Social Statistics in Action

A Canadian Introduction

Andrea M. Noack

OXFORD
UNIVERSITY PRESS

OXFORD
UNIVERSITY PRESS

Oxford University Press is a department of the University of Oxford.
It furthers the University's objective of excellence in research, scholarship,
and education by publishing worldwide. Oxford is a registered trade mark of
Oxford University Press in the UK and in certain other countries.

Published in Canada by
Oxford University Press
8 Sampson Mews, Suite 204,
Don Mills, Ontario M3C 0H5 Canada

www.oupcanada.com

The analyses presented in Chapters 3 and 12 were conducted at the Toronto Research Data
Centre (RDC) which is part of the Canadian Research Data Centre Network (CRDCN). The services and
activities provided by the Toronto RDC are made possible by the financial or in-kind support
of the SSHRC, the CIHR, the CFI, Statistics Canada, and the University of Toronto.
The views expressed in this book do not necessarily represent the CRDCN's or that of its partners.

Library and Archives Canada Cataloguing in Publication

Noack, Andrea, 1975-, author
Social statistics in action : a Canadian introduction / Andrea Noack.

Includes bibliographical references and index.
ISBN 978-0-19-901521-4 (softcover)

1. Social sciences--Statistical methods--Textbooks. 2. Textbooks.
I. Title.

HA29.N63 2018   300.1'5195   C2017-905168-7

Cover image: Magictorch/Ikon Images/Getty Images
Cover design: Sherill Chapman; Interior design: Laurie McGregor

All screen captures of IBM SPSS outputs are
reprinted courtesy of International Business Machines Corporation,
© International Business Machines Corporation

Brief Contents

Detailed Contents

PART II | Making Claims about
Populations   129

⑤ Probability, Sampling,
and Weighting   130

⑥ Making Population Estimates:
Sampling Distributions, Standard
Errors, and Confidence Intervals   157

⑦ Assessing Relationships by
Comparing Group Means:
T-Tests   195

Online Chapters

List of Boxes

Spotlight on Data

Hands-on Data Analysis

How does it look in SPSS?

Step-by-Step

Publisher's Preface

From the Publisher

Oxford University Press is delighted to present Andrea M. Noack's *Social Statistics in Action: A Canadian Introduction*. This exciting text takes a fresh and engaging approach to the study of social statistics by showing students how statistics can be used as a tool for investigating social issues. By focusing each chapter on a particular social issue and using real Canadian data to guide students through each step of the statistical procedure, *Social Statistics in Action* provides a friendly, relatable, and hands-on learning experience. This innovative approach not only shows students how statistical analyses are used in real life, it prompts them to further explore how statistics may be used to effect change for the issues that matter to them.

Key Features

Each chapter focuses on a social issue and set of research questions including young people's wages, the gendered division of labour, mental health, and racial inequality—helping students to learn the statistical procedure by demonstrating its application in everyday life.

Assessing Relationships between
Categorical Variables

Learning Objectives

In this chapter, you will learn:

- What proportionate reduction in error measures show
- How to calculate lambda and gamma measures
- The logic of chi-square tests
- How to calculate and interpret a chi-square test of independence
- How the elaboration model is used to investigate more complex relationships
- How to present cross-tabulation results

Introduction

In chapters 7 and 8, you learned how to assess a relationship between a categorical independent variable and a ratio-level dependent variable. In this chapter, you will learn how to assess relationships between two categorical variables. The first half of the chapter introduces proportionate reduction in error measures, which are used to assess the magnitude of a relationship between two categorical variables. The second half of the chapter describes how to calculate and interpret a chi-square test of independence, which is used to assess the reliability of a relationship between two categorical variables. The chapter concludes by showing how elaboration models are used to conduct multivariate analyses.

The research focus of this chapter is the relationship between racialization, contact with police, and perceptions of police. In the United States, a series of high-profile incidents have highlighted questionable police conduct towards Black people, including the shooting of unarmed suspects and deaths in police custody. Public outrage about these incidents coalesced around the "Black Lives Matter" movement, which builds on a long history of organizing by Black and racialized communities (Petersen-Smith 2015). Although the politics of race might sometimes appear less stark in Canada than in the US, racial discrimination—including discrimination against Aboriginal peoples—is also entrenched within the Canadian criminal justice system (Chan and Mirchandani 2002, Comack 2012). Most recently, debates about the use of "carding" by police forces have gained

Photo 9.1   **The Black Lives Matter movement has staged ongoing protests against police brutality and the use of "carding."**

public attention. In some Canadian cities, police officers routinely stop and question racialized people—especially racialized young men—to collect identifying information from them as they go about their daily activities (Rankin and Winsa 2012). Community leaders have spoken out about how these policing practices erode people's trust in police and in the justice system more generally (Cole 2015).

Canada's 2014 General Social Survey (GSS) on Victimization asks respondents whether they think their "local police force does a good job, an average job, or a poor job of treating people fairly" (See the "Spotlight on Data" box for more information about this survey). Since many rural communities do not have local police forces (and rely instead on the Royal Canadian Mounted Police [RCMP] or a provincial service), the analyses in this chapter only include people living in urban centres. Overall, more than two-thirds of people (68 per cent) living in Canadian urban centres say that their local police force does a good job of treating people fairly. Another 26 per cent say they do an average job, and only 7 per cent say their local police do a poor job of treating people fairly. In general, this is good news: most people think the police treat people fairly. In addition to asking about people's perceptions of police fairness, the GSS collects information about whether people have come into contact with police during the previous 12 months. The survey asks about police contact in a variety of contexts: in a public information session, because of problems with respondents' own mental health or alcohol/drug use or family members' mental health or alcohol/drug use, for a traffic violation, as a witness to a crime, or when arrested by police. Almost one-third of people living in urban centres (30 per cent) report having some contact with police in the previous

Each chapter utilizes a Canadian dataset including data from the National Graduates Survey, the Labour Force Survey, and the Canadian Community Health Survey—providing students with contemporary Canadian examples throughout.

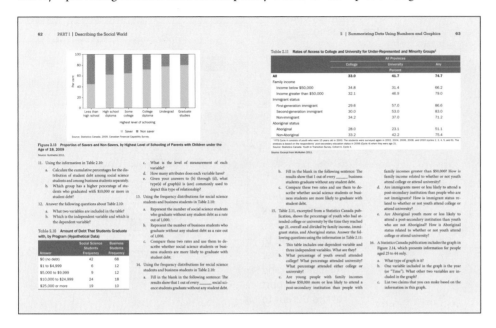

A dynamic box program provides students with a variety of examples and opportunities for hands-on learning.

"Spotlight on Data" boxes highlight a Canadian dataset, telling students when and how the data were collected and from whom, in order to show students the wide range of Canadian data available for analysis as well as where information is currently lacking.

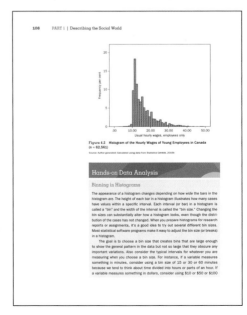

"Hands-on Data Analysis" boxes illustrate key data management skills and offer students practical advice on technical problems.

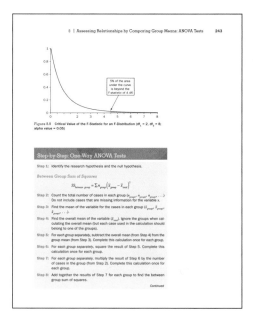

"Step-by-Step" boxes guide students through the calculation of a key formula, one step at a time.

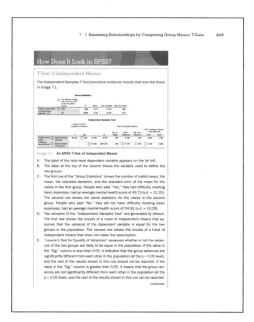

"How Does It Look in SPSS?" boxes present an SPSS output related to the topic under discussion and explain each part of the output to students.

"Best Practices in Presenting Results" sections show students how to write about and present statistical results.

Engaging learning tools enhance opportunities for critical thinking, analysis, and practice throughout.

Learning objectives at the start of each chapter provide a concise overview of the key concepts that will be covered.

Marginal glossary terms and marginal flags for key formulas reinforce learning throughout.

Figures and tables for each step of the statistical procedure guide learning and provide students with examples.

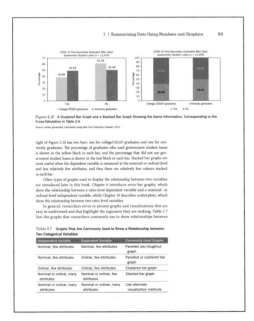

"What You Have Learned" sections provide end-of-chapter summaries.

"Check Your Understanding" sections provide conceptual questions to check students' understanding of the key concepts in the chapter.

"Practice What You Have Learned" sections provide applied practice problems that get students to engage with the statistical procedures introduced in the chapter.

"Practice Using Statistical Software (IBM SPSS)" sections provide activities for students to practice using the SPSS procedures that relate to each chapter.

"Key Formulas" list provides a reference for the formulas taught in the chapter.

Vibrant four-colour design and photos bring social statistics to life and help students connect statistics to the world around them.

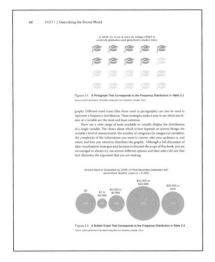

Appendices—including "A Brief Math Refresher" and a guide to "SPSS Basics," as well as an answer key to the odd-numbered "Practice What You Have Learned" problems—provide further support for students.

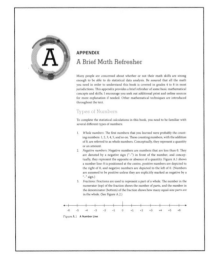

Online chapters on advanced topics—"Manipulating Independent Variables in Linear Regression" and "Logistic Regression Basics"—are available for instructors looking to teach more advanced topics.

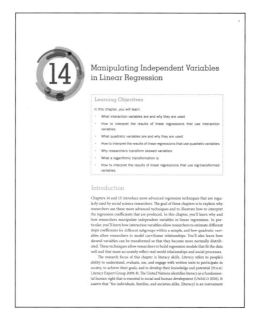

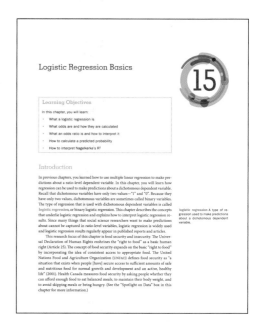

Online Supplements

Social Statistics in Action: A Canadian Introduction is supported by an outstanding array of ancillary materials for both students and instructors, all available on the companion website:

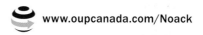 www.oupcanada.com/Noack

For Instructors

- **An instructor's manual** includes chapter overviews and suggested classroom activities.
- **A test generator** offers a comprehensive set of multiple choice, true or false, and short answer questions.
- **PowerPoint Slides** for use in classroom lectures summarize key points from each chapter and contain animations that highlight examples from the book.
- **An image bank** includes all figures, tables, examples, and formulas from the book.

For Students

- **A comprehensive student study guide** includes chapter summaries, key terms and concepts, and self-assessment quizzes designed to reinforce students' understanding of each chapter and help students prepare for tests and exams.
- **Online calculation practice activities** give students further opportunity to practice the calculations covered in the text.
- **Online interactive activities** allow students to conduct further investigation on each chapter's social issue.
- **Screencast videos** show students how to navigate SPSS and perform the SPSS procedures taught in the book.
- **SPSS datasets** for students to complete the "Practice Using Statistical Software" questions.
- Answers to the odd-numbered "Practice Using Statistical Software" questions.

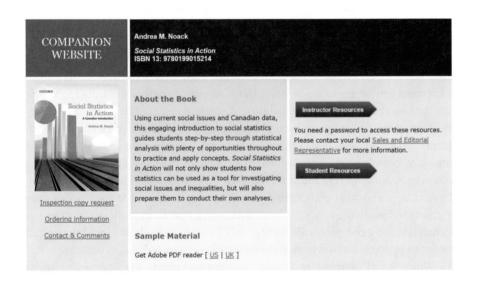

Author's Preface

When I first began teaching introductory statistics, I was dismayed by the absence of textbooks that emphasized statistics as a tool for investigating social inequalities. There are excellent books for younger students that take this approach, but few for post-secondary students, and none that focus on Canadian data. This textbook is designed to fill this gap.

In many post-secondary institutions, introductory social statistics courses are taught using a fairly technical approach. As a result, students often find it difficult to make connections between what they are learning in a statistics course and what they are learning in other courses. In this text, I aim to break down the conceptual barriers between introductory statistics and other social science courses by illustrating how statistical analysis is useful for understanding social issues, such as housing affordability, mental health, wage inequality, and more.

I firmly believe that basic statistical analysis is a skill that anyone enrolled in post-secondary education can learn. The foundational concepts needed to understand statistics are taught at the junior level in most jurisdictions. And, frankly, it is far more important to understand the *meaning* of statistical results than their calculation. The widespread availability of spreadsheets, online tools, and point-and-click statistical software means that the technical details of calculating each statistic no longer need to be central to students' learning experiences. Although I introduce formulas and illustrate how to calculate statistics throughout this text, my intention in doing so is to illustrate the underlying conceptual ideas. In large part, if you understand what each statistic is telling you, there is little need to calculate anything by hand. Most social science data—even that from small-scale surveys—is analyzed using computers. Because of this, I also describe how to read the output produced by statistical software and illustrate how to interpret it in context.

This book is designed to prompt students to become practitioners of social statistics—even if only on a small scale. As a research methods instructor, my goal is to give people the tools that they need to find out about the world for themselves. For too many students, the only data they ever see are filtered through someone else: a news reporter, a Wikipedia contributor, a government official, or another researcher. But Canada has an excellent statistical infrastructure, and a great amount of data are freely available from all levels of government as well as from other research organizations. It only takes a little bit of knowledge to be able to make use of this information. Students shouldn't have to search through outdated government reports to try to find out the percentage of young people in their province who are racialized (or looking for work or living in poverty), when they can just open up the most recent census data and generate the answer themselves—they just need the confidence to do so. Because of this practical orientation, this book includes more detailed discussions of "hands-on" data management techniques than many other texts.

Another unique feature of this text is its emphasis on writing about statistical results. Writing about numbers is a fairly specialized skill and one that many people have limited exposure to. Even people with substantial statistical training sometimes struggle to present their results in a clear and concise way. Since an important part of translating statistics into action is communicating them to others, each chapter in this book concludes with a section on presenting statistical results.

Being able to understand, manipulate, and write about data are increasingly important skills to learn. As electronic data collection becomes ubiquitous, we each generate a vast amount of data as we go through our everyday lives. Governments and corporations sift through such data to discover patterns. Many jobs require the ability to decipher and draw conclusions from information that is presented in numerical form. Taken together, these trends mean that data literacy is a key skill for understanding the communities and societies that we live in.

Ultimately, I like to think that statistics can be powerful. Although qualitative research provides much-needed nuance and depth to our understanding of people's experiences, quantitative research often has an immensity of scope that is difficult to match. When statistics are put into action, they can help to frame a debate or challenge the prevailing wisdom on a topic. Many people tend to assign more value to information that is presented using numbers. Although this valorization of numbers is often unjustified, given how difficult it is to measure the social world and given the uncertainty inherent in most estimates, it can make statistics a particularly effective tool for social advocacy. Needless to say, statistics alone do not cause social change—that's up to people. But one first step toward advocating for a more just society lies in understanding the scope and nature of injustice and inequality. I hope that this book will inspire students to seek out this knowledge and then act on it.

Acknowledgements

Many people have influenced this book and helped to make it possible. I owe a debt of gratitude to the many excellent professors of social statistics whom I have had the opportunity to learn from over the years. In particular, Michael Ornstein and Bruce Arai have each profoundly influenced my approach to statistical analysis.

I am thankful for the ongoing enthusiasm of my colleagues in the Sociology department at Ryerson University, who are, in their own ways, advocating for social change and who regularly remind me of the power of teaching as resistance; I feel as though I've found an academic home with you. In particular, my statistics co-instructors, Heather Rollwagen and Paul Moore, have helped me to both become a better teacher and hone the pedagogical approach of this book. Deidre Lam, Ryerson's Math Support Coordinator, provided invaluable assistance in shaping the Math Refresher appendix. Meghan Edwards helped to develop the practice questions and pull together all of the elements of this text.

My students over the past decade have been a constant reminder of why I love teaching statistics. They're keen to learn, they work hard to understand the material, and they will call me out when I'm being confusing. It's such a joy to join them on the journey, as the pieces fall into place and they start asking their own questions and seeking their own answers.

The title, *Social Statistics in Action,* was a burst of inspiration from the wonderful Angela Prencipe, who works as a Statistics Canada Research Data Centre Analyst. The team at Oxford University Press—particularly Tanuja Weerasooriya, Rhiannon Wong, Colleen Ste. Marie, Lisa Ball, Suzanne Clark, and Ian Nussbaum—have helped to make this book the best it could be.

My family and friends have shown exceptional patience as my time and attention have been consumed by this project. And this book simply would not exist without the ongoing support, encouragement, and endless cups of tea provided by my partner, Leanne. To simply say thank you seems woefully inadequate, so thank you, thank you, thank you.

This book has certainly been improved by the feedback and suggestions of the following reviewers, along with those who chose to remain anonymous (of course, any errors or omissions are my own):

Aaron Brauer, Concordia University
Ryan Broll, University of Guelph
Jonah Butovsky, Brock University
David Desjardins, John Abbott College
Amir Erfani, Nipissing University
Nathalie Gagnon, Kwantlen Polytechnic University
Anthony Gracey, Acadia University
Jianye Liu, Lakehead University
Kenneth MacKenzie, McGill University
Lonnie Magee, McMaster University
Michelle Maroto, University of Alberta
Natalka Patsiurko, Concordia University
Marisa Young, McMaster University

Learning to Think Statistically

Learning Objectives

In this chapter, you will learn:

- Why it is useful to learn social statistics
- How statistical analysis fits into social science research
- Some basic concepts used in statistical analysis: data, variables, attributes, and values
- How to identify a variable's level of measurement
- How statistical software organizes and displays data
- What a unit of analysis is and why it matters
- The difference between descriptive and inferential statistics
- Some strategies for writing about statistical results

Introduction

Statistical data analysis is a powerful tool that can help us to understand the world around us. Statistical evidence can also be used to advocate for social change. This book introduces you to the basic ideas of social statistics in an accessible way and shows you how data can be used to build effective arguments. **Data** is just another word for *information*. (Note that the word *data* is the plural of *datum*, which refers to a single piece of information.)

data Information.

The word *statistic* comes from the German word *statistik*, which was used to refer to the science of the state. In Britain, early statistical analysis was called "political arithmetic." These word origins give us clues about why it's important for social scientists to learn about statistics. Early forms of statistical data analysis were used by the first modern nation-states to collect information about the people they were governing (Rose 1991). Governments used statistics to keep track of births, deaths, disease, education, and income on a large scale and then used that information to manage their nation's population. A survey that collects information from every person in a population is called a census. The earliest Canadian censuses were organized around collecting information "of national interest" (Curtis 2002) and included questions about such things as agricultural production and weapon ownership in order to assess Canadians' capacity to collectively feed and defend themselves. You might be surprised to learn that Canadian census-taking has always been politicized. After the first national census was taken in 1871, both francophone and anglophone politicians made accusations that census-takers had underestimated their respective constituencies, reflecting the ongoing tensions between the French and British in founding the nation.

Today, many countries collect information about their residents using surveys administered by large national organizations or agencies. In Canada, the census and other social surveys are administered by Statistics Canada, a federal government agency. Statistics Canada is mandated to "collect, compile, analyse, abstract and publish statistical information relating to the commercial, industrial, financial, social, economic and general activities and conditions of the people of Canada" (Canada 1985, c. 15, s. 3). Over the past two decades, governments around the world have begun to rely more on data, and not just on political ideology, as the basis for decision-making. The emergence of evidence-based policy-making in Canada and elsewhere means that the results of statistical data analyses are regularly used by governments to develop and make changes to their policies and

Paul McKinnon/Shutterstock

Photo 1.1   In 2012 and 2013, protests against the federal government's cuts to research institutions and the muzzling of government scientists took place in many Canadian cities.

programs. Social scientists also use the data that Statistics Canada and other agencies collect in order to investigate social trends, provide insight into social problems, and generate evidence that policy-makers, community organizations, and social activists can use.

In this book, I illustrate how statistical analyses can be used to describe and understand social inequalities in Canada. Statistical evidence is a powerful way of showing how policies and practices have a real effect on people's lives. Good evidence, supported by community activism and organizing, can spur people, social movements, and governments into action. For instance, the stark illustration of the gender wage gap—revealed by the fact that in Canada, in 1981, women made roughly 74 cents for every dollar earned by men—prompted campaigns for pay equity and the implementation of pay equity laws in several provinces during the late 1980s. As a result, the gender wage gap has narrowed substantially over the past 30 years: in 2011, in Canada, women earned roughly 87 cents for every dollar earned by men (Morrissette, Picot, and Lu 2013).

The techniques described in this book will give you the skills that you need to generate statistical evidence about issues that you care about. You will learn how publicly available data can be used to make social inequalities more visible. As well, you will be able to generate statistics about your specific community or region, instead of relying on the generalizations of other researchers. And you will also learn how to present statistical results in a meaningful way and use them to support a larger argument. On its own, statistical evidence will not lead to social change— but accessible statistical information, combined with the experiences and stories of everyday people—can help build awareness of and engagement with important issues. As you make your way through this book, I encourage you to ask questions about the world we live in and use statistical analyses to investigate topics that are meaningful to you.

Statistical literacy is also a useful skill in the workforce. Indeed, it's essential for most professional jobs and entrepreneurial ventures. Even if you don't need statistical knowledge in your paid work, you might hold a volunteer position in an organization that needs someone with these skills. If it turns out that you like doing statistical analysis, there are many jobs that focus on doing research and analyzing data. Regardless of your plans for the future, being able to critically read and assess news articles and arguments that rely on statistics will allow you to become an informed consumer and an engaged citizen.

Features of This Book

A main feature of this book is the emphasis on using statistics to illustrate social inequalities in Canadian society. Most chapters are organized around a social science research example, such as affordable housing, wage inequality, or mental health. These research examples all rely on "real" data about people living in Canada.

Another main feature of this book is that it emphasizes the practical skills that you need to analyze data yourself, including how to deal with missing answers, grouping or combining answers, and selecting only some cases to analyze. "Hands-on Data Analysis" boxes found at the end of most chapters illustrate key data management skills. You might feel a bit anxious about learning statistics—that's okay, many people do. Like most things, though, statistics become easier the more that you practice using them. You can learn some things about data analysis from a book, but you'll also learn a lot by trying things out yourself, making mistakes, and learning how to fix them.

In order to practice analyzing data or to find out more about a research example used in a chapter, you may want to download a Statistics Canada dataset. A **dataset** is a collection of data or information, usually focused on a main topic. Most statistical datasets come in electronic form and store the information as a series of numbers. You'll learn more about computerized data later in this chapter. Every time a new dataset is used in this book, it is introduced in a "Spotlight on Data" box that provides information about the survey, its goals, how data were collected, and whom it was collected from. The "Spotlight on Data" box in this chapter describes how to access Statistics Canada data. In general, this book emphasizes statistical techniques for large datasets, such as those available from national and international agencies as well as research consortiums. Although some techniques for smaller datasets are briefly discussed, they are not the main focus.

You might be concerned about how much mathematics you will need to know in order to do statistical analyses. Be assured that the only mathematics that you will need to be successful in introductory statistics is covered at the elementary school level. (If you need a refresher, see Appendix A.) These days, most statistical analysis is done using computers. Throughout this text, I will provide mathematical formulas and explain how to use them so that you understand what is happening in the background when a computer generates a statistic. Many of the formulas in this book are accompanied by a "Step-by-Step" box that provides instructions for calculating each statistic. Simply follow the steps in order to calculate the statistic yourself.

In addition to learning how to use and calculate each statistic, you must learn to accurately interpret and present statistical information. Statistics are not meaningful on their own; they become meaningful only when they are linked to people's everyday lives or to a larger argument. Each chapter closes with a section that highlights "Best Practices in Presenting Results" and describes strategies for clearly displaying and writing about statistical information.

In the remainder of this chapter, I situate statistical analysis within the larger context of social science research and describe the main statistical traditions and approaches. I also introduce several key terms and concepts used in data analysis.

Although this chapter does not incorporate any data analyses, many of the examples relate to poverty and income inequality. In Canada, both poverty and income inequality declined from the mid-1970s to the mid-1990s, when they began to rise again rapidly (OECD 2008). Some people in Canada are more likely

dataset A collection of data or information, usually focused on a main topic.

to be living in poverty than others, including children, Aboriginal people, and single mothers. Many social advocacy groups have called on politicians and policy-makers to develop a comprehensive anti-poverty plan to address this growing social inequality. In 2009, Members of Parliament unanimously resolved to develop an immediate plan to end poverty in Canada (Canada 2009), but although individual provinces have developed anti-poverty strategies, no federal strategy has been established. In this chapter, statistical concepts are used to explore the following:

- What are the different ways of measuring poverty in Canada?
- How do the different ways of measuring poverty lead to different conclusions about how many people in Canada live in poverty?

The Research Process and Statistical Analysis

Empirical research is a type of research that relies on making direct observations of the world around us. In the social sciences, empirical research usually focuses on collecting information about people and analyzing it to make claims about the social world. Statistical analysis is only one part of the empirical research process, however, and not all empirical research includes statistical analysis. An empirical research project typically begins with a research question or something that people want to know about the world. This research question is usually informed by a theoretical perspective that shapes the researcher's understanding of how the world works. After developing a research question, most researchers will review the existing literature to find out what is already known. Then, they will decide what information they want to collect, from whom (or about whom) they will collect it, and how they will do so. Social scientists usually collect "direct observations" of the world by asking people to report on their experiences, behaviours, and attitudes in surveys, interviews, or focus groups; by conducting experiments; or by making structured observations of an environment or event. The pieces of information—or observations—that are collected using empirical research are referred to as data.

Empirical research relies on **aggregation**, which is the process of collecting and summarizing many pieces of information in order to develop conclusions. At its most basic level, data analysis involves looking for patterns and trends in information that has been collected, in order to answer a research question. Statistical data analysis is used when researchers collect information that is recorded using numbers instead of words (see the discussion of quantification later in this chapter). Once researchers have answered their research question, they present the results to other people in order to share the new knowledge they produced. The statistical techniques described in this book are used primarily in the data analysis phase of a research project, though as you will learn, decisions made throughout a research project affect the types of statistical analyses that researchers can do.

empirical research Research that relies on making direct observations of the world to generate knowledge.

aggregation The process of collecting and summarizing many pieces of information in order to develop conclusions.

Spotlight on Data

Getting Access to Canadian Data

Statistics Canada collects a substantial amount of information about people living in Canada, which is relatively easy to access. As a result of the work of the Data Liberation Initiative, an advocacy group that emerged in the mid-1990s, students and researchers at most post-secondary institutions are able to access Statistics Canada's Public Use Microdata Files (PUMFs) through their school library. Each microdata file usually contains people's answers to a survey. The file comes with a user guide that explains how the information was collected and a codebook that lists all of the questions that were asked and the possible answers. The original survey questionnaires are often available on the Statistics Canada website (www.statcan.gc.ca). Microdata files are usually either pre-formatted so that they can be opened directly by the major statistical software programs, or they come with the files that are needed to import the data into a statistical software program. Many post-secondary libraries have a data librarian, who can help you to find, download, and open microdata files.

Statistics Canada's publicly available microdata files are carefully edited to ensure that they do not contain any information that can identify a survey respondent. Researchers who want access to more detailed datasets may apply to work in one of the Statistics Canada Research Data Centres (RDCs), which are located at post-secondary institutions across the country and which have strict screening and security protocols. Most of the data used in this book are publicly available from Statistics Canada, but some information comes from datasets that are only available at an RDC.

People who are not affiliated with a post-secondary institution can also access a substantial amount of Statistics Canada information through the Government of Canada's Open Government portal (www.data.gc.ca). Many PUMFs are also freely available to members of the general public, as long as they sign a licensing agreement with Statistics Canada.

Many provincial and municipal governments also collect and distribute data through "open data" portals. As well, private polling firms, not-for-profit organizations, and research consortiums also collect information about people living in Canada. Some of these data are freely available for download from specific project websites or from larger data archives, such as the Canadian Opinion Research Archive (www.queensu.ca/cora) or the Michigan-based Inter-university Consortium for Political and Social Research (ICPSR) (www.icpsr.umich.edu).

Some Statistical Basics

Like many other fields, social statistics relies on some specialized terms and concepts. In addition, some everyday words—such as *significant*—have a different meaning for statisticians. All statistical analysis, however, relies on quantification, which is the process by which we translate ideas, behaviours, feelings, and other aspects of our everyday lives into numbers. Those numbers are then used as data in statistical analyses. Quantification can occur in many different ways. Sometimes, it is the result of simple counting: How many minutes does it take you to travel to school? How many cups of coffee have you consumed this week? Often, though, the quantifications that social scientists rely on are more complex.

Sometimes, researchers ask people to quantify for themselves; other times, researchers do it for them. For instance, a researcher might ask you to rate your level of interest in a social statistics course on a scale from 0 to 10, where 0 indicates no interest and 10 indicates a great deal of interest. You are likely to quickly assess your general feelings about statistics and/or mathematics, what you have heard from other students, and your impressions of the instructor, and then provide a number that expresses your level of interest. When you choose a number, you are engaging in the process of quantification by representing your level of interest as a numeric value. Alternatively, a researcher might ask whether you have a low, moderate, or high level of interest in a social statistics course and then assign the number 1 to people who report low interest, the number 2 to people who report moderate interest, and the number 3 to people who report high interest. In this situation, the researcher is the one engaging in quantification.

Although quantification is a simple idea, putting it into practice can quickly become complicated. Even seemingly simple counting can be challenging. Consider the question about how many minutes it takes you to travel to school. The answer might depend on traffic, on what time you travel, and on how quickly you are moving. When people answer this question, they tend to report their usual or average travel time. And, unless their travel time is very short, people also tend to round the number to the nearest five-minute value, for example, 25 minutes or 45 minutes instead of 23 or 46 minutes. This is because we have learned to not report numbers that seem too precise, especially when we are reporting an average. Similarly, your report of the number of cups of coffee you consumed this week depends on what and on how you count: Does a latte or an iced cappuccino count as a coffee? Does a double-shot of espresso count as one coffee or two? How should you include half-finished cups of coffee in your count? All of these details about what counts and how to count affect the numbers used in statistical analyses.

When concepts become more complicated, counting can become even harder. Let's use the number of people living in poverty as an example. In Canada, there is no official definition of *poverty*; instead, there are three different ways of counting people with "low income" (Statistics Canada 2015). The first way is based on what's called a low-income cut-off (LICO). A family that is below the low-income cut-off devotes a larger proportion of its income to the necessities of food, shelter, and clothing than an average family of the same size, living in a community of the same size. The

quantification The process of translating a concept, idea, behaviour, feeling, identity, or something else into a number so that it can be used as data in statistical analyses.

second way is based on what's called a low-income measure (LIM). The calculation of the LIM is more complicated, but it takes into account the number of people who live in a household and how their income compares to other households; it does not take into account where they live. The third way of measuring low income is called the market-basket measure (MBM), which is based on the cost of purchasing basic goods and services (such as food, transportation, and clothing) in a specific geographic area.

Why do these differences matter? Because you get a different estimate of the number of low-income people in Canada depending on how you count. Using the LICO method, for example, Statistics Canada estimates that there are about 3.2 million people (or 9 per cent) living in low-income families in Canada (2015). In contrast, using the LIM method, Statistics Canada estimates that there are 5.0 million people (or 14 per cent) living in low-income families. And using the MBM method, Statistics Canada estimates that there are 4.2 million people (or 12 per cent) living in low-income families. What does all of this mean? First, it means that about one in ten people in Canada have low income—so poverty isn't a particularly rare occurrence. Second, it means that, depending on how we count, there's a difference of more than a million-and-a-half people in these estimates. The way that you count, then, affects how big a problem you think that poverty is. And, if you are developing a new policy or program designed to help reduce poverty or provide assistance to low-income people, your decisions about what to do, how much it will cost, and how to measure success will all be affected by which method of counting you are using.

Practices of quantification haven't always been as prominent as they are today, however. It wasn't until the seventeenth century that people's ideas about what counts as evidence changed in a way that led to the widespread emergence of probabilistic thinking (Hacking 2006). Then, in the period between 1820 and 1840, changing social conditions in Europe and the UK led to the emergence of what scholars call an "avalanche of printed numbers"—when the production and publication of quantified measures and statistical tables became widespread (Hacking 1982). People began to think about societies as mostly stable, with a calculable chance of something happening to individuals or to members of specific social groups. These ideas prompted the expansion of the insurance industry, which used statistical analyses to determine people's chances of becoming ill, having an accident, or losing their job as the basis for establishing health, disability, and unemployment insurance.

Today, as digital devices become incorporated into more and more of our daily activities, a mind-boggling amount of quantified data is being collected. If you use a smart watch, a digital fitness/activity tracker, or health apps on your smart phone, you are generating an enormous amount of personalized data. Some enthusiasts try to use all of the quantified information that they generate about themselves to track patterns in their health and well-being. Enthusiasts of the "quantified self" movement seek to gain "self-knowledge through numbers." They hold regular conferences and present data analyses based on the information they collect about their nutritional intake, daily physical activity, heart rate, sleep cycles, and more. (Learn more about them at quantifiedself.com.)

Over the past 50 years, a great deal of quantified information has become available to use in statistical analyses. Some data come from surveys conducted by government agencies or private companies, such as marketing or public opinion research firms. Survey data are designed specifically for statistical analysis and, thus, are relatively easy for researchers to use. More recently, researchers have begun analyzing administrative data, which are collected through bureaucratic processes. For instance, data from the computerized systems used to administer social assistance (welfare) benefits is one type of administrative data. Administrative data are usually accurate because they include unbiased information on a large number of people, but such data can also be challenging to access and use, since the systems used to collect administrative data were not designed with social science research in mind. Currently, most administrative data are not publicly available and so researchers must enter into negotiations with the organization collecting the data in order to be able to use it. As more governments and government agencies adopt open data policies, however, administrative data are becoming easier to access. Some of the techniques described in this book can be used to analyze administrative data.

Over the past several years, researchers have also begun to analyze the vast amount of data that are collected by apps, search engines, and web-enabled personal and home devices. These data—which have enormous numbers of observations and are produced in or near real time—are often referred to as "big data" (Kitchin 2014). There is no specific cut-off for how many cases or observations a dataset needs in order to be considered "big," but some have millions or more. Limitations in computer processing speed and statistical software make it infeasible to use the statistical procedures described in this book with big data. Instead, researchers who analyze big data rely on techniques from the computer sciences, such as data mining and machine learning, in order to identify patterns.

Social scientists sometimes conceptually divide empirical research into two streams: quantitative approaches and qualitative approaches. Because it relies on quantification, statistical analysis is considered to be part of the quantitative research tradition. Whereas quantitative researchers typically capture and describe the social world using numbers, qualitative researchers typically capture and describe the social world using words. Both numbers and words are an imperfect way of capturing the complexity of the social world and necessarily simplify people's experiences in some way. Researchers' decisions about whether to use a quantitative or qualitative approach depends on what they think will be the best way to capture the phenomena they want to study. Today, researchers often combine both qualitative and quantitative approaches in a single project.

Quantitative data are not any more or less accurate or truthful than qualitative data. But many people still give greater weight to evidence that comes in the form of numbers than they do to quotes from an interviewee or to a researcher's careful observations. A common misconception is that numerical evidence is more "scientific" than other types of evidence. People's confidence in numbers also reveals their lack of familiarity with the complex decision-making that goes into statistical analyses. As you learn to do statistical analysis yourself, try to remain aware of the fact that any quantification will always be an imperfect representation of people's identities, behaviours, and attitudes. Although statistical evidence can be powerful, it will never capture the full complexity of people's lived experiences.

The Building Blocks of Data Analysis: Variables and Values

variable Captures some characteristic that varies across cases, such as across people, households, places, or time.

One of the most basic statistical concepts is the idea of a **variable**. A variable captures some characteristic that varies or changes—across people, families, households, different points in time, or different geographic locations. When researchers collect data by surveying people, each survey question is transformed into a variable. The opposite of a variable is a *constant*, which is a characteristic that does not change. You cannot do much data analysis with a constant, since there needs to be some variation in people's answers in order to use most statistical techniques.

independent variable (IV) The variable that captures the characteristic that is considered to be the "cause" of an outcome.

dependent variable (DV) The variable that captures the characteristic that is considered to be the "effect" or the result of whatever is captured in the independent variable.

Conceptually, researchers divide variables into **independent variables** and **dependent variables**. An independent variable captures the characteristic that is considered to be the "cause" of an outcome, whereas the dependent variable captures the characteristic that is considered to be the "effect" or the result. So, for example, a researcher who is studying the relationship between education and poverty would treat a person's level of education as the independent variable and whether or not they have low income as the dependent variable. The variable that is designated as the dependent variable must capture information about something that occurs after (or at the same time as) whatever is captured in the independent variable. Variables that capture relatively stable demographic characteristics—such as age, gender, visible minority status, ethnicity, first language spoken—can usually only be used as independent variables, since these characteristics tend to occur before everything else in time. But many variables can be used as either independent or dependent variables depending on what a researcher is studying. In the first example, low-income status was used as the dependent variable, but a researcher who is studying the relationship between poverty and whether or not people volunteer in their community would treat low-income status as the independent variable, and whether or not someone volunteers as the dependent variable.

attributes Capture the potential range of variation within a single variable.

categorical variable A variable with attributes that divide cases into groups or categories.

count variable A variable with attributes that are quantities or amounts that capture the number of times something occurs; the attributes usually include only whole numbers greater than or equal to 0.

continuous variable A variable with attributes that are quantities or amounts and that, theoretically, has an infinite number of attributes, because any fraction of an amount is a legitimate attribute.

value A number that is assigned to each attribute for the purpose of statistical manipulation.

The **attributes** of a variable represent the potential range of variation. There are two types of variable attributes. The first type divides cases (or people) into groups or categories. Variables with these types of attributes are called **categorical variables**. The second type of attributes are quantities or amounts. Variables that capture quantities or amounts can be *count variables* or *continuous variables*. **Count variables** show the number of times something occurs. Count variables usually only have attributes that are whole numbers, and there are no attributes below 0. **Continuous variables** capture quantities or amounts, but there are theoretically an infinite number of attributes because any fraction of an amount is a legitimate attribute. At the introductory level, most of the statistical techniques used with continuous variables can also be used with count variables. Count variables are treated as continuous variables throughout this book; specialized techniques for count variables are not discussed.

When the attributes of a variable divide cases into groups or categories, they are assigned numbers so that they can be used in statistical analyses. These numbers are referred to as **values**. For example, a student survey might ask, "What major are you enrolled in?" People's answers to this question become the variable "Major of study," and the attributes of the variable might include criminology, political science, psychology, sociology, and any other majors offered at the school. In order to quantify these answers for statistical analysis, numbers are assigned to each attribute (or answer): the value "1" indicates criminology students, "2" indicates political science students, "3" indicates psychology students, "4" indicates sociology students, and so on.

When the attributes of a variable capture a quantity or an amount, the number representing that amount becomes the value. So, for the earlier question about how long it takes you to travel to school, the value becomes the number of minutes you report (25 or 45). For the question about how many cups of coffee you drink each week, the value becomes the number of cups (0 or 9.5 or 12). When values are quantities, it is important to determine what unit of measurement is being used. For questions about how long something takes, for example, is the amount recorded in minutes or hours? For questions about money, is the amount recorded in dollars or thousands of dollars? Be sure to look carefully at each variable label and the data documentation in order to determine what each number actually represents.

Levels of Measurement

A variable's level of measurement is determined by its attributes. In this book, I focus on variables with three different levels of measurement: nominal-level variables, ordinal-level variables, and ratio-level variables. Nominal- and ordinal-level variables are categorical variables. Ratio-level variables can be continuous variables or count variables. (But count variables with very few attributes should be treated as ordinal-level variables in statistical analyses.) Because a variable's level of measurement determines what kinds of statistical procedures can be used with it, you must learn to correctly distinguish between them.

Nominal-level variables have attributes that divide cases into groups or categories, but the categories *do not* have an inherent order to them. In other words, each attribute can't be ranked higher or lower than any of the others based on what it represents. For each of the examples shown in Figure 1.1, the attributes can be reordered without losing coherence. In the gender identity variable, the number 1 could just as easily be associated with the label "Woman"; and the number 2, with the label "Man." In the marital status variable, there's no reason why the categories need to be in the order shown; being "Widowed" isn't inherently higher or lower than being "Divorced/separated."

nominal-level variable A variable with attributes that divide cases into groups or categories, but where the categories do not have an inherent order to them.

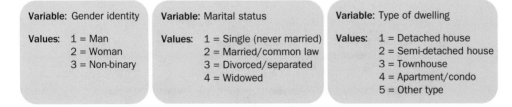

Variable: Gender identity	**Variable:** Marital status	**Variable:** Type of dwelling
Values: 1 = Man 2 = Woman 3 = Non-binary	**Values:** 1 = Single (never married) 2 = Married/common law 3 = Divorced/separated 4 = Widowed	**Values:** 1 = Detached house 2 = Semi-detached house 3 = Townhouse 4 = Apartment/condo 5 = Other type

Figure 1.1   **Examples of Nominal-Level Variables**

The values assigned to each attribute of a nominal-level variable don't need to be sequential. For instance, when Statistics Canada creates a variable to indicate the province that a person lives in, it typically uses the value 10 for Newfoundland and Labrador, 11 for Prince Edward Island, 12 for Nova Scotia, 13 for New Brunswick, 24 for Quebec, 35 for Ontario, 46 for Manitoba, 47 for Saskatchewan, 48 for Alberta, 59 for British Columbia, and 60 for Northern Canada. Using this non-sequential numbering system makes it easy to divide the country into regions using the digits in the tens columns: the Atlantic region (10 to 13), Quebec (24), Ontario (35), the Prairie region (46 to 48), British Columbia (59), and Northern Canada (60). Most of the time, people use sequential numbers for values because it's easier to remember, but this is just a matter of convenience.

dichotomous variable A special type of nominal-level variable that has only two attributes and values (0 and 1) that indicate the presence or absence of something.

Dichotomous variables are a special type of nominal-level variable. They have only two attributes and two values, numbered "0" and "1," which represent the absence (0) or presence (1) of something (see Figure 1.2). They are sometimes referred to as dummy variables. Demographic characteristics, such as visible minority status or gender, are often captured in dichotomous variables that represent the presence or absence of racialization or of being a woman. Dichotomous variables are special because in some statistical procedures, they can be used in the same way as ratio-level variables; you'll learn more about this in Chapter 12.

ordinal-level variable A variable with attributes that divide cases into groups or categories, where the categories do have an inherent order to them.

Ordinal-level variables have attributes that divide cases into groups or categories, but the categories *do* have an inherent order to them. Usually, each attribute indicates more or less of some characteristic; thus, the attributes are ordered from highest to lowest or vice versa. Figure 1.3 shows some common attributes of ordinal-level variables. Since using a food bank "Often" is more than doing so "Sometimes," the value assigned to "Often" (4) is higher than the value assigned to "Sometimes" (3), which is higher than the value assigned to "Rarely" (2). It would be confusing to list these attributes in a different order.

It doesn't matter which direction the attributes and values of an ordinal-level variable are listed in, as long as the sequence is maintained and the results are interpreted correctly. For example, the variable capturing people's opinions about government intervention into poverty assigns the value "1" to the highest level of agreement and "2" to a lower level of agreement (and so on). This is fine, as long the researcher remembers that lower values mean more agreement when they describe the results.

class intervals Used to collapse all of the possible attributes of a variable into a smaller number of groups that each represent a range of attributes.

Sometimes the attributes of ordinal-level variables are **class intervals**, which collapse all of the possible attributes of a variable into a smaller number of groups that each represent a range of attributes. The variable on the right of Figure 1.3 captures information about personal annual income using five class intervals: less

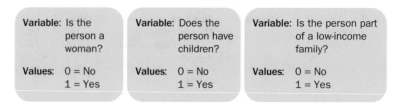

Figure 1.2   **Examples of Dichotomous Variables**

Variable: How often do you use your local food bank?	Variable: The Canadian government should do more to reduce poverty	Variable: Personal annual income from all sources, before tax
Values: 1 = Never 2 = Rarely 3 = Sometimes 4 = Often	Values: 1 = Strongly agree 2 = Somewhat agree 3 = Somewhat disagree 4 = Strongly disagree	Values: 1 = Less than $5,000 2 = $5,000 to $9,999 3 = $10,000 to $19,999 4 = $20,000 to $29,999 5 = $30,000 or higher

Figure 1.3   **Examples of Ordinal-Level Variables**

than $5,000; $5,000 to $9,999; $10,000 to $19,999; $20,000 to $29,000; and $30,000 or higher. Class intervals are typically used to simplify data collection, to simplify data analyses, or to ensure the confidentiality of people's answers. Class intervals must be exhaustive; that is, researchers must be able to place every single case into one of the intervals. In the annual income variable shown in Figure 1.3, this is accomplished making the first class interval $5,000 *or less* and making the last class interval $30,000 *or higher* so that everyone can be placed into a group, including people who earn less than $5,000 and people who earn more than $30,000.

Class intervals must also be mutually exclusive; in other words, a case cannot belong to more than one class. That is why several of the attributes in the personal income example end in "999"; the classes "$5,000 to $10,000" and "$10,000 to $20,000" are not mutually exclusive because a person earning exactly $10,000 belongs to both classes. Although ending classes with a "999" does make them mutually exclusive, it poses another problem: what class does someone who earns $9,999.50 belong to? The most common way to resolve this problem is to apply standard rounding practices: someone earning $9,999.49 is treated as if they earn $9,999 and someone earning $9,999.50 is treated as if they earn $10,000.

The distance between the values of ordinal-level variables is not always meaningful or equal. For example, in the personal income variable in Figure 1.3, the income difference between a person assigned the value "2" and the value "3" could be $10, or $500, or $9,500, depending on where each person's income falls within the class interval. So, while it's clear that people assigned a "3" make more money than people assigned a "1" or "2," we don't know how much more. Since the distance between the values of ordinal-level variables isn't equal, class intervals do not need to be the same size. In the personal income variable, the values "1" and "2" each represent a $5,000 interval, the values "3" and "4" each represent a $10,000 interval, and the value "5" potentially has a much wider interval.

Finally, **ratio-level variables** have attributes and values that are specific amounts or quantities of something. The attributes and values of ratio-level variables have an inherent order to them, and there is an equal distance between them. Figure 1.4 provides some examples of ratio-level variables, although, in order to save space, not all of the possible values are listed. A "0" value on a ratio-level variable means having none of whatever is being measured. So, a "0" value on the variable that captures hours working for pay means the person doesn't work for pay. Ratio-level variables are easy to identify because the values make sense if you

ratio-level variable A variable with attributes and values that are specific amounts or quantities of something, where the value "0" means having none of something.

Variable: How many hours do you spend working for pay in an average week?	Variable: Personal annual income from all sources, before tax, in dollars	Variable: What is your hourly wage?
Values: 0	Values: 0	Values: 10.00 or less
1	1	10.01
2	2	10.02
3	3	10.03
…	…	10.04
166	999,997	…
167	999,998	99.96
	999,999 or more	99.97
		99.98
		99.99 or more

Figure 1.4   **Examples of Ratio-Level Variables**

double them: having a value that is double that of another person means having twice the amount of something. Someone assigned the value "4" on the variable that captures hours working for pay spends twice as much time at work as someone assigned the value "2." Notice that it doesn't make sense to double the values of the ordinal-level personal income variable shown in Figure 1.3. Someone assigned the value "4" (earning $20,000 to $29,999) doesn't necessarily earn twice as much as someone assigned the value "2" (earning $5,000 to $9,999).

In many Statistics Canada datasets, some ratio-level variables have an artificial upper or lower limit to their variation. For example, people's income might be reported in exact dollars, and the values might range from $0 to $999,999. Although there are some people who earn more than $999,999, the value $999,999 is assigned to everyone with an income of a million dollars or more. This is called top-coding and is usually done to protect the confidentiality of very high income earners or others who are unusual in some way. Variables with top-coding (or bottom-coding) can still be treated as ratio-level variables in statistical procedures, but be aware that the results are affected by the artificial limit.

interval-level variable A variable with attributes and values with an inherent order to them and an equal distance between them, but where the value "0" does not indicate the absence of something.

You may also hear researchers refer to **interval-level variables**. These are similar to ratio-level variables because they have attributes and values that have an inherent order to them and an equal distance between them. Unlike ratio-level variables, however, a "0" value of an interval-level variable does not indicate the absence of something, and doubling a value does not necessarily indicate having twice as much of something. An example of an interval-level variable is a temperature scale. Temperature values have an equal distance between them, but the "0" value doesn't indicate the absence of temperature, and 20 degrees Celsius is not twice as hot as 10 degrees Celsius. Social scientists don't often use interval-level variables, and at the introductory level, most of the statistical techniques used with interval-level variables are the same as those used with ratio-level variables. In fact, many people refer to the two types of variables in combination, calling them interval-ratio level variables. As a result, I do not specifically discuss interval-level variables in this text.

Learning to identify a variable's level of measurement is key to becoming a good statistical data analyst. Many statistical techniques can only be used with ratio-level variables. Fewer statistical techniques can be used with ordinal-level variables, and fewer still can be used with nominal-level variables. As you encounter each new statistical technique in the chapters ahead, be sure to note which level of measurement a variable must have in order to use it.

Understanding Computerized Data

Most statistical data analysis is done using computers. Learning how computerized data are organized will help you to understand what you see when you use statistical software.

There are many different statistical software programs used in social science data analysis. One of the most popular is IBM's SPSS (Statistical Package for the Social Sciences). SPSS is easy to learn because most statistical procedures are accessed using drop-down menus. "How Does It Look in SPSS?" boxes, which show SPSS output and describe how to interpret it, are included throughout this book. In addition, Appendix B provides instructions for using SPSS to generate common statistics. Many post-secondary institutions provide access to SPSS in their computer labs. There is also an open-source alternative to SPSS, called PSPP (https://www.gnu.org/software/pspp). It has many features that are similar to SPSS and is available free of charge, but it requires a bit more technical knowledge to download and install. Other commonly used statistical software programs are R, Stata, and SAS (the Statistical Analysis System). Each of these programs will produce all of the statistics described in this book.

Most statistical software programs include a screen that displays the data. It typically looks similar to a spreadsheet, as shown in Figure 1.5. Each column represents a variable in the dataset, which often corresponds to a question asked in a survey. The values (numbers) in the column are those assigned to the attributes of the variable, which usually correspond to the answers to a survey question. Each row represents a single person or case, and reading across a row shows that person's answers to each of the questions in the survey. So, in Figure 1.5, the value (number) that represents the first person's answer to the first survey question goes in the yellow cell. The value that represents the first person's answer to the second survey question goes in the cell immediately to the right of the yellow cell, and so on, for all of the questions in the survey. The values that correspond to the survey answers given by the second person go in the second row, and so on, for all of the cases. Notice that the first column is labelled "ID Number." In most datasets, the first column is used to record a number that uniquely identifies each person or case. It works very much like a student ID number. If some data are missing, or an incorrect value has been recorded, researchers use the ID number to go back to the original record or survey to fix the problem.

Statistical software programs allow users to assign labels to both variables and values. For survey data, variable labels provide more information about the questions asked and the units of measurement used. Value labels show which values are assigned to each attribute of a variable. When values are amounts or quantities, value labels are typically not used, since the number itself provides the amount (e.g., "3" means three hours, "22" means twenty-two dollars, and so on). Often, you can use a variable's value labels to determine its level of measurement.

	ID Number	Answer to Question 1	Answer to Question 2	Answer to Question 3	Answer to Question 4	Answer to Question 5	Answer to Question 6	and so on . . .
Answers from Person 1 →								
Answers from Person 2 →								
Answers from Person 3 →								
Answers from Person 4 →								
Answers from Person 5 →								
Answers from Person 6 →								
Answers from Person 7 →								
Answers from Person 8 →								
Answers from Person 9 →								
Answers from Person 10 →								
and so on . . .								

Figure 1.5   **How Data Are Organized in Statistical Software**

Hands-on Data Analysis

What Are "Missing Values"?

Many people are confused by the idea of "missing values" when they first begin doing data analysis. In part, this is because missing values aren't always actually missing or blank, as the name suggests. Instead, missing values are values assigned to attributes that are not useful in statistical analysis. The decision about which attributes are not useful for a particular analysis needs to be made by the researcher.

The most common types of missing attributes are those associated with "Don't know," "Not applicable," or "Valid skip" answers. A "Don't know" answer is recorded in survey data when respondents say that they do not know the answer to a question. Sometimes, respondents legitimately don't know the answer—as in surveys of young people, who might not know their parents' exact occupation or income. Other times, people will say that they don't know the answer to a question because it's too hard to figure out, or they don't want to disclose the answer. Sometimes it is useful to analyze "Don't know" answers; for example, in election surveys it might be interesting to learn more about people who don't know who they are going to vote for (undecided voters).

"Not applicable" or "Valid skip" answers are recorded in survey data when a question doesn't apply to a person and therefore the question is not asked. This is done to avoid leaving a blank space in the dataset. A "Not applicable" or "Valid skip" attribute tells a researcher that the question wasn't just accidentally missed, it was legitimately skipped. For example, if someone does not have any children, a "Not applicable" or a "Valid skip" answer is recorded for any questions about the children's gender or age. In practice, there's usually no difference between the attributes "Not applicable" and "Valid skip"; it just depends on how the dataset was constructed.

Finally, sometimes information is truly missing from a dataset, and when that happens a blank space or a dot (.) appears in the cell instead of a number. This occurs when a value has simply not been entered for a variable and is sometimes called "system-missing" data. Usually, a researcher cannot tell why a case is missing this information.

Most statistical software programs allow researchers to designate some values of each variable as "missing values." Cases with these missing values are automatically excluded from any statistical analysis using that variable. If a variable has many cases that are missing information, be careful using it in statistical analyses since the results reflect the answers of fewer people. If you do not tell the statistical software which attributes or values of each variable should be treated as missing and excluded from the analysis, your statistical results will generally be incorrect.

Identifying the Unit of Analysis

A key decision in statistical analysis revolves around selecting or identifying the **unit of analysis**. The unit of analysis refers to what or who is in each of the cases that a researcher is analyzing. In most data used by social scientists, the unit of analysis is an individual person. That is, a person is the "unit" that information is collected about, and each row of data represents one person. Sometimes, however, social scientists collect and analyze information about households, couples, families, electoral districts, cities, or countries. Any of these, as well as other configurations, can be used as a unit of analysis.

Using different units of analysis can lead to seemingly different statistical results. Let's go back to the example of poverty or low income in Canada. Using individual people as the units of analysis, and the LICO method, Statistics Canada estimates that 9 per cent of Canadian residents have low income. But we can also investigate the number of families (not individual people) with low income. This makes sense because people who live together in families usually share expenses within the household. Using families as the unit of analysis, and the LICO method, Statistics Canada estimates that 13 per cent of Canadian families have low income (2006). Alternatively, we can also think about low-income neighbourhoods. Using neighbourhoods as the unit of analysis, and the LIM method, Statistics Canada estimates that 9 per cent of all Canadian neighbourhoods are low income (a low-income neighbourhood is one where 30 per cent or more of the residents have low income) (Statistics Canada 2013). The estimates of poverty in Canada change, then, depending on the unit of analysis. While each of these statistics is correct, they prompt slightly different interpretations of how prevalent poverty is in Canada.

unit of analysis The basic "unit" that researchers treat as a single case in their analysis.

In this book, I usually refer to individual people as the unit of analysis because this approach is both common and easy to understand. Most of the time, though, when I reference "people," I actually mean "cases." If you are analyzing data with households or countries or some other unit as cases, please just make a mental substitution.

Using Descriptive and Inferential Statistics

I want to introduce one more important distinction in statistical analysis that shapes the organization of this book. Statistical analysis is conceptually divided into two main streams: descriptive statistics and inferential statistics. **Descriptive statistics** are used to describe the characteristics of a set of data. The goal is to summarize the data in a full and complete way, by looking at the distribution of individual variables in the dataset and the relationships between variables. Descriptive statistics are used to condense information so that it can easily be conveyed to others. Descriptive statistical techniques are usually used to analyze survey data and administrative data. The first four chapters of this book focus primarily on descriptive statistics.

Inferential statistics, on the other hand, are used to make claims about a population, based on analyzing data collected from a randomly selected sample of that population. The **population** is the whole group that a researcher is interested in studying. A **sample** is a group of people (or cases) who are selected from the larger population, that is, they are a subset of the larger population. In order to generate accurate estimates or predictions from a sample, the people (or cases) in the sample must be randomly selected, using a probability sampling method. Statisticians use inferential statistics to test whether the results obtained from a sample are likely to be representative of the results in the larger population. Inferential statistics are regularly used to analyze large-scale survey data, especially those produced by government agencies and large research organizations. But inferential statistics usually cannot be used to analyze administrative data or "big data" because these data do not capture information from a randomly selected sample of cases. You will learn more about the concepts underlying inferential statistics in chapters 5 and 6, while chapters 7 and on describe inferential statistical techniques.

When data are collected from a probability sample, researchers typically use both descriptive and inferential statistical techniques. Descriptive statistics are used at the beginning of the analysis in order to summarize and find patterns in the data. Then, inferential statistics are used to assess relationships and develop models that are generalizable to the larger population.

Within inferential statistics there are two analytic traditions: frequentist and Bayesian. Frequentist statistics rely solely on the empirical data that are available and use the frequency of various outcomes to assess whether sample results are generalizable to the population. Bayesian approaches to statistics use analytic strategies developed by Thomas Bayes in the 1700s and take a different approach to probability than frequentist statistics. This book describes only frequentist statistical techniques. Most people learn frequentist approaches to statistics first. However, you may encounter academic journal articles that use Bayesian approaches.

---

descriptive statistics A series of techniques used to aggregate and summarize information (or data).

inferential statistics A series of techniques used to make estimates or predictions about a population, using information collected from a probability sample selected from that population.

population The whole group that a researcher is interested in studying or making claims about.

sample A group of people (or cases) who are selected from the larger population and about whom information is collected.

Best Practices in Presenting Results

Writing about Statistical Results

In order for statistical results to be useful, they must be translated back into everyday language. A common error made by beginning researchers is to simply list all of the numbers they have generated in their statistical analyses. The problem with this approach is that it leaves readers to draw their own conclusions, instead of highlighting the argument that a researcher wants to make. Just listing all of the numbers is the equivalent of a qualitative researcher handing someone a stack of interview transcripts and saying "Look at all these great results." The role of any researcher is to act as a translator, by identifying and highlighting the main points in the data.

In their discussion of writing about qualitative, ethnographic research, Emerson, Fretz, and Shaw (2011) introduce the idea of "excerpt-commentary units" as a way of presenting research findings and situating them in context. With some modifications, their useful strategy can be extended to quantitative research results, in what I call "statistic-commentary paragraphs." A statistic-commentary paragraph presents statistical results in the context of a larger essay or paper, or a series of statistic-commentary paragraphs can be combined together in order to produce a statistical report.

A statistic-commentary paragraph consists of four parts:

1. An *analytic point*, which is the topic sentence of the paragraph, or the main claim that you want to make with the data
2. *Orienting information*, which provides the reader with information about where the data came from and the measure(s) used
3. *Statistical results*, which are the key quantitative findings
4. An *analytic commentary*, which provides your interpretation of the statistical results and links it to a larger argument or to the existing knowledge about the topic

Here's an example of what a simple statistic-commentary paragraph might look like:

> *Analytic point*: Despite Canada's reputation as an economically prosperous nation with a strong social safety net, Canada's poverty rate is higher than many European and Scandinavian countries.
>
> *Orienting information*: Data from the 2011 Census was used to determine the number of people who live in households with an income below a low-income threshold, which is adjusted for household size. This measure is an indicator of relative poverty.
>
> *Statistical result*: Overall, more than one in ten people in Canada (13 per cent) live in households with low income. This has remained consistent over the past decade.
>
> *Analytic commentary*: Comparable measures from other countries allow us to assess how Canada fares in relation to other economically prosperous nations. Canada's poverty rate is lower than that of the United States,

which was 18 per cent (OECD 2013), but higher than many European nations, such as France (8 per cent), the Netherlands (8 per cent), and Germany (9 per cent). Smeeding, Rainwater, and Burtles (2001) suggest that Canada fares better than the US because of more careful targeting of government social programs to the poor, whereas European and Scandinavian countries simply spend more per capita on social assistance.

When you put it all together, you get this:

> Despite Canada's reputation as an economically prosperous nation with a strong social safety net, Canada's poverty rate is higher than many European and Scandinavian countries. Data from the 2011 Census was used to determine the number of people who live in households with an income below a low-income threshold, which is adjusted for household size. This measure is an indicator of relative poverty. Overall, more than one in ten people in Canada (13 per cent) live in households with low income. This has remained consistent over the past decade. Comparable measures from other countries allow us to assess how Canada fares in relation to other economically prosperous nations. Canada's poverty rate is lower than that of the United States, which was 18 per cent (OECD 2013), but higher than many European nations, such as France (8 per cent), the Netherlands (8 per cent), and Germany (9 per cent). Smeeding, Rainwater, and Burtles (2001) suggest that Canada fares better than the US because of more careful targeting of government social programs to the poor, whereas European and Scandinavian countries simply spend more per capita on social assistance.

This approach is useful because it places the statistical result in context and links it to the researcher's larger argument. You might notice that the statistical result is a relatively minor part of the paragraph. You don't necessarily need extensive statistical results to make a good argument; rather, they are just another tool that you can use to get your point across. Once you get used to writing statistic-commentary paragraphs, you can begin to group results together in order to develop a more comprehensive argument about a topic. The goal of writing about research results is to link the numerical/statistical results to real-world experiences and practices.

What You Have Learned

This chapter began by situating the practice of statistics in a historical context and explaining how statistics are used in the social sciences. Statistical data analysis is particularly useful for understanding and documenting social inequalities. The remainder of the chapter introduced some basic terms and concepts that are central to statistical data analysis. You'll put these ideas to work in the next chapter, which explains how to summarize statistical data using different types of tables and data visualization techniques.

In this chapter you also learned that there are three ways of measuring poverty in Canada: (1) using a low-income cut-off line, (2) using an internationally comparable low-income measure, and (3) using a market-basket measure. Depending on which measure is used, somewhere between 3.2 and 5.0 million people in Canada live in poverty. The estimate of how much poverty there is in Canada also changes depending on whether the unit of analysis is an individual, a family, or a neighbourhood. Further research in this area might investigate how using different measures and different units of analysis affects the description of which people are more likely to live in poverty.

Check Your Understanding

Check to see if you understand the key concepts in this chapter by answering the following questions:

1. Why is it useful to learn about social statistics?
2. What stage of the research process is statistical analysis used in?
3. What is the difference between a variable and a value?

4. What are the three commonly used levels of measurement, and what are the differences between them?
5. Why does the "unit of analysis" matter?
6. What is the difference between descriptive and inferential statistics?
7. What information goes into a statistic-commentary paragraph?

Practice What You Have Learned

Check to see if you can apply the key concepts in this chapter by answering the following questions:

1. You want to know whether or not women or men are more likely to live in poverty. What two variables will you use in a statistical analysis in order to answer this question?
2. In 1989, the federal government committed to eliminating child poverty in Canada by the year 2000. This campaign was unsuccessful. You want to investigate whether adults or children are more likely to live in poverty. What two variables will you use in a statistical analysis to find the answer to this question?
3. You are curious whether people who are not in long-term relationships are more likely to experience poverty than those who are in long-term relationships. Your dataset includes a variable for marital status, with the following attributes:

   1 - Married
   2 - Living common-law
   3 - Widowed
   4 - Separated
   5 - Divorced
   6 - Single, never married

   How would you combine these six attributes so that they divide people into two groups: people who are not in long-term relationships and those who are?

4. A researcher in a small town is investigating whether or not there is a relationship between people's income and their attitudes toward the local homeless shelter. Which variable would the researcher treat as the independent variable, and which would the researcher treat as the dependent variable?
5. You want to know how level of education is related to a person's income. You find a survey with two relevant variables: "Highest educational credential" and "Annual personal income." Which variable would you treat as the independent variable, and which would you treat as the dependent variable?
6. Researchers who study socio-economic status are interested in how people's level of education compares to their parents' level of education. Which of these variables is treated as the independent variable, and which is treated as the dependent variable?
7. You want to investigate whether people born in Canada are more likely to live in poverty than people who immigrated to Canada. You find a variable in a survey dataset that indicates whether a person is "Born in Canada" (assigned the value "1") or "Not born in Canada" (assigned the value "0"). What level of measurement does this variable have?
8. You want to know whether a person's social class is related to pursuing a post-secondary education. You conduct an informal survey of your peers and ask people to report whether their family was

"Upper-class," "Upper middle-class," "Lower middle-class," "Working-class," or "Poor" when they were growing up. When your survey question becomes a variable, what level of measurement will it have?

9. Because people's reports of social class are subjective, you also include a question in your informal survey that asks your classmates to report whether or not their family owned their own home (regardless of whether it had a mortgage). There are two possible answers: "Yes, my family owned their own home" or "No, my family did not own their own home." When your survey question becomes a variable, what level of measurement will it have?

10. A researcher is interested in learning how much support there is for a government program that would guarantee everyone a basic income. (See www.basicincomecanada.org.) The researcher conducts a survey that asks people whether they "Strongly agree," "Somewhat agree," "Somewhat disagree," or "Strongly disagree" with the idea of establishing a guaranteed basic income. When this survey question becomes a variable, what level of measurement will it have?

11. You are doing research for ACORN Canada, an organization that advocates for low-income families. (See www.acorncanada.org.) In order to learn more about grocery costs in your community, you poll the organization's members about where the cheapest place to buy food is: Walmart, No Frills, Save-On-Foods, Super C, Foodland, or some other store. When this poll question becomes a variable, what level of measurement will it have?

12. As part of your research into grocery costs, you also ask people to report how much money their family spent on groceries in the past week. You encourage people to add up their grocery receipts so that they can report the exact amount they spent in dollars and cents (for example, one family spent $90.46). When these grocery-spending reports become a variable, what level of measurement will they have?

13. You are interested in knowing how your hourly wage compares to that of other people who work for your employer. Since you know that wages are a sensitive topic for some people, you ask your co-workers to tell you whether they earn less than $12 per hour, $12 to $14.99 per hour, $15 to 19.99 per hour, or $20 or more per hour. When you turn this information into a variable, what level of measurement will it have?

14. Despite your best efforts in collecting the information described in question 13, some of your co-workers simply refused to tell you their hourly wage. You recorded their answers as "Refused to say." When analyzing the hourly wage data you collected, what is the best way to handle these "refusal" answers?

15. You want to know how affordable rental housing is in the town or city that you live in. You survey all of the people in your class who rent and ask them to tell you exactly how much they pay for their share of rent each month, in exact dollars. The amounts that people pay range from $172 per month all the way up to $769 per month. When you turn this information into a variable, what level of measurement will it have?

16. A Statistics Canada survey records the total income of every household in exact dollars. The highest recorded value is $1,000,000, and households that earn more than a million dollars are all assigned the value $1,000,000. What is this an example of?

17. A Statistics Canada survey captures how much people earn at their job each year in a variable with the following attributes:

    1 - Less than $20,000
    2 - $20,000 to $39,999
    3 - $40,000 to $59,999
    4 - $60,000 to $79,999
    5 - $80,000 to $99,999
    6 - $100,000 to $119,999
    7 - $120,000 or more
    97 - Valid skip
    98 - Don't know
    99 - Refusal

    What do the values "97," "98," and "99" indicate? How should you handle these attributes when they are used in statistical analyses?

18. Statistics Canada collects census information from each household in Canada. One person in the household is asked to provide information about the dwelling, such as the number of rooms, whether the dwelling needs repairs, and whether it is owned or rented. You want to know the percentage of low-income households whose occupants own their homes, and the percentage of low-income households whose occupants rent their homes. What is the unit of analysis for your research? (For more, see the Statistics Canada Year Book chapter

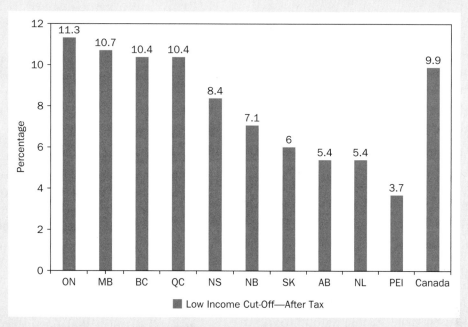

Figure 1.6   Overall Poverty Rates by Province, 2012

Source: Garner 2014.

on families, households, and housing: www.statcan.gc.ca/pub/11-402-x/2012000/chap/fam/fam-eng.htm.)

19. You use United Nations data to investigate which developed countries have the highest and lowest rates of child poverty. Each nation is asked to report the percentage of children in the country who live in low-income households. You collect information about 29 different countries. What is the unit of analysis for your research? (For more, see www.unicef-irc.org/publications/series/16/)

20. An organizer working for the BC Poverty Reduction Coalition posted the graph in Figure 1.6 on the progressive blog *Policynotes.ca*.

   a.   How is poverty measured in this graph?
   b.   What two variables are included in the graph?
   c.   Which is the independent variable, and which is the dependent variable?
   d.   What does this graph show? Which province has the highest poverty rate? Which province has the lowest poverty rate?

Practice Using Statistical Software (IBM SPSS)

The questions in this section are designed to give you practice using IBM SPSS to analyze data and interpret results. Answer these questions using IBM SPSS and the GSS27.sav dataset available from the Student Resources area of the companion website for this book. If you are using a Student Version of IBM SPSS, which is limited to 50 variables and 1,500 cases, use the GSS27_student.sav dataset instead.

1. How many cases are included in the dataset?
2. How many variables are included in the dataset?

3. Look at the answers of the person listed in the first row of the dataset. That person's "Record identification" [RECID] number is 1.

   a.   What value is recorded for the variable "Marital status of respondent" [MARSTAT]?
   b.   What attribute is associated with that value? What is that person's marital status?

4. Look at the answers of the person whose "Record identification" [RECID] number is 5.

a. What value is recorded for the variable "Annual personal income of the respondent – 2012" [INCM]?

b. What attribute is associated with that value? How much income did that person have in 2012?

5. Look at the answers of the person whose "Record identification" [RECID] number is 2.

a. What value is recorded for the variable "Total household income – 2012" [INCMHSD]?

b. What attribute is associated with that value? How much combined income did all of the people in that person's household have in 2012?

6. Look at the answers of the person whose "Record identification" [RECID] number is 10.

a. What value is recorded for the variable "Number of hours per week spent watching television" [MCR_300C]?

b. What does that value indicate? How many hours does that person spend watching television?

7. Find the variable named VBR_10.

a. What information does the variable capture?

b. What attributes does the variable have? What values are associated with each attribute?

c. What attributes/values are designated as missing?

d. What is the variable's level of measurement?

8. Find the variable named VCG_310.

a. What information does the variable capture?

b. What attributes does the variable have? What values are associated with each attribute?

c. What attributes/values are designated as missing?

d. What is the variable's level of measurement?

9. Find the variable named WHW_120C.

a. What information does this variable capture?

b. Do the values on this variable represent categories or quantities?

c. What attributes/values are designated as missing?

d. What is the variable's level of measurement?

References

Curtis, Bruce. 2002. *The Politics of Population State Formation, Statistics, and the Census of Canada, 1840–1875*. Toronto: University of Toronto Press.

Emerson, Robert M., Rachel I. Fretz, and Linda L. Shaw. 2011. *Writing Ethnographic Fieldnotes*. 2nd ed. Chicago: The University of Chicago Press.

Garner, Trish. 2014. "Latest Poverty Stats Show BC Still Has One of the Highest Poverty Rates in Canada." Policynote.ca. December 15. http.//www.policynote.ca/latest-poverty-stats-show-bc-still-has-one-of-the-highest-poverty-rates-in-canada/.

Canada. 1985. Statistics Act (R.S.C.). C. S-19. http.//laws-lois.justice.gc.ca/eng/acts/S-19/FullText.html.

———. 2009. "Edited Hansard: 40th Parliament, 2nd Session." Number 116, November 25. http.//www.parl.gc.ca/HousePublications/Publication.aspx?Language=E&Mode=1&Parl=40&Ses=2&DocId=4254820.

Hacking, Ian. 1982. "Biopower and the Avalanche of Printed Numbers." *Humanities in Society* 5 (3/4): 279–95.

———. 2006. *The Emergence of Probability: A Philosophical Study of Early Ideas about Probability, Induction and Statistical Inference*. 2nd ed. New York: Cambridge University Press.

Kitchin, Rob. 2014. "Big Data, New Epistemologies and Paradigm Shifts." *Big Data & Society* 1 (1). doi:10.1177/2053951714528481.

Morissette, René, Garnett Picot, and Yuqian Lu. 2013. *The Evolution of Canadian Wages over the Last Three Decades*. Ottawa: Statistics Canada. http://www.statcan.gc.ca/pub/11f0019m/11f0019m2013347-eng.pdf.

Organisation for Economic Co-operation and Development (OECD). 2008. *Growing Unequal? Income Distribution and Poverty in OECD Countries*. Paris: Organisation for Economic Co-operation and Development.

Organisation for Economic Co-operation and Development (OECD). 2013. "Income Distribution Database." http://www.oecd.org/social/income-distribution-database.htm.

Rose, Nikolas. 1991. "Governing by Numbers: Figuring out Democracy." *Accounting, Organizations and Society* 16 (7): 673–92.

Smeeding, Timothy, Lee Rainwater, and Gary Burtles. 2001. "United States Poverty in a Cross-National Context." Paper 151. Center for Policy Research. http://surface.syr.edu/cpr/151.

Statistics Canada, Income Statistics Division. 2015. "Low Income Lines, 2013–2014." Catalogue no. 75F0002M. http://www.statcan.gc.ca/pub/75f0002m/75f0002m2015001-eng.pdf.

Statistics Canada. 2006. "Table 109-0300, Census Indicator Profile, Canada, Provinces, Territories, Health Regions (2011 Boundaries) and Peer Groups, Every 5 Years." CANSIM (database). http://www5.statcan.gc.ca/cansim/a26?lang=eng&retrLang=eng&id=1090300&tabMode=dataTable&srchLan=-1&p1=-1&p2=9.

———. 2013. "Persons Living in Low-Income Neighbourhoods." Catalogue no. 99-014-X2011003. http://www12.statcan.gc.ca/nhs-enm/2011/as-sa/99-014-x/99-014-x2011003_3-eng.pdf.

———. 2015. "Table 206-0041 Low Income Statistics by Age, Sex and Economic Family Type, Canada, Provinces and Selected Census Metropolitan Areas (CMAs)." CANSIM (database). http://www5.statcan.gc.ca/cansim/a26?lang=eng&id=2060041.

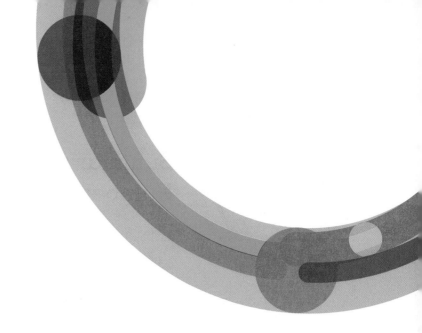

Part I

Describing the Social World

Summarizing Data Using Numbers and Graphics

Learning Objectives

In this chapter, you will learn:

- What a frequency distribution is and what it shows

- How to use ratios and rates

- How to visually display the distribution of a single variable

- What a cross-tabulation is and what it shows

- How to visually display the relationship between two variables

- Some guidelines for writing about statistical results

- How to create and format tables, graphs, and infographics

Introduction

One goal of statistical data analysis is to summarize data and present it in a way that everyone can understand, often using tables and graphs. Images are a powerful way to illustrate a problem or influence people's opinions, which is why most reports produced by community and government agencies rely heavily on well-designed graphics. On websites and social media, infographics are regularly used to present key facts. This chapter introduces techniques for summarizing one variable or the relationship between two variables. The emphasis is on displaying information from categorical variables (nominal- and ordinal-level variables), although I also discuss some techniques that can be used with ratio-level variables. More discussion of how to summarize ratio-level variables is included in chapters 3 and 4.

The research focus of this chapter is student loans—something that you may be familiar with. The total amount of student loans owed to the Canadian government is more than $15 billion, with students studying in Ontario and the Maritimes having the highest average debt loads. The Canadian Federation of Students argues that rising tuition costs and the government's use of loans as financial aid (instead of bursaries or grants) have led to the highest levels of student debt ever (CFS 2013). In 2012, students in Quebec protested for more than 100 days against proposed tuition increases. For many post-secondary students, financing their education has become a major source of concern and stress. If you have used loans to finance

FireAtDusk/iStockphoto

Photo 2.1   In the spring of 2012, students in Quebec protested for more than 100 days against potential tuition fee increases. This banner reads "Marching towards free education."

your education, I hope that the statistics in this chapter help to put your experience in context. Regardless of whether or not you have student loans, I hope that this chapter encourages you to think about why students need to borrow money to pay for school.

In this chapter, statistical analysis is used to investigate the following:

- What proportion of post-secondary students use government loans to finance their education?
- How much do students with government loans owe when they graduate?
- How has the use of loans by post-secondary students changed over time?
- What is the relationship between student loan use and tuition costs?

Frequency Distributions

The most basic way to summarize data is called a frequency distribution. In its simplest form, a **frequency distribution** provides a tally or count of how many people have each attribute of a variable. Figure 2.1 shows a simple frequency distribution that summarizes how many people responded "Yes" and "No" to a hypothetical question.

frequency distribution Shows how many cases in a dataset have each attribute of a variable, or how frequently each attribute occurs.

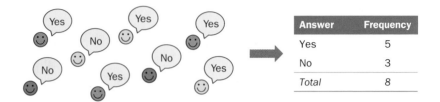

Figure 2.1   **An Illustration of a Simple Frequency Distribution**

Frequency distributions are usually reported in tables. The table in Figure 2.1 shows that five people responded "Yes," three people responded "No," and eight people in total gave an answer. In most statistical analyses, frequency tables are used to tally the answers of many more people than are shown in Figure 2.1—too many to accurately count without a computer. Although the number of people with each attribute is the starting point for every frequency table, most frequency tables also include other useful information.

Proportions and Percentages

proportion (p) Shows the fraction of people with an attribute out of a whole (out of 1).

percentage (%) Shows the proportion of people with an attribute, out of a base of 100.

Proportions show the fraction of people who do something or have some attribute, out of a whole, where the number 1 represents the whole. So, if half the number of people do something, the proportion is 0.5. If a quarter of the number of people do something, the proportion is 0.25. **Percentages** are proportions that are rescaled to be out of 100, instead of out of 1.

Let's use percentages to do some real-world data analysis. The National Graduates Survey (NGS) periodically collects information from college, CÉGEP (*Collège d'enseignement général et professionnel*, a type of college in Quebec), and university graduates across Canada, most recently from people who graduated in 2009–10. (See the "Spotlight on Data" box for more information.) Table 2.1 shows the distribution of responses to a survey question asking whether or not people used government student loans to finance their education. The first column in the table lists two attributes of the variable (Yes and No). The second column shows the frequency, or count, of people who gave each answer; these numbers are added together to find the total number of people who answered the question (5,596 + 6,809 = 12,405). The third column shows the percentage of graduates who did and did not use government student loans. In other words, it shows the proportion of people with each attribute, rescaled to be out of 100.

Table 2.1   **A Frequency Distribution with Counts and Percentages**

Did you use government student loans to finance your education?

Answer	Frequency	Percentage
Yes	5,596	45.1
No	6,809	54.9
Total	12,405	100.0

Source: Author generated; Calculated using data from Statistics Canada, 2015.

Spotlight on Data

The National Graduates Survey

Most of the data used in this chapter are from Statistics Canada's National Graduates Survey (NGS). The NGS's main goal is to collect information about the labour market experiences of Canadian post-secondary graduates (Statistics Canada 2008). Although Statistics Canada is an independent federal agency, it is organizationally situated under the umbrella of Industry Canada, whose mission is to "foster a growing, competitive, knowledge-based Canadian economy" (2014). Collecting good information about Canadians' educational activities, school-to-work transitions, and the match between people's education and their jobs is crucial to this mission.

The NGS is sent to graduates of public post-secondary educational institutions, including universities, colleges/CÉGEPs, and technical/trade schools. The survey has been administered to eight cohorts of graduates: those graduating in 1976, 1982, 1986, 1990, 1995, 2000, 2005, and 2009–10. Graduates are sent the survey two to three years after completing the requirements for their degree, diploma, or certificate; so people who graduated in 2009–10 were surveyed in 2013, people who graduated in 2005 were surveyed in 2007, people who graduated in 2000 were surveyed in 2002, and so on.

In addition to asking about labour market success, the NGS also collects information about how people paid for their education, including their use of loans, and how much debt they had accumulated at graduation. The examples in this chapter rely on the NGS questions about graduates' use of government student loans as well as loans from other sources, including banks and private loans from family or friends. Although the NGS collects information about students who have graduated from all levels of schooling, the examples in this chapter only use data from two groups of graduates: (1) those who graduated with a community college or CÉGEP certificate or diploma (labelled college/CÉGEP graduates), and (2) those who graduated with a university bachelor's degree, a university certificate below the bachelor's level, or a professional degree at the bachelor's level (labelled university graduates). Collectively, college/CÉGEP and university graduates are referred to as post-secondary graduates; the analysis in this chapter excludes graduates of technical or trade schools, as well graduates of master's or doctoral-level degree programs.

In addition, some data in this chapter are taken from the survey of Tuition and Living Accommodation Costs (TLAC) for full-time students at Canadian degree-granting institutions. This survey is sent to the administrative office of all Canadian universities and degree-granting colleges. The purpose of the survey is to collect Canada-wide information about program tuition costs, additional student fees, and residence/living accommodation costs (Statistics Canada 2014).

Table 2.1 shows that almost half (45 per cent) of the 2009–10 college/CÉGEP and university graduates who responded to the NGS used government student loans to finance their education. What this statistic means, however, depends on the argument you make. You might argue that this statistic provides evidence that post-secondary education has become unaffordable as almost half of students needed to borrow money from the government to finance their education. Alternatively, you might argue that this statistic provides evidence that government loan programs are effective as a large proportion of students are able to access them in order to finance their education. Neither argument is inherent in the statistic, which is why social science researchers need to contextualize and explain their results.

Percentages are often used to make comparisons between groups of different sizes. You probably compare your grades using this method. Imagine that you got 31 out of 42 questions correct on one test, and 23 out of 30 questions correct on a second test. (Assume that all of the questions were worth one point.) It's hard to quickly determine which test you did better on using the raw scores: you got more questions right on the first test, but it also had more questions. Converting both test scores to percentages makes them easy to compare. A quick calculation shows that you got 74 per cent of questions (the equivalent of 74 out of 100) correct on the first test and 77 per cent of questions (the equivalent of 77 out of 100) correct on the second test; thus, you can say that you did better on the second test since you got a higher proportion of questions correct.

The percentages shown in frequency tables are calculated in the same way as other percentages: find the number of people with an attribute, divide it by the total number of people who answered the question, and multiply by 100 to rescale it. For Table 2.1, the percentage of people who said yes is calculated as follows:

$$\% \text{ of people who said "Yes"} = \left(\frac{\text{number of people who said "Yes"}}{\text{number of people who answered the question}} \right) \times 100$$

$$= \left(\frac{5596}{12405} \right) \times 100$$

$$= 45.1$$

The same calculation is used to find the percentage of people who said "No," except that the number of people who said "No" is used in the numerator of the fraction instead of the number of people who said "Yes."

Let's represent the calculation for a percentage in a formula. You might not need it to calculate a percentage, but it is good to practice reading and using formulas. In the percentage formula, the F in the numerator stands for the frequency, or number, of people with the attribute of interest. In statistical notation, the letter N always stands for the total number of cases used in the calculation. In this formula the N in the denominator refers to the total number of people who answered the question.

$$\% = \left(\frac{F}{N}\right)100$$

percentage

Note that the multiplication symbol "×" is omitted from the formula. In mathematical notation, the default operation for two elements placed immediately beside one another is multiplication. In other words, unless a formula tells you to do something else with two elements, you should multiply them together. Once you have determined what each symbol in a formula stands for, just substitute in the numbers and solve for the answer by dividing, multiplying, adding, and subtracting. Be sure to use the correct order of operations: complete anything in parentheses first; then calculate exponents; then perform division, multiplication, addition, and subtraction in that order. (You may have learned to remember this order using the acronym PEDMAS or BEDMAS [Parentheses/Brackets, Exponents, Division, Multiplication, Addition, Subtraction].)

The formula for a proportion is almost the same as for a percentage; the only difference is that you do not multiply by 100:

$$p = \left(\frac{F}{N}\right)$$

proportion

As a result, it's easy to translate between proportions and percentages. If you are starting with a proportion, multiply it by 100 (or move the decimal point two places to the right) to find the percentage: so 0.75 is 75 per cent. If you are starting with a percentage, divide it by 100 (or move the decimal point two places to the left) to find the proportion: so 40 per cent becomes 0.40.

For nominal-level variables, it can be useful to look at a frequency table that lists the attributes in order from most to least popular (or the reverse). Table 2.2 shows a descending count frequency distribution, which lists the most popular attributes at the top and the least popular attributes at the bottom, while Table 2.3 shows an ascending count frequency distribution, which lists the least popular attributes at the top and the most popular attributes at the bottom. Both tables show the reasons why people did not apply for government student loans. Listing people's responses in order of popularity makes it easy to see that the most common reason that people did not apply for a government student loan was because they did not need one. The second-most common reason that people did not apply for a loan was because they did not think their (or their family's) income was low enough to qualify. The least common reason that people did not apply for a loan was because they preferred to borrow from some other source (such as a bank).

Table 2.2   **A Descending Count Frequency Distribution**

What is the main reason you did not apply for a government-sponsored student loan? (among people who did not apply for a loan to finance their post-secondary education)

Answer	Frequency	Percentage
I did not need one	3,796	63.2
I didn't think I would qualify/income was too high (self, spouse, or parents)	1,166	19.4
Another reason	371	6.2
I was not willing to borrow the amount required to go to school	291	4.8
I didn't know how to apply	184	3.1
I didn't meet the residency requirements	133	2.2
I preferred to borrow from another source	65	1.1
Total	*6,006*	*100.0*

Source: Author generated; Calculated using data from Statistics Canada, 2015.

cumulative percentage Shows the percentage of cases with an attribute or one ranked below it (with a lower value).

For ordinal- and ratio-level variables, it is better to use frequency tables that list the attributes in their original order (so avoid using ascending or descending count frequency tables). Calculating and reporting cumulative percentages for ordinal- and ratio-level variables is also often useful. A **cumulative percentage** is the percentage of cases with an attribute or one ranked below it (with a lower value). For instance, in the NGS, people were asked how much money they owed in government student loans when they graduated. Their answers are grouped into five categories: $0, $1 to $4,999, $5,000 to $9,999, $10,000 to $24,999, or $25,000 or more. Because the amounts owing are class intervals, this variable is ordinal level (and not

Table 2.3   **An Ascending Count Frequency Distribution**

What is the main reason you did not apply for a government sponsored student loan? (among people who did not apply for a loan to finance their post-secondary education)

Answer	Frequency	Percentage
I preferred to borrow from another source	65	1.1
I didn't meet the residency requirements	133	2.2
I didn't know how to apply	184	3.1
I was not willing to borrow the amount required to go to school	291	4.8
Another reason	371	6.2
I didn't think I would qualify/income was too high (self, spouse, or parents)	1,166	19.4
I did not need one	3,796	63.2
Total	*6,006*	*100.0*

Source: Author generated; Calculated using data from Statistics Canada, 2015.

ratio level). Table 2.4 shows a frequency distribution with columns for the number of students owing each amount, the percentage of students owing each amount, and the cumulative percentage of students owing each amount. The first row shows that about one in five graduates who used government student loans, or 20 per cent, owed $0 when they graduated; that is, they repaid all of their loans before graduation. The second row shows that 9 per cent owed between $1 and $4,999 when they graduated. The cumulative percentage column in the second row shows that 29 per cent of graduates who used loans owed less than $5,000 when they graduated: 20 per cent owed $0 and 9 per cent owed $1 to $4,999, which sums to 29 per cent. Almost half (45 per cent) of graduates who used loans owed less than $10,000 at graduation (20 per cent plus 9 per cent plus 16 per cent). To calculate how many people owed more than a certain amount, subtract the cumulative per cent from the total of 100 per cent. So, 55 per cent of respondents who used government student loans owed $10,000 or more when they graduated (calculated as 100 minus 45, which is the cumulative percentage of people who owe less than $10,000).

Another way to find a cumulative percentage is to add together the number of cases with each attribute or a lower attribute, and then use the sum of cases to calculate a percentage of the total. For instance, 1,525 people reported that they owed less than $5,000 in student loans (1,059 + 466 = 1,525). You can use this sum (1,525) to calculate that 29 per cent of people owed less than $5,000 in student loans, since the result of 1,525 divided by 5,265 and then multiplied by 100 is 29. Similarly, 2,344 people reported that they owed less than $10,000 in student loans (1,059 + 466 + 819 = 2,344). You can use this sum (2,344) to calculate that 45 per cent of people owed less than $10,000 in student loans, since the result of 2,344 divided by 5,265 and then multiplied by 100 is 45 (or 44.5). This method sometimes generates cumulative percentages that are slightly different from those calculated by adding together the percentages because of rounding error. Because cumulative percentages rely on whether an attribute is ranked higher or lower than other attributes of a variable, it does not make sense

Table 2.4   A Frequency Distribution with Counts, Percentages, and Cumulative Percentages

How much money did you owe in government student loans when you graduated? (among those with loans)

Answer	Frequency	Percentage	Cumulative Percentage
$0	1,059	20.1	20.1
$1 to $4,999	466	8.9	29.0
$5,000 to $9,999	819	15.6	44.5
$10,000 to $24,999	1,721	32.7	77.2
$25,000 or more	1,200	22.8	100.0
Total	5,265	100.0	

Source: Author generated; Calculated using data from Statistics Canada, 2015.

to report them for nominal-level variables since the attributes of nominal-level variables do not have an order.

The frequency distribution in Table 2.4 shows that more than half of students who use government loans (55 per cent) graduate owe $10,000 or more, and one in five (23 per cent) owe $25,000 or more. Surprisingly, 20 per cent of those who use loans report repaying them before they graduate. If you have a student loan, you can compare how much you expect to owe when you graduate to the situation of students in Canada overall.

Frequency distributions (or frequency tables) are the main tool that researchers use to summarize categorical variables. They provide a quick overview of how common each attribute is. Many researchers begin their data analysis by looking at a frequency table of each variable that they are interested in. But, for variables that have many attributes—including most continuous variables and some count variables—frequency distributions are too large to be useful. A frequency table listing the exact amount of each graduate's student loan has thousands of rows. There are better ways to summarize continuous variables and count variables with many attributes/values, which Chapter 4 will discuss.

How Does It Look in SPSS?

Frequency Distributions

The Frequencies procedure produces a table that looks like the one in Image 2.1. The variable label appears at the top of the table. Above each frequency table, SPSS prints a "Statistics" box (not shown) that lists the number of valid cases and the number of missing cases.

Debt size of government student loans at time of graduation

		⒟ Frequency	⒠ Percent	⒡ Valid Percent	⒢ Cumulative Percent
ⒶValid	$0	1059	8.4	20.1	20.1
	Less than $5,000	466	3.7	8.9	29.0
	$5,000 to less than $10,000	819	6.5	15.6	44.5
	$10,000 to less than $25,000	1721	13.7	32.7	77.2
	$25,000 or more	1200	9.5	22.8	100.0
	Total	5265	41.8	100.0	
ⒷMissing	Valid skip	6809	54.1		
	Not stated	514	4.1		
	Total	7323	58.2		
ⒸTotal		12588	100.0		

Image 2.1 An SPSS Frequency Distribution

A. The "Valid" label shows which attributes are being treated as valid (not missing). There are five valid attributes in this table. There is an indented "Total" label for the row that shows the sum of the cases with valid attributes.

B. The "Missing" label shows which attributes are being treated as missing. There are two missing attributes in this table. The "Valid skip" label indicates that the person skipped the question for a valid reason; in this example, people who did not use government student loans were not asked this question. If there is more than one missing attribute, SPSS prints an indented "Total" label for the row that shows the sum of the cases with missing attributes.

C. The "Total" label aligned on the far left indicates that this row shows the totals for all cases, regardless of whether they have a valid or a missing attribute.

D. The "Frequency" column shows the number of cases with each attribute in the dataset.

E. The "Percent" column shows the percentage of cases with each attribute in the dataset, out of the total number of cases. For example, 1,059 is 8.4 per cent of 12,588.

F. The "Valid Percent" column shows the percentage of cases with each attribute in the dataset, out of the total number of valid cases. For example, 1,059 is 20.1 per cent of 5,265. This is the percentage that is most often reported.

G. The "Cumulative Percent" column shows the percentage of cases with each attribute, or an attribute with a value lower than it, out of the total number of valid cases. For example, 29.0 per cent of the 5,265 people who answered this question owed less than $5,000 (including $0) when they graduated. SPSS prints the cumulative percent column for all frequency distributions, but it only makes sense to report this information for ordinal- and ratio-level variables.

Step-by-Step: Percentages

$$\% = \left(\frac{F}{N}\right)100$$

Step 1: Count the number of cases with the attribute or characteristic you are interested in (F).

Step 2: Count the total number of cases (N). Do not include cases that are missing information for the variable.

Step 3: Divide the number of cases with the attribute (from Step 1) by the total number of cases (from Step 2).

Step 4: Multiply the result of Step 3 by 100 to find the percentage of cases with the attribute or characteristic.

Tip: To find the proportion of cases with an attribute, stop after completing Step 3.

Ratios and Rates

Researchers use ratios and rates to illustrate how common or rare something is. Like percentages and proportions, ratios and rates standardize information so that making comparisons is easy. A **ratio** shows how the frequencies of two attributes compare directly to each other. For instance, a ratio can be used to compare how many graduates have student loans, compared to how many do not. Looking back, Table 2.1 shows that 5,596 graduates had student loans, and 6,809 did not. So, the ratio of student loan use, compared to not using student loans, is calculated as:

ratio Shows how the frequencies of two attributes compare directly to each other.

$$ratio\ of\ student\ loan\ use = \frac{number\ of\ students\ who\ used\ loans}{number\ of\ students\ who\ did\ not\ use\ loans} = \frac{5596}{6809} = 0.82$$

This ratio shows that for every 0.82 students who have a loan, 1 student does not. Ratios are often expressed using a colon between the two numbers. So, the ratio of students who used loans to students who did not use loans is 0.82:1. When ratios are smaller than one, researchers often multiply both sides of the ratio by the same number, usually a multiple of 10, to make it easier to understand. For example, when both sides of this ratio are multiplied by 10, we can say that for every 8.2 students who used loans, there were 10 students who did not use loans. When both sides of this ratio are multiplied by 100, we can say that for every 82 students who used loans, there were 100 students who did not use loans.

So, ratios are used to compare the number of people with one attribute to the number of people with a second attribute. In the ratio formula, F_1 is used to denote the frequency or number of people with the first attribute. The subscript number 1 on the F indicates that it is for the first attribute. F_2 is used to denote the frequency or number of people with the second attribute, since the F has a subscript number 2. So, the formula for a ratio is:

ratio

$$ratio = \frac{F_1}{F_2}$$

rate Shows the number of times that something occurs, relative to the number of times that it could possibly occur.

Another way that researchers summarize results is by using rates. A **rate** shows the number of times that something occurs, relative to the number of times that it could possibly occur. You are probably familiar with statistics like the fertility rate, which shows the number of live births per woman, or the mortality rate, which shows the number of deaths that occur per every 100,000 people.

Let's calculate the rate of student loan use using Table 2.1 again. The table shows that there were 12,405 students who graduated from university or college/ CÉGEP in 2009–10 and answered the survey. Of those graduates, 5,596 of them used student loans. The rate of student loan use is calculated by dividing the number of graduates who used student loans by the number of graduates overall (the number of students who could have used loans). So, the basic rate of student loan use is:

$$\text{rate of student loan use} = \frac{\text{number of students who used loans}}{\text{number of students who could have used loans}} = \frac{5596}{12405}$$

So far, this calculation is very similar to the calculation for a percentage, which is effectively a rate out of 100. Usually, though, researchers standardize rates by using division or multiplication to calculate a rate out of a denominator other than 100. The choice of what denominator (or base) to use depends on how common or rare an occurrence is. For instance, imagine that I wanted to report the number of people out of every 1,000 post-secondary graduates that used student loans. By using cross-multiplication techniques to solve for x in this equation, I can find out how many graduates out of every 1,000 used student loans (in later chapters, the notation x will be used to represent a variable, but here x represents an unknown quantity):

$$\frac{5596}{12405} = \frac{x}{1000}$$

To cross-multiply, the numerator of the first fraction is multiplied by the denominator of the second fraction, and the denominator of the first fraction is multiplied by the numerator of the second fraction. The two products are equivalent to each other, so:

$$(12405)(x) = 5596(1000)$$
$$12405x = 5596000$$
$$x = \frac{5596000}{12405}$$
$$x = 451.12$$

The product 5,596,000 on the right-hand side of the equation is divided by 12,405 to find the value of x. The calculation shows that 451 out of every 1,000 students who graduated in 2009–10 used student loans.

This method can also be used to standardize the value of the numerator instead of the denominator. Typically, the numerator is set to 1, in order to assert that something occurs 1 out of every x possible times. So, to be able to say that 1 out of every x number of students who graduated in 2009–10 used student loans, I solve for x in the following equation:

$$\frac{5596}{12405} = \frac{1}{x}$$

Again, using cross-multiplication, I know that:

$$(5596)(x) = 12405(1)$$
$$5596x = 12405$$
$$x = \frac{12405}{5596}$$
$$x = 2.22$$

This calculation shows that 1 out every 2.2 students who graduated in 2009–10 used student loans. Media reports often present rates because people understand them more intuitively. As you become more proficient at calculating and interpreting statistics, you can use these techniques to generate more complex rates, such as 2 out of every 5 people, or 7 out of every 10 people. The "Best Practices in Presenting Results" section of this chapter illustrates how specific percentages translate into commonly reported rates.

Step-by-Step: Rates

Rate with a Standardized Denominator (base)

$$\frac{F}{N} = \frac{x}{\text{base of the rate}}$$

Step 1: Count the number of cases with the attribute or characteristic you are interested in (F).

Step 2: Count the total number of cases (N). Do not include cases that are missing information for the variable.

Step 3: Decide what denominator or base you want to use in the rate. Common bases are 1,000 and 100,000.

Step 4: Multiply the number of cases with the attribute (from Step 1) by the base you are using in the rate (from Step 3).

Step 5: Divide the result of Step 4 by the total number of cases (from Step 2) to find the rate at which an attribute or characteristic occurs, out of the base that you selected.

Rate with a Standardized Numerator (1)

$$\frac{F}{N} = \frac{1}{x}$$

Step 1: Count the number of cases with the attribute or characteristic you are interested in (F).

Step 2: Count the total number of cases (N). Do not include cases that are missing information for the variable.

Step 3: Divide the total number of cases (from Step 2) by the number of cases with the attribute (from Step 1). The result shows that 1 out of every x number of cases have the attribute or characteristic.

Visualizing Data

Graphs are the main tools that social science researchers use to visually display the distribution of a variable. But other data visualization strategies are becoming

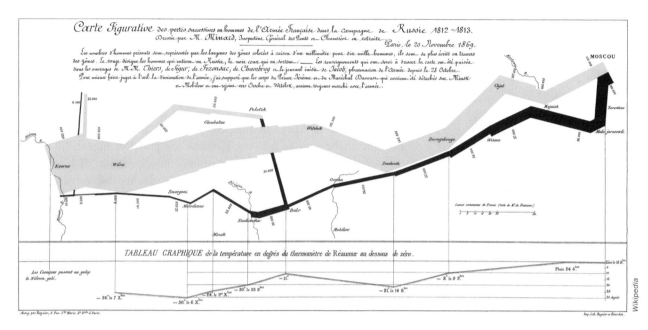

Photo 2.2   **Minard's depiction of Napoleon's advance on and retreat from Moscow.**

more prominent as a result of easy-to-use online graphics tools. Data visualization has always been a topic of interest to statistical data analysts because of the power of pictures to clearly and effectively display complex information. Edward Tufte, one of the most pre-eminent data visualization theorists, argues that graphics *reveal* data in a way that statistical computations cannot (Tufte 1999, 13, ital. orig.). Tufte identifies an 1869 map by Charles Joseph Minard, a French civil engineer, depicting Napoleon's 1812–13 advance on (and retreat from) Moscow as "probably the best statistical graphic ever drawn." (See above.) Minard's graphic uses space, size, and colour to juxtapose several types of information. The number of troops and their movements are represented by the width and location of the paths: the light brown band shows the army's advance on Moscow and the black band shows their retreat, in much smaller numbers. The graph shown below the paths links the declining number of troops in retreat to the below-zero temperatures that dropped steadily over time.

Displaying data in pictures and graphs can help researchers to better understand their own results and to convey those results to others.

Visualizing the Distribution of a Single Variable

You are probably already familiar with several common ways to display the distribution of a single variable. The graphing techniques used by data analysts rely on the same general principles as those taught in elementary schools, but the emphasis shifts to selecting the best type of graph to display the available data and modifying it so that it is easy to read. The decision about which graph is best for displaying a distribution is often related to the variable's level of measurement. In this section, I describe several ways to graph a single variable. I also introduce several data visualization techniques that are used primarily in infographics.

pie graph A graph that depicts the proportion or percentage of cases with each attribute as a portion of the area in a circle.

Pie Graphs **Pie graphs** represent the percentage of cases with each attribute as a proportion of the area of a circle. (See the left panel of Figure 2.2.) Each slice of the pie provides a visual representation of how common each attribute is relative to the others. Pie graphs are useful for displaying the distribution of dichotomous variables and nominal-level variables with relatively few attributes (usually fewer than five). Pie graphs with more than five slices are difficult to read. A popular variation of the pie graph is the doughnut graph. (See the right panel of Figure 2.2.) A doughnut graph is the same as a pie graph, but the centre of the pie is left blank (like the hole in the centre of a doughnut). This space is sometimes used to present a key finding.

bar graph A graph that depicts the proportion or percentage of cases with each attribute using the relative height or length of a bar.

Bar Graphs **Bar graphs** display the distribution of categorical variables by using the length (or height) of bars to show the percentage of cases with each attribute. Bar graphs are particularly useful for visualizing the distribution of ordinal-level variables because the order of the attributes is preserved in the order of the bars. Bar graphs can also be used to display the distribution of nominal variables with many attributes. Bar graphs of nominal-level variables should show the bars arranged from most to least common (or the

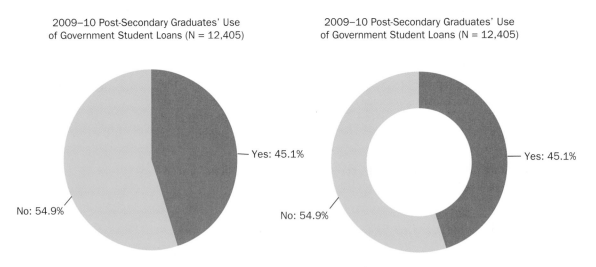

2009–10 Post-Secondary Graduates' Use of Government Student Loans (N = 12,405)

2009–10 Post-Secondary Graduates' Use of Government Student Loans (N = 12,405)

Yes: 45.1%

No: 54.9%

Yes: 45.1%

No: 54.9%

Figure 2.2   **A Pie Graph and a Doughnut Graph That Correspond to the Frequency Distribution in Table 2.1**

Source: Author generated; Calculated using data from Statistics Canada, 2015.

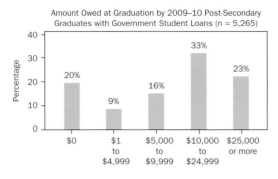
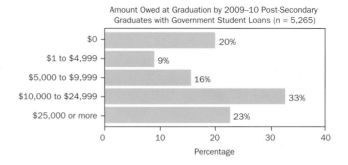

Figure 2.3 **Vertical and Horizontal Bar Graphs That Correspond to the Frequency Distribution in Table 2.4**

Source: Author generated; Calculated using data from Statistics Canada, 2015.

reverse) for easy interpretation. Bar graphs can be constructed with bars displayed either horizontally or vertically (sometimes called column graphs). (See Figure 2.3.)

Box Plots and Histograms   Two other types of graphs are sometimes used to show the distribution of a single variable: box plots and histograms. Box plots are typically used to show ordinal-level variables, and they provide information about the centre and spread of a variable. Box plots are described in Chapter 3. Histograms are used to show continuous variables (and sometimes count variables). A histogram is similar to a bar graph, but there are no gaps between the bars. In a bar graph, the space between the bars visually depicts the fact that the variable is categorical—that is, there are no legitimate answers between the specified categories or attributes. Histograms are described in Chapter 4.

Pictographs   Online data visualization tools have made it easy to use a variety of other strategies to summarize data using pictures. Some of these tools allow a user/viewer to interact with a visualization, instead of presenting a static image. One popular technique is to use **pictographs** or pictograms to visually represent a proportion or percentage. A pictograph typically converts a proportion or percentage into a rate and then illustrates that rate using coloured/shaded icons. For instance, if we look back at Table 2.1, we see that it shows that 45 per cent of 2009–10 post-secondary graduates used government student loans. This percentage can be converted into a rate out of 20: 45 per cent is equivalent to 9 out of 20. A pictograph represents this rate by showing 9 shaded icons out of a total of 20 icons. (See Figure 2.4.) The icon used in a pictograph can be changed to match the topic; in Figure 2.4, for example, I used graduation hats to indicate that I am presenting information about an education-related topic.

pictograph A data-visualization strategy that illustrates a rate using coloured/shaded icons.

Bubble Graphs   Other visualization strategies use size or area to display the differences between groups. For instance, Figure 2.5 shows how the frequency distribution in Table 2.4 can be represented using different-sized bubbles (called a bubble

In 2009–10, 9 out of every 20 college/CÉGEP &
university graduates used government student loans

Figure 2.4   **A Pictograph That Corresponds to the Frequency Distribution in Table 2.1**

Source: Author generated; Calculated using data from Statistics Canada, 2015.

graph). Different-sized icons (like those used in pictographs) can also be used to represent a frequency distribution. These strategies make it easy to see which attributes of a variable are the most and least common.

There are a wide range of tools available to visually display the distribution of a single variable. The choice about which is best depends on several things: the variable's level of measurement, the number of categories (in categorical variables), the complexity of the information you want to convey, who your audience is, and where and how you intend to distribute the graphic. Although a full discussion of data-visualization strategies and decisions is beyond the scope of this book, you are encouraged to always try out several different options and then select the one that best illustrates the argument that you are making.

Amount Owed at Graduation by 2009–10 Post-Secondary Graduates with
Government Student Loans (n = 5,265)

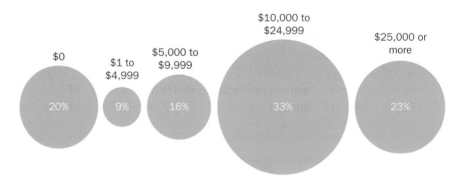

Figure 2.5   **A Bubble Graph That Corresponds to the Frequency Distribution in Table 2.4**

Source: Author generated; Calculated using data from Statistics Canada, 2015.

Displaying Change across Space or Time

Map-Based Graphics   Data visualization techniques are also useful for showing how the distribution of a variable changes across space or time. Map-based graphics allow researchers to depict how the distribution of a variable changes in relation to geography. For instance, you might be curious about whether the percentage of post-secondary students who use government student loans varies across Canada, especially since tuition is regulated at the provincial level. Figure 2.6 uses a map to show the percentage of 2009–10 university graduates from each province who used government student loans. (Data for university and college/CÉGEP students combined was not available.) Gradations in shading/colour are associated with different percentages: the province with the lowest percentage is the lightest purple and the province with the highest percentage is the darkest purple. The map shows that university graduates from New Brunswick are the most likely to use government student loans, and those from Manitoba are least likely to use student loans (information for the territories is not available) (Statistics Canada 2010).

Percentage of 2009–10 University Graduates Who Used Government Student Loans, by Province

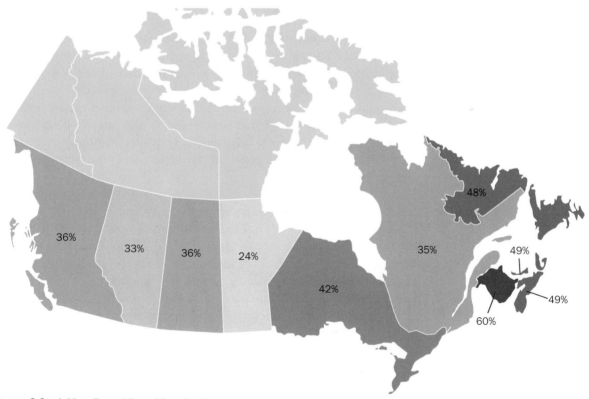

Figure 2.6   A Map-Based Data Visualization

Source: Author generated; Calculated using data from Statistics Canada, 2010.

Bar Graphs   Graphing how the distribution of a variable has changed over time lets you make claims about whether some aspect of our social world is getting better, getting worse, or holding steady. Although it's useful to know what percentage of 2009–10 post-secondary graduates used government student loans to finance their education, the obvious question is whether or not this percentage is increasing or decreasing. The bar graphs in Figure 2.7 show the percentage of graduates who borrowed money to finance their education for each year of the NGS since 1982. Because of changes to the survey questions over time, it's only possible to compare the percentage of students who borrowed money from any source to finance their education. (See the "Hands-On Data Analysis" box in this chapter.)

Although both bar graphs in Figure 2.7 show the same statistical information, the scale of the vertical axis (i.e., the y-axis) in the bottom graph has been modified to accentuate the differences over time. In the upper graph, the scale of the vertical axis ranges from 0 to 70 per cent. In the lower graph, the scale of the vertical axis ranges from 40 to 65 per cent. The upper graph emphasizes how many students overall borrow to finance their education since the bars are all quite tall. The lower graph emphasizes the change in the percentage of students who borrow over time since the differences between the heights of the bars are more pronounced. When adjusting the scale of graphs, the goal is to highlight the features of the data that are consistent with your argument, without deceiving the reader.

These graphs show that the percentage of students who used loans to finance their education grew between 1982 and 2000, and then levelled off. Students' high reliance on loans can have a negative effect on the economy, since students with high debt loads must limit their post-graduation spending in order to make loan repayments. Using data from two other Statistics Canada surveys, Luong (2010) found that post-secondary graduates with student debt were less likely to own a home, were less likely to have investment income, and had fewer assets overall than post-secondary graduates of the same age who did not have student debt.

Line Graphs   Line graphs are another way to graph change over time. Figure 2.8 shows the change in average Canadian university undergraduate tuition fees from 1980 to 2013. (Similar data for college tuition fees are not available.) The line graph shows how average undergraduate tuition fees have increased steadily over time. (You'll learn more about averages in Chapter 4.) The graph also shows that there were relatively small annual increases in average tuition fees from 1980 to 1990 but that from 1990 onward the increases were much larger. In order to make a fair comparison, I have included a second line on the graph that illustrates how price levels changed as a result of inflation during the same period. Comparing the two lines shows that average tuition fees increased on par with prices until 1990, at which point they began to rise much faster than prices overall. Schwartz and Finnie (2002) argue that the dramatic rise in Canadian university tuition fees during the

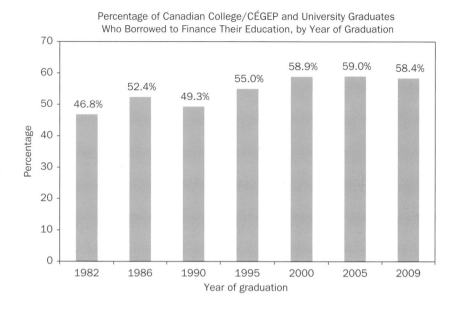

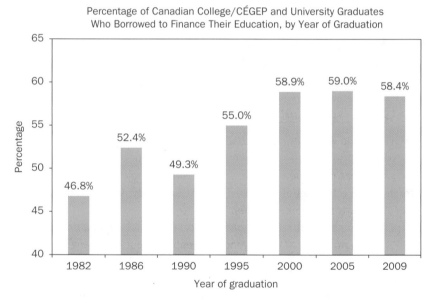

Figure 2.7   Two Versions of a Bar Graph Showing Change in a Statistic over Time

Source: Author generated; Calculated using data from Statistics Canada, 2015.

1990s was the result of stagnant levels of government funding, combined with increasing levels of student enrolment. Government grants to universities dropped from $7,569 per student in 1990 to $5,543 per student in 1997 (Schwartz and Finnie 2002). Since universities had few other sources of funding, they were forced to increase tuition fees, as well as class sizes, in response.

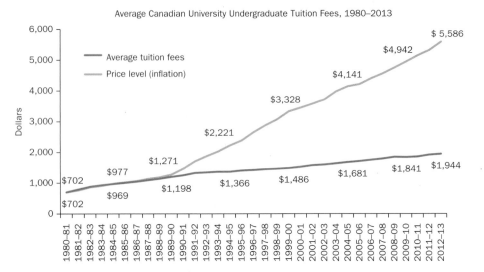

Figure 2.8   **A Line Graph Showing Change in a Statistic over Time**

Source: Author generated; Calculated using data from Statistics Canada, 2014.

Hands-on Data Analysis

Combining Variables

Researchers who work with survey data often need to modify the variables included in a dataset in order to answer their research questions. This might include combining the attributes of two or more variables to create a new variable or grouping the attributes of a variable in a different way. In this box, I explain how to combine two or more variables.

The questions about student loans used in the NGS changed over time. In 1982, graduates were asked a general question about borrowing: "Did you ever borrow money to finance any of your education?" From 1986 onwards, graduates were asked two questions about borrowing: "Did you ever borrow money to finance any of your education through the student loan program?" and "Did you ever borrow money to finance any of your education from other sources such as relatives or directly from a bank?" In order to make comparisons between the 1982 survey and later surveys, in the later survey datasets I combined people's answers to the two questions about borrowing into a single new variable indicating whether they had ever borrowed money, from any source.

All statistical software programs allow users to create new variables based on the attributes of existing variables. The key to combining variables is to think through what you want the result to be for cases with all of the different combinations of attributes. Both of the NGS variables that I combined had only two attributes: yes and no. The new variable that I created used the following logic:

if "borrowed from student loans" = yes
 or "borrowed from other sources" = yes;
 then "ever borrowed" = yes
if "borrowed from student loans" = no
 and "borrowed from other sources" = no;
 then "ever borrowed" = no

These commands are programmed into the statistical software, substituting the variable names and values for the descriptions and attributes that are written out in words. Once the new variable is created, be sure to generate a frequency distribution to make sure that the number of people with each attribute matches with what you expected. You can also generate a cross-tabulation between the new variable and each of the variables used to create it, in order to ensure that all of the cases have the correct attributes.

Writing out what you want to happen to each group of cases can help you to think through how to give a statistical software program instructions for combining variables. This can be especially helpful as your analyses become more advanced and you want to combine variables with many attributes or more than two variables.

When variables are combined, the researcher must decide what to do with cases that have attributes that are designated as missing. If a case has missing values on all of the variables that are being combined, then the case should be assigned a missing value on the new variable. If a case has a missing value on only one of the variables being combined, then it is up to the researcher whether it should be assigned a valid or missing value on the new variable. This decision depends on the meaning of the questions/variables that are being combined. What is most important is that you are able to justify the decisions that you make when combining variables and that you apply the same logic to all of the cases in a dataset.

Cross-Tabulations

Cross-tabulations, or "crosstabs" for short, are used to show the relationship between two categorical variables (nominal- or ordinal-level variables). In other words, they show the frequency distribution of a variable for different subgroups of people or cases. The frequency distribution of each subgroup is called a conditional

cross-tabulation Shows how the distribution of one variable is related to or is contingent on group membership (as defined by another variable).

distribution because it only shows the distribution for people who meet some condition—that is, they belong to a specific group. Cross-tabulations are sometimes called contingency tables because the distribution that is displayed is contingent on group membership.

The idea of "contingency" is illustrated in the frequency distributions in Table 2.5. The table on the left shows the percentage of college/CÉGEP graduates who used government student loans to finance their education. The table on the right shows the percentage of university graduates who used government student loans to finance their education. The frequency distribution of student loan use changes depending on—or contingent on—whether a person is a college/CÉGEP student or a university student (the condition). The frequency tables show that half (50 per cent) of university students used government student loans to finance their education, compared to only 39 per cent of college students. This is likely because university graduates spend longer in school: an undergraduate university degree typically takes four years, whereas many college or CÉGEP diplomas can be completed in two years. The cost of university tuition is also generally higher than college tuition, and some CÉGEP programs are free to attend.

The first step in producing a cross-tabulation is to decide which variable will be treated as the independent variable (IV), and which variable will be treated as the dependent variable (DV). For the example in Table 2.5, this means deciding which of these statements makes more sense:

A person's use of student loans (DV) depends on their level of study (IV).
Or
A person's level of study (DV) depends on their use of student loans (IV).

Here, the first statement clearly makes more sense. It's likely that the tuition fees charged by different institutions prompt students to apply for loans (or not); it's much less likely that getting a student loan (or not) influences someone's level of study. This is especially true because most student loan applications require applicants to specify which post-secondary institution they will hold their loan at. Notice that in these statements I use the phrase "depends on" instead

Table 2.5   **Two Frequency Distributions That Illustrate the Idea of Contingency**

Did you use government student loans to finance your education?

College/CÉGEP Graduates			University Graduates		
Answer	Frequency	Percentage	Answer	Frequency	Percentage
Yes	1,979	38.8	Yes	3,617	49.5
No	3,116	61.2	No	3,693	50.5
Total	*5,095*	*100.0*	*Total*	*7,310*	*100.0*

Source: Author generated; Calculated using data from Statistics Canada, 2015.

of the language of "cause" and "effect." There are very strict criteria for making claims about causality that most survey data do not meet. When researchers use cross-tabulations to investigate the relationship between two variables, they can describe the association between the variables or how the frequency distribution is contingent on group membership, but they cannot claim that one thing causes the other. There are many factors that affect whether someone uses government student loans, and we do not have enough information to definitively identify the cause of student loan use.

Cross-tabulations show the same information as frequency tables that are divided by group, like those in Table 2.5. The advantage of cross-tabulations is that they are easier to produce and read, especially if the variables have many attributes. As a general strategy, the attributes of the variable that is being treated as the independent variable should be placed in the columns of the cross-tabulation, and the attributes of the variable that is being treated as the dependent variable should be placed in the rows. Then, calculate or request "column percentages"—that is, percentages that show the distribution of the dependent variable within each column or group defined by the independent variable. This approach produces cross-tabulations that are easiest for beginning researchers to interpret.

The cross-tabulation in Table 2.6 corresponds to the frequency distributions in Table 2.5. The numbers in the column labelled "College/CÉGEP Graduates" match those in the frequency and percentage columns of the frequency table for college/CÉGEP graduates. The numbers in the column labelled "University Graduates" match those in the frequency and percentage columns of the frequency table for university graduates.

To interpret the results of a cross-tabulation, compare the percentages across each row. A helpful saying used by statistics instructors is "percentage down and compare across." In this cross-tabulation, comparing across the rows shows us that 39 per cent of college/CÉGEP graduates used government student loans, compared to 50 per cent of university graduates. Conversely, 61 per cent of college/CÉGEP graduates did not use government student loans, compared to 51 per cent

Table 2.6   A Cross-Tabulation

Did you use government student loans to finance your education?		Level of Study Graduated from in 2009–10		
		College/ CÉGEP Graduates	University Graduates	Total
Yes	Count (Frequency)	1,979	3,617	5,596
	Column % (Percentage)	38.8%	49.5%	45.1%
No	Count (Frequency)	3,116	3,693	6,809
	Column % (Percentage)	61.2%	50.5%	54.9%
Total	Count (Frequency)	5,095	7,310	12,405
	Column % (Percentage)	100.0%	100.0%	100.0%

Source: Author generated; Calculated using data from Statistics Canada, 2015.

How Does It Look in SPSS?

Cross-Tabulations

The Crosstabs procedure, with "Observed" counts and "Column" percentages selected in the Cells option, produces a table like the one in Image 2.2. The labels of the two variables in the cross-tabulation appear at the top of the table. Above each cross-tabulation, SPSS prints a "Case Processing Summary" (not shown) that lists the number and percent of valid cases, missing cases, and total cases.

Borrowed money from government for student loans * Level of studies graduated from in 2009-2010 Crosstabulation

			(F) Level of studies graduated from in 2009-2010		
			(G) College/ CEGEP graduates	(H) University graduates	(I) Total
(A) Borrowed money from government for student loans	(B) Yes (C)	Count	1979	3617	5596
		(D) % within Level of studies graduated from in 2009-2010	38.8%	49.5%	45.1%
	No	Count	3116	3693	6809
		% within Level of studies graduated from in 2009-2010	61.2%	50.5%	54.9%
(E) Total		Count	5095	7310	12405
		% within Level of studies graduated from in 2009-2010	100.0%	100.0%	100.0%

Image 2.2   **An SPSS Cross-Tabulation**

A.   The label of the row variable appears on the far left. The "Borrowed money from government" variable is placed in the rows because I am treating it as the dependent variable in this analysis.

B.   The attributes of the row variable are listed in a vertical column. Only valid attributes are shown in an SPSS cross-tabulation; rows and columns for attributes designated as missing are not printed.

C.   The "Count" label indicates that the first line in each cell shows the observed count or the number of cases with the attribute, within the groups defined by the column variable.

D.   The "% within" label indicates that the second line in each cell shows the percentage of cases with the attribute, out of the total number of cases in each column. You know that the percentages are out of the total number of cases in each column because the variable name after the "% within" label is the column variable.

E. The "Total" label indicates that these two rows show the overall column totals.

F. The label of the column variable appears above the columns. The "Level of studies" variable is placed in the columns because I am treating it as the independent variable in this analysis.

G. This column shows information for the first group defined by the column variable. The first line in the column shows that 1,979 college/CÉGEP graduates reported that "Yes" they did borrow money from the government for student loans. The second line in the column shows that 38.8 per cent of the 5,095 college/CÉGEP graduates reported that "Yes" they did borrow money from the government for student loans. The number and percentage of college/CÉGEP graduates who reported "No" they did not borrow money are reported in the third and fourth lines, respectively. The fifth line shows the total number of college/CÉGEP graduates, and the sixth line shows that the percentages sum to 100. (These lines are part of the "Total" row.)

H. This column shows the same information as G, but for the second group defined by the column variable (university graduates).

I. This column shows the overall totals for the row variable, ignoring the groups defined by the column variable. The first line in this column shows that 5,596 graduates overall reported that "Yes" they did borrow money from the government for student loans (regardless of whether they were college/CÉGEP or university graduates). The second line shows that 45.1 per cent of the 12,405 people who answered the question about using government student loans said "Yes." The information in this column will match the information in a frequency table of the row variable (as long as the column variable does not have any missing attributes that result in cases being excluded from the cross-tabulation).

of university graduates. (The percentages for university graduates sum to 101 because of rounding.) These results suggest that there is an association between students' level of study and whether or not they use government loans to finance their education.

Visualizing the Relationship between Two Variables

Much like data visualizations for one variable, the goal of using data visualizations for two variables is to convey information about a relationship in an easy-to-understand way. In this section I illustrate three general strategies for showing relationships using graphs: panelling, clustering, and stacking. I also discuss graphing two or more variables on the same chart.

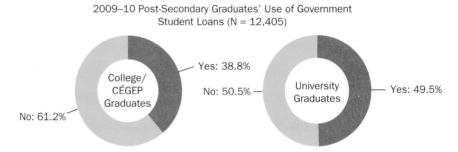

Figure 2.9   Panelled Doughnut Graphs, Corresponding to the Cross-Tabulation in Table 2.6

Source: Author generated; Calculated using data from Statistics Canada, 2015.

Panelling Techniques

Panelling techniques rely on displaying multiple graphs beside or above one another, to show the distribution of a variable for different groups of cases. They are the visual equivalent of the side-by-side frequency tables in Table 2.5. The distribution of the dependent variable is shown in each graph, and a separate graph (panel) is created for each group defined by the independent variable. For example, the panelled doughnut graphs in Figure 2.9 correspond to the frequency distributions in Tables 2.5 and 2.6. They clearly show that a larger percentage of university graduates than college/CÉGEP graduates use government student loans to finance their education, since the "Yes" slice is bigger for university graduates than college/CÉGEP graduates. Because panelling techniques use multiple graphs, they are only practical to use when the independent variable is measured at the nominal or ordinal level and has relatively few attributes (two to four).

Clustering and Stacking Techniques

Another common way to display the relationship between two variables is using clustering or stacking techniques, which are often used with bar graphs. In a clustered bar graph, each attribute of the dependent variable appears along one axis, and different coloured bars are used to display the number or percentage of people with each attribute for each group defined by the independent variable. The clustered bar graph on the left of Figure 2.10, for example, uses coloured bars to show the percentage of college/CÉGEP versus university graduates who said they did or did not use government student loans: the orange bars show the percentages for college/CÉGEP graduates and the pink bars show the same information for university graduates. Clustered bar graphs are most useful when the independent variable is measured at the nominal or ordinal level and has relatively few attributes; thus there are relatively few colours.

Stacked bar graphs show the same information as clustered bar graphs, although there are some important visual differences between them. In stacked bar graphs, each group defined by the independent variable appears along one axis. The percentage of cases with each attribute of the dependent variable is displayed using different coloured blocks that are stacked in the same bar. The stacked bar graph on the

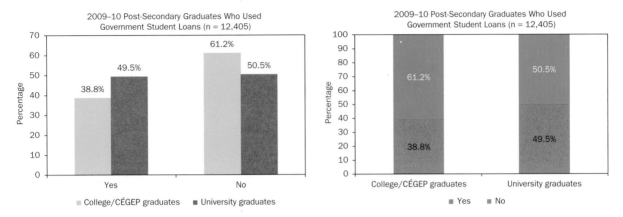

Figure 2.10   A Clustered Bar Graph and a Stacked Bar Graph Showing the Same Information, Corresponding to the Cross-Tabulation in Table 2.6

Source: Author generated; Calculated using data from Statistics Canada, 2015.

right of Figure 2.10 has two bars: one for college/CÉGEP graduates and one for university graduates. The percentage of graduates who used government student loans is shown in the yellow block in each bar, and the percentage that did not use government student loans is shown in the teal block in each bar. Stacked bar graphs are most useful when the dependent variable is measured at the nominal or ordinal level and has relatively few attributes, and thus there are relatively few colours stacked in each bar.

Other types of graphs used to display the relationship between two variables are introduced later in this book. Chapter 6 introduces error-bar graphs, which show the relationship between a ratio-level dependent variable and a nominal- or ordinal-level independent variable, while Chapter 10 describes scatterplots, which show the relationship between two ratio-level variables.

In general, researchers strive to present graphs and visualizations that are easy to understand and that highlight the argument they are making. Table 2.7 lists the graphs that researchers commonly use to show relationships between

Table 2.7   Graphs That Are Commonly Used to Show a Relationship between Two Categorical Variables

Independent Variable	Dependent Variable	Commonly Used Graphs
Nominal, few attributes	Nominal, few attributes	Panelled pie/doughnut graph
Nominal, few attributes	Ordinal, few attributes	Panelled or clustered bar graph
Ordinal, few attributes	Ordinal, few attributes	Clustered bar graph
Nominal or ordinal, many attributes	Nominal or ordinal, few attributes	Stacked bar graph
Nominal or ordinal, many attributes	Nominal or ordinal, many attributes	Use alternate visualization methods

two nominal- or ordinal-level variables. These are only guidelines, however; the best strategy for displaying a relationship also depends on the distribution of each variable. As I already noted, it's always best to try several different ways of displaying information and then select the one that most clearly illustrates your argument.

Displaying Two Variables on the Same Chart

Another way to illustrate the relationship between two (or more) variables is to graph them on the same chart. This strategy is typically used to combine line graphs, or line graphs and bar graphs. Figure 2.11 provides an example of this technique, graphing the change in average tuition fees over time and university students' borrowing for education over time in the same chart. The year (time) is displayed on the horizontal axis. The vertical axis on the left shows the scale for average tuition fees each year, which are depicted in the line graph. The vertical axis on the right shows the scale for the percentage of graduates who used government student loans to finance their education each year, which is depicted in the bar graph. Combining this information in a single chart makes it clear that there is an association between rising tuition fees and the percentage of students who borrow to pay for their education.

Despite this graph, we cannot claim that increased tuition fees have caused more student borrowing. Other factors influence both tuition fees and students' borrowing. For instance, low interest rates might prompt more people to borrow money because it won't cost them much to do so, and low interest

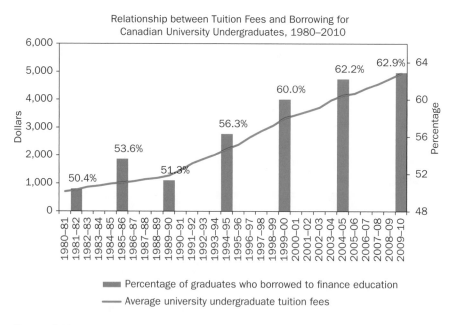

Figure 2.11   Two Variables Graphed over Time, on the Same Chart

Source: Author generated; Calculated using data from Statistics Canada, 2014 and Statistics Canada, 2015.

rates might also prompt higher tuition fees, since universities earn less from their investments when interest rates are low. Nor do these results let us make any claims about whether this trend is desirable for Canadian students or educational policy-makers; it is up to the researchers to make this argument in their analyses.

Best Practices in Presenting Results

Although tables, graphs, and other visualizations are effective tools for summarizing data, researchers also convey and interpret the results of their analysis in writing. As a general rule, a reader should be able to look at a table, graph, or other visualization and easily understand what it shows based on the titles and labels, without reading any of the accompanying text. Similarly, any written text should make sense and be meaningful without looking at the accompanying tables or graphs.

Writing about Numbers

Writing about numbers is a skill that requires practice. There are entire books dedicated solely to this topic. If you don't regularly write about numbers, you might find that it feels awkward at first. Over time, you will learn to craft sentences that fluidly incorporate numbers, without becoming repetitive. Quantitative academic journal articles can provide good examples of how to write about numbers; as you read these types of articles, pay attention to how authors have constructed the sentences that present their results. If you're having trouble writing about your results, try using the structure of statistic-commentary paragraphs, which was introduced in Chapter 1.

A common error that beginning researchers make is to overstate the level of precision in the data by using too many decimal places in the numbers they report. Technically, the percentage of post-secondary graduates who used government student loans is 45.111095 per cent. But, since this number is just an estimate based on a sample, it is not reasonable to have the level of confidence that is implied by the use of six decimal places. More importantly, knowing that 45.111095 per cent of graduates used government student loans does not really change the interpretation of the results; your ideas about student loan use likely won't change depending on whether I report that 45.111 per cent of graduates borrowed or 45.1 per cent of graduates borrowed, or even just 45 per cent of graduates borrowed.

In general, few statistical results warrant reporting more than two decimal places, and most of the time, rounding to whole numbers doesn't substantially affect their interpretation. To improve readability and make your results less intimidating, use standard rounding rules (0.500 and higher rounds upward and 0.499 and lower rounds downward) to show numbers without any decimal places, or with just one or two decimal places, depending on what is being measured. For example, if your results show dollars and cents, use either two decimal places or no decimal places since this reflects how we think about money. For each number that you report, take a moment to consider whether showing more or fewer decimals really changes the meaning—and if it doesn't particularly affect the interpretation of the results, use fewer decimals. It's also good practice to be consistent about the number of decimals you report throughout a piece of writing. One common strategy that researchers use is to report only whole numbers in written text but to present a higher level of precision (one or two decimal places) in tables.

It's inelegant to begin a number with a decimal: always write 0.67 instead of .67. Similarly, it is grammatically incorrect to begin a sentence with a number written as an Arabic numeral. It's best to avoid starting a sentence with a number, but if you must do so, the number should be written out in words, with the equivalent Arabic numeral in parentheses. For instance:

> *Incorrect:* 45% of post-secondary graduates used government student loans to finance their education.
> *Correct:* Forty-five per cent (45%) of post-secondary graduates used government student loans to finance their education.
> *Correct:* Overall, 45 per cent of post-secondary graduates used government student loans to finance their education.
> *Even better:* Slightly less than half of post-secondary graduates (45%) used government student loans to finance their education.

As illustrated in this "even better" example, it can sometimes be useful to report fractions or rates instead of percentages. This helps to give readers a better sense of the magnitude of the phenomenon that is being reported. Table 2.8 shows how some common percentages translate into fractions and rates that are easy for people to understand. You can also use this approach to report tenths, which are not listed in Table 2.8. If the percentage of people with an attribute doesn't correspond exactly with a rate, use a qualifying word like "approximately" or "about" with the closest rate, and then give the exact percentage in parentheses.

Table 2.8   How to Report Percentages as Fractions and Rates in Words

Grouping	Percentage	Fraction in Words	Rate in Words
Thirds	33	One-third of people	One out of three people
	66	Two-thirds of people	Two out of three people
Quarters	25	A quarter of people	One out of four people
	50	Half of people	One out of two people
	75	Three-quarters of people	Three out of four people
Fifths	20	One-fifth of people	One out of five people
	40	Two-fifths of people	Two out of five people
	60	Three-fifths of people	Three out of five people
	80	Four-fifths of people	Four out of five people

Finally, be careful whenever you compare two percentages in writing. There is an important distinction between a *percentage point* difference, and a *percentage* difference. A percentage point difference refers to the difference between two percentages, when each percentage is out of 100. For example, Table 2.6 shows that 39 per cent of college/CÉGEP graduates used government student loans and that 50 per cent of university graduates used government student loans. The difference between these two numbers is calculated in percentage points: 50 per cent minus 39 per cent is 11, so there is an 11 percentage point difference in student loan use between college/CÉGEP graduates and university graduates.

In contrast, a percentage difference refers to how two percentages compare to each other, using one of the percentages as the reference point for scaling the size of the difference. In other words, the percentage point difference (11 points, in this example) is divided by one of the original percentages. To find out how university graduates compare to college/CÉGEP graduates, the percentage for college/CÉGEP graduates (39) is used as a reference point. When the percentage point difference (11) is divided by the reference percentage (39), the result is 0.28, or 28 per cent. So, compared to college/CÉGEP graduates, 28 per cent more university graduates used student loans. To find out how college/CÉGEP graduates compare to university graduates, the percentage for university graduates is used as a reference point. When the percentage point difference (11) is divided by the reference percentage (50), the result is 0.22, or 22 per cent. So, compared to university graduates, 22 per cent fewer college/CÉGEP graduates used student loans.

Researchers usually report the differences between the percentages in a cross-tabulation using percentage points. To avoid confusion, always be sure to specify whether you are reporting percentage or percentage point differences:

> *Incorrect:* Fifty per cent of university graduates (50%) used government student loans to finance their education, compared to only 39 per cent of college/CÉGEP graduates, an 11 per cent difference.
> *Incorrect:* Compared to college/CÉGEP graduates, 11 per cent more university graduates used government student loans.
> *Correct:* Fifty percent of university graduates (50%) used government student loans to finance their education, compared to only 39 per cent of college/CÉGEP graduates, an 11 percentage point difference.
> *Correct:* Compared to college/CÉGEP graduates, 28 per cent more university graduates used government student loans.

Formatting Tables

Statistical software usually produces frequency tables and cross-tabulations that are too cluttered to use in professional reports and presentations. Tables should always be fully labelled, but omit information that is meaningless to the reader. The easiest way to do this is to transfer the tables produced by statistical software into a word processor for editing. Each table should have a meaningful title, such as "Post-Secondary Students' Use of Borrowing to Finance Education" (rather than simply "Table 1"), and

Table 2.9   A Table That Combines Two Frequency Distributions

Use of Government Student Loans by 2009–10 Canadian College/CÉGEP and University Graduates

Characteristic	Frequency	Percentage
Used government student loans		
Yes	5,596	45.1
No	6,809	54.9
Amount owed at graduation ($)		
0	1,059	20.1
1 to 4,999	466	8.9
5,000 to 9,999	819	15.6
10,000 to 24,999	1,721	32.7
25,000 or more	1,200	22.8

Source: Author generated; Calculated using data from Statistics Canada, 2015.

should show the percentage of people with each attribute, in addition to the frequencies, so that readers can quickly and easily make comparisons. If percentages are used without showing frequency counts, report the total number of people used in the denominator of the percentage using the notation (N = _____). Knowing the total number of cases helps readers assess how credible the results are; typically, statistics based on a large number of cases are perceived as more credible than those based on a small number of cases. In order to use space efficiently, researchers often combine several frequency distributions in a single table, listing each variable name and its attributes in the rows. Table 2.9 combines the frequency tables in Tables 2.1 and 2.4. Attributes that are designated as missing are usually not shown in formatted tables. A similar strategy can be used to combine the results of several cross-tabulations with the same independent variable.

Constructing Graphs

Statistical software programs, spreadsheets, and online tools have made it easy to produce and manipulate graphs. Following these general guidelines will help you to create professional-looking graphs. Although many of these suggestions may be common-sense ones, they often require changes to a program's default graph options.

- Include a clear and meaningful title for the graph overall and for each graph axis and legend category. For categorical variables, the group or category names should be displayed (not the value numbers). This makes the content clear to the reader.
- Report percentages, not counts. Use data labels to show the percentage of cases with each attribute directly on the graph. This makes it easier for readers to make comparisons between groups of different sizes.
- Be sure that the percentages that are displayed match the results of the corresponding frequency table or cross-tabulation.

- Report of the total number of cases (N) depicted in the graph to help the reader assess the credibility of the results. Often this is listed in parentheses after the title or in a footnote.
- Do not display any attributes or values designated as missing on graphs. If the information is not usable for your analysis, it should not be shown.
- Adjust the font size and the number of decimal places displayed so that text is legible.
- Select colour and shading options with attention to how the graph will be printed or displayed. Use greyscale options for graphs that will not be displayed in colour. Be aware of red-green colour blindness when preparing graphs that will be displayed in colour.

Creating Infographics

Researchers use infographics in order to make their statistical results accessible to a wide audience. Infographics typically present one or two statistical results using a combination of text, graphs, and other data visualization strategies. (See Figure 2.12.) The key to creating a good infographic is to use the graphic design elements to present the information in a visually interesting way. Some infographics include only a single statistic or information about a single variable; large infographics can combine four or five key statistics to develop a narrative about a topic. In general, though, infographics present less information and are less detailed than more formal reports. The text in an infographic is used primarily to draw the

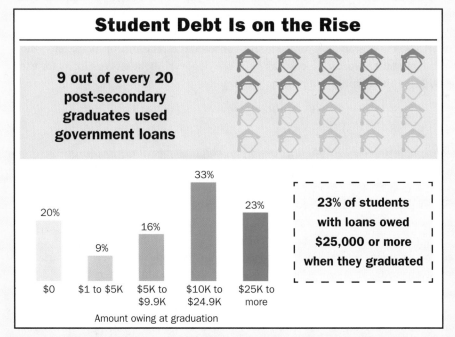

Figure 2.12 An Infographic Depicting the Information in Table 2.9 in a More Visually Interesting Way

Source: Author generated; Calculated using data from Statistics Canada, 2015.

reader's attention to a particularly interesting result. Because it doesn't take much time to read an infographic, they are particularly well-suited to sharing on social media and blog sites, and they are often used by researchers working in community organizations to raise awareness about an issue.

Online data visualization tools make it easy to create infographics, and many are free. Unless you have graphic design training, it's generally best to begin with a pre-formatted infographic template and make modifications from there.

What You Have Learned

This chapter described several different ways to summarize data. Tables and graphs are the basic tools of descriptive statistics. Frequency tables and cross-tabulations show the number and percentage of cases with some attribute. Graphs and other data visualization techniques are used to convey statistical results in an easy-to-understand way. Chapter 3 will add to these tools by introducing several techniques for describing the centre and spread of nominal- and ordinal-level variables.

The research focus of this chapter was student loans. Slightly less than half of Canadian post-secondary students who graduated in 2009–10 used government loans to finance their education. University graduates were more likely to use government student loans than were college/CÉGEP graduates. More than half of

graduates who used government student loans owed $10,000 or more at graduation, and one in five graduates who used government student loans owed $25,000 or more. Average university undergraduate tuition fees have been rising faster than the rate of inflation since 1990. Among university graduates, the number of students who borrow (from any source) to finance their education has also been steadily increasing. These findings lend support to student activists' demands to make education in Canada more affordable. Since tuition and funding for post-secondary education is regulated at the provincial level, further research in this area should consider the relationship between tuition costs and the use of government student loans for each province separately.

Check Your Understanding

Check to see if you understand the key concepts in this chapter by answering the following questions:

1. What is the difference between a percentage and a cumulative percentage? When is each used?
2. What are rates and ratios, and how are they different from each other?
3. What types of graphs or data visualizations can be used to show the distribution of a single variable?
4. How are cross-tabulations similar to and different than frequency distributions?

5. What does the saying "percentage down, compare across" for cross-tabulations mean?
6. What types of graphs or data visualizations can be used to show the relationship between two variables?
7. What is a percentage point difference? How does it compare to a percentage difference?
8. What are the features of a well-formatted table, a well-constructed graph, and a well-designed infographic?

Practice What You Have Learned

Check to see if you can apply the key concepts in this chapter by answering the following questions. Keep two decimal places in any calculations.

1. You want to learn more about how post-secondary students pay for their education. You collect information from a random sample of 50 students about whether or not their families help to pay their tuition costs. Overall, 38 people said that "Yes" their families do help to pay their tuition costs, and 12 people said that "No" their families do not help to pay their tuition costs. Show this frequency distribution in a table that includes counts and percentages.

2. Using the frequency distribution from question 1, find the proportion of students whose families help to pay their tuition costs.

3. Using the frequency distribution from question 1, represent the number of students whose families help pay their tuition costs as a rate out of 1,000.

4. Using the frequency distribution from question 1, fill in the blank in the following sentence: The results show that 1 out of every ___ students gets help from family members with their tuition costs.

5. Using the frequency distribution from question 1, show how the number of students whose families help to pay their tuition costs compares to the number of students whose families do not help to pay their tuition costs by expressing it as a ratio.

6. Create a graph that depicts the frequency distribution from question 1, either by hand, by using a spreadsheet, or by using an online data visualization program.

7. You think that people's age might be related to whether their family helps them with their tuition costs. Among the 50 students you collected information from, 16 are mature students (aged 25 or older), and the remainder are not. Among the 16 mature students, 6 said that "Yes" their families help to pay their tuition costs, and the remaining 10 said that "No" their families do not help to pay their tuition costs. Among the 34 remaining students, 32 said that "Yes" their families help to pay their tuition costs, and the remaining 2 said that "No" their families do not help to pay their tuition costs. Present this information in a cross-tabulation that displays counts and column percentages and also includes a "Total" row and a "Total" column.

8. Create a graph that depicts the different frequency distributions for mature students and for non-mature students from question 7, either by hand, by using a spreadsheet, or by using an online data visualization program.

9. A Statistics Canada publication includes the graph in Figure 2.13, which shows whether or not a family is saving money for their children's post-secondary education, in relation to the highest level of education achieved by a child's parents.

    a. What type of graph is it?
    b. What two variables are included in the graph?
    c. Which is the independent variable and which is the dependent variable?
    d. What is the level of measurement of each variable?
    e. Describe the general pattern depicted in the graph.

10. You are interested in how social science students' debt compares to business students' debt. You find the information in Table 2.10 in one of your school's publications, which shows the amount of student debt among students in a recent graduating class.

    a. Calculate the total number of students and the percentage of students with each level of debt, regardless of their field of study.
    b. Calculate the cumulative percentage of students who graduate with each level of debt, or a lower level, regardless of their field of study.
    c. What percentage of students overall graduated with less than $10,000 in debt (including people with no student debt)?
    d. What percentage of students overall graduated with $10,000 or more in debt? Explain how you calculated this percentage.

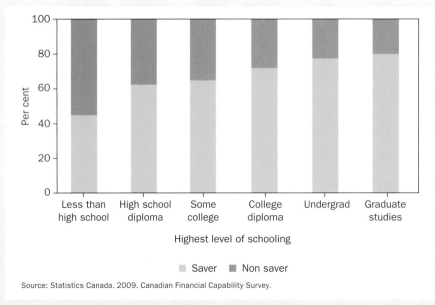

Source: Statistics Canada. 2009. Canadian Financial Capability Survey.

Figure 2.13 **Proportion of Savers and Non-Savers, by Highest Level of Schooling of Parents with Children under the Age of 18, 2009**

Source: Guilmette 2011.

11. Using the information in Table 2.10:

   a. Calculate the cumulative percentages for the distribution of student debt among social science students and among business students separately.

   b. Which group has a higher percentage of students who graduated with $10,000 or more in student debt?

12. Answer the following questions about Table 2.10:

   a. What two variables are included in the table?

   b. Which is the independent variable and which is the dependent variable?

Table 2.10 **Amount of Debt That Students Graduate with, by Program (Hypothetical Data)**

Answer	Social Science Students Frequency	Business Students Frequency
$0 (no debt)	42	68
$1 to $4,999	6	12
$5,000 to $9,999	9	12
$10,000 to $24,999	24	18
$25,000 or more	19	10

   c. What is the level of measurement of each variable?

   d. How many attributes does each variable have?

   e. Given your answers to (b) through (d), what type(s) of graph(s) is (are) commonly used to depict this type of relationship?

13. Using the frequency distributions for social science students and business students in Table 2.10:

   a. Represent the number of social science students who graduate without any student debt as a rate out of 1,000.

   b. Represent the number of business students who graduate without any student debt as a rate out of 1,000.

   c. Compare these two rates and use them to describe whether social science students or business students are more likely to graduate with student debt.

14. Using the frequency distributions for social science students and business students in Table 2.10:

   a. Fill in the blank in the following sentence: The results show that 1 out of every _____ social science students graduate without any student debt.

Table 2.11   **Rates of Access to College and University for Under-Represented and Minority Groups[1]**

	All Provinces		
	College	University	Any
		Percent	
All	**33.0**	**41.7**	**74.7**
Family income			
Income below $50,000	34.8	31.4	66.2
Income greater than $50,000	32.1	46.9	79.0
Immigrant status			
First-generation immigrant	29.6	57.0	86.6
Second-generation immigrant	30.0	53.0	83.0
Non-immigrant	34.2	37.0	71.2
Aboriginal status			
Aboriginal	28.0	23.1	51.1
Non-Aboriginal	33.2	42.2	75.4

[1] YITS Cycle A consists of youth who were 15 years old in 2000. The students were surveyed again in 2002, 2004, 2006, 2008, and 2010 (cycles 2, 3, 4, 5, and 6). The analysis is based on the respondents' post-secondary education status in 2006 (Cycle 4) when they were age 21.
Source: Statistics Canada. Youth in Transition Survey. Cohort A. Cycle 4.

Source: Excerpt from McMullen 2011.

b. Fill in the blank in the following sentence: The results show that 1 out of every _____ business students graduate without any student debt.

c. Compare these two rates and use them to describe whether social science students or business students are more likely to graduate with student debt.

15. Table 2.11, excerpted from a Statistics Canada publication, shows the percentage of youth who had attended college or university by the time they reached age 21, overall and divided by family income, immigrant status, and Aboriginal status. Answer the following questions using the information in Table 2.11:

a. This table includes one dependent variable and three independent variables. What are they?

b. What percentage of youth overall attended college? What percentage attended university? What percentage attended either college or university?

c. Are young people with family incomes below $50,000 more or less likely to attend a post-secondary institution than people with

family incomes greater than $50,000? How is family income related to whether or not youth attend college or attend university?

d. Are immigrants more or less likely to attend a post-secondary institution than people who are not immigrants? How is immigrant status related to whether or not youth attend college or attend university?

e. Are Aboriginal youth more or less likely to attend a post-secondary institution than youth who are not Aboriginal? How is Aboriginal status related to whether or not youth attend college or attend university?

16. A Statistics Canada publication includes the graph in Figure 2.14, which presents information for people aged 25 to 44 only.

a. What type of graph is it?

b. One variable included in the graph is the year (or "Time"). What other two variables are included in the graph?

c. List two claims that you can make based on the information in this graph.

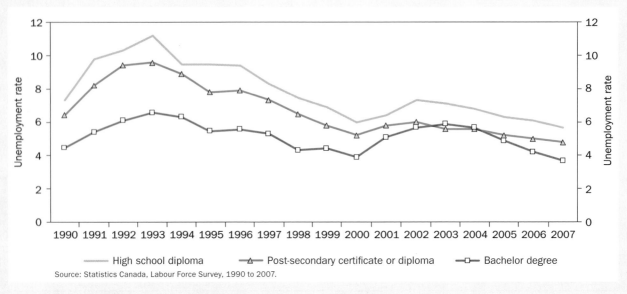

Figure 2.14 **Unemployment Rates by Level of Education, 25- to 44-Year-Olds, 1990 to 2007**

Source: Bayard and Greenlee 2009, 16.

Practice Using Statistical Software (IBM SPSS)

Answer these questions using IBM SPSS and the GSS27.sav or the GSS27_student.sav dataset available from the Student Resources area of the companion website for this book. Report two decimal places in your answers, unless fewer are printed by IBM SPSS. It is imperative that you save the dataset to keep any new variables that you create.

1. Use the Frequencies procedure to produce a frequency distribution of the variable "Sex of respondent" [SEX].

    a. How many men answered the survey? What percentage of survey respondents are men? (For this question, and going forward, assume that *male* refers to men. Statistics Canada only records people's gender presentation in its surveys.)

    b. How many women answered the survey? What percentage of survey respondents are women? (For this question, and going forward, assume that *female* refers to women.)

2. Use the Frequencies procedure to produce a descending count frequency distribution of the variable "Province of residence" [PRCODE].

    a. Which province has the most survey respondents? What percentage of respondents are from that province?

    b. Which province has the least survey respondents? What percentage of respondents are from that province?

3. The variable "Canadian shared values - Human rights" [SVR_10] shows people's answers to this question: "To what extent do you feel that Canadians share the following values? Human rights." Use the Frequencies procedure to produce a frequency distribution of this variable.

    a. What percentage of respondents say that Canadians share the value of human rights to a great extent, out of the total number of people who gave a valid answer to the question?

    b. What percentage of respondents say that Canadians share the value of human rights to a moderate extent, out of the total number of people who gave a valid answer to the question?

    c. What percentage of respondents say that Canadians share the value of human rights to either a great or a moderate extent, out of the total number of people who gave a valid answer to the question?

4. The variable "Pride - Being Canadian" [PRD_10] shows people's answers to this question: "How proud are you to be Canadian?" Use the Frequencies procedure to produce a frequency distribution of this variable.

    a. What percentage of respondents say that they are very proud to be Canadian, out of the total number of people who gave a valid answer to the question?

    b. What attributes/values are designated as missing?

5. In the *Variable View*, designate the attributes "No opinion" and "Not a Canadian citizen" as missing attributes for the variable "Pride - Being Canadian" [PRD_10], so that the variable has six missing attributes: "No opinion," "Not a Canadian citizen," "Valid skip," "Don't know," "Refusal," and "Not stated." (Do this by selecting "Range plus one optional discrete missing value" and designate the values "6" through "99" as missing.) Then, use the Frequencies procedure to produce another frequency distribution of the variable.

    a. What percentage of respondents say that they are very proud to be Canadian, out of the total number of people who gave a valid answer to the question?

    b. Explain why the percentage from question 4(a) is different than the percentage from question 5(a).

6. Use the Chart Builder tool to create a pie graph of the variable "Canadian shared values - Human rights" [SVR_10].

    a. Edit the graph so that it displays the percentage with each attribute.

    b. Edit the graph to change the colour of the most common attribute to red.

7. Use the Chart Builder tool to create a bar graph of the variable "Canadian shared values - Human rights" [SVR_10]. Display percentages on the vertical axis (y-axis).

    a. Edit the graph so that it displays the percentage with each attribute.

    b. Compare the pie graph from question 6 to the bar graph. Which displays the information more effectively?

8. Use the Crosstabs procedure to produce a cross-tabulation showing the relationship between "Place of birth of respondent - Canada" [BRTHCAN] (in the columns) and "Canadian shared values - Human rights" [SVR_10] (in the rows). Display both counts and column percentages in the cross-tabulation.

    a. What percentage of respondents overall say that Canadians share the value of human rights to a great extent?

    b. What percentage of people born in Canada say that Canadians share the value of human rights to a great extent?

    c. What percentage of people born outside of Canada say that Canadians share the value of human rights to a great extent?

    d. Use the column percentages to describe any other differences in the distribution of responses between people born in Canada and people born outside Canada.

9. Use the Chart Builder tool to create a clustered bar graph showing the relationship between "Place of birth of respondent - Canada" [BRTHCAN] and "Canadian shared values - Human rights" [SVR_10]. Display the attributes of the variable "Canadian shared values - Human rights" [SVR_10] on the horizontal axis (x-axis). Display percentages on the vertical axis (y-axis). Depict the percentage of people born in Canada and born outside Canada with each attribute using different coloured bars. Be sure that the percentages shown in the graph match those from the cross-tabulation in question 8.

    a. Edit the graph so that it shows the percentage with each attribute.

    b. Describe what the graph shows.

10. Use the Compute Variable tool and the If Cases window in the Compute Variable tool to combine the results of the two variables "Last federal election – Voted" [VBR_10] and "Last provincial election – Voted" [VBR_30] in a new variable called "Voting history" [VOTE_HISTORY]. Create five attributes in the new variable:

    1 – "People who voted in both the last federal election and the last provincial election"

2 – "People who voted in the last federal election but did not vote in the last provincial election"
3 – "People who did not vote in the last federal election but voted in the last provincial election"
4 – "People who did not vote in either the last federal election or the last provincial election"

5 - People who had a missing attribute (i.e., "Valid skip," "Don't know," "Refusal," or "Not stated") on either of the two variables

In the *Variable View*, add value labels to the new variable, and designate the final attribute as "missing information." Produce a frequency distribution of the variable "Voting history" [VOTE_HISTORY], and describe what it shows.

Key Formulas

Percentage $\% = \left(\dfrac{F}{N}\right)100$

Proportion $p = \left(\dfrac{F}{N}\right)$

Ratio $ratio = \dfrac{F_1}{F_2}$

References

Bayard, Justin, and Edith Greenlee. 2009. "Graduating in Canada Profile, Labour Market Outcomes and Student Debt of the Class of 2005." Catalogue no. 81-595-M No. 074. Ottawa: Statistics Canada. http://www.statcan.gc.ca/pub/81-595-m/81-595-m2009074-eng.pdf.

Canadian Federation of Students (CFS). 2013. "Student Debt in Canada: Education Shouldn't Be a Debt Sentence." http://cfs-fcee.ca/wp-content/uploads/sites/2/2013/11/Factsheet-2013-11-Student-Debt-EN.pdf.

Guilmette, Sylvie. 2011. "Competing Priorities—Education and Retirement Saving Behaviours of Canadian Families." Catalogue no. 81-004-X. Ottawa: Statistics Canada. http://www.statcan.gc.ca/pub/81-004-x/2011001/article/11432-eng.htm.

Industry Canada. 2014. "About Us: Mandate." November 24. http://www.ic.gc.ca/eic/site/icgc.nsf/eng/h_00018.html.

Luong, May. 2010. "The Financial Impact of Student Loans." Catalogue no. 75-001-X. Ottawa: Statistics Canada. http://www.statcan.gc.ca/pub/75-001-x/2010101/pdf/11073-eng.pdf.

McMullen, Kathryn. 2011. "Postsecondary Education Participation among Underrepresented and Minority Groups." Catalogue no. 81-004-X. Ottawa: Statistics Canada. http://www.statcan.gc.ca/pub/81-004-x/2011004/article/11595-eng.htm.

Schwartz, S., and R. Finnie. 2002. "Student Loans in Canada: An Analysis of Borrowing and Repayment." *Economics of Education Review* 21 (5): 497–512. doi:10.1016/S0272-7757(01)00041-3.

Statistics Canada. 2008. "Microdata User Guide: National Graduates Survey Class of 2005." Ottawa: Centre for Education Statistics.

——. 2010. "Table 477-0068: National Graduates Survey, Student Debt from All Sources, by Province and Level of Study." CANSIM (Database). http://www5.statcan.gc.ca/cansim/a26?lang=eng&id=4770068.

——. 2014. "Tuition and Living Accommodation Costs for Full-Time Students at Canadian Degree-Granting Institutions, 1993–2015." ODESI: Ontario Data Documentation, Extraction Service and Infrastructure. http://www.statcan.gc.ca/dli-ild/data-donnees/ftp/tlac-fssuc-eng.htm.

——. 2015. "National Graduates Survey 2013." *Public Use Microdata File.* Ottawa, ON: Statistics Canada.

Tufte, Edward Rolf. 1999. *The Visual Display of Quantitative Information.* Cheshire, CT: Graphics Press.

Describing the Centre and Dispersion of a Distribution: Focus on Categorical Variables

Learning Objectives

In this chapter, you will learn:

* How to find the mode and median of a variable

* What percentiles and quantiles are and how to calculate them

* How to find the range and interquartile range of a variable

* What box plots are and how to read them

* Some guidelines for writing about the centre and dispersion of categorical variables

Introduction

In addition to using tables, graphs, and other visual tools to summarize the distribution of a variable, researchers often find it useful to describe the centre of a variable and how cases are dispersed—or spread out—around that centre. This is especially useful when a variable has many attributes and reporting the percentage of cases with each attribute becomes cumbersome. This chapter introduces several ways to describe the centre and dispersion of nominal- and ordinal-level variables. These strategies can also be used with ratio-level variables. You'll learn additional ways to describe the centre and dispersion of ratio-level variables in Chapter 4.

The research focus of this chapter is the cost of housing and whether it is affordable. Over the past 20 years, housing in Canada has become increasingly expensive, especially in major cities, such as Vancouver and Toronto. In 2016, Vancouver was compared to cities within Canada and nine other countries and rated as the third–least affordable city for housing (countries compared were Australia, Canada, China, Ireland, Japan, Malaysia, New Zealand, Singapore, the United Kingdom, and the United States); only Hong Kong, China, and Sydney, Australia, have less affordable housing than Vancouver (Cox and Paveltich 2016). While Toronto is ranked thirteenth in the list of least-affordable major cities, its level of housing affordability has deteriorated rapidly over the past decade, partly as a result of restrictive urban land-use and planning policies (Cox and Paveltich 2016). Compared to the nine other countries that were studied, Canada's housing market is considered to be seriously unaffordable overall (Cox and Paveltich 2016).

Stephen Hui/Shutterstock

Photo 3.1   Social justice organizations in Vancouver organized the "Red Tent" campaign during the 2010 Olympic Games to highlight the need for a funded national housing strategy.

Access to adequate housing is a human right; the Universal Declaration of Human Rights asserts that "everyone has the right to a standard of living adequate for the health and well-being of himself and of his family, including food, clothing, housing and medical care and necessary social services" (United Nations 1948, Article 25). The Ontario Human Rights commission extends this position to assert that "there is an undeniable link between affordable and adequate housing and quality of life" (ONHRC 2007, 2). In 1993, Canada's National Affordable Housing Program was cancelled by the Liberal government. Since that time, advocates for people with low income and for the homeless have been petitioning the federal government to develop a national housing strategy; which was finally launched in November 2017. In 2012, a private member's bill calling for the development of a national housing strategy (Bill C-400) defined affordable housing as that which is available at a cost that does not compromise an individual's ability to meet other basic needs, including food; clothing; and access to health-care services, education, and recreational activities (the bill was ultimately defeated; Morin 2012).   The Canada Mortgage and Housing Corporation defines affordable housing as that which costs less than 30 per cent of a household's before-tax income (CMHC 2014). The CMHC includes money spent on rent or mortgage payments, property taxes, condominium fees, electricity, fuel, water, and other municipal services in its calculation of housing costs.

This chapter uses two different measures of housing affordability. The first is the total dollar amount spent on housing costs each month, which is grouped into nine categories; this provides an absolute measure of affordability. The second measure is the percentage of a household's total income that is spent on housing; this provides a relative measure of affordability. Using the CMHC's cut-off of 30 per cent as a marker of housing affordability, each household is designated as having very affordable housing (spending less than 15 per cent of income on housing), somewhat affordable housing (spending 15 to 29 per cent of income on housing), somewhat unaffordable housing (spending 30 to 44 per cent of income on housing), or very unaffordable housing (spending 45 per cent or more of income on housing). Both of these variables are measured at the ordinal level because they use class intervals to group the attributes (dollars and percentages), which are ordered. The statistical analyses in this chapter are used to investigate the following:

- What percentage of households have affordable housing?
- How much do households typically spend on housing each month?
- How is housing affordability related to household income?

Table 3.1 shows a frequency distribution of the level of housing affordability among Canadian households. The cumulative percentage column shows that housing is somewhat or very affordable for 78 per cent of households. But, about one in five households—22 per cent—have housing that is somewhat or very unaffordable.

Describing the Centre of a Variable: Mode and Median

The simplest way to describe the centre of a variable is to report the mode, or the most common answer. The **mode** of a variable is the attribute or value that occurs most frequently in the data. In other words, it is the most common answer that people give. The mode can be reported for variables with any level of measurement.

It's easy to find the mode of a variable using a frequency distribution. Simply find the row with the highest frequency or percentage. In the frequency distribution of monthly housing costs in Table 3.2, the attribute with the highest frequency is "$500–749;" 21 per cent of households report they spend this much. Thus, the mode—or the modal attribute—of the variable "Monthly housing cost" is "$500–749." In Canada, then, it is most common for households to spend $500–749 on housing each month.

mode The attribute or value of a variable that occurs most frequently in the data; the most common answer.

Table 3.1 Frequency Distribution of Housing Affordability among Canadian Households

Level of housing affordability

Category	Frequency	Percentage	Cumulative Percentage
Very affordable	722	44.5	44.5
Somewhat affordable	546	33.6	78.1
Somewhat unaffordable	205	12.6	90.7
Very unaffordable	151	9.3	100.0
Total	1,624	100.0	

Source: Author generated; Calculated using data from Statistics Canada, 2015.

Table 3.2   **Frequency Distribution of Canadian Households' Housing Costs**

Monthly housing costs

Category	Frequency	Percentage	Cumulative Percentage
Less than $250	52	3.2	3.2
$250–499	302	18.6	21.8
$500–749	342	21.1	42.9
$750–999	221	13.6	56.5
$1,000–1,249	156	9.6	66.1
$1,250–1,499	130	8.0	74.1
$1,500–1,999	210	12.9	87.0
$2,000–2,499	115	7.1	94.1
$2,500 or more	96	5.9	100.0
Total	*1,624*	*100.0*	

Source: Author generated; Calculated using data from Statistics Canada, 2015.

Sometimes a variable has more than one mode. This occurs when two (or more) attributes have exactly the same frequency or are selected by exactly the same number of people. If a variable has two modes, it is referred to as bimodal: the prefix *bi-* is used to indicate that there are two modes. If a variable has three or more modes, it is referred to as multi-modal: the prefix *multi-* is used to indicate that there are more than two modes. If a variable is bimodal or multi-modal, researchers simply report all of the modes when they describe the variable. Most of the time, though, variables are unimodal, that is, they have only one mode, or a single most common value.

Another way to describe the centre of a variable is to report the median. The **median** of a variable is the attribute or value of the case located at the middle-most point when all of the cases are arranged in order from the lowest attribute to the highest attribute (or the reverse). In other words, half of cases are at or above the median of a variable, and half of cases are at or below the median of a variable. The median is used to describe the centre of ordinal- and ratio-level variables. It's not possible to find the median of a nominal-level variable because the attributes do not have any inherent order and, thus, can't be arranged in order from lowest to highest.

If there are only a small number of cases, you can find the median of a variable by hand. The first step is to list the cases in order from the lowest attribute to the highest attribute of the variable you are finding the median of. Figure 3.1 illustrates how to list the cases in order, using a hypothetical variable with four attributes: "Strongly disagree," "Disagree," "Agree," and "Strongly agree." Figure 3.1 also shows three possible scenarios that might occur when finding the median of an ordinal-level variable. If there are an odd number of cases, as in Scenario 1, the median is the attribute of the case that is in the middle position. So, if there

median The attribute or value of the case located at the middle-most point of a variable when all of the cases are arranged in order from the lowest attribute or value to the highest attribute or value.

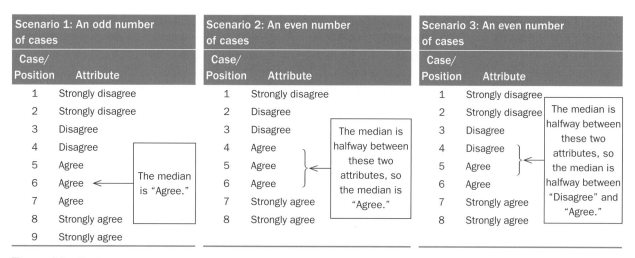

Figure 3.1   Finding the Median of a Variable When There Are a Small Number of Cases (Hypothetical Data)

are nine cases arranged in order from the lowest attribute to the highest attribute (or vice versa), the median of the variable is the attribute of the fifth case, since four cases are below it and four cases are above it. If there are an even number of cases, as in Scenario 2 and Scenario 3, the median is said to be halfway between the attributes of the cases that are in the two middle positions. So, if there are eight cases arranged in order from the lowest attribute to the highest attribute (or vice versa), the median of the variable is halfway between the attributes of the fourth and fifth cases. If the fourth and fifth cases both have the same attribute, as in Scenario 2, report that attribute as the median. In theory, though, the median doesn't need to be an attribute that occurs in the data. If the fourth and fifth cases have different attributes, as in Scenario 3, the median is reported as being halfway between those two attributes.

An easy way to find the position of the middle case is to add 1 to the total number of cases, and then divide the result by 2. For example, in Scenario 1 there are nine cases: $9 + 1 = 10$, and $10 \div 2 = 5$, so the median of the variable is the attribute of the fifth case. In Scenario 2 and Scenario 3, there are eight cases: $8 + 1 = 9$, and $9 \div 2 = 4.5$, so the median of the variable is halfway between the attributes of the fourth and fifth cases.

If there are a large number of cases, finding the median by hand is impractical because it is tedious to list the cases in order from the variable's lowest to highest attribute. Instead, the cumulative percentage column of a frequency distribution can be used to find the median of a variable. The attribute where the cumulative percentage first crosses over the 50 per cent threshold is the median of the variable. This is the equivalent of adding up all of the percentages until they become higher than 50 per cent in order to find the location of the middle case (and its attribute).

Looking back at the frequency distribution in Table 3.2, the cumulative percentage first crosses over the 50 per cent threshold at the attribute "$750–999." The cumulative percentage for the attribute "$500–749" is 43 and then jumps to 57 per cent for the category "$750–999." Since "$750–999" is the median of this variable, we can say that 50 per cent of households spend $750–999 or less on housing each month, and 50 per cent of households spend $750–999 or more on housing each month. Because the median attribute is a class interval, this claim can be simplified to say that half of households spend less than $1,000 on housing each month, and half of households spend $750 or more on housing each month.

So, for the monthly housing cost variable, the mode, or the most common attribute, is "$500–749;" and the median, or the middle-most attribute, is "$750–999." Both the mode and the median provide a way to identify the centre of this variable's distribution; it is up to the researcher to decide whether one of these attributes is a better representation of monthly housing costs. In this particular situation, even though "$500–749" is reported by the largest number of households, this attribute is located near the end of the distribution, and there are other attributes with almost as many households. As a result, the attribute "$750–999" may be a better representation of the "central" amount that Canadian households spend on housing.

Step-by-Step: Median

Step 1: Count the total number of cases (N). Do not include cases that are missing information for the variable.

Step 2: Manually or electronically arrange (or sort) the cases in order from the lowest attribute to the highest attribute of the variable. If two or more cases have the same attribute, it doesn't matter what order they are listed in.

Step 3: Label the *position* of each case in the sequence, from 1 to N, so that the case with the lowest attribute has position 1, the case with the next lowest attribute has position 2, and so on until the position of every case is labelled.

Step 4: Add 1 to the total number of cases (from Step 1).

Step 5: Divide the result of Step 4 by 2 to find the position of the case with the median attribute.

Step 6 (whole numbers): If the result of Step 5 is a whole number, find the case in the *position* that corresponds to that number; the attribute or value of that case is the median of the variable.

Step 6 (partial numbers): If the result of Step 5 is a partial number (a decimal), find the cases in the *positions* corresponding to the whole number immediately below the partial number and the whole number immediately above the partial number (so, if the result is 8.5, find the case in position 8 and the case in position 9); the median of the variable is halfway between the attributes or values of those two cases.

Spotlight on Data

The Survey of Household Spending

The Survey of Household Spending (SHS) collects information about the expenditures and income of households across Canada. It includes detailed information about how much households spend on a wide variety of items, including regular expenditures, such as rent and utilities, as well as less frequent expenditures, such as repairs and furniture (Statistics Canada 2015). The SHS also gathers data about the type and quality of people's housing and household equipment, such as appliances and electronics. The federal and provincial governments use the survey results to help them develop policies and programs; SHS results are also used in the calculation of the gross domestic product (GDP) (Statistics Canada 2015).

A key feature of the SHS is that the unit of analysis is the household, not individual people. This is because many of the expenditures that the SHS collects information about are billed at the household level. For instance, in a household with more than one person, it is difficult to calculate how much water or heat each person uses individually. Instead, each case in the SHS dataset represents information for a household of people who live together.

In 2014, the SHS collected information from 11,413 private households in Canada's 10 provinces. People living on Aboriginal reserves, members of the Canadian Forces living in military camps, people living in religious or other communal colonies, and people living full-time in institutions, such as seniors' residences, school residences, or work camps, are not surveyed (Statistics Canada 2015).

Information is collected from each selected household using computer-assisted personal interviewing (CAPI). One person from each household completes the survey on a laptop computer brought to their home by a Statistics Canada employee. With the permission of the household members, income information is taken from people's income tax files and combined to calculate the total household income. Some households are also asked to complete a detailed expenditure diary that records their daily spending over a two-week period. The overall response rate for the interview portion of the 2014 SHS was 67 per cent; that is, 67 per cent of households who were selected to participate in the survey interview actually completed it (Statistics Canada 2016). This response rate is typical for surveys that are conducted by Statistics Canada.

Households reported on different types of expenditures for different time periods (the past four weeks, the past month, the past three months, etc.). This information is annualized by Statistics Canada—in other words, it is multiplied to represent the yearly cost of each expenditure. Monthly housing costs are calculated by adding together the annual cost of rent, mortgage payments, condominium charges, property taxes, electricity, natural gas, other fuel, water and sewage,

Continued

and dividing by 12. Because household income information is collected from the income tax files that people submitted in 2014, it is only available for the 2013 calendar year (since people report their income for the previous calendar year on their taxes). In order to ensure the best match between household income and expenses, the analyses in this chapter only use data collected in January and February 2014, from people who lived in their dwelling during the entire 2013 calendar year. Since only two months of data are used, these analyses do not take into account seasonal variations in housing and utility costs. Households that reported an annual income below 0 (debt) are excluded from the analyses.

Access to SHS data from 2010 onwards is only available in a Statistics Canada Research Data Centre (RDC). In order to meet Statistics Canada's strict data-release guidelines, the values on some variables used in these analyses have been rounded, top-coded (there is an artificial upper limit), and/or bottom-coded (there is an artificial lower limit). In addition, some atypical cases (including outliers) have been suppressed or collapsed into the next nearest class interval to ensure that there are a sufficient number of cases at each data point.

Describing the Dispersion of a Variable: Percentiles and Quantiles

In addition to describing the centre of a distribution, researchers also often want to describe how the cases are spread out, or dispersed, across a variable's attributes or values. There are several strategies that researchers use to describe the dispersion or spread of a distribution; in this section I describe percentiles and quantiles, and in the following section I describe the range and the interquartile range.

One way to illustrate the dispersion or spread of ordinal- and ratio-level variables is by dividing up the cases in an ordered list and then reporting the cut-off points. The cases are ordered based on their attributes or values on a specific variable. **Percentiles** are used to divide cases based on the percentage of cases that are at or below some attribute or value on a variable. For instance, the Occupy movement used the rhetoric of "the 1 per cent" and "the 99 per cent" in reference to the unequal distribution of wealth. In 2015, people needed an annual income of at least $225,409 to be among the top 1 per cent of income earners in Canada (Younglai and Yukselir 2017). Thus, $225,409 represents the ninety-ninth percentile of income—it is the cut-off point where 99 per cent of cases are at or below that level, and 1 per cent of cases are at or above it.

Researchers more commonly use percentiles to divide cases into halves or quarters or tenths. Recall that the median of a variable divides the cases in half by identifying the cut-off point where half of cases are at or below it, and half of cases are at or above it. Thus, the median of a variable is the same as the fiftieth percentile of a variable. That is, 50 per cent of cases are at or below the median attribute or value, and fifty per cent of cases are at or above the median attribute or value.

percentiles Cut-off points used to divide cases based on the percentage of cases that are at or below some attribute or value on a variable.

The process of finding a percentile is similar to the process of finding a median. The first step is to list the cases in order from the lowest attribute to the highest attribute of the variable. Then, find the attribute or value of the case in the position that has the desired percentage of cases at or below it. For instance, to find the tenth percentile, locate the position in the ordered list where 10 per cent of cases are at or below it, and 90 per cent of cases are at or above it; the attribute or value of the case in that position is the tenth percentile.

The easiest way to locate the position in an ordered list that corresponds to a percentile is by first converting the percentage into a proportion by dividing it by 100, and then multiplying the total number of cases plus 1 by that proportion. If the result is a whole number, the attribute or value of the case located at that position in the ordered list is the cut-off point for the percentile. If the result is a partial number (a decimal), the cut-off point for the percentile is located partway between the attributes or values of the cases immediately above and below the partial number. For instance, in looking back at Scenario 1 in Figure 3.1, there are nine cases listed in order. To find the thirtieth percentile of the variable, 30 per cent is divided by 100 to get the proportion 0.3, then the proportion 0.3 is multiplied by the number of cases plus 1 ($9 + 1 = 10$), to get 3. So, the attribute or value of the case in position 3 is the cut-off point for the thirtieth percentile. Since the case in position 3 has the attribute "Disagree," the cut-off point for the thirtieth percentile is the attribute "Disagree"; in other words, 30 per cent of cases have an attribute of "Disagree" or lower. The "Step-by-Step: Percentiles" box provides more information about how to find a percentile when the result is a partial number instead of a whole number.

Step-by-Step: Percentiles

Step 1: Count the total number of cases (N). Do not include cases that are missing information for the variable.

Step 2: Manually or electronically arrange (or sort) the cases in order from the lowest attribute to the highest attribute of the variable. If two or more cases have the same attribute, it doesn't matter what order they are listed in.

Step 3: Label the *position* of each case in the sequence, from 1 to N, so that the case with the lowest attribute has position 1, the case with the next lowest attribute has position 2, and so on until the position of every case is labelled.

Step 4: Add 1 to the total number of cases (from Step 1).

Step 5: Transform the percentile that you want to find into a proportion, by dividing it by 100. (So, to find the eightieth percentile, divide 80 by 100 to get 0.80.)

Step 6: Multiply the result of Step 4 by the result of Step 5.

Continued

Step 7 (whole numbers): If the result of Step 6 is a whole number, find the case in the *position* that corresponds to that number; the attribute or value of that case is the percentile.

Step 7 (partial numbers, ordinal-level variables): If the variable is ordinal level and the result of Step 6 is a partial number (a decimal), find the cases in the *positions* corresponding to the whole number immediately below the partial number and the whole number immediately above the partial number. (So, if the result is 8.1, find the cases in position 8 and position 9.) The percentile is halfway between the attributes of those two cases.

Step 7 (partial numbers, ratio-level variables): If the variable is ratio level and the result of Step 6 is a partial number (a decimal), find the cases in the *positions* corresponding to the whole number immediately below the partial number and the whole number immediately above the partial number. (So, if the result is 8.1, find the cases in position 8 and position 9.)

Step 7.1: Subtract the *value* of the case in the position corresponding to the whole number immediately below the partial number from the *value* of the case in the position corresponding to the whole number immediately above the partial number to find the difference between the values of the two cases.

Step 7.2: Identify the partial contribution to the percentile by subtracting all of the whole numbers from the partial number that was the result of Step 6. (So, if the result of Step 6 is 8.1, subtract 8 to get 0.1; if the result of Step 6 is 135.73, subtract 135 to get 0.73.)

Step 7.3: Multiply the difference between the values of the two cases (from Step 7.1) by the partial contribution to the percentile (from Step 7.2).

Step 7.4: Add the result of Step 7.3 to the *value* of the case in the position corresponding to the whole number immediately below the partial number to find the percentile.

Note: There are several different ways to calculate a percentile. The approach described here corresponds with that used by SPSS; other calculation methods may produce different results.

The cumulative percentage column of a frequency distribution can also be used to find percentiles. The attribute or value where the cumulative percentage first crosses over the percentage corresponding with the desired percentile is the cut-off point. So, to find the cut-off point for the tenth percentile, find the attribute or value in a frequency distribution where the cumulative percentage first becomes higher than 10 per cent. Looking back at Table 3.2, which shows the distribution of monthly housing costs, the cumulative percentage first becomes higher than 10 per cent in the "$250–499" row. Thus, $250–499 is the cut-off point for the tenth percentile; 10 per cent of households spend this amount or less on housing each month, and 90 per cent of households spend this amount or more on housing each month.

Percentiles are often used to divide the cases in a dataset into equal-size groups. The cut-off points between the groups are generally referred to as **quantiles**, but specific names are used for common subdivisions. For example, if the cases are divided into four equal groups, the cut-off points are called quartiles because they divide the cases into quarters. To divide a variable into quartile groups, researchers find the twenty-fifth, fiftieth, and seventy-fifth percentile. Or, if the cases are divided into five equal groups, the cut-off points are called quintiles. To divide a variable into quintile groups, researchers find the twentieth, fortieth, sixtieth, and eightieth percentile. Quintile groups are often used in conjunction with income variables because they allow researchers to compare the situation of the poorest, that is, the bottom 20 per cent, to those in the middle and upper classes.

The SHS includes information about households' annual income from all sources, including employment, investments, and government payments (such as pensions), for all of the people in a household combined. Each household's total annual income is recorded in dollars. (I then rounded it to the nearest $1,000.) Since the values are quantities—that is, the value 1,000 means 1,000 dollars—annual household income is a ratio-level variable. The cut-off points for the percentiles (twentieth, fortieth, sixtieth, and eightieth) that divide household income into quintiles are listed in Table 3.3. The twentieth percentile of annual household income is $32,000, which indicates that 20 per cent of households have an annual income of $32,000 or less. This is the quintile group that earns the lowest income. The fortieth percentile of annual household income is $57,000, which indicates that 40 per cent of households have an annual income of $57,000 or less. The 20 per cent of households that earn $57,000 or less but more than $32,000 are the second-lowest quintile group. This same logic is used to define the minimum and maximum cut-off values for all five quintile groups.

These quintiles divide households into five groups of roughly equal size, based on their total annual income. I created a new, ordinal-level variable to indicate which income quintile group a household belongs to. (Learn how to do this in the "Hands-on Data Analysis" box in this chapter.) Table 3.4 shows the class intervals and the number of cases in each quintile group. The frequency distribution shows that each group contains roughly 20 per cent of all of the cases in the data. The groups don't contain exactly 20 per cent of cases because more than one household might have an income that is at the cut-off point. For example, there are several households with a recorded income of $32,000. Since there is no sensible way to

quantiles Cut-off points used to divide cases into roughly equal-sized groups based on their attributes or values on a variable.

Table 3.3   Cut-Offs Used to Divide Household Income into Quintile Groups
Household income before taxes

Percentile	Cut-Off Point
20th	$32,000
40th	$57,000
60th	$83,000
80th	$125,000

Source: Author generated; Calculated using data from Statistics Canada, 2015.

Table 3.4   **Frequency Distribution of a New Variable Dividing Households into Quintile Groups Based on Their Total Annual Income**

Household income before taxes, divided into quintile groups

Category	Frequency	Percentage	Cumulative Percentage
Quintile Group 1 ($0 to $32,000)	328	20.2	20.2
Quintile Group 2 ($32,001 to $57,000)	321	19.8	40.0
Quintile Group 3 ($57,001 to $83,000)	331	20.4	60.4
Quintile Group 4 ($83,001 to $125,000)	322	19.8	80.2
Quintile Group 5 (more than $125,000)	322	19.8	100.0
Total	1,624	100.0	

Source: Author generated; Calculated using data from Statistics Canada, 2015.

decide which of the households with an income of $32,000 should be put in the first quintile group and which should be put in the second quintile group, they are all placed in the first quintile group. As a result, the first quintile group includes slightly more than 20 per cent of cases, and the second quintile group includes slightly less than 20 per cent of cases.

How Does It Look in SPSS?

Mode, Median, and Percentiles

The mode and median of a variable, as well as specified percentiles, can be obtained using the Statistics option in the Frequencies procedure. The variable label appears at the top of the table. You can choose whether or not a frequency table is also printed.

Statistics

Annual household income before taxes (rounded to nearest 1,000)

N	(A)	Valid	1624
		Missing	0
Median	(B)		71000.00
Mode	(C)		32000
Percentiles	20		32000.00
	40		57000.00
	(D) 60		83000.00
	80		125000.00

Image 3.1   **SPSS Statistics (from the Frequencies Procedure) for a Ratio-Level Variable**

A. These rows show the number of cases with a valid value (or attribute) and the number of cases with a missing value (or attribute) for the variable. Only cases with valid values are used to calculate the statistics.

B. This row shows the median. Half of households have an annual income of $71,000 or lower, and half of households have an annual income of $71,000 or higher.

C. The "Mode" row shows the most common value. The annual income that is reported by the largest number of households is $32,000.

D. The "Percentiles" rows show the cut-off points where the specified percentage of cases are at or below that value. These percentiles correspond with quintiles. Twenty per cent of households have an annual income of $32,000 or less, 40 per cent of households have an annual income of $57,000 or less, 60 per cent of households have an annual income of $83,000 or less, and 80 per cent of households have an annual income of $125,000 or less.

Dividing the cases into quintile groups makes it easier to describe the dispersion, or spread, of household income in Canada. One-fifth of households (20 per cent) have an annual income of $32,000 or less, before taxes. Two-fifths of households (40 per cent) have an annual income of $57,000 or less, before taxes. We also know that about one-fifth of households (20 per cent) have an annual income of more than $125,000, before taxes. Describing the distribution of household income in this way helps us to understand how much income inequality there is in Canada. Think about which quintile group your own household income is located in: Is your household in the bottom 20 per cent? The middle 20 per cent? Or the highest 20 per cent?

One reason that social science researchers often use income quintiles is that they correspond roughly to people's ideas about social class. Although the labels are not a perfect match, you might think of the households in the first quintile group as poor, the second quintile group as working class, the third quintile group as lower middle class, the fourth quintile group as upper middle class, and the fifth quintile group as upper class. Of course, the definition of working- and middle-class families in Canada is a subject of much debate. It might be that what matters most is household income per person; even though it's cheaper for several people to live together than it is for someone to live alone, a single adult with an annual income of $32,000 wouldn't necessarily be thought of as poor, whereas a family with two or three adults and four children earning the same amount would be. Others might argue that social class isn't necessarily related to the total household income but to the amount of income that is available for discretionary spending; high-earning households that must spend everything they earn on the necessities of life might not be considered upper or middle class if they cannot afford anything that is non-essential. Some theorists argue that social class depends more on a set of shared values and culture than on income per se, whereas others would assert that social class is defined primarily by access to—or ownership of—the means of production. A full discussion of the debates around how to assess social class in Canada is beyond the scope of this book. (But to learn more, see Porter et al. [2015] and Helms-Hayes and Curtis [2015]). I just want to draw your attention to how conceptual ideas such as social class can be translated into empirical measures and to some of the challenges associated with doing so.

Hands-on Data Analysis

Recoding a Variable

Researchers often need to modify variables in order to do the statistical analyses they want to do. In the last chapter, you learned how to combine two or more variables into a single variable. In this box, I describe how to create a new variable by grouping together the attributes of another variable. There are several reasons why researchers might want to re-group the attributes in a variable. Sometimes, they want an attribute to fit better with a theoretical or conceptual idea. Other times, researchers will group the attributes of a variable into quantiles for analysis. Very often, researchers eliminate attributes with relatively few cases by grouping two or more attributes together. Many statistical procedures require that each attribute of a variable has a minimum number or percentage of cases.

The first step in recoding a variable is to produce a frequency distribution of the original variable that you plan to recode. Make note of the values or numbers that are associated with each attribute. Then, think about how you want the attributes to be grouped together in the new variable you are creating. Be sure to have a clear plan before you begin the technical process of recoding. For instance, looking back at Table 3.2, consider the frequency distribution of monthly housing costs. Let's imagine that I want to merge some of the attributes together but also want to keep some of the attributes the same so that every class interval captures a $500 range. Table 3.5 shows my plan for recoding the variable: the first and second attributes will be grouped together in the new variable, the third and fourth attributes will be grouped together in the new variable, and the fifth and sixth attributes will be grouped together in the new variable, but the seventh, eighth, and ninth attributes will remain the same as in the original variable.

Once you have decided how many attributes the new variable will have, assign each new attribute a value and a label. Then create a series of "if →then" statements that tell the statistical software how to transform the values on the original variable into values on a new variable. For the example shown in Table 3.5, I use the following series of if → then statements:

Table 3.5   Planning to Recode a Variable

Original Variable		New Variable	
Value	Attribute	Value	Attribute
1	Less than $250	1	Less than $500
2	$250–499		
3	$500–749	2	$500–999
4	$750–999		
5	$1,000–1,249	3	$1,000–1,499
6	$1,250–1,499		
7	$1,500–1,999	4	$1,500–1,999
8	$2,000–2,499	5	$2,000–2,499
9	$2,500 or more	6	$2,500 or more

- If ORIGINAL_VARIABLE = 1 then NEW_VARIABLE = 1
- If ORIGINAL_VARIABLE = 2 then NEW_VARIABLE = 1
- If ORIGINAL_VARIABLE = 3 then NEW_VARIABLE = 2
- If ORIGINAL_VARIABLE = 4 then NEW_VARIABLE = 2
- If ORIGINAL_VARIABLE = 5 then NEW_VARIABLE = 3
- If ORIGINAL_VARIABLE = 6 then NEW_VARIABLE = 3
- If ORIGINAL_VARIABLE = 7 then NEW_VARIABLE = 4
- If ORIGINAL_VARIABLE = 8 then NEW_VARIABLE = 5
- If ORIGINAL_VARIABLE = 9 then NEW_VARIABLE = 6

If the original variable is a ratio-level variable, the if → then statements can be modified to incorporate a range of values. For instance, in order to create the first quintile group in Table 3.4, I used a statement indicating that if a household's income was between $0 and $32,000, then the value of the new variable should equal 1 (the first income quintile group).

It's important to pay attention to what happens to the attributes or values that are designated as missing when you recode a variable. In some statistical software, missing attributes or values in the original variable are automatically carried over when a variable is recoded. In other statistical software, if you do not specify what to do with the attributes or values that are designated as missing in the original variable, they are transformed into blanks (or system-missing) in the new variable.

Always confirm that the new variable worked out as planned by comparing a frequency distribution of the original variable and the new variable. (See Table 3.6.) Although this step might seem tedious, it can save you from making embarrassing mistakes or encountering major problems later on.

Continued

Table 3.6   Frequency Distributions of an Original Variable and a Corresponding New Variable

Monthly housing costs (original)   Monthly housing costs (recoded)

Original Variable			New Variable		
Category	Frequency	Percentage	Category	Frequency	Percentage
Less than $250	52	3.2	Less than $500	354	21.8
$250–499	302	18.6	$500–999	563	34.7
$500–749	342	21.1	$1,000–1,499	286	17.6
$750–999	221	13.6	$1,500–1,999	210	12.9
$1,000–1,249	156	9.6	$2,000–2,499	115	7.1
$1,250–1,499	130	8.0	$2,500 or more	96	5.9
$1,500–1,999	210	12.9			
$2,000–2,499	115	7.1			
$2,500 or more	96	5.9			
Total	1,624	100.0	Total	1,624	100.0

Source: Author generated; Calculated using data from Statistics Canada, 2015.

Describing the Dispersion of a Variable: Range and Interquartile Range

range The distance between the lowest and highest attributes or values on a variable that occur in the data.

minimum/maximum The lowest and highest attributes or values on a variable that occur in the data.

In addition to percentiles and quantiles, researchers can describe the dispersion, or spread, of ordinal- and ratio-level variables by reporting the range and the interquartile range. The **range** is the distance between the lowest attribute or value and the highest attribute or value on a variable in the data. That is, the range shows the difference between the minimum attribute or value and the maximum attribute or value. The **minimum** and **maximum** are the lowest and highest attributes or values on a variable that actually occur in the data (not just those that are theoretically possible). For ordinal-level variables, the range is typically reported in words: e.g., "Monthly housing costs ranged from less than $250 to $2,500 or more," or "Respondents' answers ranged from 'Strongly disagree' to 'Somewhat agree.'" For ratio-level variables, the range is reported as a number that shows the distance between the lowest and the highest value on the variable in the dataset. For ratio-level variables, the range is calculated as the maximum value minus the minimum value.

Although the range is an easy way to report the dispersion of a variable, it doesn't provide any information about how the cases are spread out between the minimum and maximum attributes or values. Perhaps only one household spends less than $250 on housing or it could be that most households spend that little. Similarly, it could be that only one household spends $2,500 or more on housing, or it could be that almost all households spend that much. Knowing the range doesn't help to distinguish between these situations. The **interquartile range (IQR)** provides more information about where cases are located within a variable. The IQR of a variable is

interquartile range (IQR) The distance between the twenty-fifth percentile and the seventy-fifth percentile of a variable.

calculated by dividing the cases into quartile groups (four equal groups). The inter-quartile range is the distance between the twenty-fifth percentile and the seventy-fifth percentile. (See Figure 3.2.) That is, it's the range of the middle 50 per cent of the data. One advantage of using the interquartile range to describe the dispersion of a variable is that it is not affected if there are a few cases with very high or very low attributes or values.

Just like other percentiles, the twenty-fifth and the seventy-fifth percentiles of a variable can be identified using the cumulative percentage of a frequency distribution. To find the cut-off point for the twenty-fifth percentile, find the attribute or value in a frequency distribution where the cumulative percentage first becomes higher than 25 per cent; similarly, to find the cut-off point for the seventy-fifth percentile, find the attribute or value in a frequency distribution where the cumulative percentage first becomes higher than 75 per cent. Figure 3.3 illustrates how to find the interquartile range using the frequency distribution of monthly housing costs. The twenty-fifth percentile is "$500–749," since this is the row where the cumulative percentage first becomes higher than 25 per cent. One-quarter of households spend $500–749 or less on housing each month. The seventy-fifth percentile is "$1,500–1,999" since this is the row where the cumula-tive percentage first becomes higher than 75 per cent. Three-quarters of house-holds spend $1,500–1,999 or less on housing each month (and one-quarter of households spend $1,500–1,999 or more). These two attributes define the bottom and the top of the interquartile range. For ordinal-level variables, the IQR is typ-ically reported in words. For example, these data show that the middle 50 per cent of households spend between $500–749 and $1,500–1,999 on housing each month; or, more simply, the middle 50 per cent of households spend between $500 and $1,999 on housing each month.

The IQR can also be used to describe the dispersion of ratio-level variables. Because these frequency distributions can be large, it's much easier to just request the twenty-fifth and the seventy-fifth percentile using statistical software. Although the IQR technically refers to the distance between these two cut-off points, it is often more useful to simply identify the bottom and the top of the interquartile range, even for ratio-level variables.

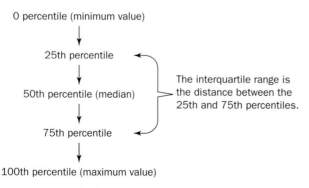

Figure 3.2   **A Depiction of the Interquartile Range**

Category	Frequency	Percentage	Cumulative Percentage	
Less than $250	52	3.2	3.2	
$250–499	302	18.6	21.8	
$500–749	342	21.1	42.9	← 25th percentile
$750–999	221	13.6	56.5	
$1,000–1,249	156	9.6	66.1	
$1,250–1,499	130	8.0	74.1	
$1,500–1,999	210	12.9	87.0	← 75th percentile
$2,000–2,499	115	7.1	94.1	
$2,500 or more	96	5.9	100.0	
Total	*1,624*	*100.0*		

The middle 50% of Canadian households spend between $500 and $1,999 on housing each month.

Figure 3.3   Using a Frequency Distribution to Find the Interquartile Range

Source: Author generated; Calculated using data from Statistics Canada, 2015.

How Does It Look in SPSS?

Minimum, Maximum, Range, and Interquartile Range

The minimum, maximum, range, and quartiles, as well as the median and mode of a variable, can be obtained using the Statistics option in the Frequencies procedure. The variable label appears at the top of the table. You can choose whether or not a frequency table is also printed.

Statistics

Monthly housing costs

N (A)		Valid	1624
		Missing	0
Median (B)			4.00
Mode (C)			3
Range (D)			8
Minimum (E)			1
Maximum (F)			9
Percentiles	25		3.00
(G)	50		4.00
	75		7.00

Image 3.2   SPSS Statistics (from the Frequencies Procedure) for a Categorical Variable

A. These rows show the number of cases with a valid value (or attribute) and the number of cases with a missing value (or attribute) for the variable. Only cases with valid values are used to calculate the statistics.

B. This row shows the median. Because this is a categorical variable, "4" refers to the attribute associated with the value 4, which is "$750–999." (The first two columns of Table 3.5 list the values and attributes of this variable.) Half of households spend $750–999 or less on housing each month, and half of households spend $750–999 or more on housing each month.

C. The "Mode" row shows the most common attribute. The attribute associated with the value "3" is "$500–749," so it is most common for households to spend $500–749 on housing each month.

D. The "Range" row shows the distance between the maximum and the minimum: 9 minus 1 is equal to 8. For ordinal-level variables, the range is usually reported in words, by stating the minimum and maximum attributes. For ratio-level variables, the range is usually reported as a number.

E. This row shows the minimum attribute of the variable that occurs in the data. The attribute associated with the value "1" is "Less than $250," so in this dataset, the lowest amount that households spend on housing is less than $250 each month.

F. This row shows the maximum attribute of the variable that occurs in the data. The attribute associated with the value "9" is "$2,500 or more," so the highest amount that households spend on housing is $2,500 or more each month.

G. The "Percentiles" rows show the cut-off points where the specified percentage of cases are at or below that attribute. These percentiles correspond with quartiles. Twenty-five per cent of households spend $500–749 or less on housing each month, 50 per cent of households spend $750–999 or less on housing each month, and 75 per cent of households spend $1,500–1,999 or less on housing each month. The interquartile range is from $500–749 to $1,500–1,499; or, more clearly, the middle 50 per cent of households spend between $500 and $1,999 on housing each month.

Using Box Plots to Show the Centre and Dispersion of a Variable

Since writing about the median, range, and IQR can sometimes be awkward, many researchers use **box plots** to visually depict these statistics. Box plots can be used to show the distribution of a single ordinal- or ratio-level variable, or they can be used to show the relationship between an ordinal- or ratio-level variable and another categorical variable. Some statistical software programs allow researchers to produce box plots for both ordinal- and ratio-level variables, but recent versions of SPSS only produce box plots for ratio-level variables. As a result, in this section the variables capturing households' monthly housing costs and the percentage of income spent on housing have been converted to ratio-level variables, rounded to

box plots Graphs that depict the median, the interquartile range, and the minimum and maximum of a variable using a box with whiskers.

the nearest $100 and 5 per cent, respectively. (These variables are also top-coded and/or bottom-coded.)

Figure 3.4 shows a box plot of households' monthly housing costs. The variable values are displayed on the vertical axis, or y-axis. The top and the bottom of the teal box are located at the twenty-fifth and seventy-fifth percentile, and thus the height of the box depicts the spread of the interquartile range. The bold horizontal line in the middle of the box shows the median. Finally, the minimum and maximum are depicted using the T-shaped bars above and below the box. These T-shaped bars are called whiskers. All of the horizontal lines in a box plot—the bottom whisker, the bottom of the box, the median, the top of the box, and the top whisker—visually divide the cases into four equal-sized groups (that is, into groups that each have the same number of cases). Roughly one-quarter of the cases are located between the bottom whisker and the bottom of the box, one-quarter are located between the bottom of the box and the median line, one-quarter are located between the median line and the top of the box, and one-quarter are located between the top of the box and the top whisker. If the height of a quartile is short, it indicates that many cases are bunched together at or around those values. If the height of a quartile is tall, it indicates that cases are more spread out across those values. The width of a box plot doesn't matter. Because of how they visually divide cases into equal-sized groups, box plots are a quick way to get information about the dispersion of a variable.

The box plot in Figure 3.4 indicates that a quarter of households spend $500 or less on housing each month, because the bottom of the box is located at $500. The height of the teal box indicates that the middle 50 per cent of households spend between $500 and $1,500 on housing each month. (This is slightly different than what is displayed in Figure 3.3 because the variable depicted here is a rounded ratio-level variable.) The top of the box shows that a quarter of households spend $1,500 or more on housing each month. Finally, the bold line in the centre of the

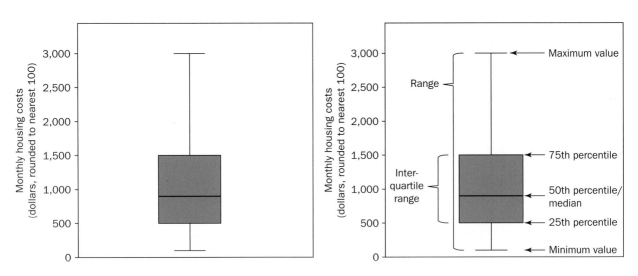

Figure 3.4   A Box Plot of Monthly Housing Costs, without and with Guidelines for Interpretation

Source: Author generated; Calculated using data from Statistics Canada, 2015.

box tells us that 50 per cent of households spend $900 or more each month on housing costs, and 50 per cent of households spend $900 or less.

Box plots are sometimes used to identify cases that do not appear to fit with the rest of the data. These cases are called "outliers" or "extreme values." When an outlier is identified in a box plot, it is depicted using a symbol (such as a dot) located above or below the minimum/maximum whiskers. (The box plots in Figure 3.4 do not show any outliers.) Each statistical software program has its own method of calculating and displaying outliers and extreme values in box plots, which is usually described in the program's "Help" files. Most software programs will display the case numbers or identification numbers of the outliers so that the researcher can assess whether the cases are truly outliers or whether they are data-entry errors or missing information that is incorrectly designated. The decision about whether to include or exclude any outlier cases is up to each researcher; you'll learn more about what might influence these decisions in later chapters.

The box plot in Figure 3.4 is easy to read because the cases are relatively evenly distributed or spread out across the variable. When cases are not as evenly distributed across a variable, box plots can appear distorted because some of the horizontal lines that divide the cases into quartiles may overlap. This occurs when more than 25 per cent of cases have the same attribute or value. Although these distortions can make box plots slightly harder to read, they still provide the same information about how cases are spread out across a variable.

Using Box Plots to Show Relationships

Box plots are also used to investigate how the dispersion of a variable is different in different groups. A separate box plot is created for each group, but they are all displayed together in a single graph. The ordinal- or ratio-level variable that the box plot shows the distribution of is typically considered the dependent variable, and the categorical variable that is used to group the cases is typically considered the independent variable. The values on the dependent variable are displayed on the vertical axis (y-axis), and the groups defined by the independent variable are labelled on the horizontal axis (x-axis). Comparing the minimums, maximums, medians, and the height of the boxes provides information about how the dispersion of cases differs in each group.

Figure 3.5 shows the relationship between households' income quintile group and the relative affordability of their housing, which is measured using the percentage of household income spent on housing. Recall that the CMHC defines affordable housing as that which costs less than 30 per cent of a household's before-tax income. In these box plots, income quintile group is treated as the independent variable and the percentage of income spent on housing is treated as the dependent variable. It's reasonable to think that the lower a household's income is, the higher the percentage of its income that is spent on housing. A comparison of the box plots for each income quintile group in Figure 3.5 supports this argument. Among households in the lowest income quintile group, half spend 35 per cent or more of their income on housing; that is, half of households have unaffordable housing. In contrast, in the highest income quintile groups, almost all households

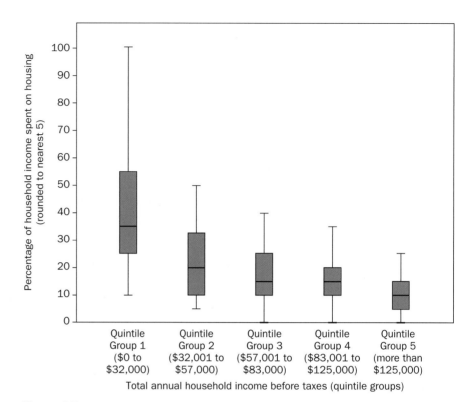

Figure 3.5   **Using Box Plots to Show the Relationship between Two Variables**

Source: Author generated; Calculated using data from Statistics Canada, 2015.

(excluding outliers not shown) spend 25 per cent or less of their income on housing; that is, they have affordable housing.

The box plots in Figure 3.5 also show that the cases in the lowest income quintile group are more spread out than those in higher income quintile groups. The middle 50 per cent of households in the lowest income quintile spend between 25 and 55 per cent of their income on housing. Some lower-income households have been able to find affordable housing, whereas others have not. (Some households in the lowest quintile report spending 100 per cent of their income on housing.) In contrast, the middle 50 per cent of households in the top income quintile only spend between 5 and 15 per cent of their income on housing. Not only do households in the highest quintile group have more income overall, they also have a much higher proportion of their income available for other necessities (such as food and transportation), as well as for non-essential spending.

Although these results might seem intuitive—of course high-income households have more affordable housing—it's valuable to show families' differential access to affordable housing with current, reliable Canadian data. Of particular concern is the prevalence of housing unaffordability among households in the lowest income quintile group. Fully 40 per cent of the households in the lowest income quintile group consist of seniors living alone. Among these low-income, unattached seniors, more than half (56 per cent) have housing that is very or somewhat unaffordable. As the Canadian government emphasizes aging-in-place, and as the number of seniors grows as a result of the aging baby-boomer cohort, attention needs to be paid to whether seniors' core housing needs are being met (CMHC 2010). For seniors on fixed incomes, the cost of property taxes, utilities, and upkeep can be expensive, even for those who own their homes. This brief statistical analysis suggests that more research is needed in this area, as affordable housing for seniors may be an emerging concern for policy-makers.

Best Practices in Presenting Results

Writing about the Centre and Dispersion of a Variable

Statistical writing—including writing about the centre and dispersion of a variable—requires a level of precision that you may not be accustomed to. Below, I illustrate some common errors that people make when writing about the mode and the median, and I show how to avoid them.

Although statistical software makes it easy to identify the mode, median, range, and IQR of a variable, the results typically require some explanation in order to be meaningful. For instance, some software programs report the mode, median, quartiles, and range for categorical variables using the value numbers instead of the value labels. It's important to determine what attribute each of these value numbers represent and to only report the value labels or the names of the categories in your writing.

One common writing error that people make is to assert that the mode is the attribute that is held by a majority of people or cases. But in order to legitimately say that a majority of cases have some characteristic, more than 50 per cent of the cases must have it. This is not always true of the mode, especially for variables with many attributes. These examples show correct and incorrect ways of reporting of the mode:

> *Incorrect*: The mode shows that the majority of households spend $500–749 on housing each month. (You know this is incorrect because the frequency distribution in Table 3.2 shows that only 21 per cent of households spend this amount.)
>
> *Correct*: The mode shows that the most common amount that households spend on housing each month is $500–749.
>
> *Even better*: It is most common for households to spend $500–749 on housing each month.

Similar wording challenges are associated with writing about percentiles or quantiles, including the median (the fiftieth percentile). Another common writing error is to report that 50 per cent of people have the median attribute. But the median is just a cut-off point: half of people are at *or below* that attribute or value. It's not true that half of people have that exact attribute or value. For example:

> *Incorrect*: The median shows that 50 per cent of households spend $750–999 on housing each month. (You know this is incorrect because the frequency distribution in Table 3.2 shows that only 14 per cent of households spend this amount.)

Correct: The median shows that 50 per cent of households spend $750–999 or less on housing each month.

Even better: Half of households spend less than $1,000 on housing each month.

Even better: Half of households spend $750 or more on housing each month.

Pay close attention to the language you use when reporting statistical results in order to make sure that your assertions are accurate and meaningful to people who aren't familiar with the data. Although writing about statistical results may seem awkward at first, it becomes easier with practice.

What You Have Learned

This chapter introduced some strategies for describing the centre and dispersion of a variable. The focus was on using these statistics with categorical variables, but all of these strategies can also be used with ratio-level variables. For nominal-level variables, researchers typically only describe the centre of the variable using the mode. For ordinal-level variables, researchers describe the centre of the variable using the mode and the median, and describe the dispersion of the variable using the minimum and maximum, the range, and the interquartile range. These statistics help to illustrate how cases are spread out around the centre of a variable. Another way of describing the dispersion of a variable is to divide it into quantile groups. This chapter also introduced box plots, which rely on quartiles to visually display the spread of a variable. All together, these are some of the main tools that researchers use to summarize and describe data. Chapter 4 introduces some additional strategies for describing the centre and dispersion of ratio-level variables.

The research focus of this chapter was housing affordability. Overall, about one in five households do not have affordable housing, as defined by the Canadian Mortgage and Housing Corporation. Half of households spend $750 or more on housing each month. Among low-income households—those with annual incomes of $32,000 or less—more than half do not have affordable housing. Two out of every five low-income households (40 per cent) are seniors who live alone. These results suggest that housing affordability may become a more pressing concern, especially as the size of Canada's senior cohort grows and as housing costs in urban areas increase. Further research in this area might investigate the relationship between housing affordability, the size of a city or town, and the availability of employment and services, in an effort to untangle the factors that contribute to high housing costs. Access to affordable housing is a human right that is not being met for all people in Canada, and thus policy interventions may be warranted.

Check Your Understanding

Check to see if you understand the key concepts in this chapter by answering the following questions:

1. What is the difference between the mode and the median?
2. What are percentiles, and why do researchers use them?
3. Why does it not make sense to report the dispersion of a nominal-level variable?
4. What strategies can you use to report the dispersion of an ordinal-level variable?
5. How are quartiles conceptually related to the median and the interquartile range?
6. What statistics are visually represented on a box plot, and how are they depicted?

Practice What You Have Learned

Check to see if you can apply the key concepts in this chapter by answering the following questions. Keep two decimal places in any calculations.

1. A local agency has received funding to build new seniors' residences. To find out what type of housing is most in demand, the agency hires you to survey people participating in seniors' programs at a local community centre. The survey results show that 36 per cent of respondents want the agency to build independent-living apartments, 32 per cent want assisted-living apartments, 14 per cent want small townhouses, and 18 per cent want housing for multi-generational families.

    a. When this survey question becomes a variable, what level of measurement will it have?
    b. How can you describe the centre of a variable with this level of measurement?
    c. What is the centre of this variable?

2. As a volunteer in a local election campaign, you go door to door to find out which issues are most important to potential voters. You ask people whether having affordable housing is "Very important," "Somewhat important," "Not very important," or "Not at all important" to them. By the end of the campaign, you collected information about affordable housing from 86 potential voters: 36 said that it is "Very important," 24 said that it is "Somewhat important," 14 said that it is "Not very important," and 12 said that it is "Not at all important."

    a. When this information becomes a variable, what level of measurement will it have?
    b. How can you describe the centre of a variable with this level of measurement?
    c. What is the centre of this variable?

3. Your student union is concerned about whether local housing is affordable for students who are living away from home, especially given rising tuition costs. The union distributes a questionnaire to a random sample of students who are not living at home and not living in residence, asking them to estimate their monthly housing costs, including utilities. Table 3.7 shows the results.

Table 3.7 **Frequency Distribution of Students' Monthly Housing Costs (Hypothetical Data)**
Approximately how much do you spend on housing costs each month (including utilities)?

Answer	Frequency	Percentage	Cumulative Percentage
$0	20	1.6	1.6
$1 to $99	43	3.4	5.0
$100 to $199	83	6.6	11.6
$200 to $299	123	9.7	21.3
$300 to $399	125	9.9	31.2
$400 to $499	187	14.8	46.0
$500 to $599	256	20.3	66.3
$600 to $699	166	13.2	79.5
$700 to $799	120	9.5	89.0
$800 to $899	97	7.7	96.7
$900 to $999	42	3.3	100.0
$1,000 or more	0	0.0	
Total	1,262	100.0	

    a. What level of measurement does this variable have?
    b. How can you describe the centre of a variable with this level of measurement?
    c. What is the centre of this variable?

4. Using the information in Table 3.7:

    a. What are two strategies you can use to describe the dispersion of a variable with this level of measurement?
    b. Describe the dispersion of this variable using each of the two strategies you identified in question 4(a).

5. By hand, draw a box plot that shows the distribution of the variable shown in Table 3.7.

6. Using the information in Table 3.7:

    a. Find the tenth percentile of the variable. Explain what it shows.
    b. Find the ninetieth percentile of the variable. Explain what it shows.

7. Using the information in Table 3.7:

    a. Find the cut-offs that divide the cases into quintile groups (the twentieth, fortieth, sixtieth, and eightieth percentiles).

    b. Use the cut-offs you found in (a) to create a new frequency distribution that groups the cases into quintile groups, based on housing costs. Your frequency distribution will be similar to that shown in Table 3.4.

8. As a follow-up to the survey of seniors described in question 1, you collect some additional information about how high housing costs influence people's lives. You ask 20 seniors about how often they worry about having enough money to pay for their home and utilities: less than once a month, once a month, a few times a month, once a week, a few times a week, or every day. Their answers are:

once a month	every day	once a week	less than once a month
a few times a month	a few times a week	less than once a month	once a month
every day	once a week	once a month	less than once a month
once a week	every day	less than once a month	a few times a week
a few times a month	less than once a month	a few times a month	every day

    a. Find the mode of this variable.
    b. Find the median of this variable.

9. Using the answers listed in question 8:

    a. Find the range of the variable.
    b. Find the interquartile range of the variable.

10. By hand, draw a box plot that shows the distribution of the answers listed in question 8.

11. Using the answers listed in question 8:

    a. Find the fortieth percentile of the variable. Explain what it shows.

    b. Find the eightieth percentile of the variable. Explain what it shows.

12. In the follow-up survey of seniors in described in question 8, you also collect information about how many years each person has lived in their current residence. Their answers are:

2 years	7 years	10 years	5 years
4 years	5 years	4 years	12 years
6 years	20 years	2 years	18 years
10 years	3 years	5 years	15 years
15 years	15 years	17 years	14 years

    a. Find the mode of this variable.
    b. Find the median of this variable.

13. Using the answers listed in question 12:

    a. Find the range of the variable.
    b. Find the interquartile range of the variable.

14. Using the answers listed in question 12:

    a. Find the fifteenth percentile of the variable. Explain what it shows.

    b. Find the sixty-fifth percentile of the variable. Explain what it shows.

15. The CMHC regularly publishes information about housing and income in Canada. The graph in Figure 3.6 uses CMHC data to compare the median incomes of homeowners and renters across time.

    a. What type of graph is it?
    b. What three variables are included in the graph?
    c. List two claims that you can make based on the information in this graph.

16. Access to affordable housing is a nationwide problem. But some provinces and municipalities are doing better than others. Use the online interactive map and database at www.rentalhousingindex.ca to investigate:

    a. What percentage of renter households in your municipality spend more than 30 per cent of their before-tax income on rent and utilities?

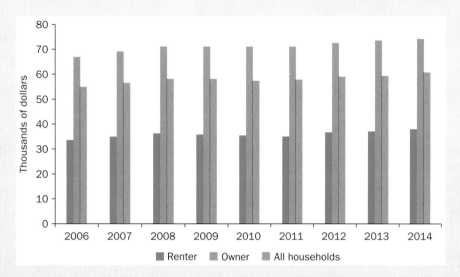

Figure 3.6   Median After-Tax Household Income, by Housing Tenure (Owner and Renter), Canada, 2006–2014 (2014 constant dollars)

Source: Statistics Canada (Canadian Income Survey) 2012–2014; Survey of Labour and Income Dynamics 2006–2011; all data are rounded to the nearest $100.

How does this compare to the province overall and to Canada overall?

b. What percentage of renter households in your municipality spend more than 50 per cent of their before-tax income on rent and utilities?

How does this compare to the province overall and to Canada overall?

c. What is the average cost of rent and utilities in your municipality? How does this compare to the province overall and to Canada overall?

Practice Using Statistical Software (IBM SPSS)

Answer these questions using IBM SPSS and the GSS27.sav or the GSS27_student.sav dataset available from the Student Resources area of the companion website for this book. Report two decimal places in your answers, unless fewer are printed by IBM SPSS. It's imperative that you save the dataset to keep any new variables that you create.

1. Use the Format option in the Frequencies procedure to produce a descending count frequency distribution of the variable "Living arrangement of respondent's household (6 categories)" [LIVARR06]. Find the mode of the variable using the frequency distribution. Explain what this statistic shows.

2. Use the Statistics option in the Frequencies procedure to find the mode of the variable "Living arrangement of respondent's household (6 categories)" [LIVARR06]. Explain what this statistic shows.

3. Produce a frequency distribution of the variable "Annual personal income of the respondent—2012" [INCM]. Find the mode and the median of the variable using the frequency distribution. Explain what these statistics show.

4. Use the Statistics option in the Frequencies procedure to find the mode and the median of the variable "Annual personal income of the respondent—2012" [INCM]. Explain what these statistics show.

5. Use the Statistics option in the Frequencies procedure to find the twenty-fifth, fiftieth, and seventy-fifth percentile of the variable "Annual personal income of the respondent—2012" [INCM]. Report the interquartile range of the variable, and explain what it shows.

6. Use the Statistics option in the Frequencies procedure to find the twentieth, fortieth, sixtieth, and eightieth percentile of the variable "Number of hours per week spent watching television" [MCR_300C]. Explain what these statistics show.

7. Use the Chart Builder tool to create a box plot of the variable "Number of hours per week spent watching television" [MCR_300C].

   a. Use the box plot to identify the twenty-fifth, fiftieth, and seventy-fifth percentile. (You can use the Statistics option in the Frequencies procedure to precisely identify these points.)

   b. Describe the interquartile range and the range (excluding outliers), using the box plot.

8. Use the Chart Builder tool to create a box plot of the variable "Number of hours per week spent watching television" [MCR_300C], divided by "Age group of respondent (groups of 10)" [AGEGR10].

   a. Which age group(s) have the lowest median? Which age group(s) have the highest median?

   b. Which age group(s) have the smallest IQR? Which age groups(s) have the widest IQR?

   c. Describe the overall pattern that appears in these box plots.

9. Use the Recode into Different Variables tool to recode the variable "Annual personal income of the respondent -2012" [INCM] into a new variable called "Annual personal income of the respondent - 2012 (recoded)" [INCM_RECODED]. The new variable should have seven attributes:

   1 - People with no income

   2 - People with an income from $1 to $19,999

   3 - People with an income from $20,000 to $39,999

   4 - People with an income from $40,000 to $59,999

   5 - People with an income from $60,000 to $79,999

   6 - People with an income of $80,000 or more

   7 - People with a "Missing" answer (i.e., "Don't know," "Refusal," or "Not stated")

In the *Variable View*, add value labels to the new, recoded variable, and designate the final attribute as missing information. Produce a frequency distribution of the original variable "Annual personal income of the respondent—2012" [INCM] and the new variable "Annual personal income of the respondent—2012 (recoded)" [INCM_RECODED] and compare them to ensure that the recoding is correct. Describe what the frequency distribution of the new, recoded variable shows.

References

Canada Mortgage and Housing Corporation (CMHC). 2010. "2006 Census Housing Series: Issue 10—The Housing Conditions of Canada's Seniors." Catalogue no. 10-021. Ottawa, ON: Canadian Mortgage and Housing Corporation. http://www.cmhc-schl.gc.ca/odpub/pdf/67201.pdf?fr=1427929957662.

———. 2014. "Housing in Canada Online: Definitions of Variables." http://cmhc.beyond2020.com/HiCODefinitions_EN.html#_Affordable_dwellings_.

Cox, Wendell, and Hugh Paveltich. 2016. "13th Annual Demographia International Housing Affordability Survey: 2017." Illinois: Demographia and Performance Urban Planning. http://www.demographia.com/dhi.pdf.

Helmes-Hayes, Richard C., and James E. Curtis, eds. 2015. *The Vertical Mosaic Revisited*. Toronto: University of Toronto Press.

Morin, Marie-Claude. 2012. *An Act to Ensure Secure, Adequate, Accessible and Affordable Housing for Canadians*. http://www.parl.gc.ca/HousePublications/Publication.aspx?Language=E&Mode=1&DocId=5391884&File=4.

Ontario Human Rights Commission (ONHRC). 2007. *Human Rights and Rental Housing in Ontario Consultation Paper*. Toronto: Ontario Human Rights Commission.

Porter, John, Wallace Clement, Jack Jedwab, Vic Satzewich, and Richard Charles Helmes-Hayes. 2015. *The Vertical Mosaic: An Analysis of Social Class and Power in Canada*. 50th anniversary edition. Toronto: University of Toronto Press.

Statistics Canada. 2015. "Survey of Household Spending 2014." http://www23.statcan.gc.ca/imdb/p2SV.pl?Function=getSurvey&Id=152606.

———. 2016. *User Guide for the Survey of Household Spending, 2014*. Catalogue no. 62F0026M. Ottawa: Statistics Canada. http://www.statcan.gc.ca/pub/62f0026m/62f0026m2016001-eng.pdf.

United Nations. 1948. "Universal Declaration of Human Rights." http://www.un.org/en/documents/udhr/.

Younglai, Rachelle, and Murat Yukselir. 2017. "Who Are Canada's 1 Per Cent and Highest Paid Workers?" *The Globe and Mail*. October 8. https://beta.theglobeandmail.com/news/canada-1-per-cent-highest-paid-workers-compare/article36383159/.

Describing the Centre, Dispersion, and Shape of a Distribution: Focus on Ratio-Level Variables

Learning Objectives

In this chapter you will learn:

- How to find the mean of a variable

- How to calculate and interpret the standard deviation (and variance) of a variable

- What z-scores are and how to calculate them

- What outliers are and how they affect the mean and standard deviation

- What histograms are and how they are different from bar charts

- What the key features of a normal distribution are

- What skew and kurtosis statistics show

- Which statistics to report for different types of variables

Introduction

This chapter introduces strategies for describing the distribution of ratio-level variables. The mean (or arithmetic average) is used to identify the centre of a distribution, and the standard deviation is used to describe the dispersion of a distribution. In the second half of this chapter, I introduce the normal distribution and its features, as well as two ways to describe the shape of a distribution: kurtosis and skew. Recall that ratio-level variables have attributes that are specific quantities or amounts, which simply become the values on the variable.

The research focus of this chapter is young people's wages, with a focus on those aged 15 to 24 who are not attending school. Fully half of young people in this age group (49 per cent) are neither full- nor part-time students. For some young people, factors such as poor grades, an unsafe or unwelcoming environment, or an inability to pay tuition fees push them away from educational institutions. For other young people, factors such as the availability of interesting or well-paying work or a desire for personal and financial independence pull them into the labour force. Among the 49 per cent of young people who are not students, four out of five (83 per cent) are in the labour force—that is, they are either employed or looking for work. Compared to older workers, young people are much more likely to work in two specific industries: retail trade and accommodation and food services.

Sorbis/Shutterstock

Photo 4.1   **Young people often work in low-wage jobs in the retail, accommodation, and food services industries.**

Young people's wages, especially in their first post-education jobs, influence their long-term personal and career trajectories, including their ability to live independently or start a family. Over the past few decades, young people have been taking longer and longer to engage in the typical transitions to adulthood, such as leaving their parents' home, entering conjugal relationships, and having children (Clark 2007). Wages that are not commensurate with the cost of living are one reason that it is taking young people longer to make these transitions. In this chapter, statistical analysis is used to discover the following:

- What is the average hourly wage for a young person who is not in school? How does that wage translate into weekly and yearly earnings?
- How much variation is there in young people's wages? How is the variation in wages related to age?
- What does the distribution of young people's wages look like? What shape does it have?

Describing the Centre of a Ratio-Level Variable: The Mean

mean ($\overline{X}$) The arithmetic average of a variable

The most common way to describe the centre of a ratio-level variable is the **mean**, or the arithmetic average. Averages are routinely reported in many different contexts. You have probably heard an instructor report the class average on a test, or calculated your own average grade on a series of assignments. Since you likely already know how to calculate an average, I'll take this opportunity to introduce some statistical notation.

Finding an average value is relatively straightforward: add together all of the values, and divide the result by the total number of cases. Let's imagine that we asked six different young people about their hourly wages. They reported earning $9 per hour, $11.50 per hour, $12 per hour, $14.50 per hour, $15 per hour, and $16 per hour. Their average hourly wage is calculated as:

$$Average\ hourly\ wage = \frac{9+11.50+12+14.50+15+16}{6}$$

$$= \frac{78}{6}$$

$$= 13$$

Like the median, there does not need to be a case in the data with the average value. The average is a theoretical construct and is not linked to a specific case. For instance, there isn't anyone who earns $13 per hour in this example.

Let's display this calculation using some different notation. The variable that captures information about hourly wages is called X. The hourly wage for the first person is X_1, the hourly wage for the second person is X_2, the hourly wage for the third person is X_3, and so on, until you reach the final person. The total number of cases is represented by the letter N. Since there are N cases, X_N represents the value on the final case. If there are 100 cases in a dataset, X_N is X_{100} or the value on the variable X for the hundredth case; if there are 5,000 cases in a dataset, X_N is X_{5000} or the value on X for the five-thousandth case. The mean or average of a variable is denoted by a horizontal bar above the letter corresponding to the variable. For example, the mean of variable X is denoted by an X with a bar above it ($\overline{X}$), and pronounced x-bar. When these symbols are used to show how to calculate a mean, we get:

$$\overline{X} = \frac{X_1 + X_2 + X_3 + X_4 + \ldots + X_N}{N}$$

This formula is a bit tedious to write out by hand, so statisticians have developed a shorthand for the numerator (or top part) of the fraction used in the formula. Instead of writing out $X_1 + X_2 + X_3 + X_4 + \ldots + X_N$, they use the upper-case Greek letter **sigma** (Σ) to indicate that the values of all of the cases should be summed together. The notation X_i is used to indicate the value on the variable X for each case in a dataset. The subscript letter i is referred to as an index, and it runs from 1 to N (unless otherwise specified). Whenever you see the sigma symbol, you need to complete the calculation that follows the symbol for each case, and then add together all of the results. Since:

$$X_1 + X_2 + X_3 + X_4 + \ldots + X_N = \Sigma X_i$$

sigma (Σ) Statistical notation used to indicate that the calculation following the symbol should be completed for each case, and the results should be summed together.

The formula for the mean is:

mean (average)

$$\bar{X} = \frac{\sum X_i}{N}$$

Let's learn about young people's wages by analyzing data from the 2016 Labour Force Survey (described in the "Spotlight on Data" box). Among employees aged 15 to 24 who are not students, the most common hourly wage (or the mode) is $12 per hour. The median hourly wage is $14 per hour. Half of young employees earn $14 per hour or less, and half earn $14 per hour or more. Finally, the average hourly wage is $15.90 per hour, slightly higher than the median and the modal hourly wage. These numbers are hard to interpret without a bit more context. One way to give these results meaning is to compare them to your own hourly wages. Do most young people in Canada earn more or less than you? Another way to give these results context is to calculate the annual income of a young person who works a full-time schedule of 40 hours each week, 52 weeks per year. On average, a young person working 40 hours a week earns $636 per week, or $33,072 per year, before taxes. In many Canadian cities, that's only slightly higher than the cut-off for the low-income (poverty) measure.

But many young people have trouble finding full-time work. Indeed, they are much more likely to be unemployed than older workers. In 2012, 14 per cent of young people in Canada were unemployed—that is, they were actively looking for work but could not find it (Galarneau, Morissette, and Usalcas 2013). This is twice the unemployment rate of people in Canada overall. As well, some young people are able to find only part-time work; in 2016, a quarter (26 per cent) of young people who were working and not in school worked part-time hours in their main job. On average, young people who were not in school worked 33.4 hours each week at their main (or only) job. Let's adjust our calculations

Step-by-Step: Means (Averages)

$$\bar{X} = \frac{\sum X_i}{N}$$

Step 1: Count the total number of cases (N). Do not include cases that are missing information for the variable.

Step 2: Add together the values of all of the cases ($\sum X_i$) to find the numerator of the mean equation.

Step 3: Divide the result of Step 2 by the total number of cases (from Step 1) to find the mean of the variable.

to take this into account. On average, a young person earning $15.90 per hour and working 33.4 hours each week earns $531 per week, or $27,615 per year. For young people with financial or housing support from their family of origin, this is a livable wage, but for young people who are living alone or with a partner, this is a relatively low income, especially for those with student debts and/or children to support.

Spotlight on Data

The Labour Force Survey

The Labour Force Survey (LFS) is one of the oldest and largest survey programs in Canada. It was initiated in 1945 to provide policy-makers with the employment and labour force information that they needed to transition to a peacetime economy after the Second World War (Statistics Canada n.d.). The survey is used to generate the employment/unemployment rate and estimates of the number of jobs created or lost each month in Canada as a whole and in each province/territory as well as in some major urban areas. Internationally, more than 100 countries conduct a labour force survey that collects similar information (ILO 2015).

The main variable used in the examples in this chapter is people's hourly wage for their main job, that is, the job that they work the most hours at. It incorporates people's base rate of pay and also includes any tips, commissions, or bonuses that they earn (Statistics Canada 2016a). An hourly wage is calculated for all employees, regardless of whether they are actually paid by the hour, but is not calculated for people who are self-employed. This is because employment earnings are not always related to hours of work for self-employed people.

The LFS collects information about people aged 15 or older. Full-time members of the Canadian Armed Forces, people who are institutionalized, and people living on reserves and in Aboriginal settlements in the provinces are not surveyed. Statistics Canada estimates that, together, these exclusions represent less than 2 per cent of the Canadian population (Statistics Canada 2016b). In recent years, the LFS has collected information about approximately 100,000 people, living in about 56,000 households, each month. Data are collected through telephone interviews or in-person visits to each household. About 90 per cent of the households who are selected to participate in the LFS actually do so (Statistics Canada 2016b). The analyses in this chapter are based on the labour force information of young people aged 15 to 24 who were formally employees (not self-employed) and not in school, collected between January and December 2016.

Describing the Dispersion of a Ratio-Level Variable: Standard Deviation and Variance

The dispersion of a ratio-level variable refers to how far the cases are spread out around the mean of that variable. The statistic that is most often used to describe the dispersion of a ratio-level variable is called the standard deviation. In this section, I describe the logic and calculation of the standard deviation, first using an example and then using a formula. The variance is the square of the standard deviation. It isn't reported as often as the standard deviation, but it is used to calculate some of the statistical tests you will learn about in later chapters.

One way to describe how far a series of cases are from their mean is by finding the average distance, or deviation, from the mean. Let's calculate the average distance from the mean by creating a table that lists each case in a row and has three columns: one with the value for each case, one with the mean of the variable (which will be the same for all of the cases), and one with the deviation from the mean for each case (the value minus the mean). Table 4.1 shows the deviation from the mean for the six cases from our hypothetical example.

Now, let's try to calculate the average deviation from the mean. To calculate the average, we sum up the numbers in the Deviation from the Mean column, and then divide the sum by the total number of cases. When we try to do this, we quickly encounter a problem: the numbers in the Deviation from the Mean column add up to 0 because of the negative numbers. When 0 is divided by 6 (since there are six cases) to find the average deviation from the mean, the result is 0. This doesn't help to explain how the cases are spread out around the mean.

One way to resolve this problem is to find the absolute deviation from the mean, that is, to ignore the negative signs of the deviations from the mean. When all of the absolute deviations from the mean are added together, ignoring the negative signs, the result is 13, and the average absolute deviation from the mean is 2.17 (13 divided by 6). Although it's possible to generate an answer this way, the absolute deviation from the mean is not commonly used by statisticians. Instead, statisticians prefer to square the deviations from the mean before adding them together. Not only does this eliminate the negative numbers (since two negatives multiplied together results in a positive number), it also assigns a bigger penalty to cases that

Table 4.1   Finding the Deviation from the Mean (Hypothetical Data)

Person (Case)	Hourly Wage ($) X_i	Variable Mean $\bar{X}$	Deviation from the Mean $(X_i - \bar{X})$
Naomi	9	13	−4
Katrina	11.5	13	−1.5
Sydney	12	13	−1
Nicolas	14.5	13	1.5
Mohammed	15	13	2
Riel	16	13	3
			column sum: 0

Table 4.2   Finding the Squared Deviation from the Mean (Hypothetical Data)

Person (Case)	Hourly Wage ($) X_i	Variable Mean $\bar{X}$	Deviation from the Mean $(X_i - \bar{X})$	Squared Deviation from the Mean $(X_i - \bar{X})^2$
Naomi	9	13	−4	16
Katrina	11.5	13	−1.5	2.25
Sydney	12	13	−1	1
Nicolas	14.5	13	1.5	2.25
Mohammed	15	13	2	4
Riel	16	13	3	9
				column sum: 34.5

are farther from the mean, since squaring the deviations makes them exponentially larger. For instance, a case that is 1 unit away from the mean still only contributes 1 to the total when the squared deviations are added together, but a case that is 10 units away from the mean contributes 100 to the total when the squared deviations are added together. The fourth column added to Table 4.2 illustrates what happens when the deviations from the mean are squared.

When the squared deviations from the mean are added together, the result is always a non-zero, positive number (the only way to get zero is if your variable is actually a constant, that is, all of the cases have exactly the same value). The sum of the squared deviations from the mean in Table 4.2 is 34.5. This result is divided by the number of cases to find the average squared deviation from the mean: $34.5 \div 6 = 5.75$. This statistic is the **variance**.

variance (S^2) The square of the standard deviation.

You might notice that the variance is much larger than the average absolute deviation from the mean. That's because the deviations from the mean were squared before they were added together to calculate the variance. To compensate for this squaring, we take the square root of the variance: $\sqrt{5.75}$ is 2.40. This statistic is the **standard deviation** from the mean. The process of squaring the deviations from the mean (to eliminate the negative numbers), finding their average (the variance), and then taking the square root of that average standardizes the deviations from the mean. Thus the standard deviation is the standardized, average deviation from the mean. Along with the variance, it is a measure of how far cases are spread out around the mean of a variable. The standard deviation has the same unit of measurement as the variable. If a variable is measured in dollars, the standard deviation is also in dollars; if a variable is measured in minutes, the standard deviation is also in minutes, and so on.

standard deviation (S) The standardized, average deviation from the mean.

The standard deviation is interpreted in comparison to the mean of a distribution, which is why researchers always report the mean and the standard deviation of a variable together. Let's return to learning about young people's wages. The average wage for young people in Canada who are not in school is $15.90 per hour, and the standard deviation is $6.06. Because the standard deviation is relatively small, compared to the mean (about a third of the size), we know that young people's hourly wages aren't widely spread out. If the standard deviation were larger than the mean, we would know that hourly wages are widely dispersed. The usefulness of the standard deviation becomes more apparent if we repeat the analysis, dividing young people into two groups: those aged 15 to 19 and those aged 20 to 24. For people aged 15 to 19, the average hourly wage is $12.56, with a standard deviation of $3.04. For people aged 20 to 24, the average hourly wage is $16.84, with a standard deviation of $6.36. These results show that the hourly wages of people aged 20 to 24 are more spread out (or more dispersed) than the hourly wages of younger people. In other words, there is more variation in wages among people aged 20 to 24 than among younger people. People in the 15- to 19-year-old group may be ghettoized in low-paying jobs with fewer opportunities to advance, such as those in retail sales or fast food restaurants. In contrast, some people in the 20- to 24-year-old group may be working in low-wage jobs, but others may have obtained a post-secondary credential and be working in professional positions with higher hourly wages. Later in this chapter, you'll learn some more ways to interpret the standard deviation statistic.

I just illustrated how to calculate the standard deviation using a table that lists each case in a row, and has four columns. The formula for the standard deviation instructs you to go through the same calculations as in Table 4.2. The formula for the standard deviation is:

standard deviation

$$S = \sqrt{\frac{\sum \left(X_i - \bar{X}\right)^2}{N}}$$

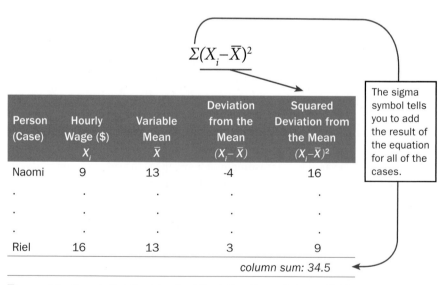

$$\sum_i (X_i - \bar{X})^2$$

Person (Case)	Hourly Wage ($) X_i	Variable Mean $\bar{X}$	Deviation from the Mean $(X_i - \bar{X})$	Squared Deviation from the Mean $(X_i - \bar{X})^2$
Naomi	9	13	-4	16
.	.	.	.	.
.	.	.	.	.
.	.	.	.	.
Riel	16	13	3	9
				column sum: 34.5

The sigma symbol tells you to add the result of the equation for all of the cases.

Figure 4.1   **Translating the Standard Deviation Formula into a Table (Hypothetical Data)**

In general, formulas that incorporate a sigma symbol are easiest to implement using a table. Each of the elements in the numerator of the fraction in the standard deviation formula correspond to a column in Table 4.2. Begin with the calculation that follows the sigma symbol, and then be mindful of the order of operations. Start with the parentheses first, calculating $(X_i - \overline{X})$ for each case, and then square the result. (See Figure 4.1.) Then, follow the instructions given by the sigma symbol and add together the result of the calculation for all of the cases.

Once you have found the numerator of the fraction in the standard deviation formula, the remainder is easy to complete. Divide the fraction numerator by the number of cases (N), and then find the square root of the result. Once the formula is broken down into its components, it becomes clear that it is just giving instructions for calculating the standardized, average deviation from the mean. The formula for the variance is identical to the standard deviation, except that you do not take the square root of the result:

$$S^2 = \frac{\Sigma \left(X_i - \overline{X} \right)^2}{N}$$

variance

Step-by-Step: Standard Deviation

$$S = \sqrt{\frac{\Sigma \left(X_i - \overline{X} \right)^2}{N}}$$

Step 1: Count the total number of cases (N). Do not include cases that are missing information for the variable.

Step 2: Find the mean or the average of the variable $(\overline{X})$.

Step 3: For *each case individually*, subtract the mean (from Step 2) from the value of the case (X_i) to find the deviation from the mean. Complete this calculation N times, once for each case.

Step 4: For *each case individually*, square the deviation from the mean (the result of Step 3). Complete this calculation N times, once for each case.

Step 5: Add together all of the squared deviations from the mean (the results of Step 4) to find the numerator of the fraction in the standard deviation equation.

Step 6: Divide the result of Step 5 by the total number of cases (from Step 1).

Step 7: Find the square root of the result of Step 6 to find the standard deviation of the variable.

Tip: To find the variance of a variable, stop after completing Step 6.

How Does It Look in SPSS?

Mean and Standard Deviation

The Means procedure produces a report that looks like Image 4.1. The variable label appears at the top of the report. Above each mean report, SPSS prints a "Case Processing Summary" (not shown) that lists the number and percent of included cases, excluded cases, and total cases.

Report

Usual hourly wages, employees only

(A) Mean	(B) N	Std. Deviation (C)
15.9049	62561	6.06440

Image 4.1   **An SPSS Means Report**

A.   This column shows the mean or average. The average hourly wage is $15.90.
B.   This column shows the number of (valid) cases used to calculate these statistics.
C.   This column shows the standard deviation. The standard deviation of hourly wages is $6.06.

The Means procedure can generate statistics for separate groups, which are defined by a second variable (usually a categorical variable). When the means report is divided by group, it looks like Image 4.2.

Report

Usual hourly wages, employees only

Age of respondent (5 year groups) (D)	Mean	N	Std. Deviation
15 to 19 (E)	12.5562	13637	3.04232
20 to 24	16.8383	48924	6.36017
Total (F)	15.9049	62561	6.06440

Image 4.2   **An SPSS Means Report, Divided by Group**

D.   The label at the top of the column shows the variable used to define the groups.
E.   The rows list each of the groups defined by the second variable, and show the mean, the standard deviation, and the number of (valid) cases for each group individually. Young people aged 15 to 19 have an average wage of $12.56, and young people age 20 to 24 have an average wage of $16.84. The standard deviation of hourly wages is $3.04 for young people aged 15 to 19, compared to $6.36 for young people aged 20 to 24.
F.   The "Total" row shows the overall mean, standard deviation, and number of cases, ignoring the groups defined by the second variable. The information in this row will match the information in a means report of the variable on its own (as long as the grouping variable does not have any missing attributes that result in cases being excluded from the means report).

Standardized Scores

The mean and standard deviation of a variable can be used to calculate a standardized score for each case, called a **z-score**. A z-score shows how far an individual case is from the mean of the variable, using standard deviations as the unit of measurement. A case with a z-score of 0 has the average value of the variable, a case with a z-score of +1 has a value that is one standard deviation above the average value, a case with a z-score of +2 has a value that is two standard deviations above the average value, and so on. Negative z-scores have a similar pattern: a case with a z-score of −1 has a value that is one standard deviation below the average value of the variable, a case with a z-score of −2 has a value that is two standard deviations below the average value, and so on. It's easy to calculate a z-score for an individual case: first, find out how far the case deviates from the mean, and then divide the result by the standard deviation. The formula for a z-score is:

> **z-score** A standardized score that shows how far an individual case is from the mean of a variable, using standard deviations as the unit of measurement.

$$Z = \frac{\left(X_i - \bar{X}\right)}{S}$$

z-score

Z-scores are used primarily to compare the relative position of two or more cases that are part of different distributions. For instance, z-scores can be used to compare the income of two employees working in different cities, where each city has a different average income and a different spread of income. By calculating how many standard deviations above or below the city average each employee's income is, you can make a fair comparison between them. Let's use z-scores to compare the wages of two young people, one from each age group. Recall that among young people aged 15 to 19 the average hourly wage is $12.56, with a standard deviation of $3.04, and that among young people aged 20 to 24, the average hourly wage is $16.84, with a standard deviation of $6.36. Now let's consider the wages of 17-year-old Soriya, who earns $15.60 per hour, and 22-year-old Matthew, who earns $20.02 per hour. Both are earning wages that are about three dollars more than the average wage for their age group. But which person is earning more in comparison to their peers? To answer this question, we compare their standardized scores. Since Soriya is part of the 15–19 age group, I use the mean and standard deviation of that group to calculate her z-score: (15.60 − 12.56) ÷ 3.04 = 1. Since Matthew is part of the 20–24 age group, I use the mean and standard deviation of that group to calculate his z-score: (20.02 − 16.84) ÷ 6.36 = 0.5. Because Soriya's z-score is higher than Matthew's, we can say that she is doing better, compared to her peers, than Matthew is. Even though Matthew earns a higher hourly wage, the average wage for 20- to 24-year-olds is higher.

Z-scores are useful for comparing the relative position of cases because they are standardized using a variable's standard deviation. The example of Soriya and Matthew illustrates how z-scores are used to compare the relative position of each case within a single variable, where each case is part of a different group. Z-scores can also be used to compare the relative position of a case on two or more variables, even if the variables are measured in different units.

Step-by-Step: Z-scores

$$Z = \frac{\left(X_i - \bar{X}\right)}{S}$$

Step 1: Find the value of the case that you want to find the z-score for (X_i).

Step 2: Find the mean or the average of the variable ($\bar{X}$).

Step 3: Find the standard deviation of the variable (S).

Step 4: Subtract the mean of the variable (from Step 2) from the value of the case (from Step 1) to find the numerator of the z-score equation.

Step 5: Divide the result of Step 4 by the standard deviation (from Step 3) to find the z-score of the case.

How Outliers Affect the Mean and Standard Deviation

Sometimes cases stand out as different because they have unusually high or unusually low values of a variable. These cases are referred to as **outliers**: cases that do not fit into the general pattern of a distribution. In other words, they are cases with attributes that are notably different than the others. In the social sciences, there is no universal rule for identifying outliers; it is up to researchers to determine whether or not a case is an outlier. If the cases with unusual values seem atypical in some way, or if it seems that there are different social processes that might be resulting in the unusual values, you might want to treat them as outliers. For example, some people in Canada earn more than a million dollars each year. These people earn much of their income by buying and selling investments, not by selling their labour power as most people do. As a result, it often makes sense to treat these high-income earners as outliers and to exclude them from analyses of income in Canada.

If a variable appears to have outliers, double-check to make sure that they don't represent missing information or data-entry errors. If the outliers seem to be legitimate values, one option is to simply exclude them from the analysis—but be sure to note this when you report the results. Another option is to include the outliers in the analysis and to describe how they affect the statistics that are presented. Some researchers present their results both with and without outliers so the reader can see their effect.

Because of how the mean and standard deviation (and variance) are calculated, they are particularly susceptible to the effect of outliers. Recall our list of six young people's hourly wages: $9 per hour, $11.50 per hour, $12 per hour, $14.50 per hour, $15 per hour, and $16 per hour; the average wage is $13 per hour. Now imagine that a young person who earns $55 per hour is added to the group. With the additional person, the average wage is $19 per hour—higher than the hourly wage of six of the seven people. In this situation, the average isn't a particularly good description of the centre of the distribution. One way to solve this problem is to calculate the average without the outlying case. Alternatively, the median hourly wage can be reported, since the median is less affected by outliers. For these seven people, the median hourly wage is $14.50—which is a better description of the centre of this distribution than the mean.

The standard deviation of young people's hourly wages for the original group of six is $2.40 (and the variance is 5.75). When the person earning $55 per hour is added to the group, the standard deviation jumps to $16.05 (and the variance is 22.93). Because the mean is higher when the seventh person is added to the group, all of the cases are farther from the mean. This effect is then compounded when the distance of each case from the mean is squared in the calculation of the variance and the standard deviation.

Because outliers affect the mean and standard deviation (and variance), which are used as the basis for many statistical calculations, it's important to look carefully at the distribution of each variable before performing more advanced statistical analyses. This can be done using frequency distributions or graphs, such as box plots or histograms.

Using Histograms to Show the Dispersion of Ratio-Level Variables

One way to investigate the distribution of a ratio-level variable is by graphing it in a **histogram**, which is like a bar graph, except that there are no gaps between the bars. Whereas bar graphs are used to display the distribution of categorical variables, histograms are used to display the distribution of continuous variables. Since continuous variables theoretically have an infinite number of values, the bars of a histogram are adjacent to one another. Histograms are sometimes also used to display the distribution of count variables that have a wide range.

The width of each bar in a histogram captures an equal-sized interval of values. This is conceptually equivalent to using class intervals of equal sizes to group the values of a ratio-level variable. For instance, Figure 4.2 is a histogram of the hourly wages of young people in Canada with a $1 interval for each bar. The height of each bar shows how common the values in the interval are. The tallest bar in Figure 4.2 shows that about 18 per cent of the young people surveyed earn an hourly wage between $11 and $12 per hour. The bar to the right of the tallest bar shows that more than 10 per cent of the young people surveyed earn an hourly wage between $12 and $13 per hour. The histogram also shows that the hourly wages of many young people are clustered near the bottom of the range (on the left of the horizontal axis) and that relatively few young people earn more than $30 per hour. If you work for pay, take a moment to figure out where your own hourly wage would be plotted in this histogram, and how your pay compares to others.

histogram A graph that shows the number or percentage of cases that have values within equal-sized class intervals (called bins) using the relative height of a bar.

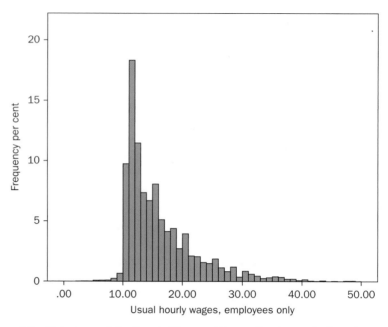

Figure 4.2 **Histogram of the Hourly Wages of Young Employees in Canada (n = 62,561)**

Source: Author generated; Calculated using data from Statistics Canada, 2016b.

Hands-on Data Analysis

Binning in Histograms

The appearance of a histogram changes depending on how wide the bars in the histogram are. The height of each bar in a histogram illustrates how many cases have values within a specific interval. Each interval (or bar) in a histogram is called a "bin" and the width of the interval is called the "bin size." Changing the bin sizes can substantially alter how a histogram looks, even though the distribution of the cases has not changed. When you prepare histograms for research reports or assignments, it's a good idea to try out several different bin sizes. Most statistical software programs make it easy to adjust the bin size (or breaks) in a histogram.

The goal is to choose a bin size that creates bins that are large enough to show the general pattern in the data but not so large that they obscure any important variations. Also consider the typical intervals for whatever you are measuring when you choose a bin size. For instance, if a variable measures something in minutes, consider using a bin size of 15 or 30 or 60 minutes because we tend to think about time divided into hours or parts of an hour. If a variable measures something in dollars, consider using $10 or $50 or $100 bins because we tend to think about money in these intervals. If a variable measures something in days, consider using 7- or 14- or 28-day bins because they translate easily into weeks.

In addition to using common intervals for bins, you can also experiment to find the bin size that makes your histogram look the best. Figure 4.3 shows the distribution of young people's hourly wages using different bin sizes. (The same distribution with $1 bin sizes is shown in Figure 4.2.) The histogram on the top left of Figure 4.3 uses $0.50 bins and has lots of unevenness in the heights of adjacent bars. Instead of showing a smooth distribution, the tops of the bars go up and down and up and down as they progress across the horizontal axis. This lack of smoothness suggests that the bin size is too small. The histogram shows more of the variation between the intervals than the general pattern in the data. (The $1 bins shown in Figure 4.2 are smoother, although there is still some up-and-down variation in the bar heights.) The $2 bins used in the histogram on the top right of Figure 4.3 are much smoother,

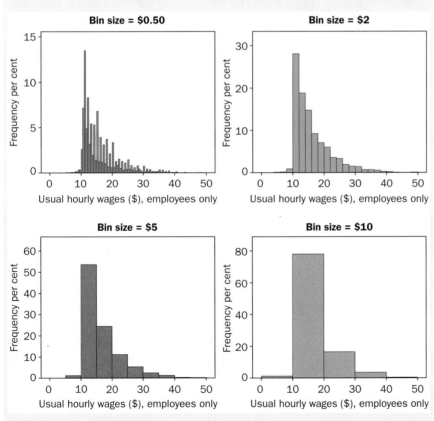

Figure 4.3   Changing Bin Sizes in a Histogram Alters the Appearance of a Distribution

Source: Author generated; Calculated using data from Statistics Canada, 2016b.

Continued

and all of the up-and-down variation in bar heights has been eliminated. The tallest bar shows that more than a quarter of young employees in Canada earn between $10 and $12 per hour, and the bar heights steadily decrease from there. When the bins become larger, as in the bottom two histograms in Figure 4.3, important details about the distribution are lost, and the histogram becomes less useful. For instance, the histograms show that more than 50 per cent of young employees in Canada earn between $10 and $15 per hour and that almost 80 per cent earn between $10 and $20 per hour. But these large bins hide the fact that many young people's wages are at the lower end of this range—between $10 and $12 per hour. Based on these bin-size experiments, it makes the most sense to use $1 or $2 bins when presenting this distribution of young employees' wages.

Like other graphs, histograms should always show the percentage of cases, not the number of cases. The horizontal axis (x-axis) scale and labels should be adjusted to correspond with the bin size so that the histogram is easy to read. There does not need to be a label for each bin, but the labels should align with the start or end of a bin so that a reader can deduce how wide each interval is. Sometimes histograms only fill a relatively small part of the chart space. If this occurs, adjust the maximum value of the scale on the horizontal axis so that the very high or very low values are not displayed.

In general, making good graphs is a bit of an art. Take the time to experiment with the scale of your axes and with different bin sizes in order to generate a graph that displays the information that you want to emphasize in your analysis.

Researchers often describe the shape of distributions by comparing them to a normal distribution—a predefined shape that is used as a point of reference in statistical analyses. In the next section, I introduce the normal distribution and its properties; you'll learn even more about the normal distribution in chapters 5 and 6.

Key Features of the Normal Distribution

normal distribution A theoretical construct that provides a reference point for describing the shape of a distribution; it is also central to the process of statistical estimation.

The **normal distribution** is central to the practice of frequentist statistics. At a basic level, a normal distribution is just a shape—you might also know it as a normal curve or a bell curve. It's sometimes called a Gaussian distribution, because one of the first people to describe the properties of a normal distribution was Carl Friedrich Gauss, a German mathematician, working in the early 1800s. In fact, Gauss and the normal distribution were so notable that they were featured on the 10 Deutschemark bill (the Deutschemark was the German currency before the Euro). Gauss wasn't alone in his discovery, however. The first known documentation of a normal distribution was written by Galileo in the seventeenth century, and French mathematician Abraham de Moivre described the normal distribution in a 1733 publication (Nicol 2010). At about the same time as Gauss, another French mathematician, Pierre-Simon Laplace, independently discovered the normal distribution and its relationship to statistical estimation (Nicol 2010). In this section, I focus on how the normal distribution serves as a reference point for describing the shape and area of a distribution. In later chapters, I describe how the normal distribution is used in statistical estimation—that is, the process of making claims about a population using data collected from a sample of that population.

A normal distribution has several key characteristics. First, it is symmetrical. In other words, it is exactly the same on both sides of the vertical line in the centre of the distribution. Second, the mean, the median, and the mode of the distribution are all the same and are located at the exact centre of the distribution. The distribution has only one mode—it is unimodal—and that mode is the centre value, where the peak of the curve is located. Because the distribution is symmetrical, that same centre value is also the median—the point at which 50 per cent of cases are below it and 50 per cent of cases are above it. Finally, the curve falls off gradually from the centre point into tails, which never actually meet the horizontal axis (x-axis), they just get closer and closer to the line. In depictions of the normal distribution, eventually the tails become so close to the x-axis that they appear to meet it to the human eye, but theoretically they never converge with the line. This shape serves as a point of reference for statisticians.

Many biological and social characteristics are approximately normally distributed. For example, people's heights are approximately normally distributed. But even when a variable is not normally distributed, researchers still describe it in comparison to a normal distribution.

Standard Deviation and the Normal Distribution

One particularly useful feature of the normal distribution is that it has a clear relationship to the standard deviation. In fact, any normal distribution can be described using only two values: its mean (which is also its median and mode) and its standard deviation. Figure 4.4 shows a standard normal distribution (sometimes called a z-distribution). It is centred on 0. In other words, the mean, the median, and the mode of the distribution are located at 0 on the horizontal axis (x-axis). In addition, each one-unit increment on the horizontal axis (x-axis) is equivalent to one standard deviation from the mean.

It's easiest to understand the relationship between the normal distribution and the standard deviation by thinking about the area that is under the curve—that is, the amount of space that there is between the line of the curve and the horizontal axis. First, imagine that the whole area is shaded in—this represents 100 per cent of the area that is under the curve. (See the top left panel of Figure 4.5.) Now, imagine a vertical line at the −1 point and at the +1 point on the horizontal axis of the standard normal distribution. If the area between the

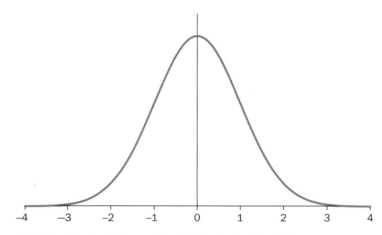

Figure 4.4   **The Standard Normal Distribution (z-distribution)**

two lines is shaded in, it takes up 68 per cent of the space. In other words, 68 per cent of the area under the curve is between −1 and +1 standard deviation from the mean. (See the top right panel of Figure 4.5.) Now, imagine a vertical line at the −2 point and the +2 point on the horizontal axis of the standard normal distribution. If the area between the two lines is shaded in, it takes up 95 per cent of the space. In other words, 95 per cent of the area under the curve is between −2 and +2 standard deviations from the mean. (More precisely, it's 1.96 standard deviations, but let's just say 2 for now; see the bottom left panel of Figure 4.5.) Finally, imagine a vertical line at the −3 point and at the +3 point on the horizontal axis of the standard normal distribution. If the area between the two lines is shaded in, it takes up 99 per cent of the space. In other words, 99 per cent of the area under the curve is between −3 and +3 standard deviations from the mean. (Again, more precisely it's 2.58 standard deviations, but let's just say 3 for now; see the bottom right panel of Figure 4.5.)

Knowing how much of the area is between plus and minus 1 standard deviation from the mean, and plus and minus 2 standard deviations from the mean helps us to describe the dispersion of a normally distributed variable. Let's return to the hypothetical example of six young people's wages. For the moment, assume that hourly wages are normally distributed. (You'll learn how to check this assumption in the next section.) The mean is $13 per hour and the standard deviation is $2.40. These statistics are substituted into the horizontal axis of a normal distribution, in place of the mean and standard deviation. (See Figure 4.6.) The shape of the distribution remains the same, but instead of a mean of 0, it has a mean of $13. The distance between the mean and the +1 point on the horizontal axis is one standard deviation, or $2.40. So, instead of +1 on the horizontal axis, I substitute in the value of the mean plus one standard deviation: $13 + $2.40 = $15.40. Instead of +2 on the horizontal axis, I substitute in the mean plus two standard deviations: $13 + $2.40 + $2.40 = $17.80. Similarly, instead of −1 on the

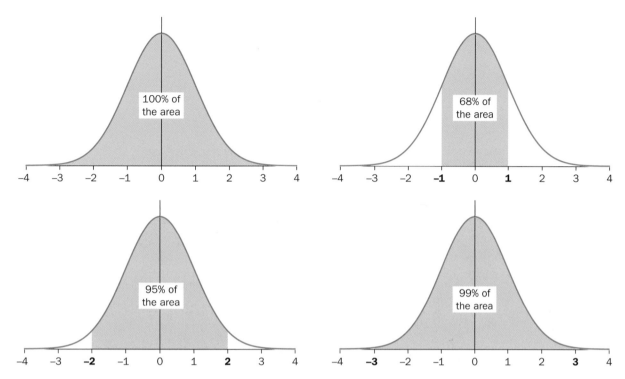

Figure 4.5   Area under the Normal Curve

horizontal axis, I substitute in the mean minus one standard deviation: $13 − $2.40 = $10.60. Instead of −2 on the horizontal axis, I substitute in the mean minus two standard deviations: $13 − $2.40 − $2.40 = $8.20. This same pattern continues for +/−3 and +/−4 standard deviations.

Then we can use what we know about the area under the normal curve to make assertions about the percentage of people in each wage range. If hourly wages are normally distributed, 68 per cent of cases will be between −1 and +1 standard deviation from the mean, because 68 per cent of the area under the normal curve is between −1 and +1 standard deviation from the mean. Based on this distribution, we expect 68 per cent of young people to earn between $10.60 and $15.40 per hour (+/− one standard deviation from the mean). Similarly, if hourly wages are normally distributed, 95 per cent of cases will be between −2 and +2 standard deviations from the mean, because 95 per cent of the area under the normal curve is between −2 and +2 standard deviations from the mean. So, we expect 95 per cent of young people to earn between $8.20 and $17.80 an hour. Our imaginary example has too few cases to be able to test these expectations, but with more cases, it becomes possible to check these results against the available data.

Figure 4.6 might also help you to better understand what z-scores show. Recall that a z-score is a standardized score that provides information about

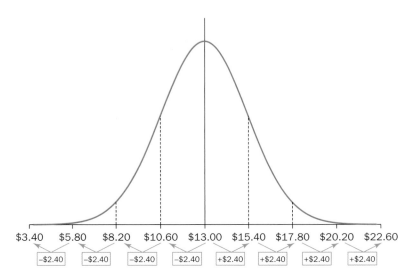

Figure 4.6   Substituting the Mean and Standard Deviation into a Normal Distribution (Hypothetical Data)

where a case is located in a distribution, by using standard deviations as the unit of measurement. The distribution in Figure 4.6 shows that a person earning $15.40 per hour has a z-score of 1 because that person's hourly wage is exactly one standard deviation above the mean of the distribution. A person earning $10.60 per hour has a z-score of –1 because that person's hourly wage is exactly one standard deviation below the mean of the distribution. Using the same logic, a person earning $17.80 per hour has a z-score of 2, and a person earning $8.20 an hour has a z-score of –2.

Finding the Area under the Normal Curve

Z-scores and the normal distribution can also be used to determine the percentage of cases that are above or below a specific value on a variable. This strategy can also be extended to find the percentage of cases that are between two values. Conceptually, this is the equivalent of identifying the percentile of an individual case.

To find the percentage of cases that are above or below a particular value on the horizontal axis of a normal distribution, you must use an "area under the normal curve" calculator or table. (One such calculator is included in a spreadsheet tool available from the Student Resources area of the companion website for this book.) These calculators/tables use calculus to determine the exact proportion of the area under the normal curve that is located to the right or to the left of a point on the horizontal axis. Such calculators and tables are widely available online and typically report the percentage of the area under the normal curve that is above or below each z-score. I'll illustrate this using our hypothetical example. Many workers' groups advocate for a $15 per hour minimum wage. To find the percentage of young people who earn $15 or less per hour, the first step is to calculate the z-score for someone making exactly $15 per hour, using the mean ($13) and the standard deviation ($2.40):

$$Z = \frac{\left(X_i - \bar{X}\right)}{S}$$

$$Z = \frac{(15 - 13)}{2.40}$$

$$= \frac{2}{2.40}$$

$$= 0.83$$

This z-score of 0.83 corresponds with what Figure 4.6 shows: someone earning $15 per hour is located just below one standard deviation above the mean ($15.40) on the horizontal axis. An "area under the normal curve" calculator tells me that 0.80 or 80 per cent of the area under the curve is located to the left of a z-score of 0.83. (See Figure 4.7.) Thus, if hourly wages were normally distributed, we would expect 80 per cent of young people to earn $15 or less per hour. Since the total area under the curve is equal to 100 per cent, it's easy to determine how many young people earn more than $15 per hour: if 80 per cent of young people earn $15 or less, 20 per cent earn more than $15 per hour.

Table 4.3 shows the area to the left and right of various z-scores. (More detailed tables are widely available.) Since the normal distribution is symmetrical, the area to the left of a negative z-score is the same as the area to the right of the corresponding positive z-score. And, of course, since a z-score of 0 is located at the centre of a normal distribution, 50 per cent of the area is to the left of it and 50 per cent of the area is to the right of it.

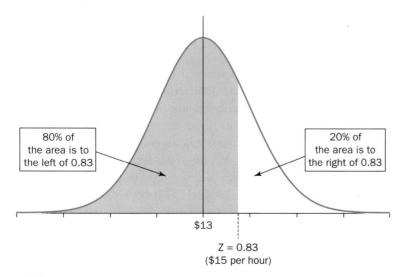

Figure 4.7 The Area under the Normal Curve Located to the Left and to the Right of a Specific Z-score

Table 4.3 **The Area under the Normal Curve Associated with Various Z-scores**

Z-score (Negative)	Area to the Left of Score	Area to the Right of Score	Z-Score (positive)	Area to the Left of Score	Area to the Right of Score
0	50%	50%	0	50%	50%
−0.25	40%	60%	0.25	60%	40%
−0.5	31%	69%	0.5	69%	31%
−0.75	23%	77%	0.75	77%	23%
−1	16%	84%	1	84%	16%
−1.25	11%	89%	1.25	89%	11%
−1.5	7%	93%	1.5	93%	7%
−1.75	4%	96%	1.75	96%	4%
−2	2%	98%	2	98%	2%
−2.25	1%	99%	2.25	99%	1%

These examples begin to illustrate why the normal distribution is so useful to statisticians. If you know that a variable is normally distributed and you know the mean and standard deviation of that variable, then you can deduce quite a bit of additional information. For instance, you know that the median and the mode will be relatively close to the mean. (In real-world data, however, they are rarely exactly the same.) You know that about two-thirds of the cases (68 per cent) will be between −1 and +1 standard deviation from the mean. And you know that about 95 per cent of the cases will be between −2 and +2 standard deviations from the mean. In addition, the mean and standard deviation can be used to calculate a z-score for any case or any value, and then the area under the normal curve can be used to determine the percentage of cases above or below that point in the distribution.

Describing the Shape of a Variable: Skew and Kurtosis

Although many ratio-level variables are approximately normally distributed, some are not. Researchers use two statistics to describe the shape of non-normal distributions: skew and kurtosis. Each provides information about how the shape of a distribution differs from a normal distribution.

skew Indicates how asymmetrical a distribution is.

The **skew** of a distribution captures how asymmetrical it is. Skewed distributions have many cases clustered on one side of the distribution and a tail that trails off to the other side. The type of skew is determined by the direction that the tail trails off in. The left panel of Figure 4.8 shows a left-skewed distribution, with a tail that trails off to the left. Left-skewed distributions are sometimes called negatively skewed distributions because the tail trails off in the negative direction on the horizontal axis. The right panel of Figure 4.8 shows a right-skewed distribution, with a tail that trails off to the right. Right-skewed distributions are sometimes called positively skewed distributions because the tail trails off in the positive direction on the horizontal axis.

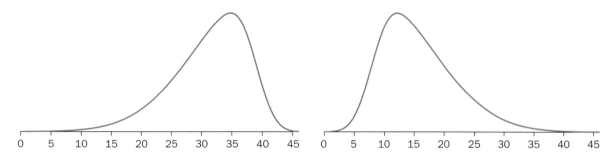

Figure 4.8 Left- and Right-Skewed Distributions

Sometimes the skew of a distribution is evident from its mean, median, and mode. In a left-skewed distribution, the mean is lower than the median, which, in turn, is lower than the mode. The mode, or the most common answer, is found at the highest point of the curve. The median, or the middle-most point of the distribution, divides the area under the curve into two equal parts and, thus, is lower than the mode because of the area contained in the tail of the curve. Finally, the mean, or arithmetic average, is typically affected by the extreme values in the tail of the distribution and, thus, is lower than the median and closer to the tail. In a right-skewed distribution, the opposite is true: the mean is higher than the median, which is in turn higher than the mode. Figure 4.9 illustrates the typical placement of mean, median, and mode in left- and right-skewed distributions.

The **kurtosis** of a distribution captures how peaked or flat the centre of a distribution is and how fat or skinny the tails of a distribution are, compared to a normal distribution. Distributions that are thinner and more peaked than a normal distribution are referred to as leptokurtic: the prefix *lepto-* comes from the Greek work *leptós*, meaning "thin." Leptokurtic distributions have tails that are thicker (or fatter) than the tails in a normal distribution. Distributions that are flatter and wider than a normal distribution are referred to as platykurtic; the prefix *platy-* comes from the Greek word *platýs*, meaning "flat." Platykurtic distributions have tails that are thinner (or skinnier) than the tails in a normal distribution. Figure 4.10 illustrates how a platykurtic and leptokurtic distributions compare to a

kurtosis Indicates how peaked or flat the centre of a distribution is and how fat or skinny the tails of a distribution are, compared to a normal distribution.

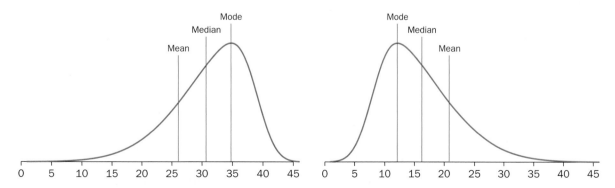

Figure 4.9 Mean, Median, and Mode in Left- and Right-Skewed Distributions

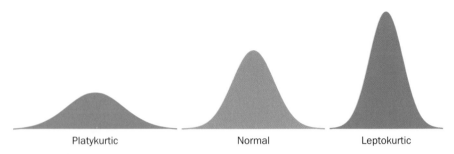

Figure 4.10   **Platykurtic and Leptokurtic Distributions, Compared to a Normal Distribution**

normal distribution. If a variable is platykurtic, it indicates that the cases are spread out across a wide range of values. If a variable is leptokurtic, it indicates that many of the cases are clustered together, instead of being widely spread out.

One way to assess skew and kurtosis is by looking at a histogram. However, statistical software can also calculate skew and kurtosis statistics. Skew statistics range from negative infinity to positive infinity, and they provide two pieces of information. First, the sign of the statistic (whether it is positive or negative) shows which direction a distribution is skewed in. A positive skew statistic indicates that the tail trails off on the right side of a distribution, or in a positive direction on the horizontal axis. A negative skew statistic, on the other hand, indicates that the tail trails off on the left side of a distribution, or in a negative direction on the horizontal axis. Second, the size of the skew statistic shows how much skew a distribution has. The skew statistic doesn't have a unit of measurement—that is, it can't be interpreted as percentages or units. Instead, researchers rely on some rough guidelines for interpreting skew statistics. A perfect normal distribution has a skew of 0. In real-world data, however, few distributions have a skew of exactly 0. But if the skew statistic is between −0.5 and +0.5, the distribution is considered approximately normal in terms of its skew—that is, it is roughly symmetrical. If the skew statistic is between −0.5 and −1 or between 0.5 and 1, the distribution is considered moderately skewed. Finally, if the skew statistic is lower than −1 or higher than 1, the distribution is considered highly skewed. Skew statistics for two or more distributions can be compared to determine which is more skewed.

Like the skew statistic, the kurtosis statistic has no unit of measurement. A normal distribution has an absolute kurtosis of 3. But most statistical software programs report "excess kurtosis," that is, they subtract 3 from the kurtosis statistic so that a kurtosis of 0 represents a normal distribution, a kurtosis less than 0 indicates a distribution that is flatter than normal (platykurtic), and a kurtosis greater than 0 indicates a distribution that is more peaked than normal (leptokurtic). The kurtosis statistic is interpreted in much the same way as the skew statistic (when it shows excess kurtosis). Like skew, in real-world data few distributions have a kurtosis of exactly 0. But if the kurtosis statistic is between −0.5 and +0.5, the distribution is considered approximately normal. If the kurtosis statistic is between −0.5 and −1 or is between 0.5 and 1, the distribution is considered moderately leptokurtic or moderately platykurtic, depending on whether the statistic is positive or negative. Finally, if the kurtosis statistic is lower than −1 or higher than 1, the distribution is considered very leptokurtic or very platykurtic. Kurtosis statistics for two or more distributions can be compared, although kurtosis statistics are only useful when distributions are not highly skewed.

Let's return to the hourly wages of young people in Canada one more time. Figure 4.11 shows how the distribution of young people's hourly wages compares to a normal distribution. The distribution has a skew of 1.8. The positive skew statistic indicates that the tail of the distribution trails off to the right (in the positive direction). Because the number is larger than 1, the distribution considered highly skewed. This is evident in the histogram, which shows that the distribution is asymmetrical and has a tail trailing off to the right. The distribution of young people's hourly wages has a(n) (excess) kurtosis of 4.2. The positive kurtosis statistic indicates that the distribution is more peaked than a normal distribution (leptokurtic). Because the number is larger than 1, the distribution is considered highly kurtotic. The histogram shows a large peak of cases in the $11 to $12 per hour range, which is more than twice the height of a normal distribution.

In practical terms, the kurtosis and skew statistics suggest that young people's hourly wages are clustered around a group of low values. Many young people in the labour market earn approximately the same hourly wage—even though they work in different jobs, industries, and regions. This result likely reflects similarities in the types of jobs that are typically available to young people, as well as that young

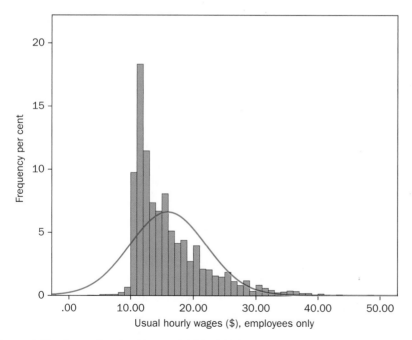

Figure 4.11 **Assessing the Shape of a Distribution Compared to a Normal Distribution**

Source: Author generated; Calculated using data from Statistics Canada, 2016b.

people tend to have similar levels of experience and/or seniority in a workplace. The skew statistic indicates that some young people have, nevertheless, managed to achieve relatively high hourly wages. The absence of a matching tail at the low end of the hourly wage distribution illustrates the effect of minimum wage laws; few young people report earning less than $10 per hour. A fuller analysis of this data might investigate whether higher levels of education or seniority are associated with higher hourly wages.

Researchers use skew and kurtosis statistics to assess whether a variable is approximately normally distributed. Although looking at a histogram provides similar information, skew and kurtosis statistics are not affected by manipulations to the binning and scale of a histogram and, thus, are a more reliable measure. In addition, skew and kurtosis statistics allow researchers to make comparisons between distributions. If a variable is approximately normally distributed—that is, it has a skew and kurtosis between −0.5 and + 0.5—then mean, standard deviation, and z-scores can be used to determine the percentage of cases with specific characteristics. If a variable is not approximately normally distributed, then the researcher should describe how the distribution differs from normal.

How Does It Look in SPSS?

Skew and Kurtosis

The skew and kurtosis, as well as measures of the centre and dispersion of a variable, can be obtained using the Statistics option in the Frequencies procedure. The variable label appears at the top of the table. You can choose whether or not a frequency table is also printed.

Statistics

Usual hourly wages, employees only

N	(A)	Valid	62561
		Missing	26857
Mean	(B)		15.9049
Median	(C)		14.0000
Mode	(D)		12.00
Std. Deviation	(E)		6.06440
Variance	(F)		36.777
Skewness	(G)		1.790
Std. Error of Skewness			.010
Kurtosis	(H)		4.213
Std. Error of Kurtosis			.020

Image 4.3   **More SPSS Statistics (from the Frequencies Procedure) for a Ratio-level Variable**

A. These rows show the number of cases with valid and missing values. Only cases with valid values are used to calculate the statistics. The missing cases represent young people who are not in school and who are not employees and, thus, do not have an hourly wage recorded.

B. This row shows the mean, or average. The average hourly wage is $15.90.

C. This row shows the median. Half of young people earn an hourly wage that is $14.00 or higher, and half of young people earn an hourly wage that is $14.00 or lower.

D. The "Mode" row shows the most common value. The hourly wage reported by the largest number of young people is $12.00.

E. This row shows the standard deviation. The standard deviation of hourly wages is $6.06.

F. This row shows the variance. The variance of hourly wages is 36.78, which is $6.0644 squared.

G. The "Skewness" row shows how skewed the distribution is and in which direction. The skew statistic of 1.790 shows that the distribution is highly right-skewed as the statistic is greater than 1 (which indicates high skew) and a positive number (which indicates right-skew). The "Std. Error of Skewness" row is printed by default. You can ignore it for now.

H. The "Kurtosis" row shows how much excess kurtosis the distribution has. The kurtosis statistic of 4.213 shows that the distribution is highly leptokurtic as the statistic is greater than 1 (which indicates high kurtosis) and a positive number (which indicates leptokurtosis). The "Std. Error of Kurtosis" row is printed by default. You can ignore it for now.

Best Practices in Presenting Results

Writing about the Centre, Dispersion, and Shape of Distributions

One hallmark of good research is a clear description of the distribution of each variable used in the analysis. Depending on a variable's level of measurement, this can include a description of the centre, dispersion, and/or shape of the distribution. Distributions can be described using text, tables, graphs, or other visual tools. The key is to use each tool in an efficient way. For instance, the distribution of a dichotomous variable can be described easily in words and doesn't necessarily warrant a graph. In contrast, the distribution of a ratio-level variable is often best illustrated using a histogram, along with a written description of key statistics. With practice, you will learn to develop a balance between text, tables, and graphs in your statistical reporting.

You should report only those statistics that are appropriate to each variable's level of measurement. Most statistical software will uncritically generate any

Table 4.4   **Statistics That Can Be Reported for Variables with Different Levels of Measurement**

		Nominal-Level Variables	Ordinal-Level Variables	Ratio-Level Variables
Centre of a Variable	Mode	Yes	Yes	Yes
	Median	–	Yes	Yes
	Mean	–	–	Yes
Dispersion of a Variable	Range	–	Yes	Yes
	Interquartile range	–	Yes	Yes
	Standard deviation	–	–	Yes
Shape of a Variable	Skew	–	–	Yes
	Kurtosis	–	–	Yes

statistic for any variable, even if it doesn't make any sense. It's up to the researcher to determine which statistics best describe each variable. Table 4.4 shows which statistics to report for variables with different levels of measurement. For nominal-level variables, only the mode can be reported. The other measures of dispersion and shape don't make sense for nominal-level variables because there is no inherent order to the attributes. If a nominal-level variable has relatively few attributes, a researcher might also show the percentage of cases in the groups, possibly in a table or graph. For ordinal-level variables, the mode and the median can be reported, as well as the range and interquartile range. But because there is not an equal distance between the attributes/values of ordinal-level variables, it doesn't make sense to describe the shape of their distribution. Finally, all of these statistics can be reported for ratio-level variables. It's up to the researcher to select which statistics are best to report: for instance, if a variable has outliers, it may be better to present the median instead of the mean and to explain why you are doing so.

As a general practice, the standard deviation is always reported with the mean. Typically, the standard deviation is shown in parentheses following the mean, using the notation "s.d." or "SD". For example, a researcher might report that the average hourly wage for young people aged 15 to 24 in Canada, who are not students, is $15.90 (s.d. = $6.06). The same convention is used in tables.

What You Have Learned

In this chapter you learned some additional strategies for describing the distribution of a ratio-level variable. The mean provides information about the centre of a distribution, and the standard deviation provides information about the dispersion of a distribution. You also learned how to describe the shape of a distribution using histograms and skew and kurtosis statistics. In addition, this chapter introduced the normal distribution and explained how it is related to the standard deviation, z-scores, skew, and kurtosis. Along with the strategies introduced in chapters 2 and 3, you now have a full range of tools for describing the centre, dispersion, and shape of distributions for variables at different levels of measurement. These are the basic tools of descriptive statistics and serve as the starting point for any statistical analyses.

The research focus of this chapter was wages for young people in Canada. Across North America, workers' advocates in many jurisdictions (including Ontario and BC) have been campaigning for a $15 per hour minimum wage, with some success in Ontario, Alberta, Los Angeles County, and New York State. Similar legislation across Canada would clearly benefit many young workers; in 2016, among young employees aged 15 to 24 who were not attending school, the most common hourly wage was $12 per hour, and half earned $14 per hour or less. On average, young people who are not in school earn $15.90 per hour (s.d. = $6.06), although the overall average masks some age-based differences: for those aged 15 to 19 the average hourly wage is $12.56 (s.d. = $3.04), compared to $16.84 (s.d. = $6.36) for those aged 20 to 24. About a quarter of young people who are not in school are not working full-time hours, even though they may want to; a young person working an average number of hours for the average hourly wage earned only $27,615 per year. This is only slightly higher than the low-income cut-off for a single person (which ranges from $16,934 to $24,600, depending on community size) (Statistics Canada 2015). Further research in this area might consider how other demographic characteristics—such as gender, racialization, immigration status, and level of education—are related to variations in young people's hourly wages.

Check Your Understanding

Check to see if you understand the key concepts in this chapter by answering the following questions:

1. When should a researcher use a median instead of an average?
2. What is the standard deviation, and how does it help you to understand the dispersion of a variable?
3. What are z-scores, and how are they related to the standard deviation?
4. What are the properties of a normal distribution?
5. How are z-scores related to the standard normal distribution?
6. What is a histogram, and how is it different from a bar chart?
7. How does changing the bin size affect the appearance of a histogram?
8. What do skew and kurtosis statistics tell you about the shape of a distribution?

Practice What You Have Learned

Check to see if you can apply the key concepts in this chapter by answering the following questions. Keep two decimal places in any calculations.

1. You work for a youth employment agency. The agency collects information about how many weeks it takes clients to find work after they start using the agency's services, as well as about their starting wage. The information collected from 10 recent clients is summarized in Table 4.5.

   a. Find the median number of weeks that it took these 10 clients to find a job.
   b. Find the average (mean) number of weeks that it took these 10 clients to find a job.

Table 4.5   Information Collected from 10 Recent Clients of a Youth Employment Agency (Hypothetical Data)

Person	Weeks until Finding a Job	Starting Wage (Hourly)
Derek	5	$12.50
Renée	2	$15.00
Zahra	2.5	$23.40
Sarunja	10	$12.80
Jenodini	6	$11.75
Sarah	6	$34.00
Markus	3	$14.50
Olivia	5.5	$12.75
Abdullah	8	$18.00
Alison	5	$13.00

2. Using the information in Table 4.5, find the standard deviation of the number of weeks that it took clients to find a job. Begin by making a table similar to Table 4.2.

3. One more client—Jennifer—has just found a job after spending 35 weeks looking for work. You're excited to hear the news, so you add Jennifer to the list in Table 4.5 and re-do your analysis.

   a. Find the median number of weeks that it took these 11 clients to find a job.

   b. Find the average (mean) number of weeks that it took these 11 clients to find a job.

   c. Which statistic (the mean or the median) most accurately reflects how long it takes people to find a job using the agency's services?

4. The executive director of the agency knows that it's also important for clients to get well-paying jobs. She asks you to analyze clients' starting wages. Using the information in Table 4.5:

   a. Find the median starting wage among the original 10 clients.

   b. Find the average (mean) starting wage among these 10 clients.

   c. Which statistic (the mean or the median) most accurately reflects clients' starting wages?

5. Using the information in Table 4.5, find the standard deviation of clients' starting wage.

6. Using the mean and standard deviation you calculated in questions 4(b) and 5, find the z-score of Zahra's starting wage.

7. The agency has collected information from hundreds of clients over the past year, and it shows that the number of weeks that it takes to find a job is normally distributed. Among all of the agency's clients in the past year, it took an average of 10 weeks (s.d. = 2.5 weeks) to find work.

   a. Using what you know about the normal distribution, how long does it take 68 per cent of the agency's clients to find a job?

   b. How long does it take 95 per cent of the agency's clients to find a job?

8. You're particularly interested in the situation of one client, named Steve. It took Steve 6.25 weeks to find a job.

   a. Using the mean and standard deviation in question 6, find the z-score for Steve.

   b. Use the z-score to find the percentage of the agency's clients who took less time than Steve to find a job and the percentage of the agency's clients who took more time than Steve to find a job.

9. You are preparing a report for the agency about the different job search experiences of young people with and without a post-secondary educational credential. The agency's information shows that young people without a post-secondary educational credential took an average of 12 weeks to find a job (s.d. = 4 weeks) and that young people with a post-secondary educational credential took an average of 3.5 weeks to find a job (s.d. = 1 week). What can you write in your report about the differences between these two groups?

10. You decide you want to compare Steve's situation to that of his friend Mustafa. It took Steve 6.25 weeks to find a job, and it took Mustafa 6 weeks. However, Steve has a post-secondary educational credential, and Mustafa does not.

   a. Calculate the z-scores for both Steve and Mustafa, using the mean and standard deviation of people who have the same educational credentials as them, from question 9.

   b. Compare the two z-scores and determine who was able to find a job more quickly, relative to others with the same educational credential.

11. The agency has also collected information about hundreds of clients' starting wages over the past year. This data is not normally distributed: the distribution has a skew of 1.7 and an (excess) kurtosis of 2. Describe the shape of the histogram.

12. You analyze the distribution of the starting wages for people with and without a post-secondary educational credential separately for your report. For people *without* a post-secondary educational credential, the distribution of starting wages has a skew of 0.6 and a kurtosis of –1.4. For people *with* a post-secondary educational credential, the distribution of starting wages has a skew of 3 and a kurtosis of 4.5. Describe how the shape of the two distributions are different, and explain how you can interpret these differently shaped distributions in relation to hourly wages.

13. Data from the 2011 National Household Survey (NHS) show that, on average, employees aged 15 to 24 who are enrolled in post-secondary education earn $11,446 in wages each year (s.d. = $10,835). The median annual wage is $9,000, and the modal wage is $5,000.

    a. Interpret the mean, the median, and the mode of this distribution in terms of what they tell us about post-secondary students' annual wages. How do these statistics compare to typical costs for post-secondary students?

    b. What does the standard deviation of this variable indicate about the distribution of post-secondary students' annual wages?

    c. Based on these statistics, will the distribution of post-secondary students' annual wages be left-skewed or right-skewed?

14. The histogram in Figure 4.12 shows the distribution of wages for employees aged 15 to 24 who are enrolled in post-secondary education:

    a. What bin size is used in this histogram?

    b. Based on the information in this histogram, how much do post-secondary students most commonly earn each year? (Provide a range using the bin size.) Approximately what percentage of post-secondary students earn this amount?

    c. How would you describe the shape of this distribution, compared to the normal curve?

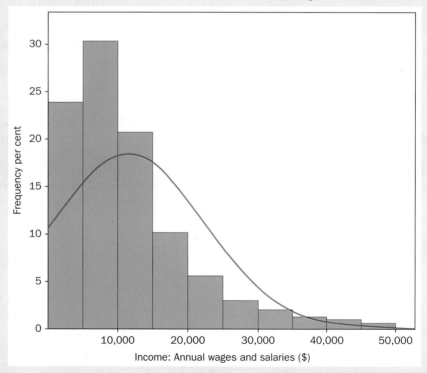

Figure 4.12   **Annual Wages for Employees Aged 15 to 24 Who Are Enrolled in Post-Secondary Education**

Source: Author generated; Calculated using data from the National Household Survey (NHS) 2011.

Table 4.6 Youth Unemployment Rate, Canada, by Selected Demographic Group, 2011 or 2012 (%)

Characteristic	15 to 19 Years—2011	20 to 24 Years—2011
Aboriginal identity	25.9	22.6
Non-Aboriginal identity	19.5	14.4
Immigrant	24.9	16.9
Non-immigrant	19.3	14.4
Visible minority	26.7	17.7
Not a visible minority	18.6	14.1

Characteristic	15 to 24 Years—2012
With a disability	25.9
Without a disability	15.3

Note: Data for "with a disability" or "without a disability" are not available by five-year age groups (15 to 19 years and 20 to 24 years).

Sources: Aboriginal identity and immigrant status: Statistics Canada, "2011 National Household Survey," Data Tables, Labour Force Status by Aboriginal identity, immigrant status and visible minority status; Presence of disability: Statistics Canada, "2012 Canadian Survey on Disability."

Source: Rajotte 2014, 24.

15. A recent report produced by the House of Commons Standing Committee on Finance highlights issues related to employment and unemployment among young people in Canada. Table 4.6 is excerpted from that report.

a. Which group of young people has the highest unemployment rate? Which group has the lowest unemployment rate?

b. Describe the trends in unemployment rates, as they relate to age, Aboriginal identity status, immigrant status, visible minority status, and disability status. What general pattern is illustrated by the information in this table?

16. Table 4.7, excerpted from a Statistics Canada publication, shows how Canada compares with other Organisation for Economic Co-operation and Development (OECD) countries in terms of employment and unemployment for young people who are not enrolled in school. Young people are divided into groups: (1) those enrolled in an educational institution, regardless of whether they are employed, (2) those who are employed, and not enrolled in an educational institution, and (3) those who are neither employed nor in education. This third group is then divided into those who are unemployed—that is, they are actively looking for work but cannot find it; and those who are not in the labour force—that is, they are not employed or looking for work.

a. How does the percentage of young people in Canada who are enrolled in an educational institution compare to the other countries listed?

Table 4.7 Employment and Education Status of Youth Age 15 to 29 in Selected OECD Countries

| | In Education[1] | Employed | Not Employed or in Education (NEET) | | |
			Total NEET	Unemployed	Not in Labour Force
			%		
Total age 15 to 29					
Canada	42.8	43.9	13.3	5.7	7.6
France	44.0	40.5	15.6	9.0	6.6
Germany	52.4	36.0	11.6	5.5	6.1
Italy	45.3	33.5	21.2	7.5	13.7
United Kingdom	40.4	43.9	15.6	7.0	8.6
United States	45.7	37.4	16.9	6.7	10.2

1. Includes students with and without jobs.

Source: Organisation for Economic Co-operation and Development (OECD) database, 2009.

Source: Excerpt from Marshall 2012, 6.

b. How does the percentage of young people in Canada who are employed compare to the other countries listed?

c. How does the percentage of young people in Canada who are unemployed compare to the other countries listed?

d. How do you think these three things—the rate of young people in education, who are employed, and who are unemployed—are all related to each other?

Practice Using Statistical Software (IBM SPSS)

Answer these questions using IBM SPSS and the GSS27.sav or the GSS27_student.sav dataset available from the Student Resources area of the companion website for this book. Report two decimal places in your answers, unless fewer are printed by IBM SPSS. It is imperative that you save the dataset to keep any new variables that you create.

1. The variable "Number of paid hours worked per week - All jobs" [WKWEHRC] shows how many paid hours people work at all of their jobs, among people who are working for pay. (People who are not working for pay are assigned the attribute "Valid skip".) Use the Means procedure to find the mean and the standard deviation of this variable. Explain what these statistics show.

2. Use the Means procedure to find the mean and the standard deviation of the variable "Number of paid hours worked per week - All jobs" [WKWEHRC], divided by "Sex of respondent" [SEX]. Explain what these statistics show about the differences between men and women.

3. Use the Statistics option in the Frequencies procedure to find the mean, the median, the skew, and the kurtosis of the variable "Number of paid hours worked per week - All jobs" [WKWEHRC].

   a. What is the median of the variable? What information does it provide? How does the median compare to the mean?

   b. What is the kurtosis of the variable? What does it indicate about the shape of the distribution?

   c. What is the skew of the variable? What does it indicate about the shape of the distribution?

4. Use the Chart Builder tool to create a histogram of the variable "Number of paid hours worked per week - All jobs" [WKWEHRC]. Display percentages on the vertical axis (y-axis).

   a. Adjust the bin sizes so that the histogram provides a good representation of the data.

   b. Add a normal curve to the histogram to use as a reference.

   c. Describe the shape of the distribution of this variable, using the histogram and the kurtosis and skew statistics from question 3.

5. The variable "Number of close friends" [SCF_100C] shows people's answers to this question: "How many close friends do you have (that is, people who are not your relatives, but who you feel at ease with, can talk to about what is on your mind, or call on for help)?" Use the Means procedure to find the mean and the standard deviation of this variable. Explain what these statistics show.

6. Use the Means procedure to find the mean and the standard deviation of the variable "Number of close friends" [SCF_100C], divided by "Sex of respondent" [SEX]. Explain what these statistics show about the differences between men and women.

7. Use the Statistics option in the Frequencies procedure to find the mean, the median, the mode, the standard deviation, the range, the skew, and the kurtosis of the variable "Number of close friends" [SCF_100C].

   a. What are the mean, median, and mode of the variable? What information do they provide about the centre of the variable?

   b. What are the standard deviation and range of the variable? What information do they provide about the dispersion of the variable?

   c. What is the kurtosis of the variable? What does it indicate about the shape of the distribution?

   d. What is the skew of the variable? What does it indicate about the shape of the distribution?

8. Use the Chart Builder tool to create a histogram of the variable "Number of close friends" [SCF_100C]. Display percentages on the vertical axis (y-axis).

    a. Adjust the bin sizes so that the histogram provides a good representation of the data.
    b. Adjust the scale of the horizontal axis (x-axis) so that the histogram takes up more of the chart space.

    c. Add a normal curve to the histogram to use as a reference.
    d. Describe the shape of the distribution of this variable, using the histogram and the kurtosis and skew statistics from question 6.

Key Formulas

Mean

$$\bar{X} = \frac{\sum X_i}{N}$$

Standard deviation (using population data)

$$S = \sqrt{\frac{\sum\left(X_i - \bar{X}\right)^2}{N}}$$

Variance (using population data)

$$S^2 = \frac{\sum\left(X_i - \bar{X}\right)^2}{N}$$

Z-score

$$Z = \frac{\left(X_i - \bar{X}\right)}{S}$$

References

Clark, Warren. 2007. "Delayed Transitions of Young Adults." *Canadian Social Trends* 84: 14–22.

Galarneau, Diane, René Morissette, and Jeannine Usalcas. 2013. "What Has Changed for Young People in Canada?" Catalogue no. 75006X. Ottawa: Statistics Canada.

International Labour Organization (ILO). 2015. "Labour Force Surveys." http://www.ilo.org/dyn/lfsurvey/lfsurvey.home.

Marshall, Katherine. 2012. "Youth Neither Enrolled nor Employed." Perspectives on Labour and Income. Ottawa: Statistics Canada. http://www.statcan.gc.ca/pub/75-001-x/2012002/article/11675-eng.pdf.

National Household Survey (NHS). 2011. *Public Use Microdata File.* Statistics Canada: Ottawa, ON.

Nicol, Adelheid A. M. 2010. "Normal Distribution." In *Encyclopedia of Research Design*, edited by Neil Salkind. California: Sage. http://knowledge.sagepub.com/view/researchdesign/n276.xml.

Rajotte, James. 2014. "Youth Employment in Canada: Challenges and Potential Solutions." *Report of the Standing Committee on Finance.* Ottawa: House of Commons, 41st Parliament, Second Session. http://www.parl.gc.ca/content/hoc/Committee/412/FINA/Reports/RP6658485/finarp06/finarp06-e.pdf.

Statistics Canada. n.d. "History of the Labour Force Survey." http://www23.statcan.gc.ca/imdb-bmdi/document/3701_D7_T9_V1-eng.pdf.

———. 2015. "Table 206-0094: Low Income Cut-Offs (LICOs) Before and After Tax by Community and Family Size in Current Dollars." CANSIM (Database). http://www5.statcan.gc.ca/cansim/a26?lang=eng&id=2060094.

———. 2016a. "Guide to the Labour Force Survey." Catalogue no. 71-543-G. http://www.statcan.gc.ca/pub/71-543-g/71-543-g2016001-eng.pdf.

———. 2016b. "Labour Force Survey 2016." http://www23.statcan.gc.ca/imdb/p2SV.pl?Function=getSurvey&Id=331692.

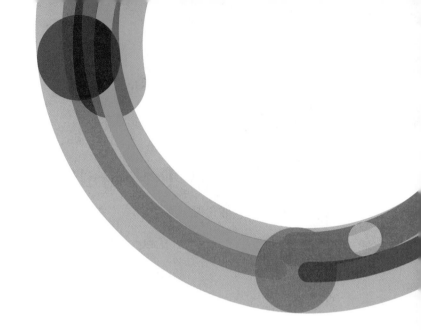

Part II

Making Claims about
Populations

Probability, Sampling, and Weighting

Learning Objectives

In this chapter you will learn:

- The basic concepts of probability
- How probabilities relate to the normal distribution
- Some strategies for probability sampling
- What weights are and why they are used
- How to create a standardized weight
- What goes in the methodology section of a survey research report

Introduction

In Chapter 1, you learned that statistical analysis is conceptually divided into two main streams: descriptive statistics and inferential statistics. Descriptive statistics are used to summarize and condense information while inferential statistics are used to make claims about a population, using data collected from a randomly selected sample of that population. Until this point, this book has emphasized descriptive statistical techniques. This chapter and the next set out the conceptual background that you need to understand inferential statistics; later chapters will discuss the specific techniques and tests used in inferential statistics.

As you learned in Chapter 1, the population is the whole group that a researcher is interested in studying and making claims about. A sample is a subset of the population that information is collected from. In this chapter, I introduce some of the theoretical ideas that researchers rely on when they generalize from samples to populations, including the logic of probability and probability sampling. I also introduce the concept of weighting and explain how it affects statistical analyses. Although introductory statistics textbooks often omit a discussion of weighting, it is a crucial aspect of most hands-on data analysis. If appropriate weights are not used, the statistics produced by most sample survey data are incorrect—that is, they don't accurately represent the population that a researcher wants to describe.

Unlike the previous chapters, this chapter does not have a specific research focus. Instead, in the "Spotlight on Data" box I discuss the people who are routinely excluded from survey research in Canada.

Probability: Some Basic Concepts

Probability refers to the chance that an event or outcome will occur. Probabilities range between 0 and 1, where 0 indicates that there is absolutely no chance that the outcome will occur, and 1 indicates that the outcome will definitely occur. The closer a probability is to 1, the more likely it is that something will occur; the closer a probability is to 0, the less likely it is to occur. A probability of 0.5 indicates that something is just as likely to happen as it is to not happen. Probabilities are usually denoted using the letter p in statistical notation. It might seem confusing that probabilities use the same symbol as proportions, but probabilities actually do show a proportion: specifically, the proportion of the time that an outcome is expected to occur. Like proportions, probabilities are often reported as percentages; for instance, a weather forecast might report a 40 per cent chance of rain or a 70 per cent chance of snow.

probability (p) The chance that an outcome or event will occur; probabilities range from 0 to 1.

Let's review some basic probability concepts. Simple probability relies on the idea that each possible outcome has an equal chance of occurring. The most common probability example is that of a coin toss, using a fair coin with two sides: one "heads" and one "tails." When you toss that coin, the probability of obtaining heads is 0.5 or 50 per cent, and the probability of obtaining tails is also 0.5 or 50 per cent. The probability of a specific outcome occurring is calculated by dividing the number of outcomes of interest by the total number of possible outcomes. For the coin toss example, the probability of getting heads is:

$$p(heads) = \frac{number\ of\ sides\ on\ a\ coin\ with\ heads}{number\ of\ sides\ on\ a\ coin}$$

$$= \frac{1}{2} \ or \ 0.5$$

Another example is a roll of a standard dice with six sides, which are numbered from one through six using dots on each side. To find the probability of rolling a side with a specific number, such as a 3, a similar calculation is used:

$$p(3) = \frac{number\ of\ sides\ on\ a\ dice\ with\ exactly\ 3\ dots}{number\ of\ sides\ on\ a\ dice}$$

$$= \frac{1}{6} \ or \ 0.167$$

So, for any roll of a standard dice, the probability of rolling a 3 is 0.167, or 17 per cent. In other words, every time you roll a dice, there's a 17 per cent chance of rolling a 3.

This idea can be extended to find the probability that one of several outcomes will occur. To do this, the number of all of the outcomes of interest are summed in the numerator of the probability calculation. For instance, in a roll of a standard dice, to find the probability of rolling one of two numbers, such as a 3 *or* a 5, the numerator becomes:

$$p(3 \text{ or } 5) = \frac{\text{number of sides on a dice with exactly 3 dots} + \text{number of sides on a dice with exactly 5 dots}}{\text{number of sides on a dice}}$$

$$= \frac{1+1}{6}$$

$$= \frac{2}{6} \text{ or } 0.333$$

So, in any roll of a standard dice, the chance of rolling *either* a 3 or a 5 is 0.33, or 33 per cent. In other words, every time you roll a dice, there's a one in three chance (or a 33 per cent chance) of having a 3 or a 5 come up. In general, the probability of an event occurring is calculated as:

probability

$$probability \left(outcome \ of \ interest\right) = \frac{number \ of \ outcomes \ of \ interest}{number \ of \ possible \ outcomes}$$

joint probability The chance that two or more outcomes will occur, either simultaneously or in sequence.

You can also calculate the probability of more than one event occurring. This is called a **joint probability**. To find the joint probability of two events occurring, the probability of the first event is multiplied by the probability of the second event. Since the probability of tossing a coin and getting heads once is 1/2 or 0.5, then the probability of getting heads twice in a row is:

$$p(heads, heads) = \frac{\text{number of sides on a coin with heads}}{\text{number of sides on a coin}} \times \frac{\text{number of sides on a coin with heads}}{\text{number of sides on a coin}}$$

$$= \frac{1}{2} \times \frac{1}{2}$$

$$= \frac{1}{4} \text{ or } 0.25$$

Every time you flip a coin twice in a row, you have a 0.25 probability or a 25 per cent chance of getting two heads. Similarly, since the probability of rolling a dice and getting a 3 once is 1/6 or 0.167, the probability of rolling a 3 twice in a row is:

$$p(3,3) = \frac{\text{number of sides on a dice with exactly 3 dots}}{\text{number of sides on a dice}} \times \frac{\text{number of sides on a dice with exactly 3 dots}}{\text{number of sides on a dice}}$$

$$= \frac{1}{6} \times \frac{1}{6}$$

$$= \frac{1}{36} \text{ or } 0.028$$

So, every time that you roll a dice twice in a row, you have a 0.028 probability or a 3 per cent chance of rolling a 3 both times. The same approach is used to find the probability of two events occurring at simultaneously. If you roll two dice at the same time, the probability of getting two 3s is also 0.028 or 3 per cent. If you think about it, when you roll two dice, one dice will almost always land first, so the probability of rolling two 3s at the same time is the same as the probability of rolling a 3 twice in a row (the rolls are just very close together). This approach to calculating joint probability is also used when the two outcomes occur at exactly the same time.

In each joint probability example used so far, the probability of getting the outcome of interest the second time is exactly the same as it was the first time—that is, it is an independent event. The events are considered independent because the outcome of the second coin toss or dice roll is not affected by the outcome of the first coin toss or dice roll. But sometimes the probability of a second (or subsequent) event in a joint probability changes depending on the outcome of the first event. For example, consider the probability of selecting a card of a specific suit, from a standard deck of 52 cards. A standard deck has four suits (hearts, diamonds, clubs, and spades) with 13 cards each. So, in one draw from a deck, the probability of selecting a heart is 13 divided by 52, or 0.25, or 25 per cent.

Now let's find the probability of selecting two hearts in a row. The probabilities change slightly depending on *whether or not I return the card to the deck* after the first draw. If I *do* return the card to the deck, then the probability of selecting a heart in the second draw doesn't change:

$$probability\ (heart, heart) = \frac{number\ of\ hearts\ in\ a\ deck}{number\ of\ cards\ in\ a\ deck} \times \frac{number\ of\ hearts\ in\ a\ deck}{number\ of\ cards\ in\ a\ deck}$$

$$= \frac{13}{52} \times \frac{13}{52}$$

$$= \frac{169}{2704}\ or\ 0.063$$

But, if I *do not* return the card to the deck before the second draw, the chance of selecting a heart in the second draw changes slightly. The total number of cards in the deck is reduced by one, and, if I selected a heart in the first draw, the total number of hearts in the deck is reduced by one. So, the probability of selecting a heart twice in a row becomes:

$$probability\ (heart, heart) = \frac{number\ of\ hearts\ in\ a\ deck}{number\ of\ cards\ in\ a\ deck} \times \frac{number\ of\ hearts\ in\ a\ deck}{number\ of\ cards\ in\ a\ deck}$$

$$= \frac{13}{52} \times \frac{12}{51}$$

$$= \frac{156}{2652}\ or\ 0.059$$

The probability of selecting two hearts in a row is reduced from 0.063 (6.3 per cent) to 0.059 (5.9 per cent) if I don't return the card to the deck after the first draw. This is because the second event, or second draw, is a dependent event. In other words, the probability of the event changes depending on the outcome of the first event. Regardless of whether the second or subsequent event is independent or dependent, however, the same general approach is used to calculate the probability of two events occurring:

joint probability

$$probability\left(outcome\ of\ interest, outcome\ of\ interest\right)$$
$$= \frac{number\ of\ outcomes\ of\ interest}{number\ of\ possible\ outcomes} \times \frac{number\ of\ outcomes\ of\ interest}{number\ of\ possible\ outcomes}$$

theoretical probability A probability that is established by dividing the number of outcomes of interest by the total number of possible outcomes.

All of the probabilities described so far are **theoretical probabilities**. A theoretical probability is the chance that an event or outcome will occur, calculated by dividing the number of outcomes of interest by the total number of possible outcomes. Researchers use theoretical probabilities to calculate the likelihood of a person in the population being selected to be part of a sample.

The probability of something occurring can also be tested empirically. For example, you can toss a coin 10 times and count how many times you get heads in order to determine the probability of the outcome. When the probability of an outcome is determined using an empirical test, it is called an **observed probability**, or an empirical probability. Observed probabilities are established by conducting multiple trials—that is, multiple coin tosses or dice rolls—to determine how often a specific outcome occurs. For instance, if you toss a coin 10 times, it isn't unusual to get 4 heads; thus the observed probability of getting a head is 4/10 or 0.4. Similarly, if you roll a dice 10 times and get a 3 twice, the observed probability of rolling a 3 is 2/10 or 0.20.

observed probability A probability that is established by conducting multiple empirical trials to determine how often a specific outcome occurs.

law of large numbers A law stating that the larger the number of trials used to establish the observed probability of an outcome, the closer the result will be to the theoretical probability of that outcome.

In general, the larger the number of trials used in an empirical test, the closer the observed probability will be to the theoretical probability. So, after 10 coin tosses, the observed probability of getting heads may not match exactly with the theoretical probability of getting heads (0.5). But, after 1,000 coin tosses, the observed probability will be closer to 0.5, and after 10,000 coin tosses, you are likely to get an observed probability very close to 0.5. This is how the **law of large numbers** is related to probability. The idea of observed probability will help you understand some of the material in Chapter 6, which describes what happens if a researcher selects multiple samples from the same population, a process that mimics the use of multiple trials to establish an observed probability.

Probabilities, Frequency Distributions, and the Normal Distribution

Let's consider how probabilities relate to frequency distributions and the normal distribution. In Chapter 2, you learned about frequency distributions, which show the number and percentage of people with each attribute for a specific variable.

Step-by-Step: Probabilities

$$probability\left(outcome\ of\ interest\right)=\frac{number\ of\ outcomes\ of\ interest}{number\ of\ possible\ outcomes}$$

Step 1: Identify the total number of outcomes of interest.

Step 2: Identify the total number of possible outcomes. The total number of possible outcomes is equal to the total number of outcomes of interest plus the total number of outcomes that are not of interest.

Step 3: Divide the total number of outcomes of interest (from Step 1) by the total number of possible outcomes (from Step 2) to find the probability of the outcome of interest.

Step-by-Step: Joint Probabilities

$$probability\left(outcome\ of\ interest,\ outcome\ of\ interest\right)$$
$$=\frac{number\ of\ outcomes\ of\ interest}{number\ of\ possible\ outcomes}\times\frac{number\ of\ outcomes\ of\ interest}{number\ of\ possible\ outcomes}$$

Step 1: For the first event, identify the number of possible outcomes of interest.

Step 2: For the first event, identify the total number of possible outcomes. The total number of possible outcomes is equal to the total number of outcomes of interest plus the total number of outcomes that are not of interest.

Step 3: For the second event, identify the number of possible outcomes of interest. This might be different from the first event; it depends on whether the second event is affected by the first event.

Step 4: For the second event, identify the total number of possible outcomes. This might also be different from the first event; again, it depends on whether the second event is affected by the first event.

Step 5: Multiply the number of possible outcomes of interest for the first event (from Step 1) by the number of possible outcomes of interest for the second event (from Step 3).

Step 6: Multiply the total number of possible outcomes for the first event (from Step 2) by the total number of possible outcomes for the second event (from Step 4).

Step 7: Divide the result of Step 5 by the result of Step 6 to find the probability of the outcome of interest for both events.

Frequency distributions can also be thought of as probability distributions: they show the observed probability (or empirical probability) that a randomly selected person will have a specific attribute. So, for instance, in the last chapter I investigated the labour force outcomes of young people aged 15 to 24 who were not enrolled in school. Table 5.1 shows a frequency distribution of the employment status for this group.

If we think about this frequency distribution as an observed probability distribution, we can calculate the probability that a randomly selected young person in Canada will be unemployed: take the number of outcomes of interest and divide it by the total number of outcomes. To calculate the probability of a young person's being unemployed, divide the total number of outcomes of interest (9,737) by the total number of outcomes (89,418) to get 0.109. In other words, for a randomly selected young person in Canada who is not in school, the probability of being unemployed is 0.109 or 11 per cent. A similar calculation can be used to find the probability that a randomly selected young person in Canada is either unemployed *or* not in the labour force: add together 9,737 and 14,918, and divide the sum by 89,418, to get 0.276. So, for a randomly selected young person in Canada who is not in school, the probability of being *either* unemployed or not in the labour force is 0.276, or 28 per cent.

The normal distribution can also be thought of as a probability distribution. When a variable is normally distributed, the area under the normal curve can be used to calculate the probability that an event will occur for any randomly selected case. In the section of Chapter 4 describing z-scores and how they relate to the area under the curve of a normal distribution, we calculated the (hypothetical) proportion of young employees expected to earn $15 or less per hour: using the mean and standard deviation, we found that a wage of $15 per hour corresponds to a z-score of 0.83, and we then determined that 80 per cent of the area under the normal curve is located to the left of the z-score of 0.83. This result can also be interpreted as a probability: a randomly selected young employee has a 0.80 probability (or 80 per cent chance) of earning $15 or less per hour.

There are some important theoretical implications when we think about the normal distribution as a probability distribution. Most of the cases in a normal distribution are located close to the mean, or the centre of the distribution. Fewer cases are located far from the mean, or in the tails of the distribution. As a result, if you randomly select a single case from a normal distribution, you are much more likely

Table 5.1   A Frequency Distribution of Young People's Labour Force Status (LFS 2016)

Answer	Frequency	Percentage
Employed (employee or self-employed)	64,763	72.4
Unemployed	9,737	10.9
Not in the labour force	14,918	16.7
Total	*89,418*	*100.0*

Source: Author generated; Calculated using data from Statistics Canada. 2016a.

to select a case that is close to the mean and much less likely to select a case that is far from the mean. And, if you know the mean and standard deviation of the normal distribution, you can use the area under the curve to determine the probability of randomly selecting a case with a specific value (or one that is higher/lower). These ideas are central to statistical inference and will be discussed further in Chapter 6.

Sampling

Probabilities are also related to the process of sampling in survey research. A **census** is a survey that collects information from every person (or case) in a population—in other words, it does not take a sample. Many countries conduct a census to collect nationwide information about how many people live in each region, as well as the age and sex, and other characteristics, of those people. National censuses are usually administered by a central government agency that is mandated and funded to collect population information. In Canada, a major census of population is conducted by Statistics Canada every ten years (known as a decennial census), and a smaller census is conducted five years later (known as a quinquennial census). The first national decennial Canadian census was taken in 1871, and the first national quinquennial census was taken in 1956; following this pattern, decennial censuses in Canada are taken on the years ending in "1," and quinquennial censuses are taken in years ending in "6."

Governments typically use census information to inform policy decisions. Some countries also use census population counts to adjust the boundaries of political districts. In Canada, the Constitution specifies that federal electoral districts must be reviewed and redistributed after each decennial census. The redistribution of electoral districts based on the 2011 census led to the creation of 30 new seats in the House of Commons as a result of population changes: 15 in Ontario, 6 in Alberta, 6 in British Columbia, and 3 in Quebec (Government of Canada 2012).

Although censuses are extremely accurate, they are also expensive and time consuming to conduct. As a result, most survey research relies on collecting information from only a sample of the population. Social science researchers conceptually divide sampling strategies into two types: **non-probability sampling** and **probability sampling**. Non-probability sampling relies on techniques such as convenience sampling (collecting information from people who volunteer or who are easy to recruit) or snowball sampling (collecting information from people who are suggested by other participants). Data collected from non-probability samples can be summarized using descriptive statistics but cannot be used to make generalizations about a larger population.

In contrast, probability sampling relies on the principle of random selection. The defining feature of probability sampling is that every person (or case) in the population has a known, non-zero chance of being randomly selected into the sample. When researchers use probability samples, they rely on the laws of probability to make generalizations about the larger population. The inferential statistical techniques described in the remainder of this book all require data collected from probability samples.

census A survey that collects information from every case in a population, rather than from a sample of cases.

non-probability sample A sample where it is not possible to calculate the chance that a case from the population will be selected into the sample; some cases in the population may have no chance of being selected into the sample.

probability sample A sample where each case in the population has a known, non-zero chance of being randomly selected.

representative sample A sample that accurately represents the diversity of the population that it was selected from.

sampling error The error that inevitably results from collecting information from a sample of cases as opposed to the total population.

bias Systematic error, leading to the consistent overestimation or underestimation of a population parameter; sampling bias occurs when a sample is systematically different from the population in some way.

Probability sampling helps to ensure that the sample of people whom information is collected from represents the diversity of the larger population. The goal is to select a **representative sample**. Any mismatch between the characteristics of the sample and the characteristics of the population that results from sampling is referred to as **sampling error**. Sampling error is inevitable—the characteristics of a randomly selected sample will never perfectly match the characteristics of the larger population. In Chapter 6, you will learn how researchers account for sampling error by calculating the amount of uncertainty associated with the statistics they generate from sample data. Although it's not possible to avoid sampling error, researchers work very hard to avoid **bias**, or systematic error. Sampling bias occurs when information is collected from a sample of cases that are systematically different from the population in some way.

Sample Statistics and Population Parameters

When researchers use probability sampling techniques, the results produced by sample data are referred to as "statistics" whereas the true results in the population are referred to as "parameters." (This is easy to remember since *sample* and *statistic* both begin with the letter *s* and *population* and *parameter* both begin with the letter *p*.) The goal is to use the sample statistics to estimate the population parameters. Sometimes a "hat" symbol (^), or a caret, is used to indicate that something is an estimate in statistical notation. With a few exceptions, I do not use this notation for estimates in this text because doing so can be confusing. All of the examples in this book rely on sample survey data; thus, all of the population parameters are estimates.

In this book, I use upper-case letters to denote population elements and lower-case letters to denote sample elements. So N represents the number of people or cases in the population, and n represents the number of people or cases in a sample. Up until this point, the formulas have all used upper-case notations, which assume that information was collected from the entire population. In more formal statistical notation, lower-case Greek letters are used to denote population parameters: the lower-case Greek letter mu (μ; pronounced "mew") is used for the population mean, the lower-case Greek letter sigma (σ) is used for the population standard deviation, and σ^2 is used for the population variance. I generally do not use the Greek letter notations in this book, but be aware of these equivalencies when you look at formulas in other textbooks or online.

Types of Probability Sampling

There are several ways to select a probability sample. The easiest, however, is **simple random sampling**, which relies on generating a list of every person (or case) in the population, assigning each case a number, and then randomly selecting numbers (usually using a computer) to choose who will be included in the sample. Simple random sampling is the equivalent of taking the name of every person in the population, putting it into a hat, mixing it well, and then randomly drawing names from

simple random sampling A sampling method that relies on listing each case in the population, assigning each case a number, and then randomly selecting numbers to select a sample.

the hat to select a sample. In a simple random sample, every person in the population has an equal chance of being selected into the sample.

Another way to select a probability sample is **stratified sampling**, which relies on dividing the population into groups, or strata, prior to selecting a sample. Each group or stratum is defined by a characteristic that is important to the research. Then, a sample is randomly selected from within each group or stratum. Stratified sampling is typically used when some analytically important groups have a population that is much smaller than other groups. Simple random sampling might not select enough cases from the small groups for researchers to be able to draw conclusions from the results. In Canadian surveys, the population is typically stratified by province before selecting a sample. This is because the number of people who live in each province varies substantially and researchers want to be sure that a sample includes enough people from every province. Of the approximately 35.2 million people who live in Canada, about 13.4 million live in Ontario, about 4.6 million live in British Columbia, but only about 143,000 people live in PEI. If simple random sampling were used to select people from the Canadian population, people living in Ontario would be much more likely to be selected than would people living in PEI. And, depending on how many cases are selected, it's possible that no one from PEI would be in the sample. To avoid this, each province is often treated as a stratum that cases are randomly selected from; that way, people from every province will appear in the sample.

When researchers use stratified sampling, they often select the same number of people from each stratum so that they capture a diversity of cases within each group. As a result, cases from different strata have different probabilities of being selected into the sample. For instance, if researchers sample 10,000 people from each province, roughly 1 out of every 1,340 people in Ontario are chosen (10,000 out of 13.4 million), 1 out of every 460 people in British Columbia are chosen (10,000 out of 4.6 million), and 1 out of every 14 people from PEI are chosen (10,000 out of 143,000). It's okay for different cases to have different probabilities of selection, as long as that probability can be calculated by a researcher and is not zero. If someone has no chance of being selected into a probability sample, then by definition they are not part of the population.

The probability sampling method used most often by large agencies, such as Statistics Canada, is called **cluster sampling**. This type of sampling relies on dividing the population into groups, or clusters, and then randomly selecting among the clusters. Unlike stratified sampling, the clusters do not need to be analytically important, and researchers choose between clusters (not within them). Cluster sampling is primarily used when it is not possible to generate a list of the population. For instance, imagine that a researcher wants to select a random sample of elementary school students in Canada. There is no central list of elementary school students, but the researcher can make a list of all of the school boards in Canada. Each school board is considered a cluster, and the researcher can randomly select among them. Then, the researcher can survey every elementary school student enrolled in each selected school board. Cluster sampling is often less expensive to implement—especially if clusters are related to geography—which is why many large agencies use it. Like stratified sampling, cluster sampling can lead to cases having different probabilities of being selected into the sample, especially if the clusters are different sizes.

stratified sampling A sampling method that relies on dividing the population into analytically important strata or groups and then randomly selecting a sample from each stratum.

cluster sampling A sampling method that relies on dividing the population into groups, or clusters, and then randomly sampling among the clusters.

multi-stage sampling A process that involves sequentially sampling at different levels in order to select a final sample.

In a **multi-stage sampling** design, a sample is selected at several different levels, one after another. Often, a different sampling method is used at each level. Let's think again about selecting a sample from the population of elementary school students in Canada. In the first stage, the researcher can use a cluster sample to select among the school boards in Canada. But instead of collecting information from every elementary school student in a school board, they can conduct a second stage of sampling and choose a random sample of elementary schools from within each selected school board. Then, in a third stage, the researcher can randomly select a sample of classes from within each selected school. At any stage, the sample can be stratified based on the characteristics a researcher thinks are important. In multi-stage sampling, a researcher can establish the probability that an individual case will be selected by multiplying the independent probabilities of selection at each stage (just like calculating the joint probability of two independent events). Most contemporary surveys conducted by large statistical agencies use a multi-stage sampling design.

Estimating Variation Using a Sample

Even when probability sampling is used, the characteristics of a sample will never perfectly match those of the whole population. Researchers know that samples always tend to slightly underestimate the amount of variation within a population. Although a detailed explanation of why this occurs is beyond the scope of this book, this phenomenon makes sense when you think about it: the variation—or the amount of difference—within a sample *must* be less than the variation within a population because the sample is only a portion of the whole population. In other words, there will always be less diversity among the cases in a sample than there will be among all of the cases in a population.

Fortunately, it's possible to compensate for this underestimation. When the variance and standard deviation are calculated using sample data, dividing by $n - 1$ (instead of by N) results in less biased population estimates. Thus, there are actually two formulas for the standard deviation and the variance: one to use with data collected from the whole population, and one to use with data collected from a sample of the population. In Chapter 4, you learned to calculate the population standard deviation and the population variance. The formulas for the sample standard deviation and the sample variance are very similar:

standard deviation

Sample data: $s = \sqrt{\dfrac{\sum(x_i - \bar{x})^2}{n-1}}$ Population data: $S = \sqrt{\dfrac{\sum(X_i - \bar{X})^2}{N}}$

variance

Sample data: $s^2 = \dfrac{\sum(x_i - \bar{x})^2}{n-1}$ Population data: $S^2 = \dfrac{\sum(X_i - \bar{X})^2}{N}$

The denominator in the fraction of the two formulas for samples is changed from N to $n - 1$. The letters are also changed to lower case to denote that the information is from a sample. To illustrate the effect of this change, let's return to the hypothetical example from Chapter 4 of six people's hourly wages. We began calculating the standard deviation by finding the deviation of each individual case from the mean, squaring them, and adding the results together. The sum of the squared deviations from the mean, which becomes the numerator of the fraction in the standard deviation equation, was 34.5. To calculate the population standard deviation, we divided 34.5 by 6 (because there were six cases) and found the square root of the result: 2.40. Now imagine these six cases are a sample selected from a larger population. To calculate the sample standard deviation, we divide 34.5 by 5 (because $6 - 1 = 5$) and find the square root of the result: 2.63. In this example, using $n - 1$ in the denominator of the equation results in a substantially higher standard deviation because there are only a small number of cases. With larger samples, the use of $n - 1$ in the denominator of the fraction often makes little difference. Most statistical software uses the sample standard deviation and sample variance formulas by default because it's rare for researchers to collect information from an entire population.

Spotlight on Data

Who Is Excluded from Canadian Surveys?

In the "Spotlight on Data" boxes so far, you may have noticed that there are groups of people who are routinely excluded from surveys. Depending on the survey, Statistics Canada excludes residents of institutions, members of the armed forces, Aboriginal people living on-reserve, and residents of the three Canadian territories (the Northwest Territories, Nunavut, and Yukon) from the survey population. Since they are not part of the survey population, these people are not included in the sample, and the survey results cannot be generalized to them.

The exclusion of residents of institutions, such as those in long-term-care homes or prisons, is a practical operational decision by Statistics Canada. It notes the difficulties "associated with the practical implications of sampling and interviewing residents of institutions (for example, access within prisons or interviewing people that are very ill) that would make their inclusion operationally problematic" (Statistics Canada 2014a, 19). In many institutions, residents do not have private telephone lines; moreover, gaining access to conduct in-person surveys may also be difficult. In long-term-care homes, some residents may have cognitive or physical limitations that detract from their ability to be effective research participants. Although basic administrative information is available about people living in institutions—their numbers, age, and sex—it is not necessarily

Continued

collected or summarized at a national level. Statistics Canada provides a similar rationale for excluding members of the Armed Forces from many surveys, "since many of these persons live in locations that are not accessible for the purposes of conducting the [survey], such as naval vessels, military camps, and barracks. Many of them may be stationed in other countries" (Statistics Canada 2014a, 19). Statistics Canada also notes that the Department of National Defence collects administrative data about this group. Although these exclusions make sense from an operational standpoint, they also imply that the experiences of people in these excluded groups do not need to be considered. Though administrative data can provide information about people's basic demographic characteristics, it usually does not capture people's attitudes, behaviours, and self-assessments. As a result, only limited information is available about topics such as the self-rated health of inmates, the political orientation of active members of the Armed Forces, or the social networks of people living in long-term-care homes, and whether these characteristics are changing over time.

Aboriginal people living on reserves are also routinely excluded from Canadian surveys. In part, this is because of practical difficulties accessing these communities: many reserves are geographically remote or still use a post-office-box system, which makes it difficult to identify dwelling addresses. In other communities, band councils do not give Statistics Canada staff permission to access their territories. The refusal to grant access stems from concerns that Aboriginal people have raised about Statistics Canada's practices, as well as government practices more generally. One concern relates to how Statistics Canada asks questions about Aboriginal identity and whether such identity is defined based on Aboriginal ancestry, self-identification as a Aboriginal person, or legal identity as a "status Indian" under the Constitution Act (Walter and Andersen 2013, Bailey 2008). Others note that Statistics Canada's definition of such concepts as education, work, income, housing, and family are conceptually different from those used by Aboriginal people (Saku 1999, Swimmer and Hennes 1993). For the census in particular, some Aboriginal communities understand enumeration as an activity of a colonizing settler government that is trying to subsume Aboriginal people into the nation through administrative processes, instead of recognizing the sovereignty of the First Nations. These tensions are exacerbated by the Canadian government's history of using administrative processes, such as registration under the Indian Act, to effectively circumscribe and discriminate against Aboriginal peoples. As a result, the census has substantially undercounted the number of Aboriginal people in Canada over time (Wright 1993, Bailey 2008). Many other national surveys provide a poor understanding of on-reserve living conditions, either because they do not collect information from people living on-reserve or because the questions asked are not meaningful in the context of an Aboriginal world view. In particular, Canada's Truth and Reconciliation Commission (TRC) noted that the available information about health indicators for Aboriginal peoples is out of date and that the lack of accessible data about how the health of

Aboriginal people compares to that of non-Aboriginal people means that these issues "receive less public, media, and political attention" (TRC 2015, 161). Although Statistics Canada uses the Aboriginal People's Survey—which was developed in partnership with several national Aboriginal organizations—to collect some in-depth information about Aboriginal people, in recent years it also has excluded Aboriginals living on-reserve (in the provinces). In contrast, the First Nations Information Governance Centre (www.fnigc.ca), which was established by the Assembly of First Nations, conducts surveys of on-reserve and northern First Nations communities that incorporate both Western and traditional understandings of key concepts.

People living in Canada's three territories—Nunavut, the Yukon, and the Northwest Territories—are also regularly excluded from survey populations. This is primarily because of cost and access issues: only a small proportion of Canada's residents live in the territories, it can be difficult to obtain accurate dwelling counts, and some communities are accessible only during part of the year. Nonetheless, this exclusion means we know little about the particular challenges faced by Canada's northern residents and further exacerbates the information gap about Aboriginal people, since many northern residents are either Inuit or Aboriginal peoples. For example, the exclusion of people living in the territories from the Survey of Household Spending (described in Chapter 3) results in limited knowledge about how exorbitantly high food prices in some northern communities influence other household spending. Whenever you analyze data that are collected by other people or agencies, note who is excluded from the population and consider how this might affect the results.

Finally, the questions that are asked (or not asked) in a survey make some social groups invisible in statistical analyses. Even though individuals from these groups are included in the sample, and may respond to a survey, the questions asked and the allowed responses effectively erase their presence. For example, most Statistics Canada surveys include information about the sex of a respondent, using a variable that has two attributes: male and female. The existence of intersex people, transgender people, or those with a non-binary sex/gender identity is neither acknowledged nor captured in the data. In addition, this question prioritizes biological sex and does not acknowledge the social construction of gender: men are assumed to be males and women are assumed to be females. In fact, in Statistic Canada's telephone and face-to-face surveys, interviewers usually record a person's sex, based on their gender presentation, without asking the person; given this practice, throughout this book I treat the variable "Sex" as though it were gender. It's certainly not possible to complete a statistical analysis of the diversity of gender identities using this information. In other surveys, questions about group membership simply go unasked. For instance, the Labour Force Survey (described in Chapter 4) does not ask whether a person is a member of a racialized group (or a "visible minority"), making it impossible to analyze the relationship between racialization and wages.

Sergei Bachiakov/Shutterstock

Photo 5.1   **The Idle No More movement calls on people to honour Indigenous sovereignty and to recognize the nation-to-nation relationship between First Nations and the federal government.**

Weighting in Sample Surveys

weights Multipliers assigned to each case that make the sample represent the larger population more accurately.

Most sample surveys use weights. **Weights** are multipliers assigned to each case that make the sample represent the larger population more accurately. In the previous section, you learned that cases can have different probabilities of being selected into a sample. If cases with different probabilities of selection are treated as equivalent in the final sample, the result is a biased representation of the population. The answers of people who are less likely to be selected into the sample should count more when we make generalizations about the population overall. Conversely, the answers of people who are more likely to be selected into the sample should count less when we make generalizations about the population overall. Weights that account for each case's probability of selection assign higher multipliers to people with a low probability of being selected into the sample, and lower multipliers to people with a high probability of being selected into the sample.

In addition to weights that compensate for the probability of selection, many statistical agencies also use weights to make the total number of cases in a sample match with the population size. Statistics Canada describes population weights like this:

When a probability sample is used, the principle behind estimation is that each person selected in the sample represents (in addition to himself/herself) several other persons not in the sample. For example, in a simple random sample of 2 per cent of the population, each person in the sample represents 50 persons in the population (himself/herself and 49 others). The number of persons represented by a given respondent is usually known as the weight or weighting factor. (Statistics Canada 2011, 13)

One major concern for survey researchers is non-response. Many people who are selected into a sample and asked to complete a survey don't actually do so. They may be too busy, or they aren't interested in the topic, or they have concerns around privacy, or they simply don't want to. This is of particular concern because the people who do not participate are not a random subset of the people who are sampled. There are clear demographic trends related to research participation. In general, women are more likely to participate than men, seniors are more likely to participate than young people, and people living in rural areas are more likely to participate than people living in urban areas. These differences can make the results biased.

Researchers use a strategy called **post-stratification** to compensate for people's unequal participation in research. Post-stratification relies on using information about the population that is taken from an outside source, such as a census. Researchers identify several key characteristics of the population, and then assign weights that make the overall demographic characteristics of the sample match with the overall demographic characteristics of the larger population. Statistics Canada typically uses post-stratification to weight survey samples in relation to people's age, sex, and province of residence. Practically speaking, people from demographic groups who are more likely to respond to surveys (such as seniors) have a slightly lower weight assigned to their answers, and people from demographic groups that are less likely to respond to surveys (such as young people) have a slightly higher weight assigned to their answers. Fortunately, in most survey datasets produced by large organizations or agencies, the weights (multipliers) that account for the probability of selection, for the population size, and for post-stratification are combined into a single weight variable.

post-stratification The process of assigning weights to a sample so that it matches the population on key characteristics, such as age and sex.

You probably have some idea about how weighting works because in many courses, instructors give different weights to different assignments. In a course where each assignment has the same weight, your mark on each assignment makes exactly the same contribution to your final grade. But some instructors assign more weight to a final cumulative course assignment. In a course with two assignments weighted 20 per cent each, and a final assignment weighted 60 per cent, your mark on the third assignment has much more influence on your final grade than your mark on the first two assignments. In fact, your mark on the third assignment has three times the influence on your final grade than your mark on the first assignment because it has three times the weight.

Weights for survey data work in much the same way: cases with higher weights have more influence on a statistic than cases that with lower weights. An example will help to illustrate this. Table 5.2 shows five hypothetical people's weekly hours of work. To find the unweighted average, add up the weekly hours of work for each person (they sum to 75) and divide the result by 5 (because there are 5 people), to get 15 hours a week.

Table 5.2   Steps in Calculating a Weighted Average (Hypothetical Data)

Person	Weekly Hours of Work	Weight	Weighted Contribution
Samantha	10	0.5	5.00
Ryan	15	1	15.00
Abigail	12	1.5	18.00
Amy	16	2	32.00
Suraj	22	3	66.00
Sum	75	8	*136.00*

The weight variable (in the third column of Table 5.2) assigns a multiplier to each case, indicating whether it should contribute more or less to the statistical result. Adding up the weights shows that these five cases are weighted to be equivalent to eight people. To find the weighted average, multiply each person's weekly hours of work by the weight; this becomes the weighted contribution of each case to the statistic. Cases with weights less than 1 have a weighted contribution that is less than their original contribution (their original weekly hours of work). Cases with weights more than 1 have a weighted contribution that is more than their original contribution. Add together the weighted contributions (they sum to 136) and divide by the weighted number of cases (the weight variable sums to 8) to find the weighted average: 17 hours a week. The weights change the contribution of each case to the final statistic.

Let's look at Table 5.3 to see how weighting affects a frequency distribution from Canada's 2014 General Social Survey (which is described in Chapter 8). For convenience, I show the distribution of gender, but weighting affects all variables in a similar way. The first two columns in Table 5.3 show the unweighted distribution of gender collected directly from the sample. The sample includes 33,089 people: 15,134 men and 17,955 women. Women make up 54 per cent of the sample. The middle two columns of Table 5.3 show the distribution of gender when weights are used. There are two major differences compared to the unweighted distribution. First, the number of cases is inflated to match the population size (29 million people). Second, the percentage of men and women matches the gender distribution found in the census. (Typically, 51 per cent of the Canadian population are women, given women's slightly longer lifespans.) Using the weights results in a frequency distribution that is a more accurate representation of the Canadian population.

The final two columns of Table 5.3 show the distribution of gender when a **standardized weight** is used. Standardized weights retain the *relative* contribution of each case but have an average multiplier of 1 so that the weighted number of cases matches the unweighted sample size. So, a case with a high (unstandardized) weight, compared to other cases, will also have a high standardized weight; similarly, a case with a low (unstandardized) weight, compared to other cases, will also have a low standardized weight. But since the average standardized weight is 1, the total number of cases used to calculate each statistic appears to be the same as the original sample size. (In general, standardized variables have a mean of 0 and include both positive and negative values. However, this is not true of standardized weight variables,

standardized weights Versions of weights that retain the relative contribution of each case but that have an average multiplier of 1 so that the weighted number of cases matches the unweighted sample size.

Table 5.3   A Frequency Distribution Using No Weights, Using (Unstandardized) Weights, and Using Standardized Weights, All from the Same Sample

Answer	Unweighted Sample		Using Weights		Using Standardized Weights	
	Frequency	Percentage	Frequency	Percentage	Frequency	Percentage
Male	15,134	45.7	14,542,093	49.4	16,351	49.4
Female	17,955	54.3	14,886,347	50.6	16,738	50.6
Total	33,089	100.0	29,428,440	100.0	33,089	100.0

Source: Author generated; Calculated using data from Statistics Canada, 2016b.

which have a mean of 1 and cannot include negative values.) Using standardized weights is the equivalent of using the weights that account for the probability of selection and post-stratification, but not for the population size. You can see how this works by comparing the three pairs of columns in Table 5.3. When the standardized weights are used, the frequency distribution shows that the total number of cases (33,089) is the same as the total number of cases in the frequency distribution of the unweighted sample. However, the percentage column of the frequency distribution when the standardized weights are used matches with the percentage column of the frequency distribution when the (unstandardized) weights are used. That is, the percentage of men and women in the frequency distribution when the standardized weights are used match the actual percentage of men and women in the population. Because weight variables are not standardized in the same way as other variables, standardized weights are sometimes called *relative weights*, or *adjusted weights*, to avoid any potential confusion. (See the "Hands-on Data Analysis" box to learn how to create a standardized weight from an unstandardized weight.)

Hands-on Data Analysis

Creating a Standardized Weight Variable

Survey datasets often include a weight variable, which must be used in order to generate accurate population estimates. If the weights inflate the number of cases in the sample so that it matches the number of people in the population, some statistical software programs do not calculate statistical tests accurately. To avoid this problem, researchers can create and use standardized weights in their data analysis. Some statistical software programs can make this adjustment automatically.

If a dataset has an unstandardized weight variable, it's easy to create a standardized weight variable. First, find the mean or the average of the weight variable (when the data are unweighted or the weights are "turned off"). If the mean of the weight variable is 1, or very close to 1 (such as 1.001 or 0.999), it is already standardized and you do not need to do anything further.

Continued

To compute a standardized weight variable, divide the unstandardized weight for each case by the mean of the unstandardized weight variable:

$$standardized\ weight = \frac{weight\ for\ each\ case}{mean\ of\ the\ weight\ variable}$$

After you compute the standardized weight variable, it will be displayed like any other variable in the dataset. Table 5.4 shows the unstandardized weight and the standardized weight for 10 example cases. The standardized weights are much smaller than the unstandardized weights, and some are less than 1. But the relative size of the weights remains the same—that is, cases with higher unstandardized weights have higher standardized weights, and cases with lower unstandardized weights have lower standardized weights.

Table 5.4   A Comparison of Unstandardized and Standardized Weights for 10 Hypothetical Cases

Person	Unstandardized Weight	Standardized Weight
Myla	4,311.04	2.36
Fareed	1,211.28	0.66
Hassan	3,794.18	2.08
Mia	1,960.80	1.07
Sylvia	1,267.65	0.69
Kevin	1,889.30	1.04
Madison	2,963.48	1.62
Sean	1,209.12	0.66
Aisha	1,872.24	1.03
Theo	4,373.10	2.40

To check that the standardized weight variable has been computed correctly, find its mean, which should be 1 (when the data are unweighted or the weights are "turned off"). After confirming that the standardized weight variable is correct, save the dataset so that the new variable will always be available.

Some statistical software programs allow you to designate a weight variable that is used in all subsequent calculations. Other programs use specialized commands to generated weighted statistics and require you to specify the weight variable each time. Researchers use standardized weights to compensate for the fact that some statistical software programs do not distinguish between the actual sample size (33,089) and the weighted sample size (29 million), and erroneously use the weighted sample size as the "n" for statistical estimation (which you will learn about in Chapter 6). Up until this point, all of the real-world examples in this book have used standardized weights, and I will continue to do so. As a result, the statistical results in each chapter accurately represent the Canadian population.

Some datasets contain more than one weight variable. For example, some surveys have a "person weight" variable and a "household weight" variable. In this situation, use the weight variable that corresponds to your unit of analysis. If you are analyzing people and reporting individual characteristics, such as age, sex/gender, or personal income, use the person weight variable, or a standardized version of the person weight variable. If you are analyzing households and you are reporting household characteristics, such as the number of children in a household or household income, use the household weight variable, or a standardized version of the household weight variable. Sometimes datasets also contain additional weight variables to use when analyzing particular groups of survey questions (modules). When only some of the people in a sample are asked a group of questions (usually to reduce costs), the cases have a different probability of being selected and, thus, have different weights. If you are unsure about which weight to use, consult the survey user guide. If a sample survey dataset contains a weight variable, though, it is crucial that you use it in any statistical analyses. If you do not use the provided weight variable(s), the results are guaranteed to be inaccurate.

Misconceptions about Weighting

Using weights can sometimes be contentious. People without any statistical knowledge may find it difficult to see how one person's answers represent those of hundreds of other people: How is it possible that there are hundreds of other people who are exactly the same as the person who answered the survey? Of course, there aren't hundreds of people who are exactly the same as the survey respondent, but when people's answers are aggregated across a large sample, the differences even out. Other people accuse statisticians of lying or of mathematical voodoo when they use weights; they perceive the use of weights as a form of manipulating the "truth" of the data.

Unequal weights can be a particularly troublesome idea; it bothers some people to discover that one person's information counts less than another person's. The use of unequal weights appears to be fundamentally opposed to the basic principles of equality and democracy that many people cherish, characterized in phrases such as "one person, one vote." People generally don't agree with the idea that wealthy people should have more votes or more input on who governs than people who are not wealthy. So why are people who are poor, who tend not to respond to surveys, assigned a weight that makes their information worth more than the information collected from others? The obvious reason is because assigning a higher weight to some people results in more accurate estimates of the population than if everyone were assigned an equal weight. Social activists often argue that equality cannot be achieved by treating everyone equally; historical and social context must be accounted for. A similar principle applies to survey weights: providing a fair representation of everyone does not mean that every respondent's answers are assigned the same weight; the context of their selection and social location must be accounted for.

Best Practices in Presenting Results

What Goes in a Methodology Section?

primary data collection Collecting information for the explicit purpose of answering a specific research question.

In studies where researchers engage in **primary data collection**, that is, collecting data for the explicit purpose of answering a specific research question, any report of the results usually includes a "methodology" section. This section typically describes the research design, data collection, sampling, data entry, and analysis. The goals of the methodology section for a primary data study are to provide other researchers with enough information so that they can conduct a similar study themselves and to illustrate the limits of the conclusions that can be drawn from the collected data.

secondary data analysis The analysis of data that were not specifically collected to answer researchers' questions and were often not collected by the researchers themselves.

Many statistical analyses rely on analyzing secondary data, that is, data that were not specifically collected to answer researchers' questions and were often not collected by the researchers themselves. Research that involves **secondary data analysis** still requires a methodology section whenever results are reported, but it tends to be shorter than for primary data studies. This is because researchers can highlight the key aspects of the methodology and then direct readers to the survey user guide or other documentation for more detailed information about how data were collected and processed.

For studies that involve analyzing secondary survey data, the methodology section should report the following:

- The name of the survey and the organization that collected it
- The survey's goal or objective; this is especially important if you are using the data to answer a research question that is not explicitly related to the survey topic
- The population of the survey, including a statement of any exclusions; if you are using a subsample of the population for your specific analysis, this should be noted
- A general description of the sampling procedures
- How and when information was collected
- The statistical analysis techniques that were used
- The statistical software used to complete the data analysis
- A statement that weights were used and, if appropriate, a note about the characteristics used to post-stratify

Depending on the research, the methodology section might also include information about the main variables used in the analyses and the survey questions used to generate those variables, as well as any information about combining or recoding variables. Providing this basic information gives readers the context they need to interpret the statistical results and to understand the limitations of the analysis. Writing a clear methodology section to accompany your statistical analyses also shows the reader that you have used the secondary data in an appropriate way.

What You Have Learned

This chapter began by introducing the basic logic of probability. This logic underlies the practice of probability sampling, which involves strategies such as simple random sampling, stratified sampling, or cluster sampling, in one or more stages. Researchers use these strategies to select representative samples. You also learned about weights, which researchers use to make the sample more accurately represent the larger population. The final section of the chapter briefly described the elements that should be included in a "methodology" section for studies that involve analyzing secondary data.

This chapter also discussed the systematic exclusion of some groups of people from sample surveys in Canada. There is limited quantitative data available about some social groups, either because they are excluded from survey populations or because of the failure to ask survey questions that make them visible in data. These gaps in our knowledge often make it more difficult to advocate for members of these social groups since they may be invisible to policy-makers and since we, collectively, know less about them.

Check Your Understanding

Check to see if you understand the key concepts in this chapter by answering the following questions:

1. What is the difference between a theoretical probability and an observed probability?
2. How do you calculate the theoretical probability that an event will occur?
3. How do you calculate a joint probability?
4. How are frequency distributions related to probabilities?
5. What is the relationship between z-scores and probabilities?
6. What is the difference between probability sampling and non-probability sampling?
7. What are some different ways to select a probability sample?
8. What are weights? What do they account for?
9. What is the difference between an unstandardized weight and a standardized weight?
10. What is post-stratification, and how is it related to people's willingness to participate in research?
11. Who is typically excluded from the population of Canadian surveys, and why?

Practice What You Have Learned

Check to see if you can apply the key concepts in this chapter by answering the following questions. Keep two decimal places in any calculations.

1. You have a bag of 10 marbles: 2 are green, 4 are blue, and 4 are red. If you close your eyes and randomly pick one marble from the bag:
   a. What is the probability that the marble you pick will be red?
   b. What is the probability that the marble you pick will be red or green?
2. If you roll two standard dice, each with six sides that are numbered from 1 through 6 using dots on each side:
   a. What is the probability of rolling a 6 on both dice?
   b. What it the probability of rolling an even number on both dice?
3. A standard deck of cards has 52 cards divided into four suits (hearts, diamonds, clubs, and spades) with

13 cards each (ace, 2, 3, 4, 5, 6, 7, 8, 9, 10, jack, queen, and king). If you randomly draw one card from a standard deck:

a. What is the probability of drawing a 7?
b. What is the probability of drawing a face card (a jack, queen, or king)?

4. You randomly draw two cards in a row from a standard deck:

a. What is the probability of selecting an ace in the first draw?
b. If you return the selected card to the deck, what is the probability of selecting an ace in the second draw?
c. If you select an ace in the first draw and do not return the card to the deck, what is the probability of selecting an ace in the second draw?

5. Your largest class has 120 people enrolled in it, including 22 mature students. If you are randomly assigned one classmate to interview, what is the probability that the classmate will not be a mature student?

6. There are 18,500 students enrolled at your school. The student union sends a survey to a random sample of 2,000 of them. What is the probability that any one student you encounter on campus will have been selected to complete the survey?

7. You survey students enrolled in an introductory social science course about their classroom participation. There are 600 students enrolled in all sections of the course. You use simple random sampling to select 100 students to survey. What is the probability that a student in the course will be included in the sample?

8. The following year you repeat the survey described in question 7, with a new group of students enrolled in the introductory social science course. The previous results had suggested that students' classroom participation was influenced by the course instructor. This year, there are three sections of the course: one taught by Professor Sundar, with 200 students; one taught by Professor Martin, with 100 students; and one taught by Professor Roja, with 300 students. You randomly select 50 students from each of the three sections to survey:

a. What sampling method are you using?
b. What is the probability of being included in the sample for students in each of the three sections?

9. Statistics Canada provides the information in the box below about sampling in the National Graduates Survey (described in Chapter 2). In your own words, explain how cases are selected to be part of the sample.

10. Among the population of post-secondary students who are employees, the number of hours people work each week is normally distributed. On average, post-secondary student employees work 18 hours per week. Working 21 hours per week is associated with a z-score of 0.25.

a. What is the probability that a randomly selected post-secondary student employee will work 21 hours a week or more?

Sampling

This is a sample survey with a cross-sectional design.

The National Graduates Survey uses a stratified simple random sample design. The sample selection of graduates within strata is done without replacement and using a systematic method.

Three variables are used for stratification: geographical location of the institution, level of certification and field of study. There are 13 geographical locations: the 10 provinces and the three northern territories. There are five levels of certification: trade/vocational certificate or diploma (Quebec only), college diploma, bachelor's degree, master's degree, and doctorate. The "trade/vocational" level in Quebec pertains to programs typically offered outside the post-secondary sector. Finally, there are 12 fields of study: categories 01 to 12 of the primary groupings of the Classification of Instructional Programs (CIP).

The combination of these three variables makes for a possibility of 636 strata in total. However, there are not graduates in every possible strata and, therefore, the final number of strata created was 434.

Source: Statistics Canada 2014b, 16.

b. What is the probability that a randomly selected post-secondary student employee will work 21 hours a week or less?

11. Among the population of post-secondary student employees described in question 10, working 7.5 hours per week is associated with a z-score of –1.00.

   a. What is the probability that a randomly selected post-secondary student employee will work 7.5 hours a week or more?
   b. What is the probability that a randomly selected post-secondary student employee will work 7.5 hours a week or less?

12. You use simple random sampling to select 20 students enrolled in your major who are also employees. You ask them how many hours they work each week. Their answers are as follows:

21 hours	8 hours	20 hours	38 hours	5 hours
40 hours	4 hours	15 hours	17 hours	30 hours
9 hours	8 hours	32 hours	22 hours	20 hours
22 hours	15 hours	12 hours	12 hours	10 hours

   a. Calculate the mean number of hours that students work each week.
   b. Calculate the standard deviation of the number of hours that students work each week, using the formula for the standard deviation of a sample.

13. Statistics Canada routinely excludes First Nations people who live in reserve or northern communities from their survey population. The First Nations Information Governance Centre (FNIGC) is dedicated to conducting surveys in these communities, including the First Nations Regional Health Survey (FNRHS, or RHS), which has existed for more than 20 years. To date, two phases of the RHS have been published: Phase 1 in 2002/03 and Phase 2 in 2008/10. (Data from Phase 3 is scheduled for release in 2018.) Figure 5.1 shows how the results of these two surveys compare to the results of the 2006 Census for the general Canadian population.

   a. How does the household income of people living in First Nations communities differ from that of the general Canadian population?

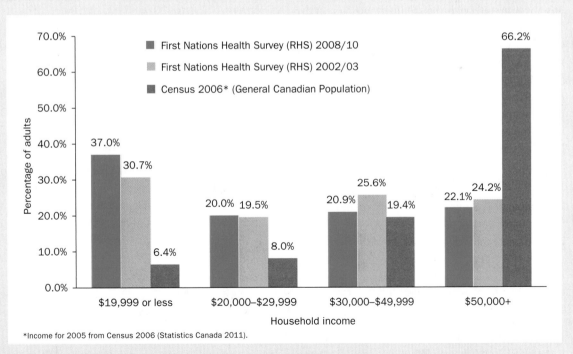

*Income for 2005 from Census 2006 (Statistics Canada 2011).

Figure 5.1   **Household Income Levels, RHS 2008/10 (n = 7,431), RHS 2002/03, and Census 2006 Households**

Source: First Nations Information Governance Centre 2012, 54.

Estimation

The estimation of population characteristics from a sample survey is based on the premise that each sampled household represents a certain number of other households in addition to itself. This number is referred to as the survey weight, and the weighting process involves computing the weight assigned to each household. There are a number of steps in that process. First, each household is given an initial weight equal to the inverse of its selection probability. A few adjustments are later applied to the interview weights.

The interview weights are first adjusted to take into account the households that did not answer the questionnaire. They are then adjusted so that selected survey estimates agree with aggregates or estimates from independent auxiliary sources. The first source is the number of persons by age group and the number of households by household size from population estimates produced by the Demography Division using data from the 2006 Census. Annual estimates of the number of persons in eight age groups (0–6, 7–17, 18–24, 25–34, 35–44, 45–54, 55–64, and 65+) are used at the provincial level and two age groups (0–17 and 18+) at the census metropolitan area level. For the number of households, the weights are calibrated to the annual provincial estimates for three household size categories (one, two, and three or more persons).

Source: Statistics Canada 2015, 7–8.

b. In general, how might the results of surveys that exclude information from on-reserve and northern First Nations communities be biased?

14. Statistics Canada provides the information in the box above about weighting the interview data in the Survey of Household Spending (described in Chapter 3). In your own words, explain what Statistics Canada takes into account when it creates weights for this survey.

15. A researcher collects employment information from a random sample of post-secondary students. Since each student had a different probability of being selected into the sample, they each have a different weight assigned to their answers. The results, including the weight assigned to each case, are listed in Table 5.5.

   a. Calculate the unweighted average of the number of hours worked per week.
   b. Calculate the weighted average of the number of hours worked per week.

Table 5.5 Students' Weekly Hours of Work and Hourly Wages, with Weights (Hypothetical Data)

Person	Hours Worked per Week	Hourly Wage ($)	Weight
Sanjay	20	11.00	1.5
Kareem	30	15.00	2
Marissa	10	21.50	0.5
Lin	15	18.50	1
Juan	50	14.00	3
Ernesto	5	16.00	0.25

16. Using the information in Table 5.5:

   a. Calculate the unweighted average of hourly wages.
   b. Calculate the weighted average of hourly wages.

Practice Using Statistical Software (IBM SPSS)

Answer these questions using IBM SPSS and the GSS27. sav or the GSS27_student.sav dataset available from the Student Resources area of the companion website for this book. Report two decimal places in your answers, unless fewer are printed by IBM SPSS. It is imperative that you save the dataset to keep any new variables that you create.

1. Create a standardized weight variable by following these steps:

   a. Use the Means procedure to find the mean of the variable "Person weight" [WGHT_PER]. Complete this step with the weight off.

   b. Use the Compute Variable tool to create a new, standardized weight variable, called "Standardized person weight" [STD_WGHT]. Make the new, standardized weight variable equal to "Person weight" [WGHT_PER], divided by its mean (keep/use all of the decimals of the mean).

   c. Use the Means procedure to find the mean of the variable "Standardized person weight" [STD_WGHT]. The mean should equal 1; if it does not, re-do steps (a) and (b).

2. Use the Sort Cases tool to sort the cases by "Person weight" [WGHT_PER].

   a. Find the case with the lowest value on the "Person weight" [WGHT_PER] variable. How do the values on the "Person weight" [WGHT_PER] and the "Standardized person weight" [STD_WGHT] variables compare for that case?

   b. Find the case with the highest value on the "Person weight" [WGHT_PER] variable. How do the values on the "Person weight" [WGHT_PER] and the "Standardized person weight" [STD_WGHT] variables compare for that case?

3. Investigate the effects of weighting by comparing the results of three frequency distributions of the variable "Sex of respondent" [SEX].

   a. With the weight off, produce a frequency distribution of "Sex of respondent" [SEX].

   b. Use the Weight Cases tool to weight the data by "Person weight" [WGHT_PER]. Then, produce another frequency distribution of "Sex of respondent" [SEX].

   c. Use the Weight Cases tool to weight the data by "Standardized person weight" [STD_WGHT]. Produce another frequency distribution of "Sex of respondent" [SEX].

   d. Describe the similarities and differences between the three frequency distributions of "Sex of respondent" [SEX] that are produced when different weights are used.

4. Investigate the effects of weighting by comparing the results of three frequency distributions of the variable "Place of birth of respondent - Canada" [BRTHCAN].

   a. With the weight off, produce a frequency distribution of "Place of birth of respondent - Canada" [BRTHCAN].

   b. Use the Weight Cases tool to weight the data by "Person weight" [WGHT_PER]. Then, produce another frequency distribution of "Place of birth of respondent - Canada" [BRTHCAN].

   c. Use the Weight Cases tool to weight the data by "Standardized person weight" [STD_WGHT]. Produce another frequency distribution of "Place of birth of respondent - Canada" [BRTHCAN].

   d. Describe the similarities and differences between the three frequency distributions of "Place of birth of respondent - Canada" [BRTHCAN] that are produced when different weights are used.

Key Formulas

Probability

$$probability\:(outcome\:of\:interest) = \frac{number\:of\:outcomes\:of\:interest}{number\:of\:possible\:outcomes}$$

Joint probability

$$probability\:(outcome\:of\:interest, outcome\:of\:interest)$$
$$= \frac{number\:of\:outcomes\:of\:interest}{number\:of\:possible\:outcomes} \times \frac{number\:of\:outcomes\:of\:interest}{number\:of\:possible\:outcomes}$$

Standard deviation (using sample data)	$s = \sqrt{\dfrac{\sum\left(x_i - \bar{x}\right)^2}{n-1}}$

Standard deviation (using population data)	$S = \sqrt{\dfrac{\sum\left(X_i - \bar{X}\right)^2}{N}}$

Variance (using sample data)	$s^2 = \dfrac{\sum\left(x_i - \bar{x}\right)^2}{n-1}$

Variance (using population data)	$S^2 = \dfrac{\sum\left(X_i - \bar{X}\right)^2}{N}$

Standardized weight	$standardized\ weight = \dfrac{weight\ for\ each\ case}{mean\ of\ the\ weight\ variable}$

References

Bailey, Sue. 2008. "First Nations Assails Census Results." *Toronto Star*, January 27. http://www.thestar.com/news/canada/2008/01/27/first_nations_assails_census_results.html.

First Nations Information Governance Centre (FNIGC). 2012. "First Nations Regional Health Survey (RHS) 2008/10: National Report on Adults, Youth and Children Living in First Nations Communities." Ottawa. http://fnigc.ca/sites/default/files/First%20Nations%20Regional%20Health%20Survey%20(RHS)%202008-10%20-%20National%20Report.pdf.

Canada. 2012. "Redistribution of Federal Electoral Districts." http://www.redecoupage-federal-redistribution.ca/.

Saku, James. 1999. "Aboriginal Census Data in Canada: A Research Note." *Canadian Journal of Native Studies* 19 (2): 365–79.

Statistics Canada. 2011. "Cycle 24: Time-Stress and Well-Being Public Use Microdata File Documentation and User's Guide." Catalogue no. 12M0024X.

———. 2014a. "Guide to the Labour Force Survey." Catalogue no. 71–543–G. http://www.statcan.gc.ca/pub/71-543-g/71-543-g2014001-eng.pdf.

———. 2014b. "Microdata User Guide 2013 National Graduate Survey (Class of 2009–2010)." Centre for Education Statistics.

———. 2015. "User Guide for the Survey of Household Spending, 2013." Catalogue no. 62F0026M, 1. http://www.statcan.gc.ca/pub/62f0026m/62f0026m2015001-eng.pdf.

———. 2016a. "Labour Force Survey 2016." http://www23.statcan.gc.ca/imdb/p2SV.pl?Function=getSurvey&Id=331692.

———. 2016b. "General Social Survey—Victimization (GSS) 2014." http://www23.statcan.gc.ca/imdb/p2SV.pl?Function=getSurvey&SDDS=4504.

Swimmer, Eugene, and David Hennes. 1993. *Inuit Statistics: An Analysis of the Categories Used in Government Data Collections.* Report for the Royal Commission on Aboriginal People. April 5. http://publications.gc.ca/collections/collection_2016/bcp-pco/Z1-1991-1-41-12-eng.pdf.

Truth and Reconciliation Commission of Canada (TRC). 2015. *Honouring the Truth, Reconciling for the Future: Summary of the Final Report of the Truth and Reconciliation Commission of Canada.* http://epe.lac-bac.gc.ca/100/201/301/weekly_acquisition_lists/2015/w15-24-F-E.html/collections/collection_2015/trc/IR4-7-2015-eng.pdf.

Walter, Maggie, and Chris Andersen. 2013. *Indigenous Statistics: A Quantitative Research Methodology.* Walnut Creek, CA: Left Coast Press.

Wright, Robert. 1993. "Using Census Data to Examine Aboriginal Issues: A Methodological Note." *Canadian Journal of Native Studies* 13 (2): 291–307.

Making Population Estimates: Sampling Distributions, Standard Errors, and Confidence Intervals

6

Learning Objectives

In this chapter you will learn:

* What a sampling distribution is

* Why the central limit theorem is central to statistical estimation

* How to calculate the standard error of means and proportions

* How to calculate 95 per cent confidence intervals for means and proportions

* What the relationship is between confidence intervals and the margin of error

* How to write about confidence intervals

Introduction

This chapter introduces one of the most important ideas in social statistics: **estimation**. Statistical estimation is the process of making inferences about a population parameter using information collected from a random sample of that population. Chapter 5 introduced this idea, but this chapter develops it more fully through a discussion of sampling distributions, the central limit theorem, and confidence intervals. A key feature of statistical estimation is that it is associated with a level of uncertainty; in other words, you can never be absolutely confident that the sample reflects the larger population. In order to acknowledge the uncertainty associated with population estimates, researchers generally talk about the "probability," "likelihood," or "chance" that something will occur.

The research focus of this chapter is the gendered division of domestic labour. In a capitalist economy, domestic labour tends to be less valued because it is usually unpaid and, thus, not directly associated with an economic cost. Nonetheless, domestic labour—such as cooking and cleaning—is crucial to our everyday personal and social well-being. Without domestic labour, people would be unable to effectively participate in the paid labour force. The majority of unpaid domestic labour is done by women. Gender roles position women as caregivers and assign them responsibility for the private sphere of the home. Even after

estimation The process of making inferences about a population parameter using information collected from a random sample of that population.

many women began participating in the paid labour force, the gendered distribution of domestic labour did not change substantially (Hochschild and Machung 2012). Instead, women who work for pay end up doing a "double day" or "second shift" of work: first their paid labour, and then their unpaid domestic labour. In addition, the types of domestic labour that women and men take responsibility for tend to be different (Luxton and Corman 2001). Women are more likely to do domestic tasks that require attention every day or multiple times a day, such as meal preparation, dishes, laundry, and other cleaning. In contrast, men are more likely to do domestic tasks that require attention weekly or occasionally, such as household repairs or yard work. In this chapter, statistical analysis is used to answer the following:

- Is there a gender gap in the amount of time spent doing domestic labour?
- Do women who work for pay do a "double day"? How is working for pay related to the amount of time that women and men spend doing domestic labour?
- How are people's living arrangements related to the amount of time that women and men spend doing domestic labour?

KatarzynaBialasiewicz/iStockphoto

Photo 6.1   **Many women who work for pay still do a "second shift" of domestic labour when they get home.**

The Sampling Distribution of a Mean

The fundamental problem associated with sampling is this: whenever researchers select a sample, they have only one sample out of all of the possible samples that could be selected from a population. By sheer chance, each sample will have a slightly different composition of cases and, thus, produce slightly different statistics. Some samples will produce statistics that are very close to the actual population parameters, while others will produce statistics that are farther away from the actual population parameters.

When researchers conduct studies using a single sample from a population, the question then becomes this: How close to the actual population parameters are the statistics produced by that single sample? This would be easy to determine if researchers knew the population parameters. But if that were the case, they wouldn't need to select a sample at all because they would already have the answer. Fortunately, there are some well-proven mathematical techniques that can help researchers make an educated guess about the relationship between their sample statistics and the population parameters. Some of these ideas may seem complex at first, but once you get the hang of it, you'll understand the fundamental principle of inferential statistics. In order to explain more clearly, I'm going to start by using some simulated (made-up) data, before turning to a real-world example.

To begin, I create a dataset of 15,000 cases that represent an entire population. Imagine that the simulated population are the residents of a small city, such as Prince Rupert, BC; Steinbach, MB; or Gander, NL. The dataset has a variable capturing how much time each person in the population spent doing domestic labour the previous day. In this simulated population, the average is exactly one hour (or 60 minutes), with a standard deviation of 30 minutes. Some people spent as little as 1 minute doing domestic labour the previous day, and some people spent as much as 3 hours (180 minutes).

Next, I select a simple random sample of 200 cases from the simulated population, as if I were a researcher collecting information from a random sample of people living in the city. In the first sample I select, the average (mean) amount of time spent doing daily domestic labour is 57 minutes, with a standard deviation of 28.5 minutes. Because this is a simulation, I know that the average of 57 minutes is close to the population average but not exactly the same. (Of course, in an actual study, I wouldn't know the population average.)

Now, I do it a second time. Starting again with the 15,000 cases in the simulated population, I select another simple random sample of 200 cases. The second sample is completely independent of the first one—that is, being selected into the first sample did not affect the chance of being selected into the second sample. As it happens, a few of the cases that were randomly selected into the second sample are also in the first sample, but most of the cases are not. In the second sample, the average (mean) amount of time spent doing daily domestic labour is 67 minutes, with a standard deviation of 30 minutes. That's a much higher average than in the first sample and higher than in the population. By chance, the second sample includes several people who spent a lot of time doing domestic labour the previous day, which makes the average much higher.

Table 6.1   **Statistics from Three Samples from a Simulated Population**

Sample	Mean	Standard Deviation
Sample 1	57	28.5
Sample 2	67	30.0
Sample 3	63	31.5

Now I select a third sample. Starting with the 15,000 cases in the simulated population, I randomly select another 200 cases and find that the average amount of time spent doing daily domestic labour is 63 minutes, right between the averages of the first two samples. The third sample has a standard deviation of 31.5 minutes. (See Table 6.1.)

I repeat this process to select 20 samples from the simulated population and find the average for each sample. Figure 6.1 shows a histogram of those 20 sample averages (or sample means). In five of the samples, the average is between 61 and 62. In four of the samples, the average is between 59 and 60. The average in most of the samples is close to the actual population average, but some averages are farther away. (And, again, we usually wouldn't know the population average.)

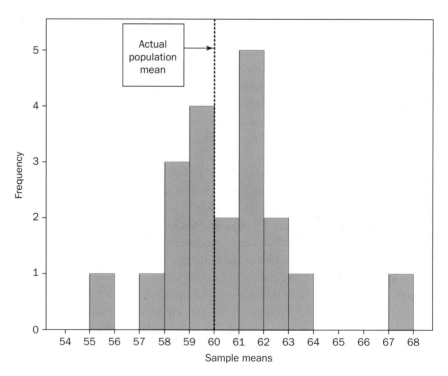

Figure 6.1   **A Histogram of the Means (Averages) of 20 Samples from the Same Population (Simulated Data)**

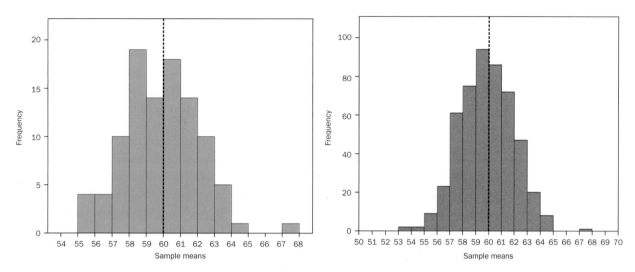

Figure 6.2 A Histogram of the Means (Averages) of 100 Samples and 500 Samples from the Same Population (Simulated Data)

Now that you have the idea, I'll continue to randomly select even more samples. (It's a good thing that this is only a simulation; if I were really sampling respondents from a small city over and over again, people might be getting irritated by now.) The distribution of a statistic for every possible sample of the same size that could theoretically be selected from a population is called a **sampling distribution**. Figure 6.2 shows a histogram of 100 sample means and a histogram of 500 sample means. These histograms are an approximation of the sampling distribution of means for this population; each histogram plots the mean from multiple samples. The true, theoretical sampling distribution includes the mean of every possible sample of 200 cases that can be selected from the population of 15,000 people.

sampling distribution The distribution of a statistic produced by every possible unique sample of the same size that can theoretically be selected from a population.

The Central Limit Theorem

Notice that the distribution in the right panel of Figure 6.2 has a familiar shape: the histogram of sample means looks like it is close to a normal distribution. In fact, as more and more samples are selected from the same population, and the averages are graphed, the distribution becomes closer and closer to a normal distribution. In theory, every sampling distribution is approximately normally distributed. And, crucially, that normal distribution is centred on the population parameter; in this example, the normal distribution is centred on the population mean (60 minutes). This phenomenon is the result of a statistical principle called the **central limit theorem**, which states that, given a large enough number of random samples of the same size selected from a population, the distribution of a statistic calculated from each of those samples (such as the mean) will approximate a normal distribution; the theorem also states that that distribution will be centred on the population parameter. The central limit theorem applies when the sample size is 100 cases or more; the sampling

central limit theorem A theorem that states that the distribution of a statistic from a large number of same-sized random samples from a population will be approximately normal, and centre on the population parameter.

distribution of smaller samples is a slightly different shape. One way to think about sampling distributions is as the result of an observed probability test with many, many trials; each new sample that is selected is equivalent to another toss of the coin or roll of the dice, and, as the number of trials increases, the observed probability converges with the theoretical probability. Because of the law of large numbers, with a sufficiently large number of samples the observed centre of the sampling distribution converges with the population parameter.

An important feature of the central limit theorem is that it does not require any assumptions about the shape of a distribution in the population. In the simulated population of 15,000 people, the distribution of time spent doing domestic labour might be left-skewed or right-skewed, or platykurtic or leptokurtic. Regardless, if I select every possible sample of 200 cases from that population and graph the mean of each of those samples, the histogram will approximate a normal curve and be centred on the population mean.

The Standard Error of a Mean

Let's review what has been established so far. First, we know that if multiple samples are randomly selected from a population and the mean of each of the samples is graphed in a histogram, the result is approximately a normal distribution. Second, we know that the centre of the normal distribution of sample means is located at the population mean. But we also know more because we know some things about the normal distribution (described in Chapter 4). As you might recall, the area under the normal curve shows that 68 per cent of cases are within one standard deviation of the mean, 95 per cent of cases are within two standard deviations of the mean, and 99 per cent of cases are within three standard deviations of the mean.

Since the sampling distribution of means is approximately normally distributed, then, roughly 68 per cent of samples have a mean within one standard deviation of the population mean; roughly 95 per cent of samples have a mean within two standard deviations of the population mean; and roughly 99 per cent of samples have a mean within three standard deviations of the population mean. Figure 6.3 shows how this applies to the distribution of the means of the samples from my simulation. The distribution of the means of the 500 samples selected from the population is approximately normally distributed, and it centres on 60 minutes (the population mean). And, by using the 500 means as values, I calculated that they have a standard deviation of two minutes. So, because of what we know about a normal distribution, we expect that 68 per cent of samples have a mean within one standard deviation of the population mean; that is, they have a mean between 58 and 62 minutes. Similarly, we expect that 95 per cent of samples have a mean within two standard deviations of the population mean; that is, they have a mean between 56 and 64 minutes. Most of the time, however, researchers select only one sample from a population and so they cannot use this method to calculate the standard deviation of the means in a sampling distribution.

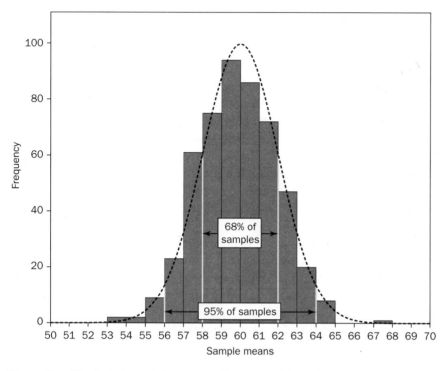

Figure 6.3 The Relationship between a Histogram of Sample Means and the Normal Distribution (Simulated Data)

In general, the standard deviation of a sampling distribution is called the standard error. And, crucially, the standard error can be estimated using *only information from the sample*. For the standard error (*se*) of a mean, the calculation incorporates the sample standard deviation (*s*) and the sample size (*n*). The subscript $\bar{x}$ notation shows that this is the formula for the standard error of a mean:

standard error (se) The standard deviation of a sampling distribution; it is estimated using only information from a sample.

$$se_{\bar{x}} = \frac{s}{\sqrt{n}}$$

standard error of a mean

It makes sense that the standard error is related to the sample standard deviation and the sample size. The sample standard deviation is an indicator of how much variation there might be in the population. If the sample standard deviation is low, there likely isn't much variation in the population that the sample was selected from; if the sample standard deviation is high, there is likely quite a lot of variation in the population that the sample was selected from. Imagine I randomly select a sample of 200 people from a city, and each person reports doing exactly 59 or 60 minutes of domestic labour the previous day. In this highly unlikely situation, the sample standard deviation would be close to 0. As a result, I would be fairly confident that most people in the city spent about the same amount of time doing domestic labour, and thus I would be more confident about using the sample to estimate the population parameter. Now, imagine I randomly select a sample of 200 people from a city and people report doing widely varying amounts of domestic labour the previous day, ranging from 1 minute to 10 hours (600 minutes). The sample standard deviation would be high, and I wouldn't be as confident making assertions about the population parameter using the sample. It's even easier to understand why having a larger sample size makes researchers more confident in their estimates. The larger the sample size, the more of the population it includes. As the sample size increases, it becomes a better and better representation of the population because more cases from the population are included in the sample.

Let's return to the first sample of 200 cases that I selected from the simulated population. In a typical research project, it would be the only sample that was selected. The sample mean is 57 minutes and the standard deviation is 28.5 minutes. Only the standard deviation and the number of cases are used to estimate the standard error of a mean. This calculation shows that the standard error of the mean is estimated to be 2.0—the same as the actual standard deviation of the sample means in Figure 6.3 (although it won't always be an exact match):

$$se_{\bar{x}} = \frac{s}{\sqrt{n}}$$

$$se_{\bar{x}} = \frac{28.5}{\sqrt{200}}$$

$$= \frac{28.5}{14.1}$$

$$= 2.0$$

There are other formulas used to estimate the standard error of other statistics, but they all rely on the same general logic. Frankly, though, the standard error isn't that useful on its own. It provides a general indication of how well a sample statistic matches with a population parameter: statistics with smaller standard errors are considered to be more precise estimates than those with larger standard errors. But the main use of the standard error is to calculate confidence intervals.

Confidence Intervals

Let's review again. When social science researchers conduct a study, they usually randomly select a single sample from the population. They want to find a population parameter—such as the average amount of time that people spend doing domestic labour—but they don't have information from every person in the

Step-by-Step: Standard Error of a Mean

$$se_{\bar{x}} = \frac{s}{\sqrt{n}}$$

Step 1: Count the total number of cases (n). Do not include cases that are missing information for the variable.

Step 2: Find the standard deviation of the variable (s).

Step 3: Find the square root of the total number of cases (from Step 1) to find the denominator of the standard error of a mean equation.

Step 4: Divide the standard deviation (from Step 2) by the result of Step 3 to find the standard error of the mean of the variable.

population. They know, however, that if they select all of the possible random samples of the same size from a population, the sampling distribution of a statistic is approximately normal, centred on the population parameter, and its standard deviation is equivalent to the standard error. Because the sampling distribution is approximately normally distributed, the researchers also expect that 68 per cent of the samples selected from a population will produce a statistic within one standard error of the population parameter and that 95 per cent of the samples selected from a population will produce a statistic within two standard errors of the population parameter. As well, the researchers can estimate the size of the standard error using information from the single sample that they selected.

In Chapter 5, you learned that the normal distribution can also be thought of as a probability distribution. A sampling distribution, which is normally distributed, shows all of the samples that can possibly be selected from a population; researchers randomly select one of those samples. Because the sampling distribution is normally distributed, researchers are more likely to select a sample that produces a statistic that is close to the centre of the distribution (which is near the population parameter). They are less likely to select a sample that produces a statistic that is located in the tails of the distribution (which is far from the population parameter). Indeed, a statistic produced by any one, randomly selected sample has a 0.68 probability (or a 68 per cent chance) of being within one standard error of the population parameter. Similarly, a statistic produced by any one, randomly selected sample has a 0.95 probability (or a 95 per cent chance) of being within two standard errors of the population parameter.

Before going further, I need to make a minor correction. For ease, I have said that 95 per cent of the area under the normal curve is between −2 and +2 standard deviations from the mean. In other words, 95 per cent of samples selected from the population will produce a statistic that is between −2 and +2 standard errors from the mean. To be more precise, though, 95 per cent of the area under the normal curve is actually between −1.96 and +1.96 standard deviations from the mean. Similarly, 95 per cent of samples selected from the population will produce a statistic that is between −1.96 and +1.96 standard errors from the mean. In much the same way, I have also said that 99 per cent of the area under the normal curve is between −3 and +3 standard deviations from the mean; to be more precise, 99 per cent of the area under the normal curve is actually between −2.58 and +2.58 standard deviations from the mean. This precision is important going forward.

So, more precisely, 95 per cent of the samples selected from a population will produce a statistic within −1.96 and +1.96 standard errors of the actual population parameter. Conversely, a statistic produced by any one, randomly selected sample has a 95 per cent chance of being within −1.96 and +1.96 standard errors of the actual population parameter. This range is referred to as a **95 per cent confidence interval (95% CI)**. A 95 per cent confidence interval shows the range that the population parameter is estimated to be within, based on information from a sample selected from that population. There's a 95 per cent chance that the population parameter will actually be within this range. To illustrate, let's return to the simulation example. The first sample selected from the population has a mean of 57 minutes, a standard deviation of 28.5 minutes, and a standard error of 2 minutes. The lower bound of the 95 per cent confidence interval is calculated using the formula:

95 per cent confidence interval (95% CI) Shows the range that a population parameter is estimated to be within, based on information from a sample selected from the population; there is a 95 per cent chance that this is correct.

$$Lower\ Bound\ of\ the\ 95\%\ CI_{\bar{X}} = \bar{x} - 1.96\left(se_{\bar{x}}\right)$$

$$Lower\ Bound\ of\ the\ 95\%\ CI_{\bar{X}} = 57 - 1.96(2.0)$$
$$= 57 - 3.92$$
$$= 53.08$$

The upper bound of the 95 per cent confidence interval is calculated using almost the same formula, except that 1.96 multiplied by the standard error is added to the mean instead of subtracted from it:

$$Upper\ Bound\ of\ the\ 95\%\ CI_{\bar{X}} = \bar{x} + 1.96\left(se_{\bar{x}}\right)$$

$$Upper\ Bound\ of\ the\ 95\%\ CI_{\bar{X}} = 57 + 1.96(2.0)$$
$$= 57 + 3.92$$
$$= 60.92$$

Thus, the 95 per cent confidence interval for the mean of the first sample in the simulation example ranges from 53.08 to 60.92. We are 95 per cent confident that the average amount of time spent doing daily domestic labour in the population

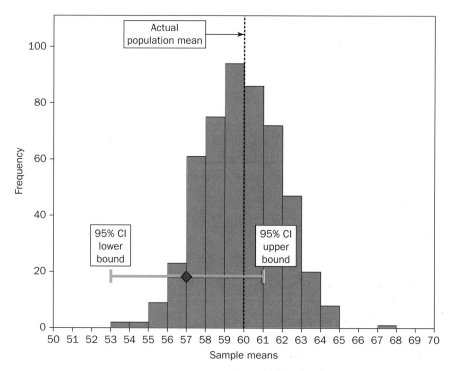

Figure 6.4 How the 95 Per Cent Confidence Interval for the Mean of a Single Sample Compares to an Approximation of the Sampling Distribution (Simulated Data)

is between 53.08 and 60.92 minutes. Because this is a simulation, I can illustrate the relationship between this 95 per cent confidence interval, the approximation of the sampling distribution, and the population parameter. Figure 6.4 shows the distribution of the average amount of time spent doing daily domestic labour in 500 samples selected from the simulated population. The brown diamond shows the average amount of time spent doing domestic labour in the first sample that was selected. The green whiskers to the left and the right of the diamond show the span of the 95 per cent confidence interval. We are 95 per cent confident that the population average is within this range; here, you can see that this is indeed correct, since the actual population mean of 60 is between the left and right whiskers.

The sample mean will always be located exactly in the middle of the 95 per cent confidence interval because of how it is calculated. The same amount (1.96 multiplied by the standard error of the mean) is both added to and subtracted from the mean to obtain the confidence interval. In fact, often the formulas for the upper and lower bounds of the 95 per cent confidence interval are condensed into a single formula, using a combined "+/−" sign instead of the "−" sign for the lower bound and the "+" sign for the upper bound. The combined "+/−" sign indicates that the calculation needs to be done twice: once using the subtraction sign to find the lower bound, and once using the addition sign to find the upper bound:

95% confidence
interval for a mean

$$95\% \ CI_{\bar{x}} = \bar{x} \pm 1.96\left(se_{\bar{x}}\right)$$

Now let's find the 95 per cent confidence interval for the mean of the second sample selected from the simulated population. In that sample of 200 cases, the mean amount of time spent doing domestic labour is 67 minutes, with a standard deviation of 30 minutes. The standard error of the mean is estimated as:

$$se_{\bar{x}} = \frac{s}{\sqrt{n}}$$

$$se_{\bar{x}} = \frac{30}{\sqrt{200}}$$

$$= \frac{30}{14.1}$$

$$= 2.1$$

The standard error is used to find the 95 per cent confidence interval for the mean of the second sample:

$$95\% \ CI_{\bar{x}} = \bar{x} \pm 1.96\left(se_{\bar{x}}\right)$$

$$95\% \ CI_{\bar{x}} = 67 \pm 1.96(2.1)$$

$$= 67 \pm 4.1$$

$$= 62.9 \ and \ 71.1$$

So, the 95 per cent confidence interval for the mean of the second sample ranges from 62.9 to 71.1. Figure 6.5 shows how this confidence interval compares to the distribution of the means of the 500 samples selected from the simulated population, and to the population parameter.

In this instance, the actual population mean is not within the confidence interval. This is why it is referred to as a 95 per cent confidence interval, and why we are only 95 per cent confident that the population parameter is within the upper and lower bounds. Five per cent of the time, or for 5 per cent of samples selected from the population, the 95 per cent confidence interval will not include the population parameter. Figure 6.5 shows that the second sample selected was relatively unusual, though in an actual study there would be no way to know this. Thus, when researchers report 95 per cent confidence intervals, they are relatively confident that the population parameter is within that range, but they can't be absolutely sure.

Social science researchers most commonly report 95 per cent confidence intervals. But the same approach can be used to obtain other confidence intervals. For instance, since 99 per cent of the area under a normal curve is within −2.58 and +2.58 standard deviations from the mean, a 99 per cent confidence interval is

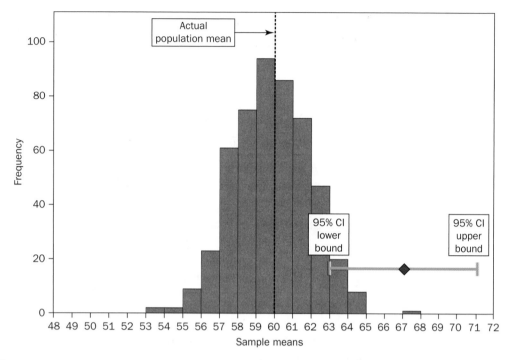

Figure 6.5 How the 95 Per Cent Confidence Interval for the Mean of Another Sample Compares to an Approximation of the Sampling Distribution (Simulated Data)

calculated by multiplying the standard error of a mean by 2.58, and then adding and subtracting the result from the mean to find the upper and lower bounds of the confidence interval. In fact, z-scores and the associated area under the normal curve can be used to calculate any confidence interval. For example, to calculate a 90 per cent confidence interval, find the z-score where 5 per cent of the area under the normal curve is to the right of it, and the z-score where 5 per cent of the area under the normal curve is to the left of it. Since the normal curve is symmetrical, these two z-scores will be the same number, only one will be positive and one will be negative. This captures the middle 90 per cent of a sampling distribution by identifying the point at which 5 per cent of the area is in *each* tail (and thus 10 per cent of the total area is in both tails). An "area under the normal curve" calculator indicates that 5 per cent of the area is to the right of +1.65 and 5 per cent of the area is to the left of −1.65. So, to find a 90 per cent confidence interval, the standard error of a mean is multiplied by 1.65 and then the result is added to and subtracted from the mean to find the upper and lower bounds.

Confidence Intervals in Action

Now that you understand sampling distributions, standard errors, and confidence intervals, let's investigate the amount of time that people in Canada spend doing domestic labour, using data from the 2010 General Social Survey

Step-by-Step: 95 Per cent Confidence Intervals for Means

$$95\% \ CI_{\bar{x}} = \bar{x} \pm 1.96\left(se_{\bar{x}}\right)$$

Step 1: Find the mean of the variable ($\bar{x}$).

Step 2: Find the standard error of the mean of the variable ($se_{\bar{x}}$).

Step 3: Multiply the standard error of the mean (from Step 2) by 1.96.

Step 4: Complete two calculations to find the upper and lower bound of the confidence interval.

> **Step 4.1:** Subtract the result of Step 3 from the mean (from Step 1) to find the lower bound of the confidence interval.
>
> **Step 4.2:** Add the result of Step 3 to the mean (from Step 1) to find the upper bound of the confidence interval.

on Time Use (GSS). (See the "Spotlight on Data" box for more information.) The GSS asks people to record exactly how many minutes they spent doing various activities on the previous day. This included time spent doing meal preparation and cleanup, baking, laundry, sewing/mending, interior and exterior cleaning of the home, interior and exterior home maintenance and repairs, home improvements, vehicle maintenance, gardening and yard/grounds maintenance, taking care of pets or houseplants, packing or unpacking groceries or luggage, household management or financial administration, and any other domestic activities. The amount of time spent doing all of these activities was combined to calculate the total number of minutes each person spent doing unpaid domestic labour on the previous day.

This information from the sample can be used to estimate the average amount of time spent doing daily domestic labour in the population. A first look at the distribution in the sample showed there were some outliers: people who reported spending 18 to 20 hours a day doing domestic labour. This seems as though it might be either an unusual situation or a reporting error. To better understand the distribution of time spent on domestic labour by most people in Canada, the 0.2 per cent of survey respondents who reported spending more than 12 hours a day doing domestic labour are omitted from this analysis. People who report doing no domestic labour the previous day are also excluded. So, for the cases used in this analysis, the minimum amount of time spent doing daily domestic labour is 1 minute, and the maximum is 720 minutes.

Table 6.2 shows statistics that describe the distribution of time spent doing daily domestic labour. The median is 105 minutes, or an hour and 45 minutes.

Table 6.2   Statistics Describing the Distribution of Time Spent Doing Daily Domestic Labour (Minutes)

Statistic	Value
Minimum	1
Maximum	720
Median	105.00
Mean	144.60
Std. Deviation	132.21
Std. Error of Mean	1.20
95% CI Lower Bound	142.24
95% CI Upper Bound	146.96
Skewness	1.43
Kurtosis	1.88
Number of Cases	*12,075*

Source: Author generated; Calculated using data from Statistics Canada, 2015.

Among the people in the sample who did domestic labour the previous day, half spent an hour and 45 minutes or less, and half spent an hour and 45 minutes or more. The average amount of time spent doing daily domestic labour in the sample is 145 minutes, or 2 hours and 25 minutes. The standard deviation of 132 minutes, which is relatively large compared to the mean, shows that there is quite a bit of variation in the amount of time that people spend doing daily domestic labour. The upper and lower bounds of the 95 per cent confidence interval show that the average amount of time spent doing daily domestic labour in the population is likely to be between 142 and 147 minutes. Although this variable has a large standard deviation, because the sample size is large, the 95 per cent confidence interval is still relatively small.

How Does It Look in SPSS?

Standard Errors and Confidence Intervals

The Explore procedure produces descriptive statistics that look like those in Image 6.1. Above the descriptives, SPSS prints a "Case Processing Summary" that shows the number and percent of valid cases, missing cases, and total cases. You can choose whether "Statistics," "Plots," or both are printed.

Continued

Descriptives

				Statistic	Std. Error
Total duration (in minutes) of domestic work activities (A)	(B)	Mean		144.60	(C) 1.203
	(D)	95% Confidence Interval for Mean	Lower Bound	142.24	
			Upper Bound	146.96	
	(E)	5% Trimmed Mean		131.43	
	(F)	Median		105.00	
	(G)	Variance		17479.934	
		Std. Deviation		132.212	
	(H)	Minimum		1	
		Maximum		720	
		Range		719	
	(I)	Interquartile Range		165	
	(J)	Skewness		1.434	.022
	(K)	Kurtosis		1.882	.045

Image 6.1   **Statistics from the SPSS Explore Procedure**

A.   The variable label appears on the far left.

B.   This row shows the mean or average in the sample in the "Statistic" column. On average, people in this sample spent 144.60 minutes doing domestic labour each day.

C.   This row shows the estimate of standard error of the mean in the "Std. Error" column. The estimated standard error of the mean is 1.20.

D.   These rows show the lower and upper bounds of the 95 per cent confidence interval for the mean in the population. We are 95 per cent confident that, in the population, the average amount of time spent doing domestic labour each day is between 142.24 and 146.96 minutes.

E.   This row shows what the mean in the sample would be if the 5 per cent of cases with the highest values and the 5 per cent of cases with the lowest values were excluded from the calculation. The trimmed mean is not usually reported, but if it is substantially different from the mean, the variable may have outliers that warrant investigation.

F.   This row shows the median in the sample. Half of people in this sample spend 105 minutes or less doing domestic labour each day, and half of people in this sample spend 105 minutes or more doing domestic labour each day.

G.   These rows show estimates of the variance and standard deviation in the population. The standard deviation of daily time spent doing domestic labour is 132.21 minutes, which is relatively large compared to the mean. There is likely a wide dispersion (or spread) of time spent doing domestic labour in the population.

H.   These rows show the lowest and highest values in the sample, and the difference between them. The lowest amount of time spent doing domestic labour reported in this sample is 1 minute, and the highest amount of time is 720 minutes; thus, the range is 719 minutes (720 − 1). I excluded outliers in the distribution.

I. The "Interquartile Range" row shows the distance between the twenty-fifth percentile and the seventy-fifth percentile in the sample; that is, it shows the range of the middle 50 per cent of cases. In this sample, the twenty-fifth percentile is 45 minutes and the seventy-fifth percentile is 210 minutes, and so the interquartile range is 165 minutes (210 − 45). (Percentiles are not produced by this procedure.)

J. The "Statistic" column of this row shows the skew of the distribution in the sample. The skew statistic of 1.434 shows that the distribution is highly skewed to the right. The "Std. Error" column of this row shows the standard error of the skew statistic, which is calculated using the sample size. The standard error can't be used to calculate confidence intervals for a skew statistic. But a common strategy is to divide the skew statistic by its standard error: $1.434 \div 0.022 = 65.182$. If the result is greater than 1.96, then it is likely that the distribution is skewed in the population; since 65.182 is higher than 1.96, it is likely that the distribution of time spent doing domestic labour is skewed in the population.

K. The "Statistic" column of this row shows the excess kurtosis of the distribution in the sample. The kurtosis statistic of 1.882 shows that the distribution is highly leptokurtic. The "Std. Error" column of this row shows the standard error of the kurtosis statistic, which is calculated using the sample size and the standard error of the skew. The standard error can't be used to calculate confidence intervals for a kurtosis statistic. But a common strategy is to divide the kurtosis statistic by its standard error: $1.882 \div 0.045 = 41.822$. If the result is greater than 1.96, then it is likely that the distribution is kurtotic in the population; since 41.882 is higher than 1.96, it is likely that the distribution of time spent doing domestic labour is kurtotic in the population.

Using Confidence Intervals to Compare Group Means

Confidence intervals can be used to assess whether different groups are likely to have different averages in the population. For instance, do men and women spend different amounts of time doing domestic labour each day, on average? To make comparisons between groups using 95 per cent confidence intervals, calculate the range of the confidence interval for the mean of each group separately. To compare men and women, calculate the mean amount of time spent doing daily domestic labour and the 95 per cent confidence interval for women alone, and then the mean amount of time spent doing daily domestic labour and the 95 per cent confidence interval for men alone. If the ranges of the two confidence intervals overlap, then the groups could have exactly the same mean in the population. If the ranges of the two confidence intervals do not overlap, then the groups are likely to have different means in the population.

   Table 6.3 shows statistics describing the distribution of time spent doing daily domestic labour divided by gender. The results show that the average amount of time

Spotlight on Data

The General Social Survey on Time-Stress and Well-Being

Each year since 1985, Statistics Canada has conducted a General Social Survey (GSS). The objective of the GSS is to collect information on social trends, to monitor changes in Canadians' living conditions and well-being, and to provide information about social policy issues (Statistics Canada 2011). Each year, Canada's GSS focuses on collecting in-depth information about a single topic. The GSS cycles through a series of main topics, typically collecting information about each of them once every five years. GSS topics include families, caregiving and receiving, social identity, victimization, giving and volunteering, and time use.

This chapter relies on data from the 2010 General Social Survey on Time-Stress and Well-Being. A key feature of this survey is the use of a time diary that asks respondents to track their activities on the previous day and to record the amount of time spent doing each activity. The survey allows respondents to record multiple activities during the same time period in order to understand multi-tasking. Because survey respondents are randomly selected to be called each day, the "previous day" that people report their time use for is effectively a random selection and, thus, can be generalized to represent the daily activities of people living in Canada.

The 2010 GSS collected information from residents of Canada aged 15 and older, excluding people living in the three territories and people living in institutions (Statistics Canada 2011). Data were collected using computer-assisted telephone interviewing (CATI). The sample was selected in two stages. The first stage used stratified random sampling: the country was divided into 27 strata, in order to ensure sufficient representation from each province and from both urban and rural areas. A list of land-line telephone numbers was obtained for all provinces and then assigned to each of their geographic strata. Next, a simple random sample of telephone numbers was selected from each of the strata. In the second stage, one person over the age of 15 was randomly selected from each household to complete the survey. People with cellular phone service only were excluded from the sample because interviews were not conducted on cellular phones.

The overall response rate for the 2010 GSS was 55 per cent. The data are weighted to compensate for the probability of selection, to adjust the population size to the size of each stratum, and to ensure that the sample represents the known age and sex distribution for each province.

Table 6.3   Statistics Describing the Distribution of Daily Domestic Labour, by Sex/Gender (Minutes)

Statistic	Men	Women	Overall
Minimum	2	1	1
Maximum	720	705	720
Median	80.00	120.00	105.00
Mean	124.73	160.58	144.60
Std. Deviation	125.43	135.33	132.21
Std. Error of Mean	1.71	1.65	1.20
95% CI Lower Bound	121.38	157.34	142.24
95% CI Upper Bound	128.08	163.82	146.96
Skewness	1.70	1.27	1.43
Kurtosis	2.96	1.36	1.88
Number of Cases	*5,383*	*6,692*	*12,075*

Source: Author generated; Calculated using data from Statistics Canada, 2015.

that men spend doing daily domestic labour, in the population, is likely between 121 and 128 minutes (slightly more than two hours). The average amount of time that women spend doing daily domestic labour, in the population, is likely between 157 and 164 minutes (about two-and-a-half hours). Because these confidence intervals do not overlap with one another, it's likely that, on average, women in Canada spend more time doing daily domestic labour than men. Using the upper bound of the confidence interval for men and the lower bound of the confidence interval for women, a conservative estimate is that women do an average of about 30 minutes more domestic labour each day than men; this translates into an extra three-and-a-half hours of domestic labour each week, or about 15 hours each month.

In this example, the width of the confidence interval is roughly the same for each group (about seven minutes). But sometimes groups will have confidence intervals of different widths. Recall that the confidence interval depends on the standard error, which is estimated using the sample standard deviation and the sample size. When one group has a wider confidence interval than other groups, either that group has a substantially larger standard deviation (in other words, there is more variation within the group), or a substantially smaller sample size than the others, or both. Table 6.3 shows that the standard deviations for men and women are roughly the same size, although there is slightly more variation among women. It also shows that there are slightly more women included in the analysis (because people who do not do any domestic labour are excluded, and these are more likely to be men).

Error-Bar Graphs

Error-bar graphs are used to display confidence intervals for different groups in a population. Error-bar graphs make it easy to assess whether confidence intervals overlap, especially when there are many groups. Since we are relatively confident

error-bar graph A graph that shows a sample statistic and the associated 95 per cent confidence interval for one or more groups.

How Does It Look in SPSS?

Standard Errors and Confidence Intervals for Groups

The Explore procedure can generate statistics for groups, which are defined by a second variable (typically a categorical variable). The descriptive statistics that are produced look like those in Image 6.2. Above the descriptives, SPSS prints a "Case Processing Summary" that shows the number and percent of valid cases, missing cases, and total cases for each group individually. You can choose whether "Statistics," "Plots," or both are printed for each group.

Descriptives

Sex of respondent.				Statistic	Std. Error
Total duration (in minutes) of domestic work activities (A)	(C) Male	Mean		124.73	1.710
		95% Confidence Interval for Mean	Lower Bound	121.38	
			Upper Bound	128.08	
		5% Trimmed Mean		110.62	
		Median		80.00	
		Variance		15733.938	
	(D)	Std. Deviation		125.435	
		Minimum		2	
		Maximum		720	
		Range		718	
		Interquartile Range		144	
		Skewness		1.701	.033
		Kurtosis		2.959	.067
	Female	Mean		160.58	1.654
		95% Confidence Interval for Mean	Lower Bound	157.34	
			Upper Bound	163.82	
		5% Trimmed Mean		148.34	
		Median		120.00	
		Variance		18314.096	
	(E)	Std. Deviation		135.330	
		Minimum		1	
		Maximum		705	
		Range		704	
		Interquartile Range		165	
		Skewness		1.274	.030
		Kurtosis		1.355	.060

Image 6.2   **Statistics from the SPSS Explore Procedure, Divided by Group**

A.   The variable label appears on the far left.
B.   The label at the top of the column shows the variable used to define the groups.
C.   The attributes of the grouping variable are listed in a vertical column. Only valid attributes are shown. Attributes designated as missing are not printed.

D. These rows show descriptive statistics for the cases in the first group. See the "How Does It Look in SPSS?" box earlier in this chapter for a description of each row. The sample statistics are for only men in this sample, and the population estimates are for only men in the population.

E. These rows show descriptive statistics for the cases in the second group (the same sequence of rows is printed for every group). See the "How Does It Look in SPSS?" box earlier in this chapter for a description of each row. The sample statistics are for only women in this sample, and the population estimates are for only women in the population. The results can be compared between groups.

that, on average, women spend more time doing daily domestic labour than men, let's investigate whether working for pay makes any difference. Figure 6.6 compares the average time spent doing daily domestic labour for four groups: men who work for pay, women who work for pay, men who do not work for pay, and women who do not work for pay (with an adjusted scale on the y-axis). The brown dot in the centre of each error bar shows the mean of that group in the sample. The whiskers above and below the dot show the upper and lower bound of the 95 per cent confidence interval, that is, they show the range that the mean of each group is likely to be within in the

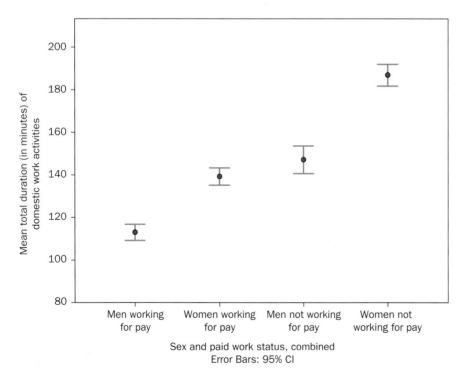

Figure 6.6   An Error-Bar Graph Showing 95 Per Cent Confidence Intervals for Four Groups

Source: Author generated; Calculated using data from Statistics Canada, 2015.

larger population (with 95 per cent confidence). If the whiskers for one group overlap with the whiskers for another group, the groups could have the same average in the population. If the whiskers for one group do not overlap with the whiskers for another group, it's likely that the two groups have different averages in the population.

Figure 6.6 shows that men who work for pay are likely to spend less time doing daily domestic labour, on average, than women who work for pay. Because the whiskers for the two groups do not overlap, the two group averages are probably different in the larger population. Similarly, men who do not work for pay likely spend less time doing daily domestic labour, on average, than women who do not work for pay. Notice that the whiskers for women who work for pay and men who do not work for pay overlap: the average amount of time spent doing daily domestic labour could be the same for these two groups in the larger population. These results starkly illustrate the effect of gender roles and support the notion that working women still do a second shift of domestic labour.

The usefulness of error-bar graphs becomes apparent as the number of groups increase. For instance, we might want to account for people's living arrangements. Maybe more women than men live alone, and so women are the only ones responsible for domestic labour in their households. The panelled error-bar graph in Figure 6.7 shows the results when the four groups from Figure 6.6 are divided into those who live

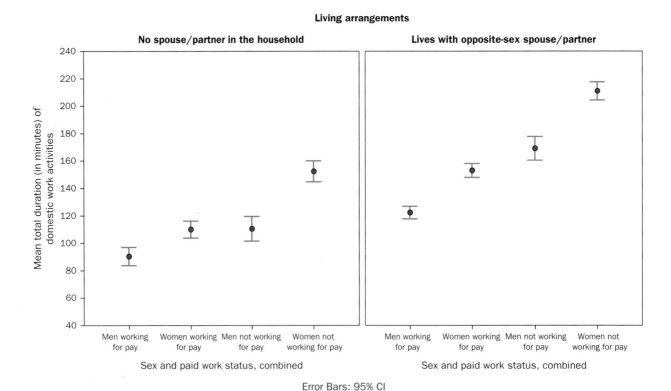

Figure 6.7   A Panelled Error-Bar Graph Showing 95 Per Cent Confidence Intervals for Eight Groups

Source: Author generated; Calculated using data from Statistics Canada, 2015.

with an opposite-sex spouse/partner, and those who do not (fewer than 0.5 per cent of respondents reported living with a same-sex spouse/partner). The results show a clear pattern. People who live with a partner are likely to spend more time doing daily domestic labour, on average, than those who do not, for each of the groups. The group with the highest average are women who do not work for pay and who live with an opposite-sex spouse/partner; we are 95 per cent confident that the mean of this group is between 204 to 218 minutes in the population (about three-and-a-half hours per day). The group with the lowest average are men who work for pay and who do not live with a spouse/partner; we are 95 per cent confident that the mean of this group is between 84 and 97 minutes a day in the population (about an hour-and-a-half per day).

At this point, you might be feeling a bit frustrated, especially if you are a woman who is working for pay or who is planning to work for pay in the future. It seems like not much has changed, despite decades of feminist activism. So before moving on, let's investigate whether the average amount of time that women and men spend doing daily domestic labour has changed over time. The first year that the GSS collected information about time use was in 1986. The error-bar graph in Figure 6.8 suggests that the difference between the average amount of time spent by men on daily domestic labour and the average amount of time spent by women

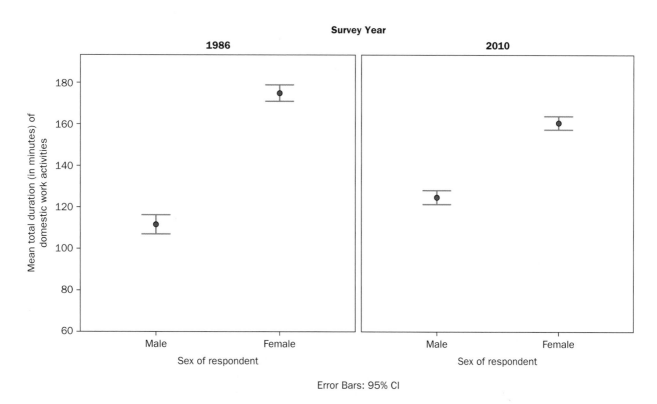

Figure 6.8 **A Panelled Error-Bar Graph Showing 95 Per Cent Confidence Intervals for Time Spent Doing Daily Domestic Labour, by Gender, in 1986 and 2010**

Source: Author generated; Calculated using data from Statistics Canada, 2015.

on daily domestic labour is shrinking, a little bit. Compared to 1986, in 2010 men spent slightly more time doing domestic labour, on average, and women spent slightly less time doing domestic labour, on average. The 95 per cent confidence intervals, shown using error bars, suggest this change is likely occurring in the population and is not just because of the samples that were randomly selected in each year.

Standard Errors and Confidence Intervals for Proportions

So far, I have described sampling distributions, standard errors, and confidence intervals for means (averages). But these ideas also extend to other statistics. In this section, I show how these concepts are applied to proportions or percentages. Recall that proportions show the fraction of people who do something out of a base of 1 and that percentages show the fraction of people who do something out of a base of 100. To translate from a proportion to a percentage, multiply by 100 or move the decimal point two places to the right; to translate from a percentage to a proportion, divide by 100 or move the decimal point two places to the left.

Let's investigate the sampling distribution of a proportion using the simulation example from the beginning of this chapter. For simplicity, I'll discuss a dichotomous variable, though the same principles apply to variables with more than two attributes. In the simulated population, everyone did at least one minute of domestic labour the previous day. Now let's consider how many people also did paid work, that is, they did both paid work and unpaid domestic work on the previous day. In the dataset representing the simulated population of 15,000 people, exactly 40 per cent of people did paid work the previous day.

In the first sample of 200 people I selected from the simulated population, 41.5 per cent did paid work the previous day (a proportion of 0.415). Once again, the percentage of people in the sample who did paid work is close to the population percentage, but not exactly the same. In the second sample of 200 people that I selected, 46 per cent did paid work the previous day (a proportion of 0.460). As you know, I continued to select 500 samples from the population. Then, I created another histogram—this one shows the percentage of people who did paid work in each sample. The top panel of Figure 6.9 shows the distribution of the percentages (or proportions) of people who did paid work in 500 samples selected from the same population. Because of the central limit theorem and the law of large numbers, the sampling distribution of proportions is approximately normally distributed and centred on the actual proportion in the population. And, because the sampling distribution is approximately normally distributed, roughly 95 per cent of samples will produce a proportion that is within 1.96 standard errors of the proportion in the population. Conversely, the proportion produced by any one randomly selected sample has a 95 per cent chance of being within 1.96 standard errors of the proportion in the population.

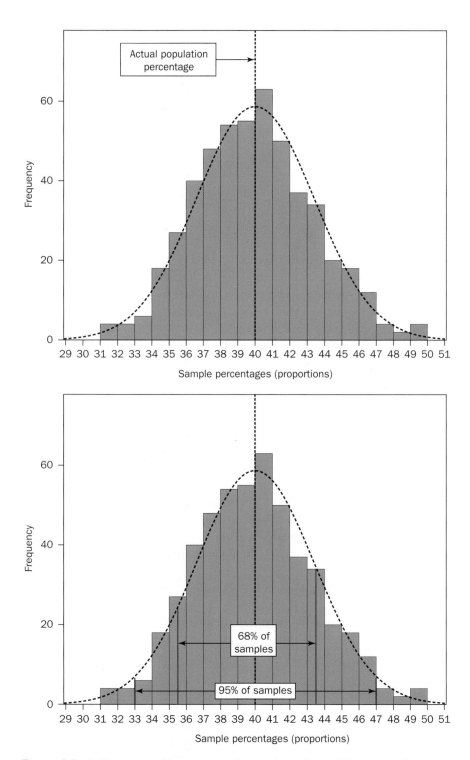

Figure 6.9   A Histogram of Percentages (Proportions) from 500 Samples from the Same Population, Compared to a Normal Distribution (Simulated Data)

The formula used to calculate the standard error of a proportion is slightly different than that for the standard error of a mean. Instead of relying on the standard deviation (the amount of variation), it accounts for how common the attribute is in the sample as well as the sample size. The formula for the standard error of a proportion is:

standard error of a proportion

$$se_p = \sqrt{\frac{p(1-p)}{n}}$$

The subscript p notation on the se indicates that this is the standard error of a proportion, instead of a mean. The p notation on the right-hand side of the equation refers to the proportion, expressed as a decimal (not the percentage). The sample size is accounted for by the n in the denominator. In general, the larger the denominator, the smaller the standard error will be.

The calculation in the numerator of the fraction accounts for how common or rare the attribute is in the sample and, thus, how common or rare it is likely to be in the population. The highest possible value of $p(1-p)$ occurs when half of the cases in a sample have an attribute: the proportion is 0.5, and $p(1-p)$ is equal to 0.25. For less common attributes, the value of $p(1-p)$ is smaller: if only 10 per cent of cases in a sample have an attribute, the proportion is 0.1, and $p(1-p)$ is equal to 0.09. For more common attributes, their opposite is rarer and, thus, the value of $p(1-p)$ is also smaller: if 90 per cent of cases in a sample have an attribute, the proportion is 0.9 and $p(1-p)$ is also equal to 0.09. When an attribute is very rare (or very common) the numerator is very small, which results in a smaller standard error. In practice, there are several alternative ways to estimate the standard error of a proportion. The method used here is conceptually equivalent to the standard error of the mean, but it performs poorly for small samples and for proportions lower than 0.1 or greater than 0.9.

In the first sample of 200 cases I selected from the simulated population, the percentage of people who did paid work was 41.5 per cent, or a proportion of 0.415. Thus, the standard error of the proportion is estimated to be 0.032:

$$se_p = \sqrt{\frac{p(1-p)}{n}}$$

$$se_p = \sqrt{\frac{0.415(1-0.415)}{200}}$$

$$= \sqrt{\frac{0.415(0.585)}{200}}$$

$$= \sqrt{\frac{0.243}{200}}$$

$$= \sqrt{0.001}$$

$$= 0.032 \ or \ 3.2\%$$

The standard error of a proportion is used primarily to calculate 95 per cent confidence intervals, in exactly the same way as for a mean:

$$95\% \ CI_p = \ p \pm 1.96 \left(se_p \right)$$

95% confidence interval for a proportion

$$95\% \ CI_p = 0.415 \pm 1.96 \left(0.032 \right)$$

$$= 0.415 \pm 0.063$$

$$= 0.352 \ and \ 0.478 \left[35.2\% \ and \ 47.8\% \right]$$

The 95 per cent confidence interval shows that the proportion of people who did paid work in the population is likely to be between 35.2 per cent and 47.8 per cent. Because this is a simulation, we know that this is indeed correct: 40 per cent of people in the simulated population did paid work. Like all confidence intervals, though, researchers can never be absolutely sure that the population proportion will be within the interval they calculate using sample data; in 5 per cent of samples the confidence interval will not include the population proportion.

Let's go back to investigating domestic labour in Canada, and estimate the proportion of people who did any domestic labour on the previous day. Overall, about four out of every five people (78.7 per cent) reported doing domestic labour the previous day (see Table 6.4).

Now, let's add a gender dimension: Table 6.4 shows the confidence intervals for the proportion of men and the proportion of women who did domestic labour on the previous day. The results show that, in the population, the percentage of men who did domestic labour is likely between 70.0 per cent and 72.1 per cent, and the percentage of women who did domestic labour is likely between 85.3 per cent and 86.9 per cent. Since these two confidence intervals do not overlap, it is likely that a higher proportion of women than men do domestic labour each day.

Error bars can be added to bar graphs to compare the 95 per cent confidence intervals for multiple groups. Figure 6.10 shows the proportion of people who

Table 6.4   Percentage of People Who Do Daily Domestic Labour, by Sex/Gender

Statistic	Men	Women	Overall
Percentage	71.1	86.1	78.7
Std. Error of Percentage	0.5	0.4	0.3
95% CI Lower Bound	70.0	85.3	78.1
95% CI Upper Bound	72.1	86.9	79.3
Number of Cases	*7,695*	*7,795*	*15,390*

Source: Author generated; Calculated using data from Statistics Canada, 2015.

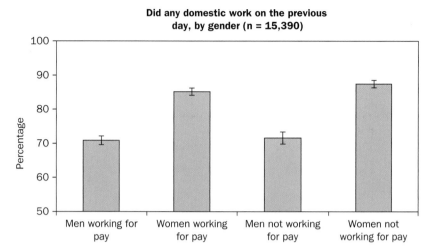

Did any domestic work on the previous day, by gender (n = 15,390)

Figure 6.10   Bar Graph with Error Bars Showing 95 Per Cent Confidence Intervals
Source: Author generated; Calculated using data from Statistics Canada, 2015.

did domestic labour the previous day, depending on their gender and whether they work for pay. The height of each bar shows the proportion of people who did domestic labour in the sample, and the error bars show the range that the proportion is likely to be within in the larger population. (The y-axis is adjusted to make the differences easier to see.) The results suggest that women are more likely to do daily domestic labour than men. For men, working for pay does not seem to be related to doing domestic labour. The overlapping error bars for men who are working for pay and men who are not working for pay show that, in the population, the proportion of men from each group who did daily domestic labour could be the same. But for women, working for pay does appear to be related to doing domestic labour. The proportion of women working for pay who did domestic labour the previous day is likely smaller than the proportion of women who do not work for pay. (The error bars for the two groups are close to one another, but do not overlap.)

Step-by-Step: Standard Error of a Proportion

$$se_p = \sqrt{\frac{p(1-p)}{n}}$$

Step 1: Count the total number of cases (n). Do not include cases that are missing information for the variable.

Step 2: Find the proportion of cases with the attribute you are interested in (p), expressed as a decimal (out of 1).

Step 3:　Subtract the proportion (from Step 2) from 1.

Step 4:　Multiply the proportion (from Step 2) by the result of Step 3 to find the numerator of the fraction in the standard error of a proportion equation.

Step 5:　Divide the result of Step 4 by the total number of cases (from Step 1).

Step 6:　Find the square root of the result of Step 5 to find the standard error of the proportion.

Step-by-Step: 95 Per Cent Confidence Intervals for Proportions

$$95\%\ CI_p = p \pm 1.96\left(se_p\right)$$

Step 1:　Find the proportion of cases with the attribute you are interested in (p), expressed as a decimal (out of 1).

Step 2:　Find the standard error of the proportion (se_p), expressed as a decimal (out of 1).

Step 3:　Multiply the standard error of the proportion (from Step 2) by 1.96.

Step 4:　Complete two calculations to find the upper and lower bound of the confidence interval.

　　Step 4.1:　Subtract the result of Step 3 from the proportion (from Step 1) to find the lower bound of the confidence interval.

　　Step 4.2:　Add the results of Step 3 to the proportion (from Step 1) to find the upper bound of the confidence interval.

Hands-on Data Analysis

Finding the Confidence Interval for a Proportion Using a Dichotomous Variable

Statistical software programs do not consistently produce confidence intervals for proportions. In part, this is because there is more than one way to calculate the confidence interval for a proportion. But the procedure used to generate the confidence interval for a mean will produce a confidence interval for a proportion that is consistent with the approach used in this text.

Continued

In order to use the Means procedure to produce the confidence interval for a proportion, an attribute in a categorical variable must be recoded into a new, dichotomous variable. People who have the attribute are assigned the value "1", and people who do not have the attribute (but still have a valid value on the categorical variable) are assigned the value "0". For example, Table 6.5 shows a frequency distribution for "Did you do any domestic work the previous day," where people who did not do any domestic labour the previous day are assigned the value "0" and people who did do domestic labour the previous day are assigned the value "1".

Table 6.5   Frequency Distribution for Doing Domestic Labour, Recoded as a Dichotomous Variable

Did you do any domestic work the previous day?

Answer	Frequency	Percentage
No (0)	3,280	21.3
Yes (1)	12,110	78.7
Total	*15,390*	*100.0*

Source: Author generated; Calculated using data from Statistics Canada, 2015.

Once the attribute is recoded into a dichotomous variable, find the mean, the standard error of the mean, and 95 per cent confidence interval for the mean of that dichotomous variable. Table 6.6 shows these statistics for the dichotomous variable shown in Table 6.5. Notice that the mean is the same as the proportion of people who did domestic labour and that the standard error and confidence interval correspond with the results in the final column of Table 6.4.

To find the 95 per cent confidence interval for the proportion of people in the "No" group, create a second, new dichotomous variable, where people who did not do any domestic labour the previous day are assigned the value "1" and people who did do domestic labour the previous day are assigned the value "0". Then, find the mean, the standard error of the mean, and the 95 per cent confidence interval for the mean of that second dichotomous variable. If a categorical variable has more than two attributes, you must create a new, dichotomous variable for each attribute to find the 95 per cent confidence interval for that proportion. Nonetheless, this approach makes it easy to find the standard error and 95 per cent confidence interval for a proportion without any manual calculations.

Table 6.6   The Mean, Standard Error of the Mean, and 95 Per Cent Confidence Interval for the Mean of the Dichotomous Variable Shown in Table 6.5

Statistics	Value
Mean	0.787
Std. Error of Mean	0.003
95% CI Lower Bound	0.781
95% CI Upper Bound	0.793

Source: Author generated; Calculated using data from Statistics Canada, 2015.

The Margin of Error

Most academic researchers who report statistics provide the standard error or the confidence interval for each estimate. But in news media and polling reports, this information is often presented slightly differently. Typically, reports intended for the general public contain a statement like this: "The margin of error for a sample of this size is +/−3 percentage points, 19 times out of 20." This statement is just another way to describe a 95 per cent confidence interval. The **margin of error** (+/−3 percentage points in this example) is equivalent to the standard error multiplied by 1.96. In other words, it's the amount that is both added to and subtracted from a sample statistic to obtain the upper and lower bounds of the confidence interval. The qualifying phrase "19 times out of 20" is equivalent to 95 per cent. (If you score 19/20 on a test, you score 95 per cent.)

> **margin of error** Specifies the distance above and below a statistic that a population parameter is likely to be within.

Since the margin of error is just a less technical way to report a 95 per cent confidence interval, it is interpreted in exactly the same way. For instance, one poll shows that 30 per cent of people only do laundry when they've run out of underwear (CBC News 2008), and the poll has a margin of error of 3.1 percentage points, 19 times out of 20. Thus, we are 95 per cent confident that, in the population, the percentage of people who wait to do laundry until they have run out of underwear is between 26.9 per cent and 33.1 per cent (30 − 3.1 = 26.9; 30 + 3.1 = 33.1).

Some polling firms report **credibility intervals** instead of confidence intervals. Credibility intervals are the equivalent of confidence intervals but rely on a Bayesian approach to statistics as opposed to a frequentist approach. (Only the frequentist approach is covered in this book.) Credibility intervals are typically reported when polling firms use a sample that is not randomly selected from the population, as in panel surveys. Much like margins of error, though, credibility intervals measure the degree of uncertainty in the estimates produced by a sample.

> **credibility interval** The equivalent of a confidence interval or a margin of error for researchers using a Bayesian statistical approach.

Best Practices in Presenting Results

Writing about Confidence Intervals

It can be tricky to write about confidence intervals. Technically, in 95 per cent of samples (of the same size) selected from a population, confidence intervals show the range that a population parameter will be within. One key aspect of this explanation is that the confidence interval is for 95 per cent of *samples,* not 95 per cent of *people* within a single sample. Earlier in this chapter, we found that the average amount of time that people spend doing daily domestic labour is 145 minutes, with a 95 per cent confidence interval ranging from 142 to 147 minutes. Here are some examples of incorrect and correct ways to report this information:

> *Incorrect:* The confidence interval shows that 95 per cent of people in the population spend between 142 and 147 minutes doing domestic labour each day.

Incorrect: People spend an average of 142 to 147 minutes doing daily domestic labour, 95 per cent of the time.

Correct: We are 95 per cent confident that in the population, the average amount of time people spend doing domestic labour each day is between 142 and 147 minutes.

Correct: We estimate that people spend an average of 145 minutes doing domestic labour each day (95% CI: 142–147).

The first incorrect example confuses 95 per cent of samples with 95 per cent of people in the population. Although the wording of the second incorrect example—which says that the statistic occurs 95 per cent of the time—is commonly used, it is especially confusing in the context of an example that has to do with time. This wording could imply that, on average, each person spends 142 to 147 minutes doing daily domestic labour for 95 per cent of the year, and then goes on vacation for the remaining 5 per cent of the year. In contrast, the correct examples express the level of confidence in the population estimate without making either of these errors.

What You Have Learned

This chapter introduced several key concepts in inferential statistics: sampling distributions, the central limit theorem, standard errors, and confidence intervals. Each concept was first illustrated using a simulated example, with known population parameters, before using it to analyze the amount of time people spend doing domestic labour. You learned how to calculate the standard error and a 95 per cent confidence interval for both means and proportions. The final section of the chapter described some strategies for accurately writing about confidence intervals.

The research focus of this chapter was the gendered division of domestic labour. The results suggest that women are still more likely to do domestic labour than men and that, on average, women spend more time than men doing domestic labour each day. There is some evidence, however, that this time gap between women and men may be shrinking: men appear to be spending more time doing daily domestic labour, on average, than they did in 1986. Women who work for pay seem to spend less time doing domestic labour, on average, than women who do not work for pay but still more than men who work for pay. These findings support the idea that many women continue to do a double day or second shift of work. People who live with an opposite-sex spouse or common-law partner also seem to spend more time doing domestic labour, on average, than their non-partnered peers. Further analyses could assess how the presence of children in a household influences the amount of time spent doing domestic labour for both men and women. Taken together, however, these results suggest that despite women's entrance into the paid labour force and the gains of the feminist movement, gender inequalities are still firmly entrenched in the domestic sphere.

Check Your Understanding

Check to see if you understand the key concepts in this chapter by answering the following questions:

1. What is a sampling distribution?
2. What is a standard error? How is it related to a standard deviation and to a sampling distribution?
3. How do the properties of a normal distribution help in estimating population parameters?
4. What does a 95 per cent confidence interval show?
5. What do error-bar graphs show, and how are they typically used?
6. How does calculating the standard error and confidence intervals differ between means and proportions?
7. What is a margin of error, and how is it related to a confidence interval?

Practice What You Have Learned

Check to see if you can apply the key concepts in this chapter by answering the following questions. Keep two decimal places in any calculations, unless otherwise specified.

1. A team of researchers is investigating post-secondary students' time use. They survey a random sample of 785 students enrolled in Canadian post-secondary institutions. On average, students report doing 300 minutes of domestic labour per week (s.d. = 93 minutes).

    a. Calculate the standard error of the mean of time spent doing domestic labour.
    b. Calculate the 95 per cent confidence interval for the mean of time spent doing domestic labour.
    c. Explain what the 95 per cent confidence interval for the mean shows. In the population of post-secondary students in Canada, what is the average amount of time spent doing domestic labour likely to be?

2. Using the information from question 1:

    a. Calculate the 99 per cent confidence interval for the mean of time spent doing domestic labour.
    b. Determine whether the 99 per cent confidence interval for the mean is wider or smaller than the 95 per cent confidence interval for the mean you calculated in question 1(b).
    c. In your own words, explain the reason for the result of question 2(b).

3. Many post-secondary students who move away from home to attend school find it challenging to manage domestic responsibilities. In the random sample described in question 1, among the 400 students who live with parents or relatives, the average amount of time spent doing domestic labour each week is 262 minutes (s.d. = 97 minutes). Among the 385 students who live independently, the average amount of time spent doing domestic labour each week is 340 minutes (s.d. = 83 minutes).

    a. Calculate the standard error of the mean of time spent doing domestic labour for each of the two groups.
    b. Calculate the 95 per cent confidence interval for the mean of time spent doing domestic labour for each of the two groups.
    c. Compare the 95 per cent confidence intervals for the means. In the population of post-secondary students in Canada, is there likely a difference in the average amount of time spent doing domestic labour between the two groups? Explain how you determined your answer.

4. Either by hand or using a spreadsheet program, create an error-bar graph that displays the results of question 3.

5. In the research project described in question 1, post-secondary students were asked whether or not they consider themselves feminists. In the sample

of 785 people, 365 said that yes, they consider them-selves feminists. Keep four decimal places in the cal-culations for this question.

a. Find the proportion of students in the sample who consider themselves feminists.

b. Calculate the standard error of this proportion.

6. You want to know what percentage of post-secondary students in Canada consider themselves feminists. Still keeping four decimal places in your calcula-tions, use the results of question 5 to:

a. Calculate the 95 per cent confidence interval for the proportion of students who consider them-selves feminists.

b. Transform the 95 per cent confidence interval for the proportion into percentages.

c. Explain what the results show. In the population of post-secondary students in Canada, what is the percentage of students who consider them-selves feminists likely to be?

7. You think that students' areas of study might be related to whether or not they consider themselves feminists. In the sample, among the 350 students in social science programs, 116 said they consider them-selves feminists. Among the 200 students in natural science programs, 96 said they consider themselves feminists. Keep four decimal places in the calcula-tions for this question.

a. Find the proportion of social science students who consider themselves feminists and the pro-portion of natural science students who con-sider themselves feminists.

b. Calculate the standard errors of these two proportions.

8. Still keeping four decimal places in your calcula-tions, use the results of question 9 to:

a. Calculate the 95 per cent confidence interval for the proportion of social science students who consider themselves feminists and the 95 per cent confidence interval for the proportion of natural science students who consider them-selves feminists.

b. Transform the 95 per cent confidence intervals for the two proportions into percentages.

c. Explain what the results show. In the population of post-secondary students in Canada, is there likely a difference in the percentage of students enrolled in the two areas of study who consider themselves feminists? Explain how you deter-mined your answer.

9. Either by hand or using a spreadsheet program, create a bar graph comparing the percentage of social science students and the percentage of natural science students who consider themselves feminists. Include error bars showing the 95 per cent confi-dence intervals on the graph.

10. A poll commissioned by the Canadian Medical Association (2015) shows that three in five people (63 per cent) say their family is not in a good position, financially or otherwise, to care for older family members if they need long-term health care. The information accompanying the poll points out that *"the sample of n = 2,008 Canadian adults is accurate to within +/− 2.5 percentage points, 19 times out of 20. The data was weighted by region, age, and gender to ensure that the sample accurately reflects the population according to Census data."* Given this information, what percentage of people in the Canadian population are likely to say they are not in a good position to care for older family members?

11. Table 6.7, excerpted from a Statistics Canada publi-cation, compares the demographic characteristics of people who are responsible for providing care to seniors (caregivers) to those who are not (non-caregivers). Only people aged 45 and older are in-cluded in the analyses.

a. Use the 95 per cent confidence intervals for the percentages (proportions) to determine whether caregivers are more likely to be women or men.

b. Use the 95 per cent confidence intervals for the percentages (proportions) to determine whether non-caregivers are more likely to be women or men.

c. Given your answers to (a) and (b), what claims can you make about how caregivers differ from non-caregivers in the population of people aged 45 and older?

Table 6.7   Percentage Distribution of Selected Characteristics of Caregivers and Non-Caregivers, Household Population Aged 45 or Older, Canada Excluding Territories, 2008/2009

| | Caregiver | | | Non-Caregiver | | |
| | | 95% Confidence interval | | | 95% Confidence interval | |
	%	From	To	%	From	To
Total	**100.0**	—	—	**100.0**	—	—
Sex						
Men	43.2	41.3	45.1	50.1	49.0	51.2
Women	56.8	54.9	58.7	49.9	48.8	51.0
Age group						
45 to 54	39.8	37.7	42.0	35.9	34.6	37.2
55 to 64	32.8	31.2	34.4	26.5	25.6	27.4
65 to 74	17.1	16.1	18.0	19.2	18.6	19.8
75 or older	10.3	9.6	11.1	18.4	17.9	19.0
Marital status						
Married/common-law	77.9	76.3	79.4	71.3	70.0	72.5
Widowed/separated/ divorced/single	22.1	20.6	23.7	28.7	27.5	30.0

Source: 2008/2009 Canadian Community Health Survey–Healthy Aging.

Source: Excerpt from Turner and Findlay 2012, 23.

12. Looking at the information in Table 6.7:

   a. Use the 95 per cent confidence intervals to determine whether the percentage (proportion) of caregivers who are in married/common-law relationships is likely to be the same as or different than the percentage (proportion) of non-caregivers who are in married/common-law relationships.

   b. Use the 95 per cent confidence intervals to determine whether the percentage (proportion) of caregivers who are widowed/separated/divorced/single is likely to be the same as or different than the percentage (proportion) of non-caregivers who are widowed/separated/divorced/single.

   c. Given your answers to (a) and (b), what claims can you make about how caregivers differ from non-caregivers in the population of people aged 45 and older?

13. Looking at the information in Table 6.7:

   a. Use the 95 per cent confidence intervals for the percentages (proportions) to determine how the age distribution of caregivers compares to the age distribution of non-caregivers.

   b. Given your answers to (a), what claims can you make about how caregivers differ from non-caregivers in the population of people aged 45 and older?

14. The Organisation for Economic Co-operation and Development (OECD) regularly publishes information about well-being in its member countries. Figure 6.11 shows the average amount of time that full-time workers in OECD countries devote to leisure and personal care each day: that is, sleeping, eating, exercising, attending to hygiene, socializing with friends and family, and travelling associated with these activities.

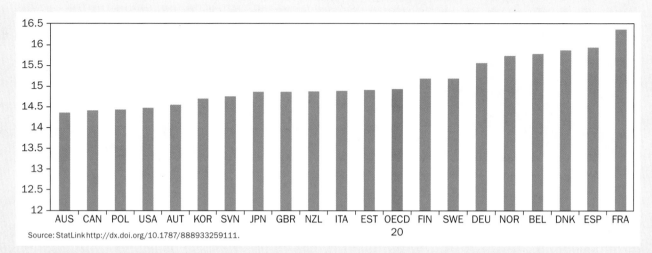

Source: StatLink http://dx.doi.org/10.1787/888933259111.

Figure 6.11   Time Devoted to Leisure and Personal Care; Average Hours per Day, People in Full-Time Employment, Latest Available Year

Source: OECD 2015, 76.

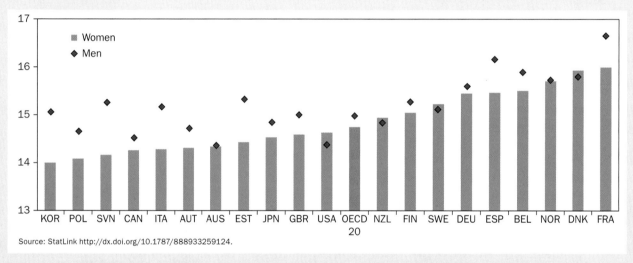

Source: StatLink http://dx.doi.org/10.1787/888933259124.

Figure 6.12   Time Devoted to Leisure and Personal Care for Men and Women; Average Hours per Day, People in Full-Time Employment, Latest Available Year

Source: OECD 2015, 78.

a. How does workers' average daily leisure and personal care time in Canada compare to other countries? What might explain these results?

b. In which countries do full-time workers devote the most amount of time to leisure and personal care each day, on average? In which countries do full-time workers devote the least amount of time to leisure and personal care each day, on average? What might explain these results?

15. Figure 6.12 shows the same information as Figure 6.11, but divided by gender.

a. How does women workers' average daily leisure and personal care time in Canada compare to other countries? What might explain these results?

b. How does the *difference* between men's and women's average daily leisure and personal care time in Canada compare to the *difference* between men and women in other countries? In which countries do men devote less time than women to leisure and personal care each day, on average?

Practice Using Statistical Software (IBM SPSS)

Answer these questions using IBM SPSS and the GSS27.sav or the GSS27_student.sav dataset available from the Student Resources area of the companion website for this book. Weight the data using the "Standardized person weight" [STD_WGHT] variable you created following the instructions in Chapter 5. Report two decimal places in your answers, unless fewer are printed by IBM SPSS. It is imperative that you save the dataset to keep any new variables that you create.

1. Use the Explore procedure to find the mean, standard error, and 95 per cent confidence interval for the mean of "Number of close friends" [SCF_100C].

   a. What is the mean of the variable? In the sample, what is the average number of close friends that people have?

   b. What is the 95 per cent confidence interval for the mean? In the population, what is the average number of close friends likely to be?

2. Use the Explore procedure to find the mean, standard error, and 95 per cent confidence interval for the mean of "Number of close friends" [SCF_100C] divided by "Sex of respondent" [SEX].

   a. In the sample, how does the mean number of close friends among men compare to the mean number of close friends among women?

   b. How do the 95 per cent confidence intervals for the mean number of close friends among men and women compare to each other? In the population, are men and women likely to have the same number of close friends, on average, or not? Explain how you determined your answer.

3. Use the Chart Builder tool to create an error-bar graph showing the mean and 95 per cent confidence interval for the mean of "Number of close friends" [SCF_100C] divided by "Sex of respondent" [SEX]. Describe the information that is displayed in the graph and how it reflects the statistics from question 2.

4. Use the Explore procedure to find the mean, standard error, and 95 per cent confidence interval for the mean of "Number of close friends" [SCF_100C] divided by "Age group of respondent (groups of 10)" [AGEGR10].

   a. In the sample, which age groups have a higher (or lower) number of close friends, on average? Describe the overall pattern of the relationship between the two variables.

   b. In the population, which age groups are likely to have a higher (or lower) number of close friends, on average? Explain how you determined your answer.

5. Use the Chart Builder tool to create an error-bar graph showing the mean and 95 per cent confidence interval for the mean of "Number of close friends" [SCF_100C] divided by "Age group of respondent (groups of 10)" [AGEGR10]. Describe the information that is displayed in the graph and how it reflects the statistics from question 4.

6. The variable "Volunteer work - 12 months" [VCG_300] shows people's answers to this question: "In the past 12 months, did you do unpaid volunteer work for any organization?" Use the Recode into Different Variables tool to recode this variable into a new variable in which the "No" attribute is associated with the value "0" and the "Yes" attribute is associated with the value "1". Call the new variable "Volunteer work - 12 months (recoded)" [VCG_300_RECODED]. Be sure that cases with attributes designated as missing information in the original variable are designated the same way in the new variable. Produce frequency distributions of the original variable and the new variable, and compare them to be sure that the recoding is correct.

7. Use the Frequencies procedure to produce a frequency distribution of "Volunteer work - 12 months" [VCG_300]. Then, use the Means procedure to find the mean of "Volunteer work - 12 months (recoded)" [VCG_300_RECODED]. Compare the percentage of people who volunteered in the past 12 months to the mean of the recoded variable. Explain how these two statistics are related.

8. Use the Explore procedure to find the mean, standard error, and 95 per cent confidence interval for the mean of "Volunteer work - 12 months (recoded)" [VCG_300_RECODED].

   a. Using the mean, determine the proportion (or percentage) of people in the sample who volunteered the past 12 months.

   b. Using the 95 per cent confidence interval for the mean, determine what the proportion (or percentage) of people in the population who volunteered in the past 12 months is likely to be.

9. Use the Explore procedure to find the mean, standard error, and 95 per cent confidence interval for the mean of "Volunteer work - 12 months (recoded)" [VCG_300_RECODED] divided by "Sex of respondent" [SEX].

   a. Using the mean, determine the proportion (or percentage) of men in the sample who volunteered in the past 12 months and the proportion

(or percentage) of women in the sample who volunteered in the past 12 months.

   b. Using the 95 per cent confidence intervals for the means, determine what the proportion (or percentage) of men in the population who volunteered in the past 12 months is likely to be and what the proportion (or percentage) of women in the population who volunteered in the past 12 months is likely to be. In the population, is it likely that the same proportion (or percentage) of men and women volunteered in the past 12 months?

Key Formulas

Standard error of the mean	$se_{\bar{x}} = \dfrac{s}{\sqrt{n}}$
95 per cent confidence interval for a mean	$95\% \; CI_{\bar{X}} = \bar{x} \pm 1.96\left(se_{\bar{x}}\right)$
Standard error of a proportion	$se_p = \sqrt{\dfrac{p(1-p)}{n}}$
95 per cent confidence interval for a proportion	$95\% \; CI_P = p \pm 1.96\left(se_p\right)$

References

Canadian Medical Association. 2015. "2015 National Report Card: Canadian Views on a National Seniors' Health Care Strategy." Ottawa, ON: Ipsos Reid Public Affairs. https://www.cma.ca/En/Lists/Medias/cma-national-report-card-2015.pdf.

CBC News. 2008. "Canadians Come Clean about Dirty Laundry: Poll." September 15. http://www.cbc.ca/news/canadians-come-clean-about-dirty-laundry-poll-1.744689.

Hochschild, Arlie Russell, and Anne Machung. 2012. *The Second Shift: Working Families and the Revolution at Home*. New York: Penguin Books.

Luxton, Meg, and June Shirley Corman. 2001. *Getting by in Hard Times: Gendered Labour at Home and on the Job*. Toronto: University of Toronto Press.

Organisation for Economic Co-operation and Development (OECD). 2015. *How's Life? 2015*. Paris: OECD Publishing. http://www.oecd-ilibrary.org/economics/how-s-life-2015_how_life-2015-en.

Statistics Canada. 2011. "Cycle 24: Time-Stress and Well-Being Public Use Microdata File Documentation and User's Guide." Catalogue no. 12M0024X.

———. 2015. "General Social Survey, Cycle 24, 2010: Time-Stress and Well-Being, Main File." *Public Use Microdata File*. Ottawa, ON: Statistics Canada.

Turner, Annie, and Leanne Findlay. 2012. "Informal Caregiving for Seniors." Health Reports 82-003-X. Ottawa, ON: Statistics Canada. http://www.statcan.gc.ca/pub/82-003-x/2012003/article/11694-eng.htm.

Assessing Relationships by Comparing Group Means: T-Tests

Introduction

In Chapter 6, you learned how information collected from a sample is used to make population estimates. This chapter develops these ideas further, by introducing some strategies for assessing whether or not there is a relationship between two variables in a population, using sample data. I begin by introducing a series of concepts related to hypothesis testing, including the idea of statistical significance. In the remainder of the chapter, I illustrate how to test a hypothesis about the relationship between a categorical independent variable with two attributes and a ratio-level dependent variable. In Chapter 8, you'll learn how to test a hypothesis about the relationship between a categorical independent variable with more than two attributes and a ratio-level dependent variable.

The research focus of this chapter is mental health or mental well-being. Several high-profile media campaigns have raised awareness about the importance of mental health as well as the effects of mental illness. In 2007, the federal government established the Mental Health Commission of Canada as an independent, non-profit organization mandated to improve the mental-health system and to change the attitudes and behaviours of Canadians around mental-health issues. In 2012, the commission launched a strategy that sets out priorities for improving Canadians' mental health and well-being. Mental health is not merely the absence of mental illness. The World Health Organization (WHO 2014) defines mental health as "a state of well-being in which an individual realizes his or her own abilities, can cope with the normal stresses of life, can work productively, and is able to make a contribution to his or her community." Mental health is related to social justice because experiencing disadvantage and discrimination often affect a person's mental well-being. For instance, Canada's Mental Health Strategy identifies several groups of Canadians who routinely face challenges in regard to mental well-being, including First Nations, Inuit, and Métis people as well as residents of northern or remote communities, immigrants and refugees, and those with a non-normative gender identity or sexual orientation (Mental Health Commission of Canada 2012).

Mental health depends on a combination of social, psychological, and biological factors. The analyses in this chapter emphasize the social determinants of mental health, that is, the structural factors that can affect people's mental health. Researchers who take a social determinants of health perspective argue that people with the same biological or psychological risk factors have a wide range of mental-health outcomes because they are influenced by their social environments and interactions. In particular, the experience of socio-economic disadvantage, or poverty, is associated with lower levels of mental well-being (WHO 2014). The World Health Organization notes that poor mental health is often related to stressful work conditions, gender discrimination, social exclusion, risks of violence, and human rights violations (2014). In contrast, factors like having a sense of belonging, enjoying good relationships and good physical health, feeling in control of one's life, and possessing good problem-solving skills can help to promote mental well-being (Mental Health Commission of Canada 2012).

The statistical analyses in this chapter use data from the mental-health module of the 2012 Canadian Community Health Survey (CCHS). The main measure of mental health is a person's "positive mental health" score. (See the "Spotlight on Data" box in this chapter for more information.) For readability, I refer to this "positive mental-health score" as a "mental-health score" throughout this chapter. The highest possible mental-health score is 70 and the lowest is 0; the average mental-health score of people in the sample is 54.3. Because the CCHS has a very large sample, the 95 per cent confidence interval is very narrow. Based on this sample, we are 95 per cent confident that the average mental-health score in the population is between 54.2 and 54.4. The median mental health score is 56.0, that is, half of respondents have this score or higher, and half have this score or lower. The most common mental-health score in the sample is also 56.0. The standard deviation of 10.8 indicates that there is a moderate spread around the mean, and respondents' actual mental-health scores range from 0 to 70. The distribution is slightly left-skewed with more people clustered around the higher scores, reflecting the overall positive mental health of many people in Canada. This generally positive result should be interpreted with some caution, however, since some of the people who tend to have poorer mental health—such as Aboriginal people living on reserves or people living in institutions—are not surveyed. In this chapter, statistical analysis is used to investigate the following question:

- Do people who have difficulty meeting their basic expenses have better or worse mental health than people who do not have difficulty meeting their basic expenses?

Spotlight on Data

The Canadian Community Health Survey (Mental-Health Component)

The Canadian Community Health Survey (CCHS) collects information about Canadians' health and health-care use. The purpose of the survey is to support health monitoring and surveillance programs and to inform health policy at the municipal, provincial, and national levels. The CCHS was first launched in 2001, and data were collected every two years until 2007; in 2007 the survey was re-designed so that data are collected on an ongoing basis and released once a year.

The examples in this chapter rely on data from a series of questions on mental health that are occasionally included in the CCHS and were last asked in 2012. The aim of these supplementary questions is to gather information about who is affected by mental disorders and who has positive mental health. The questions also ask about people's access to and utilization of mental-health services. Everyone who participated in the CCHS in 2012 was asked about their mental health, regardless of whether they reported a mental-health problem or not.

The CCHS population is people aged 12 or older who live in a private dwelling in one of Canada's ten provinces or three territories (although only people living in the ten largest communities in Nunavut are surveyed). People living on reserves and other Aboriginal settlements, full-time members of the Canadian forces, and residents of institutions are excluded (Statistics Canada 2015). People living in two of Quebec's health regions, whose populations are primarily Aboriginal peoples, are also excluded. Multi-stage cluster sampling is used to select potential respondents. First, geographic clusters are selected within each province. Then, a systematic sample of dwellings is selected from within each cluster, and, finally, one member of each household is randomly selected to complete the survey. Overall 69 per cent of the people who were selected to participate in the 2012 mental-health component of the CCHS actually did so: 25,113 individuals completed this component.

Most surveys were completed in-person using computer-assisted personal interviewing (CAPI), where an interviewer brings a laptop to a person's house. CAPI is especially good for collecting information about a potentially sensitive topic like mental health because respondents can enter their answers directly into the computer, without disclosing them to the interviewer, and because there are not the same security concerns as with online surveys. CCHS data are weighted to represent the population in regards to age, sex, and province of residence.

Continued

The main variable used in the statistical analysis in this chapter is people's "Positive mental-health score." This variable was created by combining the responses to the following questions:

In the past month, how often did you feel:
. . . happy?
. . . interested in life?
. . . satisfied with your life?
. . . that you had something important to contribute to society?
. . . that you belonged to a community (like a social group, your neighbourhood, your city, your school)?
. . . that our society is becoming a better place for people like you?
. . . that people are basically good?
. . . that the way our society works makes sense to you?
. . . that you liked most parts of your personality?
. . . good at managing the responsibilities of your daily life?
. . . that you had warm and trusting relationships with others?
. . . that you had experiences that challenge you to grow and become a better person?
. . . confident to think or express your own ideas and opinions?
. . . that your life has a sense of direction or meaning to it?

(Statistics Canada 2011)

The first three items measure emotional well-being, while the remaining items measure positive functioning, including psychological and social well-being (Statistics Canada 2014a, Keyes 2009). Respondents were asked how often they had each of these feelings: "Every day," "Almost every day," "Two or three times a week," "Once a week," "Once or twice a month," or "Never in the past month." Each response is assigned a score ranging from 5 (feeling something every day) to 0 (never feeling something). The response scores for the 14 questions are added together to find the positive mental-health score for each respondent.

Making and Testing Hypotheses

The first chapter of this book introduced the idea of a research question, that is, the general question that a researcher is trying to answer. In order to use statistical analysis to answer a research question, it needs to be restructured into one or more testable hypotheses. A **hypothesis** is a statement of the expected relationship between two or more variables. It's an educated guess that researchers make about what they expect to find, framed in the context of the specific variables included in the data. Establishing a clear hypothesis and then testing it is a cornerstone of the scientific method. Social scientists who engage in statistical analysis strive to set out hypotheses based on established theories and then test (and re-test) those hypotheses using empirical data. In chapters 7 to 11 of this book, I focus on testing hypotheses with only two variables, called bivariate hypotheses.

hypothesis A statement of the expected relationship between two or more variables.

Quantitative researchers conceptually divide hypotheses into two types: non-directional hypotheses and directional hypotheses. A **non-directional hypothesis** asserts that there is a relationship between two variables but does not specify the direction of the relationship. Here is an example of a non-directional hypothesis: "There is a relationship between people's income status and their level of mental health" or, more simply, "People's income status is related to their level of mental health." This hypothesis is non-directional because it leaves open the possibility that people with low income can have either better or worse mental health than people with high income. In contrast, a **directional hypothesis** specifies the direction of the expected relationship between the two variables. Here is an example of a directional hypothesis: "People with low income will have lower levels of mental health than people with high income." This hypothesis is directional because it only tests whether people with low income have worse mental health than people with high income and does not leave open the possibility that people with low income can have better mental health than people with high income.

Researchers' decisions about when to use directional or non-directional hypotheses depend on the theories that inform their hypotheses. In general, researchers prefer to use non-directional hypotheses because they make it possible to detect unexpected results. Researchers use directional hypotheses only in very specific situations—such as in formal experiments or when they know quite a lot about a topic.

When researchers set out bivariate hypotheses, they typically treat one variable as the independent variable and one variable as the dependent variable. Recall that an independent variable captures the characteristic that is considered to be the "cause" of an outcome whereas the dependent variable captures the characteristic that is considered to be the "effect," or the result. However, social science researchers usually cannot make claims about causality—that's why the words *cause* and *effect* appear in quotation marks above. In order for researchers to assert that there is a **causal relationship** between two or more variables, three conditions must be met. First, there must be a statistical relationship between the variables. Second, the causal variable must occur before the outcome variable in time. These first two conditions are relatively easy to meet if a researcher makes careful choices. But the third condition is that there cannot be another factor that affects both the causal variable and the outcome variables that is unaccounted for. This third condition is very difficult to meet without having experimental data since it's rarely possible to account for all of the other factors that affect the variables of interest. Researchers who do not have experimental data typically frame their hypotheses as "relationships" and "associations" and avoid the language of "cause" and "effect."

In the remainder of this section, I describe how researchers can assess the relationship between two variables set out in a hypothesis by considering two things: (1) the magnitude of the relationship and (2) the reliability of the relationship.

Assessing the Magnitude of a Relationship

The magnitude of a relationship refers to the size or the strength of a relationship. When one variable in a relationship is categorical, the simplest way to report the size of the relationship is to report the difference that group membership makes: Does it make a little bit of difference or a big difference? For a relationship between

non-directional hypothesis A statement of the expected relationship between variables that does not specify the direction of the relationship.

directional hypothesis A statement of the expected relationship between variables that specifies the direction of the relationship.

causal relationship A statistical relationship where a causal variable occurs before an outcome variable in time and where there are no other unaccounted for factors influencing both the causal and outcome variables.

measures of effect size Measures that show how much of an effect an independent variable has on a dependent variable, often in a standardized way.

measures of association Measures that show the strength of a relationship between variables, usually summarized in a single number.

positive relationship A relationship where higher values or attributes on one variable are associated with higher values or attributes on another variable (or lower values or attributes on one variable are associated with lower values or attributes on another variable).

negative relationship A relationship where higher values or attributes on one variable are associated with lower values or attributes on another variable (or vice versa).

two categorical variables, this might involve reporting how the percentage of cases with each attribute differs between groups (as in the discussion of cross-tabulations in Chapter 2). For a relationship between a categorical variable and a ratio-level variable, this might involve reporting the difference between group means (as in some of the examples in Chapter 4). For relationships between some types of variables, there are formal **measures of effect size**, which provide standardized information about the magnitude of a relationship.

In addition to reporting differences between groups and effect sizes, there are also a series of formal **measures of association** that researchers sometimes use to describe the magnitude of a relationship. Measures of association are statistics that measure the strength of a relationship between variables, usually summarized in a single number. Some measures of association also provide information about the direction of a relationship. A **positive relationship** between two variables occurs when a higher values or attributes on one variable are associated with higher values or attributes on the other variable (or a lower values or attributes on one variable are associated with lower values or attributes on the other variable). In contrast, a **negative relationship** between two variables occurs when higher values or attributes on one variable are associated with a lower values or attributes on the other variable (or vice versa). Reporting the direction of a relationship makes sense only when the attributes of both variables have an order, that is, they are measured at the ordinal or ratio level. Decisions about which measure of association to use in an analysis depend on the level of measurement for each variable and, sometimes, on the number of groups or attributes in a categorical variable. The upcoming chapters introduce several common measures of association used by social science researchers.

Comparing Means

This chapter focuses on statistical techniques for assessing relationships between a categorical independent variable with two attributes and a ratio-level dependent variable. The simplest way to assess the magnitude of this type of relationship is to compare the two group means.

Let's use this strategy to investigate how people's mental health is related to economic disadvantage. The CCHS asks people this yes/no question: "With your current household income, do you have any difficulty meeting basic expenses such as food, shelter, and clothing?" About one in eight people (12 per cent) say that they have difficulty covering their basic expenses with their current household income. Since this question only allows people to give a yes or no answer, the results are captured in a categorical variable with two attributes. The ratio-level dependent variable is a person's mental-health score, described in the introduction to this chapter and in the "Spotlight on Data" box.

As I reported in the introduction to this chapter, among people overall, the average mental health score is 54. To assess the magnitude of the relationship between income adequacy and mental health, let's compare the average mental-health score for people who have difficulty meeting their basic expenses and those who do not. Table 7.1 shows the means and standard deviations for each group. Among

Table 7.1   Comparing Two Group Means

Group	Frequency	Percentage	Mental-Health Score Mean	Mental-Health Score Std. Deviation
Has difficulty meeting basic expenses	2,626	11.6	49.7	13.35
Does not have difficulty meeting basic expenses	20,095	88.4	54.9	10.29
Total/Overall	*22,721*	*100.0*	*54.3*	*10.82*

Source: Author generated; Calculated using data from Statistics Canada, 2014b.

people who have difficulty meeting their basic expenses the average mental-health score is 49.7, whereas among people who do not have difficulty meeting their basic expenses the average mental-health score is 54.9.

It's up to the researcher to decide whether or not the size of the difference between group means is big enough to matter in a real-world context. For example, an average difference of $2 in yearly wages isn't enough to make a substantial difference in someone's life, but an average difference of $2 in hourly wages is. For a mental-health score that ranges from 0 to 70, an average difference of 5.2 points seems like it might reflect a substantial difference in mental well-being.

Effect Size: Cohen's d

In addition to using your common sense to assess whether the size of the difference between group means is large enough to matter, there is a statistic that can help you to make this assessment. **Cohen's d** is a measure of effect size; it standardizes the size of an effect by dividing the difference between the group means by the standard deviation of the variable in the sample overall. In other words, Cohen's d shows how large the difference between means is, compared to the overall variation in the sample. Cohen's d is conceptually similar to a z-score, which shows how far a particular case is from the mean, using standard deviations as the unit of measurement. (See Chapter 4.)

Cohen's d is easy to calculate: find the difference between the two group means, and divide it by the standard deviation of the variable overall (this is sometimes called the "pooled standard deviation"). Remember that $\bar{x}$ is used to denote the mean or the average. In the Cohen's d formula, the subscript number 1 on the first $\bar{x}$ in the numerator shows that it refers to the mean of the cases in the first group, and the subscript number 2 on the second $\bar{x}$ in the numerator shows that it refers to the mean of the cases in the second group. The s in the denominator denotes the standard deviation; because it has no subscript, it refers to the overall standard deviation.

Cohen's d (d) A measure of effect size that standardizes the size of an effect by dividing the difference between the group means by the standard deviation of the variable in the sample overall.

$$d = \frac{\bar{x}_1 - \bar{x}_2}{s}$$

Cohen's d

A Cohen's d of 1 indicates that the difference between the group means is the same size as the standard deviation of the variable. Cohen (1988) proposes a series of guidelines for interpretation:

- 0.2 is a small effect (or a weak relationship).
- 0.5 is a medium effect (or a moderate relationship).
- 0.8 is a large effect (or a strong relationship).

The sign (positive or negative) of Cohen's d doesn't matter and can be ignored; it simply shows whether the larger group mean was subtracted from the smaller group mean or vice versa.

Let's calculate Cohen's d to assess the magnitude of the relationship between income adequacy and mental-health scores. Using the information in Table 7.1, Cohen's d is calculated as:

$$d = \frac{\bar{x}_1 - \bar{x}_2}{s}$$

$$d = \frac{49.7 - 54.9}{10.8}$$

$$= \frac{-5.2}{10.8}$$

$$= -0.48$$

The effect size is −0.48, or a medium effect. The difference between the average mental-health scores of people who can and who cannot meet their basic expenses is approximately half the size of the standard deviation of the variable in the sample overall. In other words, there is a moderate relationship between whether or not people can meet their basic expenses and mental-health scores. Cohen's d provides researchers with a way of quantifying effect size by comparing it to established benchmarks, but it is just another way to assess the relative size of the difference between group means. Saying that something has a small, medium, or large effect, or saying that something has a weak, moderate, or strong relationship, is only meaningful if you can explain it in relation to people's everyday social contexts.

Step-by-Step: Cohen's d

$$d = \frac{\bar{x}_1 - \bar{x}_2}{s}$$

Step 1: Find the mean of the variable for the cases in the first group $(\bar{x}_1)$.

Step 2: Find the mean of the variable for the cases in the second group $(\bar{x}_2)$.

Step 3: Find the standard deviation of the variable (s). Ignore the groups when calculating the standard deviation (but each case used in the calculation should belong to one of the two groups).

Step 4: Subtract the mean of the cases in the second group (from Step 2) from the mean of the cases in the first group (from Step 1) to find the numerator of the Cohen's d equation.

Step 5: Divide the result of Step 4 by the standard deviation (from Step 3) to find Cohen's d.

Assessing the Reliability of a Relationship

Reliability refers to whether or not a researcher is confident that a relationship found in a sample exists in the larger population. The reliability of a relationship is typically assessed using a **statistical significance test**. Tests of statistical significance estimate the likelihood of randomly selecting a sample with the observed relationship, if no relationship exists in the larger population that the sample is selected from. This likelihood is typically expressed as a probability (p) and is thus sometimes called a **p-value**. Each different test of statistical significance is associated with a specific probability distribution. In Chapter 5, you learned that the normal distribution can be thought of as a probability distribution, and you also learned how to use the area under the normal curve to estimate the probability of a randomly selected case having a value above or below a particular cut-off point. Tests of statistical significance mirror this approach, by using the area under a distribution to estimate the probability of randomly selecting a sample with an observed relationship, from a population in which a relationship does not exist.

A variety of tests are used to determine the statistical significance of a relationship; the choice of which test to use depends on the level of measurement of the variables and, sometimes, on the number of groups in a categorical variable. Despite the different strategies used to calculate different tests of statistical significance, however, they all rely on the same general logic, and p-values are all interpreted in the same way.

Significance testing relies on a logic of falsification. That is, the goal of a statistical significance test is to *disprove* a hypothesis. This might seem counterintuitive since researchers are usually interested in trying to prove that there *is* a relationship between variables in the population. To compensate, after setting out a **research hypothesis**, statisticians establish a **null hypothesis**, which is the hypothesis that they set out to disprove. A null hypothesis is simply a statement that there is no relationship between the variables of interest in the population (or that the relationship is not in the expected direction, for directional hypotheses). In other words, the null hypothesis is the opposite of the research hypothesis. Typically, the notation H_1 is used to denote the research hypothesis, and H_0 is used to denote the null hypothesis. To illustrate, here are the null hypotheses associated with the non-directional and directional research hypotheses from earlier in this chapter:

> Research hypothesis (H_1): There is a relationship between people's income status and their level of mental health, in the population.

statistical significance test Estimates the likelihood of randomly selecting a sample with the observed relationship (or one of greater magnitude), if no relationship exists in the population that the sample was selected from.

p-value (p) Shows the probability of randomly selecting a sample with the observed relationship (or one of greater magnitude), if no relationship exists in the population that the sample was selected from.

research hypothesis (H_1) A statement of an expected relationship between two or more variables in the population.

null hypothesis (H_0) A statement that there is no relationship between two or more variables in the population.

Null hypothesis (H$_0$): There is no relationship between people's income status and their level of mental health, in the population.

Research hypothesis (H$_1$): People with low income will have lower levels of mental health than people with high income, in the population.

Null hypothesis (H$_0$): People with low income will not have lower levels of mental health than people with high income, in the population.

Once researchers have established a null hypothesis, they set out to disprove it, in an attempt to implicitly prove their research hypothesis. So, instead of trying to prove that there *is a relationship* between income status and level of mental health, a researcher tries to prove that there is *not no relationship* between income status and level of mental health in the population. Of course, researchers can never definitively prove or disprove any hypothesis using statistical significance testing and sample data. The best that they can do is to determine whether a hypothesis is *likely* to be true in the population.

There are two major types of errors that researchers strive to avoid when they assess the reliability of a relationship. The first type of error occurs when no relationship actually exists between two or more variables in the population, but a researcher claims that it likely does exist, based on an analysis of a random sample of cases from that population. This is called a **type I error**. You might think of a type I error as a "false positive," that is, claiming that there probably is a relationship when there really isn't. The second type of error occurs when the opposite happens: a relationship actually exists between two or more variables in the population, but a researcher claims it likely doesn't exist, based on an analysis of a random sample of cases from that population. This is called a **type II error**. You might think of a type II error as a "false negative," that is, claiming that there probably is not a relationship when there really is. Figure 7.1 illustrates the conceptual difference between type I and type II errors. Of course, the ideal situation—as indicated by the happy faces in Figure 7.1—is for a researcher to correctly predict whether or not a relationship exists in the population, based on the sample data.

type I error When no relationship exists between two or more variables in the population, but a researcher claims that it likely does exist.

type II error When a relationship exists between two or more variables in the population, but a researcher claims that it likely does not exist.

		In the population:	
		No relationship exists	A relationship exists
Based on sample data, the researcher decides:	It's likely that no relationship exists	:)	TYPE II ERROR false negative
	It's likely that a relationship exists	TYPE I ERROR false positive	:)

Figure 7.1   Understanding Type I and Type II Errors

P-values, which I introduced earlier, show the probability of making a type I error. In other words, they show the probability of randomly selecting a sample that shows a relationship between two or more variables, if no relationship exists in the larger population. Social science researchers typically try to obtain a p-value of less than 0.05 before they assert that a relationship is likely to exist in the population. That is, they are willing to accept up to a 5 per cent chance of making a type I error (claiming that there is likely a relationship when there really isn't one). The choice to use a p-value of less than 0.05 as a threshold is entirely arbitrary. Researchers in disciplines such as medicine or the natural sciences often use the more stringent p-value thresholds of less than 0.01 or less than 0.001.

The p-value that researchers establish as the threshold for their statistical tests is called an **alpha value**. The alpha value is sometimes denoted using the Greek letter alpha (α). If a test of statistical significance results in a p-value that is lower than the alpha value, then the researcher rejects the null hypothesis, and asserts that there is likely a relationship between the variables that they are testing in the population. If a test of statistical significance results in a p-value that is higher than (or equal to) the alpha value, then the researcher fails to reject the null hypothesis and asserts that there may not be a relationship between the variables that they are testing in the population.

alpha value (α) The threshold that researchers establish for their p-value.

Recall that each test of statistical significance is associated with a probability distribution. In other words, the statistic that is produced by each test has a known distribution. Once researchers have established an alpha value, they can work backwards to determine the **critical value** for each test statistic. The critical value of each test statistic is the cut-off point where the probability of making a type I error matches the alpha value. So, for an alpha value of 0.05, the critical value of a test statistic is the cut-off point where there is only a 5 per cent chance (or 0.05 probability) of randomly selecting a sample that would produce a test statistic of that size or higher, if no relationship exists in the larger population. This critical value is established by finding the point on the horizontal axis of a probability distribution that has only 5 per cent of the area under the curve beyond it. Figure 7.2 shows some examples of probability distributions associated with different tests of statistical significance and their critical values. Because each distribution has a different shape, the critical value is different for each one, even though they all use an alpha value of 0.05. A researcher who obtains a test statistic that is larger than the critical value for the probability distribution can assert that there is less than a 5 per cent chance of selecting a sample with the observed relationship, if no relationship exists in the population.

critical value For a test statistic, the cut-off point where the probability of making a type I error matches the alpha value.

The size and shape of the probability distributions associated with different tests of statistical significance change depending on the number of cases in the sample and the number of groups being compared. In this chapter and those that follow, you'll learn more about how to calculate and interpret test statistics that use these probability distributions.

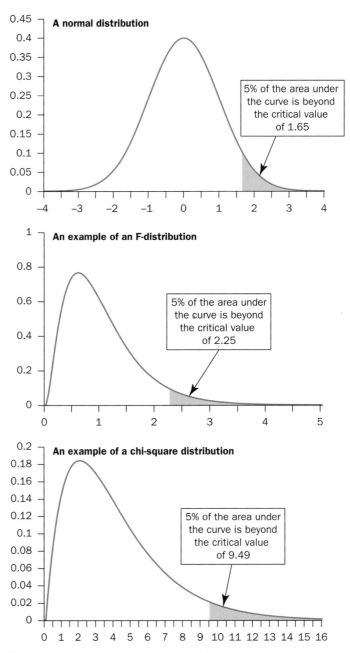

Figure 7.2   The Critical Values of Some Probability Distributions

The Logic of Statistical Significance Tests

As you know, in most studies, researchers collect information from a single, random sample of cases from a population. The particular sample that a researcher selects is only one of many possible samples that could be selected from a population. Just by chance, a researcher might randomly select a sample that shows a relationship between two variables, even if that relationship doesn't actually exist in the larger population. For quantitative researchers, a central question then becomes this: How likely are they to randomly select a sample that shows a relationship between two variables, if no relationship actually exists in the larger population? This is the question that tests of statistical significance answer.

In Chapter 6, you learned how sampling distributions are used to estimate means and proportions in a population. In this section, I use a similar approach to show how a sampling distribution of the *differences* between group means is used to assess the reliability of sample data. This logic is central to testing the relationship between a categorical independent variable with two attributes and a ratio-level dependent variable; other tests of statistical significance rely on a similar (but not identical) logic.

To illustrate, I'll use another simulated dataset that represents an entire population, although the imaginary population is smaller this time—only 100 people. It might help to think of this population as a particular group of people, such as all of the residents of a long-term-care home or all of the users of a community centre. Imagine that everyone in the population completed a mental-health assessment and was assigned a mental-health score ranging from 0 to 100, where 0 indicates that a person has very poor mental health and 100 indicates that a person has excellent mental health. Overall, the average mental-health score in the simulated population is 63 out of 100 (s.d. = 18). About half of the people in the population also have low income.

Now let's investigate whether people with low income have the same average mental-health score as people with high income in the simulated population. Income status is treated as the independent variable, and mental-health score is treated as the dependent variable. The research hypothesis is that income status is related to mental health in the population; that is, the average mental-health score for the low-income group is different than the average mental-health score for the high-income group. The null hypothesis is that there is no relationship between income status and mental health in the population. Another way to state the null hypothesis is to say that the group means are equal in the population; that is, the average mental-health score for the low-income group is the same as the average mental-health score for the high-income group.

To begin, I select a random sample of 20 people from the simulated population. The sample includes 10 people with low income, who have an average mental-health score of 66.3, and 10 people with high income, who have an average mental-health score of 74.4. When the mean of the low-income group is subtracted from the mean of the high-income group, the difference is 8.1. (See Table 7.2.) In an actual research study, this would be the only sample that I select.

But this first sample is only one of many possible samples that can be selected from the population. So, just to see what happens, I select another random sample of 20 people from the simulated population. As in the Chapter 6 simulation, being chosen for the first sample does not affect the chance of being chosen for the second sample. In the second sample, the people with low income have an average

Table 7.2   Mental-Health Score Statistics from Three Samples Randomly Selected from a Simulated Population

	High-Income Group			Low-Income Group			Difference between Group Means
	Mean	Std. Dev.	n	Mean	Std. Dev.	n	
Sample 1	74.4	15.7	10	66.3	18.7	10	8.1
Sample 2	65.4	18.5	9	65.4	22.8	11	0.0
Sample 3	62.4	24.7	7	66.6	17.8	13	−4.2

mental-health score of 65.4, and the people with high income also have an average mental-health score of 65.4, so there's no difference between the group means. I repeat the process a third time, and in the third sample, people with low income have an average mental-health score of 66.6, and people with high income have an average mental-health score of 62.4. When the mean of the low-income group is subtracted from the mean of the high-income group, the difference is −4.2. This time, the mean difference is a negative number because the low-income group has a higher average mental-health score than the high-income group.

By now, you can probably guess where this is going. I continue to select random samples of 20 people from the imaginary population until I have 200 samples. For each sample, I find the difference between the mean of the high-income group and the mean of the low-income group. When the differences between the means in each sample are graphed in a histogram, it forms a familiar shape. (See Figure 7.3.) This histogram is an approximation of a sampling distribution of mean differences. Theoretically, a sampling distribution of mean differences is the distribution of the differences between the group means produced by every possible unique sample of the same size that can be selected from a population. Like other sampling distributions, the sampling distribution of mean differences is centred on the population parameter. This histogram of mean differences is centred on 0; in many samples there is no difference or only a very small difference between the group means. In fact, in the original simulated population, the two group means are exactly the same.

This histogram helps to illustrate some general principles. First, if the group means are equal in the population, the group means are approximately equal in most of the samples selected from that population. But, even if the group means are equal in the population, some samples have group means that are very unequal. As a result, it becomes useful to think about the probability—or chance—of selecting a particular sample from the sampling distribution of mean differences that is produced when the population parameter is 0. If the group means are equal in the population, then a researcher has a higher chance of selecting a sample with approximately equal group means (a sample from the centre of the sampling distribution) and a lower chance of selecting a sample with very unequal group means (a sample from the tails of the sampling distribution).

Now, let's reverse this logic to consider a situation where a researcher randomly selects a single sample but doesn't know anything about the population. If the group

sampling distribution of mean differences A theoretical distribution of the differences between the group means produced by every possible unique sample of the same size that can be selected from a population.

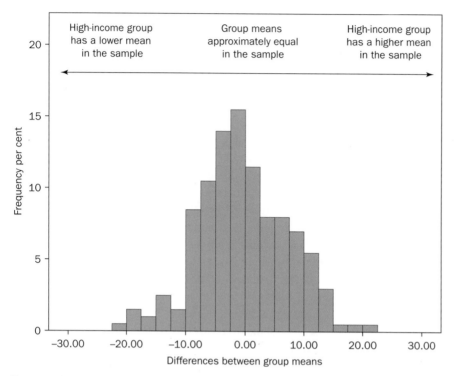

Figure 7.3   A Histogram of the Differences between Group Means in 200 Samples Randomly Selected from the Same Population (Simulated Data)

means in the sample are approximately equal, then it's more likely that the sample was selected from a population in which the group means are equal. If the group means in the sample are very unequal, then it's less likely that the sample was selected from a population in which the group means are equal.

The question then becomes this: How big a difference does there need to be between the group means in the sample for a researcher to assert that there is likely a difference between the group means in the population? The sampling distribution in Figure 7.3 shows that the mean differences in the first and third sample I selected from the simulated population (8 and −4) are actually relatively common, even though the group means are the same in the population. And, answering the question "How big a difference is big enough?" is made even harder because variables are measured in different units. If people's mental health were measured using a scale of 0 to 10, or 0 to 70, instead of a scale from 0 to 100, the size of the difference between group means would vary accordingly.

The T-Test of Independent Means

The **t-test of independent means** is a statistical significance test that allows researchers to determine whether the difference between two group means in a sample is large enough to assert that there is likely a difference between the two group means in the

t-test of independent means A statistical significance test used to determine whether the difference between two group means in a sample is large enough to assert that there is likely a difference between the group means in the population.

population. It is called a test of "independent means" (or an independent-samples t-test or a two-sample t-test) because the cases in the two groups are independent of each other; they're not linked or paired in any way. The calculation of a t-statistic starts with finding the difference between the group means. But, to compensate for the fact that variables have different units of measurement, the size of the difference between the two group means is standardized by dividing it by the standard error (which is the standard deviation of the sampling distribution). In other words, the t-statistic shows how far the difference between means in a sample is from the centre of a sampling distribution of mean differences when the population parameter is 0, using standard errors as the unit of measurement.

Because the histogram of mean differences in Figure 7.3 is from a simulation, I can calculate the actual standard deviation of the sample means: it's 7.7. If this histogram were a complete sampling distribution, a sample in which the average mental-health score of the high-income group was 7.7 points higher than the average mental-health score of the low-income group would produce a t-statistic of 1, since 7.7 is one standard error (or standard deviation) above the mean. Similarly, a sample in which the average mental-health score of the high-income group was 15.4 points higher than the average mental-health score of the low-income group would produce a t-statistic of 2, since 15.4 is two standard errors above the mean. A sample in which the average mental-health score of the high-income group was 7.7 points lower than the average mental-health score of the low-income group would produce a t-statistic of −1, since 7.7 is one standard error below the mean, and so on.

As you learned in Chapter 6, the standard error (which is the standard deviation of a sampling distribution) can be estimated using only information from the sample. You learned to calculate the standard error of a mean using the standard deviation and the number of cases in the sample. The t-statistic uses a similar approach. There are actually two ways to calculate a t-statistic. One way uses the standard deviation and number of cases in the sample overall to estimate the standard error. This approach assumes that each of the two groups have the same variance (or the same standard deviation) in the population. Since this assumption is not always met, I don't describe this approach in this book. The second way uses the standard deviation and the number of cases in each group separately (instead of pooled together) to estimate the standard error. This approach does not rely on any assumptions about the group variances in the population. When this second approach is used, the t-statistic formula uses subscripts to denote the statistics for different groups: so s_1 is the standard deviation for group one and s_2 is the standard deviation for group two, and so on. The exponents are interpreted in the usual way, so s_1^2 is the squared standard deviation of group one (or the variance of group one) and s_2^2 is the squared standard deviation of group two (or the variance of group two). The formula is:

T-statistic

$$t = \frac{\overline{x}_1 - \overline{x}_2}{\sqrt{\frac{s_1^2}{n_1} + \frac{s_2^2}{n_2}}}$$

The numerator of the t-statistic equation is simply the difference between the group means. The denominator provides an estimate of the standard error of the sampling distribution and is used to standardize the difference between the group means. Using this equation, let's calculate the t-statistic for the first sample I selected from the simulated population. In the first sample, the 10 people in the high-income group had an average mental-health score of 74.4 (s.d. = 15.7), and the 10 people in the low-income group had an average mental-health score of 66.3 (s.d. = 18.7). (See Table 7.2.) The group means, standard deviations, and number of cases are substituted into the formula for the t-statistic as follows:

$$t = \frac{\bar{x}_1 - \bar{x}_2}{\sqrt{\dfrac{s_1^2}{n_1} + \dfrac{s_2^2}{n_2}}}$$

$$t = \frac{74.4 - 66.3}{\sqrt{\dfrac{15.7^2}{10} + \dfrac{18.7^2}{10}}}$$

$$= \frac{8.1}{\sqrt{\dfrac{246.49}{10} + \dfrac{349.69}{10}}}$$

$$= \frac{8.1}{\sqrt{24.65 + 34.97}}$$

$$= \frac{8.1}{\sqrt{59.62}}$$

$$= \frac{8.1}{7.72}$$

$$= 1.05$$

If the population parameter—the difference between the group means in the population—was 0, the sampling distribution of mean differences would centre on 0. The t-statistic for this difference between group means is 1.05. This indicates that a sample with a difference this big between two group means would be located roughly one standard error above the centre of the sampling distribution of mean differences, if the population parameter were 0. This matches with what we know from the simulation: the standard deviation of the mean differences in the histogram was 7.7, and in this calculation, the estimate of the standard error is also 7.7. So, a sample with a 7.7 point difference between group means has a t-statistic equal to 1 (or −1). In this first sample I selected, the difference between the group means is 8.1 (slightly higher than 7.7), and so the t-statistic is slightly higher than 1. In fact, if you take the difference between the group means (8.1 points) and divide it by the standard error (7.7) you also get the t-statistic of 1.05.

Statisticians interpret the t-statistic of 1.05 in relation to a **t-distribution**. The histogram in Figure 7.3 shows the distribution of mean differences for 200 random

t-distribution A probability distribution used to determine the likelihood of randomly selecting a sample with the observed difference between group means if the group means are equal in the population.

samples of 20 cases selected from a population in which the group means are equal. If I selected every possible unique sample of 20 cases from the population, the sampling distribution of mean differences would form a symmetrical shape called a t-distribution. The shape of the t-distribution changes slightly depending on the sample size. For large samples (more than 500 cases), the t-distribution is approximately the same shape as the normal distribution. For smaller samples, the t-distribution is a bit wider, with slightly fatter tails than the normal distribution. Figure 7.4 shows how the shape of a t-distribution for small samples compares to the normal distribution.

The area under the t-distribution is used to determine the probability that a researcher will random select a sample with the observed difference between the group means if the sampling distribution of mean differences centres on 0 (that is, if the group means are equal in the population). The t-statistic is measured in standard errors, which are shown on the horizontal axis of the sampling distribution of mean differences. The amount of the area under the curve that is beyond the value of the t-statistic on the horizontal axis shows the probability of randomly selecting a sample with a mean difference of the observed size or larger, if the group means are equal in the population (that is, if the null hypothesis is true).

An example will help to clarify. Figure 7.5 shows a t-distribution for the first sample in the simulated example and highlights the point on the horizontal axis that corresponds to the calculated t-statistic of 1.05. The orange shaded areas show the percentage of the area under the curve that is beyond the t-statistic of 1.05. The shaded areas show the percentage of samples that can be randomly selected from the population that will produce a t-statistic of 1.05 or higher, if the group means are equal in the population. By the same logic, the shaded areas show the probability that any one randomly selected sample will produce a t-statistic of 1.05 or higher, if the group means are equal in the population. Notice that the area above 1.05 *and* the area below −1.05 is shaded orange. This is because the sign (positive or negative) associated with the t-statistic depends only on which

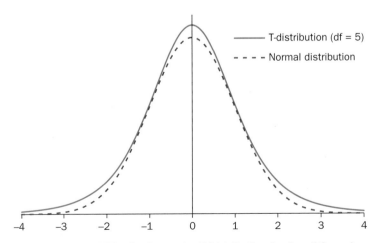

Figure 7.4   **The Normal Distribution and a T-Distribution for Small Samples**

group was used as group 1 and which group was used as group 2 in the calculation. I arbitrarily decided to use the high-income earners as group 1 and the low-income earners as group 2; if that choice were reversed, the t-statistic would be –1.05. (Go back and try it if you like.) When the two shaded areas under the curve in Figure 7.5 are added together, they take up 32 per cent of the total area under the curve. This shows that I have a 32 per cent chance of randomly selecting a sample with an 8.1 point difference (or greater) between the average mental-health scores of the two groups if the group means are the same in the population. The p-value is thus 0.32 (since 32 per cent is equivalent to 0.32). In other words, I have about a one in three chance of randomly selecting a sample with these results, even if the null hypothesis is true.

Earlier in this chapter, you learned about alpha values and critical values. For a researcher using an alpha value of 0.05, the critical value of a t-statistic is the cut-off point on the horizontal axis of the t-distribution where there is only a 5 per cent chance of randomly selecting a sample with the observed difference (or a larger difference) between the group means if the null hypothesis were true. So, instead of starting with a specific t-statistic and finding the percentage of samples that produce a t-statistic of that size or higher, you start with a specific percentage of samples (5 per cent) and find the corresponding cut-off point for the t-statistic. Because the t-distribution is symmetrical, and the sign of the t-statistic is arbitrary, the critical value is actually the point on the horizontal axis with 2.5 per cent of samples above it, and the matching negative point on the horizontal axis with 2.5 per cent of samples below it.

Recall that the shape of the t-distribution changes slightly depending on the sample size. Mathematically, these differences in sample size are reflected in the

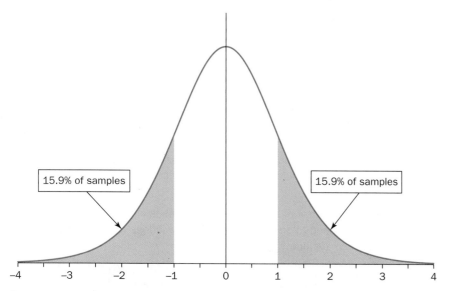

Figure 7.5   Finding the Probability of Randomly Selecting a Sample with a T-Statistic of 1.05 or Higher Using a T-Distribution

degrees of freedom (df) Account for the number of free parameters in an equation.

number of **degrees of freedom** associated with a distribution. Degrees of freedom account for the number of "free" or unconstrained parameters in an equation. For statistics that rely on means, the number of degrees of freedom is often equivalent to the number of cases minus 1. One is subtracted from the number of cases because if you know a mean, the score of the final case can always be calculated from the scores of the other cases; it is no longer a "free" parameter. For example, if I know that the average of three mental-health scores is 74 and that the first score is 67 and the second score is 82, I can calculate the third score by working backward from the average. The sum of all of the scores will be 222 (74×3—the average multiplied by the number of cases); the first two scores add up to 149, so the third score must be 73. For the version of the t-statistic that assumes the two groups have equal variances in the population (which is not described in this book), the number of degrees of freedom is typically the number of cases in the first group minus 1, plus the number of cases in the second group minus 1; that is, it's the sample size minus 2. For the version of the t-statistic that does not assume the two groups have equal variances in the population (which is described in this chapter), calculating the degrees of freedom is slightly more complicated. The formula and process are described in the following "Step-by-Step" box. (We'll return to the idea of degrees of freedom again in chapters 8 and 9.)

Step-by-Step: T-Test of Independent Means

Step 1: Identify the research hypothesis and the null hypothesis.

T-Statistic

$$t = \frac{\bar{x}_1 - \bar{x}_2}{\sqrt{\frac{s_1^2}{n_1} + \frac{s_2^2}{n_2}}}$$

Step 2: Count the total number of cases in first group (n_1) and the total number of cases in the second group (n_2). Do not include cases that are missing information for the variable x.

Step 3: Find the mean of the variable for the cases in the first group $\left(\bar{x}_1\right)$ and the mean of the variable for the cases in the second group $\left(\bar{x}_2\right)$.

Step 4: Find the variance of the variable for the cases in the first group $\left(s_1^2\right)$ and the variance of the variable for the cases in the second group $\left(s_2^2\right)$. You may need to square the standard deviations (s_1 and s_2) to find the variances $\left(s_1^2 \text{ and } s_2^2\right)$.

Step 5: Subtract the mean of the cases in the second group from the mean of the cases in the first group (both from Step 3) to find the numerator of the t-statistic equation.

Step 6: Divide the variance of the cases in the first group (from Step 4) by the number of cases in the first group (from Step 2).

Step 7: Divide the variance of the cases in the second group (from Step 4) by the number of cases in the second group (from Step 2).

Step 8: Add together the results of Step 6 and Step 7.

Step 9: Find the square root of the result of Step 8 to find the denominator of the t-statistic equation.

Step 10: Divide the result of Step 5 by the result of Step 9 to find the t-statistic.

Degrees of Freedom

$$df_t = \frac{\left(\dfrac{s_1^2}{n_1} + \dfrac{s_2^2}{n_2} \right)^2}{\dfrac{\left(s_1^2 / n_1 \right)^2}{\left(n_1 - 1 \right)} + \dfrac{\left(s_2^2 / n_2 \right)^2}{\left(n_2 - 1 \right)}}$$

Step 11: Square the result of Step 8 to find the numerator of the degrees of freedom equation.

Step 12: Square the result of Step 6.

Step 13: Subtract 1 from the number of cases in the first group (from Step 2).

Step 14: Divide the result of Step 12 by the result of Step 13.

Step 15: Square the result of Step 7.

Step 16: Subtract 1 from the number of cases in the second group (from Step 2).

Step 17: Divide the result of Step 15 by the result of Step 16.

Step 18: Add together the result of Step 14 and Step 17 to find the denominator of the degrees of freedom equation.

Step 19: Divide the result of Step 11 by the result of Step 18 to find the degrees of freedom. If the result is a decimal, round to the nearest whole number.

Statistical Significance

Step 20: Determine the alpha value you want to use for the t-test. The most common alpha value used in the social sciences is 0.05.

Step 21: Determine the critical value of the t-statistic, using the alpha value you selected in Step 20, for a t-distribution with the degrees of freedom you found in Step 19. If you are using an alpha value of 0.05 and the degrees of freedom are greater than 500, the critical value is +/−1.96. If you are using a different alpha value and/or have fewer degrees of freedom, find the critical value using a printed table of the critical values of the Student's t-distribution, an online calculator, or the spreadsheet tool available from the Student Resources area of the companion website for this book.

Continued

Step 22: Compare the t-statistic (from Step 10) to the critical value (from Step 21). If the t-statistic is beyond the critical value, the result is statistically significant: you can reject the null hypothesis and state that the group means are likely to be different in the population. If the t-statistic is not beyond the critical value, the result is not statistically significant: you fail to reject the null hypothesis and state that you are not confident that the group means are different in the population.

Figure 7.6 shows the critical value of the t-statistic, using an alpha value of 0.05, for a t-distribution with 10 degrees of freedom and a t-distribution with 500 degrees of freedom (which is the same as a normal distribution). For a t-distribution with 10 degrees of freedom, the point on the horizontal axis with 2.5 per cent of the area under the curve beyond it is +/–2.23. In other words, we need to obtain a t-statistic higher than 2.23 or lower than –2.23 to obtain a p-value of less than 0.05. If a sample produces a t-statistic that is higher than 2.23 or lower than –2.23, there is less than a 5 per cent chance that the sample was selected from a population where the group means are equal. Similarly, for a t-distribution with 500 degrees of freedom, the point on the horizontal axis with 2.5 per cent of the area under the curve beyond it is +/–1.96. In other words, we need to obtain a t-statistic higher than 1.96 or lower than –1.96 to obtain a p-value of less than 0.05. If a sample produces a t-statistic that is higher than 1.96 or lower than –1.96, there is less than a 5 per cent chance that the sample was selected from a population where the group means are equal. After 500 degrees of freedom, the shape of the t-distribution doesn't change much, and so the critical value of +/–1.96 can be used for most large samples. Since the shape of the t-distribution converges with the shape of the normal distribution after 500 degrees of freedom, this area between the two critical values (–1.96 and +1.96) corresponds with 95 per cent of the area under the normal curve (and 5 per cent of the area remains in the two tails).

You might be wondering how to find the critical value of the t-statistic, given that the shape of the t-distribution changes depending on the degrees of freedom (or sample size). One way is to use an online or interactive calculator designed for this purpose. (One such calculator is included in a spreadsheet tool available from the Student Resources area of the companion website for this book.) Another way is to use a printed, static table that lists the critical values of the t-statistic for commonly used alpha values and different degrees of freedom. Table 7.3 lists the critical value of the t-statistic, using an alpha value of 0.05, for t-distributions with various degrees of freedom. Notice that the critical value decreases as the degrees of freedom increase. Most often, though, researchers rely on statistical software to calculate the p-value associated with each t-statistic.

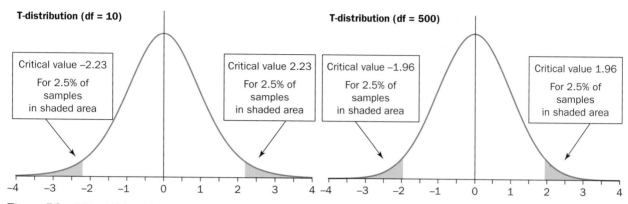

Figure 7.6 Critical Value of the T-Statistic for T-Distributions with Different Degrees of Freedom (alpha value = 0.05)

Ultimately, t-statistics and t-distributions are used to determine the probability of randomly selecting a sample with the observed difference between the group means, if the group means are equal in the population. If a researcher obtains a t-statistic that is lower than the critical value, or that is associated with a p-value of less than 0.05, there is less than a 5 per cent chance that the sample was selected from a population in which the group means are equal. In this situation, a researcher can assert that the sample was more likely to be selected from a population in which the group means are different. In essence, the critical value of a t-statistic provides a standardized answer to this question: "How much of a difference between means is big enough to say that there is likely a difference between the group means in the population?"

Let's return to the real-world data on mental-health scores and income adequacy to see t-statistics in action. Recall that in the CCHS sample overall, the average mental-health score was 54.3. Among the 2,626 people who had difficulty meeting their basic expenses, the average mental-health score was 49.7 (s.d. = 13.3), and among the 20,095 people who did not have difficulty meeting their basic expenses, the average mental-health score was 54.9 (s.d. = 10.3). Based on this sample,

Table 7.3 Critical Values of the T-Statistic for Distributions with Various Degrees of Freedom, Using an Alpha Value of 0.05

Degrees of Freedom	Critical Value of the T-Statistic	Degrees of Freedom	Critical Value of the T-Statistic
1	12.71	10	2.23
2	4.30	50	2.01
3	3.18	100	1.98
4	2.78	200	1.97
5	2.57	500	1.96

it seems that people who do not have adequate income have lower mental-health scores, on average, than those who do not.

A t-test makes it possible to assess whether this difference between the group means in the sample is large enough to confidently assert that there is likely a difference between the group means in the population. The research hypothesis is that income adequacy is related to mental health in the population; that is, the average mental-health scores for people who have difficulty meeting basic expenses and for people who do not have difficulty meeting basic expenses are different in the population. The null hypothesis is that income adequacy is not related to mental health in the population; that is, the average mental-health scores for the two groups are the same in the population. To calculate the t-statistic, the two group means, the standard deviations, and the number of cases in the sample are substituted into the formula, to obtain a t-statistic of 19.22 (or −19.22):

$$t = \frac{\bar{x}_1 - \bar{x}_2}{\sqrt{\dfrac{s_1^2}{n_1} + \dfrac{s_2^2}{n_2}}}$$

$$t = \frac{49.73 - 54.92}{\sqrt{\dfrac{13.35^2}{2626} + \dfrac{10.29^2}{20095}}}$$

$$= \frac{-5.19}{\sqrt{\dfrac{178.22}{2626} + \dfrac{105.88}{20095}}}$$

$$= \frac{-5.19}{\sqrt{0.068 + 0.005}}$$

$$= \frac{-5.19}{\sqrt{0.073}}$$

$$= \frac{-5.19}{0.27}$$

$$= -19.22$$

This t-statistic is compared to the critical value. Because the sample is so large, the t-distribution has the same shape as a normal distribution; thus, using an alpha value of 0.05, the critical value of the t-statistic is +/−1.96. Since 19.22 is well beyond 1.96, it would be very unlikely to randomly select this sample if there is no difference between the average mental-health scores of people who have difficulty meeting their basic expenses and those who do not, in the population. As a result, we can reject the null hypothesis and assert that there is likely a relationship between income adequacy and mental health in the population.

How Does It Look in SPSS?

T-Test of Independent Means

The Independent-Samples T-Test procedure produces results that look like those in Image 7.1.

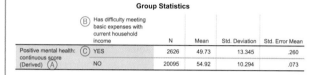

Group Statistics

	Has difficulty meeting basic expenses with current household income	N	Mean	Std. Deviation	Std. Error Mean
Positive mental health: continuous score (Derived) Ⓐ	Ⓒ YES	2626	49.73	13.345	.260
	NO	20095	54.92	10.294	.073

Ⓑ label on "Has difficulty meeting basic expenses with current household income"

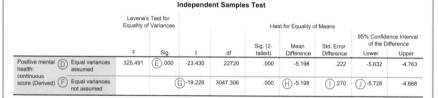

Independent Samples Test

		Levene's Test for Equality of Variances		t-test for Equality of Means						95% Confidence Interval of the Difference	
		F	Sig.	t	df	Sig. (2-tailed)	Mean Difference	Std. Error Difference		Lower	Upper
Positive mental health: continuous score (Derived) Ⓕ	Ⓓ Equal variances assumed	325.491	Ⓔ .000	-23.430	22720	.000	-5.198	.222		-5.632	-4.763
	Equal variances not assumed			Ⓖ -19.226	3047.306	.000	Ⓗ -5.198	Ⓘ .270		Ⓙ -5.728	-4.668

Image 7.1   **An SPSS T-Test of Independent Means**

A.   The label of the ratio-level dependent variable appears on the far left.

B.   The label at the top of the column shows the variable used to define the two groups.

C.   The first row of the "Group Statistics" shows the number of (valid) cases, the mean, the standard deviation, and the standard error of the mean for the cases in the first group. People who said "Yes," they had difficulty meeting basic expenses, had an average mental-health score of 49.73 (s.d. = 13.35). The second row shows the same statistics for the cases in the second group. People who said "No," they did not have difficulty meeting basic expenses, had an average mental-health score of 54.92 (s.d. = 10.29).

D.   Two versions of the "Independent Samples Test" are generated by default. The first row shows the results of a t-test of independent means that assumes that the variance of the dependent variable is equal for the two groups in the population. The second row shows the results of a t-test of independent means that does not make this assumption.

E.   "Levene's Test for Equality of Variances" assesses whether or not the variances of the two groups are likely to be equal in the population. If the value in the "Sig." column is less than 0.05, it indicates that the group variances are significantly different from each other in the population (at the $p < 0.05$ level), and the rest of the results shown in this row should *not* be reported. If the value in the "Sig." column is greater than 0.05, it means that the group variances are not significantly different from each other in the population (at the $p < 0.05$ level), and the rest of the results shown in this row can be reported.

Continued

F.   The results in the "Equal Variances Not Assumed" row match with the version of the t-statistic described in this chapter. The slight differences in the results are due to rounding.

G.   The result in the "t" column matches the t-statistic calculated in this chapter. The "df" column shows the degrees of freedom, which are used to determine the shape of the t-distribution that the t-statistic is evaluated against. The "Sig." column shows the p-value associated with this t-statistic. It shows the likelihood of randomly selecting a sample with the observed difference (or greater) between the group means, if there is no difference between the group means in the population. This p-value indicates that there is less than a 0.1 per cent chance of selecting this sample from a population in which the group means were the same ($p < 0.001$).

H.   The "Mean Difference" column shows the size of the difference between the two group means. There is a 5.2 point difference between the average mental-health score of people who have difficulty meeting basic expenses and people who do not ($49.7 - 54.9 = -5.2$). The sign (positive or negative) of the difference doesn't matter and can be ignored.

I.   The "Std. Error Difference" column shows an estimate of the standard error; it's an estimate of the standard deviation of the sampling distribution of mean differences.

J.   The "95% Confidence Interval of the Difference" columns show the lower and upper bounds of the mean difference that is likely to exist in the population. We are 95 per cent confident that the mean difference in the population is between 4.7 and 5.7 points. Since the upper and lower bounds of the confidence interval do not cross over 0, the difference between the group means is unlikely to be 0 in the larger population.

One-Tailed and Two-Tailed Tests

two-tailed significance tests Used to assess non-directional hypotheses; they are used most of the time.

So far, I have only discussed **two-tailed significance tests**. When a significance test relies on a symmetrical probability distribution, a two-tailed test takes into account the percentage (or proportion) of samples that produce a test statistic that is either higher than a positive value *or* lower than the corresponding negative value. For example, the critical values shown in Figure 7.6 and the left panel of Figure 7.7 account for the proportion of samples in *both* tails of the symmetrical t-distribution, added together. Thus, these are critical values for two-tailed t-tests. Two-tailed tests are used for non-directional hypotheses. Using both tails of the distribution to determine the critical value of a t-statistic indicates that the researcher is willing to consider the possibility that the first group has a mean that is either higher or lower than the second group.

one-tailed significance tests Used to assess directional hypotheses; they should be used only when there is a clear argument for why a relationship can go in only one direction.

When researchers have directional hypotheses, they use **one-tailed significance tests**. One-tailed t-tests allow researchers to only assess whether the first group has a higher mean than the second group but not whether it has a lower mean. As a result, one-tailed significance tests should be used only when a researcher is confident that the opposite relationship cannot exist. The critical value for a one-tailed t-test is lower than

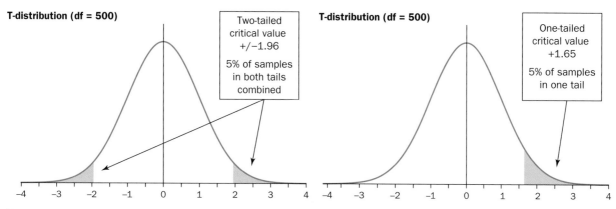

Figure 7.7   Critical Values for a Two-Tailed and One-Tailed T-Test (alpha value = 0.05)

the critical value for a two-tailed t-test for a sample of the same size. This is because, if a researcher is using an alpha value of 0.05, the 5 per cent of samples are not split between the two tails. Instead, they are in only one tail of the distribution, as shown in the right panel of Figure 7.7. In general, researchers should use two-tailed significance tests unless there is a clear and compelling argument for why a relationship can go in only one direction. Most of the time, researchers use two-tailed significance tests, in order to avoid missing a relationship that goes in an unexpected direction.

Other Types of T-Tests

The t-test of independent means described in this chapter is just one test within a larger family of t-tests. For instance, as I noted, there is an alternate version of the t-test of independent means that is used if a researcher is confident that the groups have equal variances in the population. In addition, there is a t-test for dependent means, sometimes called a paired t-test. The t-test for dependent means is used when data are paired or linked in some way, such as in experimental situations with a pre-test measure and a post-test measure. The calculation of the t-test for dependent means takes into account the fact that differences between the group means reflect changes within cases, and not differences between cases. If you are analyzing linked or paired data, be sure to do some additional research on how to conduct and interpret t-tests for dependent means.

Best Practices in Presenting Results

Writing about differences between means is relatively straightforward. To do so, clearly state the mean of each group and the differences between them. This might include a description of the effect size (Cohen's d). Always show the standard deviation for each mean that you report, as this allows readers to see the differences in variation between the groups, in addition to the differences between the means. Another common strategy is to present the 95 per cent confidence intervals for each mean, in either a table or graph. Confidence intervals allow the readers to assess which group means are likely to be different from the others, and how. And, unless you have experimental data, avoid using words such as *cause* and *effect*. Instead, point to the *association* or the *relationship* between the variables in the population.

Writing about Tests of Statistical Significance

It can take some practice to get used to writing about the results of statistical significance tests. First, to avoid confusing readers, the word *significant* should only ever be used to indicate that something is statistically significant. Do not use *significant* as a synonym for *important*; instead, use words like *substantial* or *meaningful* to highlight important differences. In general, tests of statistical significance should not be reported as if they are substantial findings on their own. Instead, report the results of statistical significance tests in parentheses, like a citation, following a claim about the likely presence or absence of a relationship in the population. Cite the statistical significance test as "proof" of your claim, in place of citing the academic literature. Each citation should include the test statistic, the relevant degrees of freedom, and the associated p-value (or an indication that the p-value is less than the alpha value the researchers are using). For example, a researcher reporting the results of the t-test of independent means for the CCHS example in this chapter might write this: "People who have difficulty meeting their basic expenses with their current household income have significantly lower mental-health scores, on average, than those who do not have difficulty meeting their basic expenses with their current household income ($t = -19.22$, df = 3,047, $p < 0.001$)."

When results are not statistically significant, researchers have several choices. If there is good evidence based on theory and the academic literature to suggest that there *should* be a relationship between the variables being tested, they can report that contrary to this expectation, no significant relationship was found. This might take the form of a statement such as, "Based on this sample data, there does not appear to be a relationship between these characteristics in the larger population." This statement should be followed by a citation that includes the non-significant test statistic, the relevant degrees of freedom, and the associated p-value. Nonetheless, researchers may still describe the relationship in the sample, as long as they make it clear that the relationship is not likely to exist in the population. Alternatively, if some results are not statistically significant and are incidental or secondary to the main research question, researchers may choose to omit them from their reporting.

Finally, it's important to once again acknowledge that tests of statistical significance are probability based. Researchers can never definitively claim that a relationship does or does not exist in a population based on the results of a statistical significance test using sample data. They only can assert that there is "likely" to be a relationship or that a relationship is "estimated" to be a certain size in the population. To improve the readability of results, many researchers do not use the language of likelihood when reporting every result, but you should strive to include this language wherever possible.

What You Have Learned

This chapter began by introducing hypothesis testing and distinguishing between directional and non-directional hypotheses. You learned that researchers assess the relationship between two variables by considering two things: the magnitude of the relationship and the reliability of the relationship. Cohen's d is a measure of effect size that is used to assess the magnitude of a relationship between a categorical independent variable and a ratio-level dependent variable. In the context of assessing the reliability of a relationship, you learned about research and null hypotheses, and statistical significance tests. Several key concepts, such as type I and type II errors, alpha values, and critical values were introduced. You then learned to calculate and interpret a t-test of independent means, which is used to assess the reliability of the relationship between

an independent variable with two attributes and a ratio-level dependent variable. This chapter concluded by presenting some guidelines for writing about means comparisons and tests of statistical significance.

The research focus of this chapter was the relationship between economic well-being and mental well-being. The results of the mental-health component of the Canadian Community Health Survey show that people who have difficulty meeting their basic expenses are likely to have lower levels of positive mental health, on average. These results demonstrate how structural inequalities are related to people's mental well-being. Reducing economic inequalities, and ensuring that all are able to meet their basic living expenses, may help to ensure improved mental health for all people living in Canada.

Check Your Understanding

Check to see if you understand the key concepts in this chapter by answering the following questions:

1. What is the difference between a directional hypothesis and a non-directional hypothesis?
2. What is the difference between a research hypothesis and a null hypothesis?
3. What information does Cohen's d provide? How is it interpreted?

4. What does it mean to say that something is "statistically significant"?
5. How is a t-statistic related to a sampling distribution of mean differences?
6. What do the results of a t-test of independent means show?
7. What is the difference between a one-tailed test of statistical significance and a two-tailed test of statistical significance?

Practice What You Have Learned

Check to see if you can apply the key concepts in this chapter by answering the following questions. Keep two decimal places in any calculations.

1. Your school's counselling centre is investigating whether students' anxiety is related to non-educational stressors. They administer an anxiety inventory—a series of questions designed to measure anxiety—to a random sample of 60 students from the school population. Each student who completes the inventory is assigned an anxiety score,

ranging from 0 to 50, where 0 indicates that the student has no anxiety and 50 indicates that the student has very high anxiety. A counsellor decides to investigate whether having paid employment is related to students' anxiety.

a. Determine which variable should be treated as independent and which should be treated as dependent.
b. State a non-directional hypothesis that the counsellor can test.

c. State a directional hypothesis that the counsellor can test.

d. State the null hypothesis for the non-directional and the directional hypotheses you identified in (b) and (c).

2. For the research described in question 1, the average anxiety score for students in the sample is 32.9 (s.d. = 13.5). For the 25 students who have paid employment, the average anxiety score is 37 (s.d. = 10); for the 35 students who do not have paid employment, the average anxiety score is 30 (s.d. = 14).

a. Describe the magnitude of the relationship between anxiety scores and having paid employment in the sample, that is, the size of the difference between the means.

b. Calculate Cohen's d. Would you describe the effect as small, medium, or large?

3. The counsellor decides to assess the relationship between anxiety scores and having paid employment in the population by testing a non-directional hypothesis.

a. Calculate the t-statistic.

b. Calculate the degrees of freedom of the t-statistic.

c. Using an alpha value of 0.05, the critical value of a t-statistic with these degrees of freedom is +/–2.00. Based on your answer to (a), state whether you reject or fail to reject the null hypothesis. Is there likely to be a relationship between having paid employment and anxiety scores in the school population?

4. A month later, school administrators decide to take a census. They have every student at the school complete the anxiety inventory. (Assume that every student in the population participates.) The school census results show that there is no difference between the average anxiety scores of students who have paid employment and those who do not. Given the results of question 3, determine whether the counselling centre research made an accurate conclusion. If not, has a type I or type II error been made?

5. The counselling centre decides to focus specifically on the experiences of first-year students. There are 18 first-year students included in the sample: 10 have paid employment and 8 do not. The anxiety scores of first-year students who have paid employment are as follows:

22	41	26	38	29
36	19	24	21	16

The anxiety scores of first-year students who do not have paid employment are as follows:

35	13	34	28	25
42	17	46		

a. Find the mean and the standard deviation of the anxiety scores of the 18 first-year students overall.

b. Find the mean and the standard deviation of the anxiety scores of the first-year students who have paid employment.

c. Find the mean and the standard deviation of the anxiety scores of the first-year students who do not have paid employment.

6. Using the results of question 5:

a. Describe the magnitude of the relationship between having paid employment and anxiety scores among first-year students in the sample, that is, the size of the difference between the means.

b. Calculate Cohen's d. Would you describe the effect as small, medium, or large?

7. A counsellor decides to assess the relationship between having paid employment and anxiety scores in the population of first-year students by testing a non-directional hypothesis. Using the scores listed in question 5:

a. Calculate the t-statistic.

b. Calculate the degrees of freedom of the t-statistic.

c. Using an alpha value of 0.05, the critical value of a t-statistic with these degrees of freedom is +/–2.16. Based on your answer to (a), state whether you reject or fail to reject the null hypothesis. Is there likely to be a relationship between having paid employment and anxiety scores in the population of first-year students at the school?

8. You want to know whether being a student is related to young people's mental health. You use data from the 2012 CCHS to compare the average positive

mental-health scores of people aged 20 to 24 who are currently attending school and those who are not.

a. Determine which variable should be treated as independent and which should be treated as dependent.
b. State a non-directional hypothesis that you can test.
c. State a directional hypothesis that you can test.
d. State the null hypothesis for the non-directional and the directional hypotheses you identified in (a) and (b).

9. Overall, people aged 20 to 24 who responded to the 2012 CCHS have an average positive mental-health score of 53.3 (s.d. = 11.2). The average among those who are currently attending school is 54.3 (s.d. = 10.2; n = 770), and the average among those who are not currently attending school is 52.7 (s.d. = 11.8; n = 1,063).

a. Describe the magnitude of the relationship between school attendance and positive mental-health scores in the sample, that is, the size of the difference between the means.
b. Calculate Cohen's d. Would you describe the effect as small, medium, or large?

10. You decide to assess the relationship between school attendance and positive mental-health scores in the population of people aged 20 to 24 in Canada by testing a non-directional hypothesis. Using the information in question 9:

a. Calculate the t-statistic.
b. Using an alpha value of 0.05, the critical value of a t-statistic with these degrees of freedom is +/−1.96. Based on your answer to (a), state whether you reject or fail to reject the null hypothesis. Is there likely to be a relationship between school attendance and positive mental-health scores in the population of people aged 20 to 24 in Canada?

11. The Mental Health Commission of Canada has an initiative called Opening Minds (OM) which attempts to address stigmas around mental health. As part of this initiative, researchers showed health-care providers and students in health care a one-hour DVD of a play that illustrates the stigma that people with mental-health issues face. Before and after watching the DVD, participants were given a survey that included a scale that measured attitudes and intentions toward people with mental illness (called the OMS-HC). Scores on the scale can range from 12 to 60, with lower scores indicating less stigma.

a. State a non-directional hypothesis that the researchers can test.
b. State the null hypothesis for the hypothesis you identified in (a).

12. In addition to surveying the health-care providers and students before and after watching the DVD, the researchers also asked participants to complete a follow-up survey one month later. Table 7.4 shows the average scores on the OMS-HC (the scale measuring attitudes toward people with mental illness) for each group before watching the DVD (pre-test), immediately after watching the DVD (post-test), and one month after watching the DVD (follow-up).

a. Describe the magnitude of the difference between the average pre-test, post-test, and follow-up

Table 7.4 **OMS-HC Scores across All Three Time Points by Participant Type: Practicing Health-Care Providers and Students**

	Pre-Test Score (95% CI)	Post-Test Score (95% CI)	Follow-Up Score (95% CI)	T-Test (Mean Change from Baseline to Follow-Up)
Practicing health-care providers (n = 22)	28.8 (27.9 − 29.7)	25.7 (24.6 − 26.8)	24.7 (23.4 − 26.0)	t(21) = 3.81 p = .001
Students (n = 20)	27.9 (26.9 − 29.0)	27.6 (26.6 − 28.6)	28.0 (26.7 − 29.3)	t(19) = 0.07 p = .943

Source: Knaak, Hawke, and Patten 2013, 10.

scores for the practicing health-care providers. Is this change an indicator of increased or reduced stigma?

b. Interpret the results of the t-test comparing the average pre-test scores to the average follow-up scores for health-care providers. *Note: These results are from a paired t-test, but the interpretation of the p-value is the same as in an independent samples t-test.*

13. Using the information in Table 7.4:

a. Describe the magnitude of the difference between the average pre-test, post-test, and follow-up scores for the students in health care. Is this change an indicator of increased or reduced stigma?

b. Interpret the results of the t-test comparing the average pre-test scores to the average follow-up scores for the students in health care. *Note: These results are from a paired t-test, but the interpretation of the p-value is the same as in an independent samples t-test.*

14. Given your answers to questions 12 and 13:

a. Determine which group experienced more change in attitudes after watching the DVD. Explain how you determined your answer.

b. Can this be considered a causal relationship? Explain why or why not.

15. World Mental Health day is held on 10 October of each year. In 2016, the OECD published an infographic in recognition of this day, part of which is shown in Figure 7.8.

a. What two types of graphs are used in this infographic?

b. In addition to the information in the two graphs, this infographic presents four other pieces of statistical information. What are they?

16. Using the infographic in Figure 7.8:

a. Identify the main message of the left panel on the "Mental health of young people."

b. Identify the main message of the right panel on "Treatment and outcomes."

c. List two strengths of this infographic.

d. Make one suggestion for how this infographic could be improved.

World Mental Health Day 2016
Good mental health throughout life

Mental health of young people

 Mental illness starts early. The median age of onset for any mental illness is 14 years-old and for anxiety disorders as low as 11 years-old

Disadvantages early in life have long-term implications:

- Poorer educational outcomes
- Higher risk of dropping out of school
- Bigger problems finding work after school

Percentages of young people leaving school early

No mental illness	14% leave early
Moderate mental illness	20% leave early
Severe mental illness	26% leave early

Treatment and outcomes

Treatment for mild and moderate mental illness is often unavailable, or patients have to wait a long time, or face high costs.

 23%   Only 23% of people with a severe mental disorder, and less than 10% of people with a moderate disorder, are in specialist treatment, e.g. with a psychiatrist or psychologist.

Undertreatment contributes to poor outcomes. In the typical OECD country, people with bipolar disorder or schizophrenia have a mortality rate 4-6 times higher than the general population.

Improvements are needed: effective diagnosis and coordination by GPs, access to psychological therapies, and good community services should be secured.

Figure 7.8   Mental Health Facts and Figures from the OECD

Source: Excerpt from OECD World Mental Health Day Infographic 2016.

Practice Using Statistical Software (IBM SPSS)

Answer these questions using IBM SPSS and the GSS27.sav or the GSS27_student.sav dataset available from the Student Resources area of the companion website for this book. Weight the data using the "Standardized person weight" [STD_WGHT] variable you created following the instructions in Chapter 5. Report two decimal places in your answers, unless fewer are printed by IBM SPSS. It is imperative that you save the dataset to keep any new variables that you create.

1. Use the Means procedure to determine how the average "Number of close friends" [SCF_100C] that people have is related to "Sex of respondent" [SEX] in the sample.

   a. Describe the magnitude of the relationship, that is, the size of the difference between the means in the sample.

   b. Use the statistics produced by SPSS to calculate Cohen's d. Would you describe the effect as small, medium, or large?

2. Use the Independent Samples T-Test procedure to assess whether the average "Number of close friends" [SCF_100C] that people have is related to "Sex of respondent" [SEX] in the population.

   a. State a non-directional research hypothesis for this relationship.

   b. State the null hypothesis associated with the non-directional research hypothesis you identified in (a).

   c. Interpret the t-statistic and its associated p-value in relation to the null hypothesis you identified in (b). In the population, is there likely to be a relationship between people's sex/gender and their number of close friends?

3. Compare the output and answers for questions 1 and 2 in "Practice Using Statistical Software" in this chapter to the output and answers for question 2 in the same section in Chapter 6.

   a. Which information displayed in the output from the Explore procedure, the Means procedure, and the Independent Samples T-Test procedure is exactly the same?

   b. Which information displayed in the output from the Explore procedure, the Means procedure, and the Independent Samples T-Test procedure is unique to each one?

   c. Does your answer to question 2(c) in this chapter correspond with your answer to question 2(b) in Chapter 6? Explain why or why not.

4. The variable "Trust people in general" [PCT_10] shows people's answers to this question: "Generally speaking, would you say that most people can be trusted or that you cannot be too careful in dealing with people?" Use the Means procedure to determine how the average "Number of close friends" [SCF_100C] that people have is related to this variable in the sample.

   a. Describe the magnitude of the relationship, that is, the size of the difference between the means.

   b. Use the statistics produced by SPSS to calculate Cohen's d. Would you describe the effect as small, medium, or large?

5. Use the Independent Samples T-Test procedure to assess whether, in the population, the average "Number of close friends" [SCF_100C] that people have is related to their general orientation towards trusting people ("Trust people in general" [PCT_10]).

   a. State a non-directional research hypothesis for this relationship.

   b. State the null hypothesis associated with the non-directional research hypothesis you identified in (a).

   c. Interpret the t-statistic and its associated p-value in relation to the null hypothesis you identified in (b). In the population, is there likely to be a relationship between people's general orientation towards trusting people and their number of close friends?

6. Use the Means procedure to determine whether, in the sample, the average "Number of close friends"

[SCF_100C] that people have is related to their highest level of education (using "Education - Highest degree (4 categories)" [DH1GED]).

a. Describe the magnitude of the relationship, that is, the size of the difference between the means.

b. Describe the overall pattern of the relationship between level of education and number of close friends. Is the relationship positive or negative?

7. Use the Independent Samples T-Test procedure and the cut-point method of defining groups to assess whether, in the population, the average "Number of close friends" [SCF_100C] that people have is related to whether or not they have a post-secondary education (using "Education - Highest degree (4 categories)"

[DH1GED]). People whose highest level of education is a post-secondary diploma (value 3) or a university degree (value 4) have a post-secondary education.

a. State a non-directional research hypothesis for this relationship.

b. State the null hypothesis associated with the non-directional research hypothesis you identified in (a).

c. Interpret the t-statistic and its associated p-value in relation to the null hypothesis you identified in (b). In the population, is there likely to be a relationship between whether or not people have a post-secondary education and their number of close friends?

Key Formulas

Cohen's d	$d = \dfrac{\bar{x}_1 - \bar{x}_2}{s}$
T-statistic (unequal variances)	$t = \dfrac{\bar{x}_1 - \bar{x}_2}{\sqrt{\dfrac{s_1^2}{n_1} + \dfrac{s_2^2}{n_2}}}$
Degrees of freedom of the t-statistic (unequal variances)	$df_t = \dfrac{\left(\dfrac{s_1^2}{n_1} + \dfrac{s_2^2}{n_2}\right)^2}{\dfrac{\left(s_1^2/n_1\right)^2}{(n_1 - 1)} + \dfrac{\left(s_2^2/n_2\right)^2}{(n_2 - 1)}}$

References

Keyes, Corey L. M. 2009. "Atlanta: Brief Description of the Mental Health Continuum Short Form (MCH-SF)." http://www.sociology.emory.edu/ckeyes/.

Knaak, Stephanie, Lisa Hawke, and Scott Patten. 2013. "That's Just Crazy Talk: Evaluation Report." Ottawa: Mental Health Commission of Canada. http://www.mentalhealthcommission.ca/sites/default/files/Stigma_OM_Thats_Just_Crazy_Talk_Evaluation_Report_ENG_0.pdf.

Mental Health Commission of Canada. 2012. "Changing Directions, Changing Lives: The Mental Health Strategy for Canada." Calgary: Mental Health Commission of Canada. http://strategy.mentalhealthcommission.ca/pdf/strategy-images-en.pdf.

Organisation for Economic Co-operation and Development (OECD). 2016. "World Mental Health Day 2016 (Infographic)." *Mental Health Systems in OECD Countries.* http://www.oecd.org/health/mental-health-systems.htm.

Statistics Canada. 2011. "Canadian Community Health Survey (CCHS)—Mental Health, Questionnaire." November 30. http://www23.statcan.gc.ca/imdb-bmdi/instrument/5105_Q1_V3-eng.pdf

———. 2014a. "Canadian Community Health Survey (CCHS)—Mental Health Public Use Microdata File: Derived Variable Specifications."

———. 2014b. "Canadian Community Health Survey, 2012: Mental Health Component." *Public Use Microdata File*. Ottawa, ON: Statistics Canada.

———. 2015. "Canadian Community Health Survey—Annual Component (CCHS)." April 20. http://www23.statcan.gc.ca/imdb/p2SV.pl?Function=getSurvey&Id=164081.

World Health Organization (WHO). 2014. "Mental Health: Strengthening Our Response." Fact Sheet 220. http://www.who.int/mediacentre/factsheets/fs220/en/.

8

Assessing Relationships by Comparing Group Means: ANOVA Tests

Learning Objectives

In this chapter, you will learn:

*   The logic of ANOVA tests

*   How to calculate and interpret a one-way ANOVA test

*   What post-hoc tests show and how to interpret them

*   How to write about ANOVA test results

Introduction

In Chapter 7, you learned how to assess the magnitude and reliability of a relationship between a categorical independent variable with two attributes and a ratio-level dependent variable. But social science researchers are often interested in independent variables that have more than two attributes or groups. For instance, you might be interested in analyzing the differences between people living in different regions of Canada, or between people with different ethnic backgrounds, or between people in different income quintiles. This chapter extends the ideas introduced in Chapter 7 by illustrating how to assess the magnitude and reliability of a relationship between a categorical independent variable with more than two attributes and a ratio-level dependent variable. In Chapter 9, you will learn how to test a hypothesis about the relationship between two categorical variables, and in Chapter 10 you will learn how to test a hypothesis about the relationship between two ratio-level variables.

Like Chapter 7, this chapter focuses on mental health and well-being, using data from the mental-health module of the 2012 Canadian Community Health Survey (CCHS). Once again, the main measure of mental health is a person's "positive mental health score" or, more briefly, their "mental-health score." (See the "Spotlight on Data" box in Chapter 7 for more information on how this was measured.) The highest possible mental-health score is 70, and the lowest is 0. This chapter continues to investigate the relationship between mental health and economic disadvantage by finding out the following:

- How is employment status related to people's mental health?
- How is household income related to people's mental health?

Means Comparison in Action

Let's begin by considering how employment status is related to mental health, among people aged 15 to 64 in Canada. Employment status is an indicator of economic advantage or disadvantage because it reflects a person's level of economic independence. Employment status is also related to social integration. In a longitudinal study of four OECD countries (Australia, Canada, Switzerland, and the UK), people who were not working had worse mental health, generally, than those who were working (Llena-Nozal 2009). It is also well established that unemployment is associated with poor mental-health outcomes (Paul and Moser 2009). In the CCHS, employment status is divided into three categories: (1) people who are employees (they work for someone else), (2) people who are self-employed (they operate their own business), and (3) people who were not employed during the past 12 months. Among working-aged people in Canada, almost two-thirds (64 per cent) are employees. About one in eight (12 per cent) are self-employed. The vast majority of self-employed people work alone, that is, they do not have other employees (Statistics Canada 2015). Finally, a quarter (25 per cent) of working-age people are not employed, for a variety of reasons: they might be caring for children or other family members, they might be unable to work, they might be looking for work but unable to find it, or they may have retired early.

As you learned in Chapter 7, the starting point for assessing a relationship between a categorical independent variable and a ratio-level dependent variable is comparing the group means. The size of the differences between the group means helps to describe the magnitude of the relationship. Table 8.1 lists the average mental-health scores for people who are employees, for people who are self-employed, and for people who are not employed. People who are self-employed have the highest average mental-health score: 56.0 (s.d. = 9.2). The average mental-health score for employees is only 1.5 points lower: 54.5 (s.d. = 10.3). But the average mental-health

Table 8.1   Comparing Three Group Means

Group	Frequency	Percentage	Mental-Health Score Mean	Std. Deviation
Employee	12,203	63.6	54.5	10.3
Self-employed	2,275	11.8	56.0	9.2
Not employed	4,724	24.6	52.2	12.7
Total/Overall	*19,202*	*100.0*	*54.1*	*10.9*

Source: Author generated; Calculated using data from Statistics Canada, 2014.

Table 8.2   **Calculating Three Effect Sizes**

Comparison	Group Means	Difference between Group Means	Standard Deviation (Overall)	Cohen's d (Effect Size)
Employee vs. self-employed	54.5 vs. 56.0	1.5	10.9	0.14
Employee vs. not employed	54.5 vs. 52.2	2.3	10.9	0.21
Self-employed vs. not employed	56.0 vs. 52.2	3.8	10.9	0.35

Source: Author generated; Calculated using data from Statistics Canada, 2014.

score for people who are not employed is 52.2 (s.d. = 12.7), 3.8 points lower than the average for the self-employed group and 2.3 points lower than the average for employees. You might notice that the overall mean and standard deviation of mental-health scores (overall mean = 54.1; overall s.d. = 10.9) is slightly different from those reported in Chapter 7; this is because this analysis excludes people older than 64, as well some people in the sample do not have an employment status recorded, and they are also excluded.

It may be difficult to assess whether a difference of two or three points in a person's mental-health score is meaningful. Cohen's d is used to help characterize the size of the effect. Remember that Cohen's d shows how the differences between group means compared to the standard deviation in the sample overall. In this example, the overall standard deviation is 10.9. The 1.5 point difference between the group means for employees and the self-employed is divided by 10.9, to get a Cohen's d of 0.14. (See Table 8.2.) The 2.3 point difference between the group means for employees and people who are not employed is likewise divided by 10.9, to get a Cohen's d of 0.21. The same approach is used for the difference between the group means for the self-employed and people who are not employed, to get a Cohen's d of 0.35.

Using the guidelines for interpreting Cohen's d described in Chapter 7, being an employee compared to not being employed has a small effect on mental-health scores, and being self-employed compared to not being employed also has a small effect on mental-health scores.

One-Way ANOVA Tests

In the previous chapter, you learned how to use a t-test of independent means to assess the reliability of a relationship between an independent variable with two attributes and a ratio-level dependent variable. When a categorical independent variable has more than two attributes, however, researchers must use a different strategy. One option is to use a series of t-tests to assess how each group compares to every other group. But, when the number of categories is large, this becomes tedious.

Instead, researchers typically use an ANalysis Of VAriance, or ANOVA, test to determine the reliability of a relationship between a categorical independent

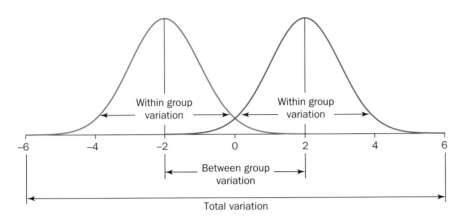

Figure 8.1 **The Logic of Between and Within Group Variation**

variable and a ratio-level dependent variable. The simplest type of ANOVA test is called a **one-way ANOVA test**. It is used to assess whether two or more group means are likely to be different from each other in the population, using sample data. For large samples with only two groups, the results of a one-way ANOVA test mirror the results of a t-test of independent means.

Conceptually, the one-way ANOVA test relies on assessing variation (or variances). The total variation in the sample is divided into two: the variation between the groups and the variation within the groups. Figure 8.1 shows the distribution of an imaginary dependent variable for two groups in a sample (this same idea extends to three or more groups). The teal curve on the left shows the distribution for the first group, and the purple curve on the right shows the distribution for the second group. The vertical line in the centre of each distribution represents the mean of that group. The variation *between* the groups captures the distance between the group means. The variation *within* the groups captures how far the cases in each group are spread out around their group mean. Finally, the total variation captures how far all of the cases in the sample are spread out around the sample mean, regardless of which group they belong to.

Using this conceptual framework, the main question becomes this: Is there more variation between the groups than there is within the groups? If so, group membership is likely related to people's values on the dependent variable. In contrast, if there is more variation within the groups than between the groups, then group membership likely isn't related to people's values on the dependent variable.

Figure 8.2 illustrates these two potential situations. In the top panel, the between group variation is larger than the within group variation; thus, there is no area where the two groups overlap. In this situation, a researcher is more confident that group membership is related to the dependent variable. In the bottom panel, the between group variation is smaller than the within group variation; thus, there is a large area where the two groups overlap. In this situation, a researcher is less confident that group membership is related to the dependent variable. Notice that in both panels of Figure 8.2, the group means are the same (−2 and +2). The

one-way ANOVA test A statistical significance test used to assess whether the means of two or more groups are likely to be different from each other in the population, using sample data.

The between group variation is larger than the within group variation:

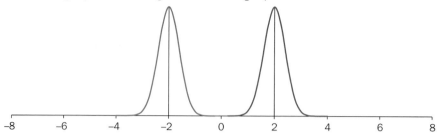

The between group variation is smaller than the within group variation:

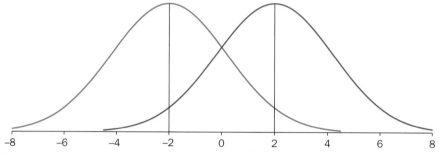

Figure 8.2   Two Patterns of Between and Within Group Variation

difference between the two situations is the amount of variation around the mean for each of the groups. This illustrates how ANOVA tests use the amount of variation to make inferences about whether there are differences between group means in the population. In fact, the ANOVA test statistic is a ratio that compares the between group variation to the within group variation.

Calculating One-Way ANOVA Tests

As you might expect, one-way ANOVA tests rely on assessing the amount of variation within a sample. In general, the variation is measured by summing together squared distances from a mean; the result is called the "sum of squares" (abbreviated SS). This approach should be familiar from the standard deviation and variance calculations, which also capture the amount of variation. The total sum of squares in a sample (SS_{total}) is equal to the between group sum of squares ($SS_{between\,group}$) plus the within group sum of squares ($SS_{within\,group}$):

total sum of squares (version 1)

$$SS_{total} = SS_{between\ group} + SS_{within\ group}$$

I'll use a tiny, hypothetical dataset to illustrate how to calculate each of the three sum of squares components (although ANOVA tests are typically used with larger samples). In this hypothetical example, the independent variable is employment status, which divides the sample into three groups: employees, the self-employed, and people who are not employed. The dependent variable is a person's mental-health score, which potentially ranges from 0 to 100. The research hypothesis is that there is a relationship between employment status and mental-health scores in the population, that is, the average mental-health score for at least one of the groups is different from the others. The null hypothesis is that there is no relationship between employment status and mental-health scores in the population. Or, another way to state the null hypothesis is to say that the group means are equal in the population; that is, the average mental-health scores for all three groups are equal in the population. Table 8.3 lists each hypothetical person's employment status and mental-health score.

Let's begin by finding the within group sum of squares for this sample because the formula is very similar to those that you have already learned. The within group sum of squares captures the distance of each case from the mean of its group. These distances are then squared and summed together (thus the name "sum of squares"). The formula for the within group sum of squares is:

$$SS_{within\ group} = \Sigma \left(x_i - \bar{x}_{group} \right)^2$$

within group sum of squares

Recall that the sigma (Σ) symbol means that you need to complete the calculation that follows the symbol for each case, and then add together all of the results. The x_i symbol represents the value of each individual case, and the $\bar{x}_{group}$ symbol represents the mean of the group that the case belongs to. The easiest way to calculate the within group sum of squares is to create a table such as the one in Figure 8.3. The first three columns are the same as Table 8.3. The fourth column

Table 8.3   A Sample of Hypothetical Cases

Person (Case)	Employment Status (Group)	Mental-Health Score
Manroop	Employed	80
Stephanie	Employed	56
Katya	Employed	78
Ramez	Employed	82
Runa	Self-employed	58
Sharon	Self-employed	84
Novera	Self-employed	76
Mac	Self-employed	82
Jodie	Not employed	57
Kristen	Not employed	42
Jullian	Not employed	63

$$\frac{\Sigma(x_i - \overline{x}_{group})^2}{}$$

Person (Case)	Employment Status (Group)	Mental-Health Score x_i	Mean of the Group the Case Belongs to $\overline{x}_{group}$	Deviation from Group Mean $(x_i - \overline{x}_{group})$	Squared Deviation from Group Mean $(x_i - \overline{x}_{group})^2$
Manroop	Employed	80	74	6	36
Stephanie	Employed	56	74	−18	324
Katya	Employed	78	74	4	16
Ramez	Employed	82	74	8	64
Runa	Self-employed	58	75	−17	289
Sharon	Self-employed	84	75	9	81
Novera	Self-employed	76	75	1	1
Mac	Self-employed	82	75	7	49
Jodie	Not employed	57	54	3	9
Kristen	Not employed	42	54	−12	144
Jullian	Not employed	63	54	9	81
				column sum:	1,094

The sigma symbol tells you to add the result of the equation for all of the cases.

Figure 8.3   Translating the Within Group Sum of Squares Formula into a Table (Hypothetical Data)

shows the mean of the group that the person belongs to. For the employed group, which includes the first four people, the mean is 74 ([80 + 56 + 78 + 82] ÷ 4 = 74). For the self-employed group, which includes the next four people, the group mean is 75 ([58 + 84 + 76 + 82] ÷ 4 = 75). For the not employed group, which includes the three remaining people, the group mean is 54 ([57 + 42 + 63] ÷ 3 = 54). The fifth column shows the difference between each person's mental-health score and the mean of the group. Notice that if you sum the differences in this column, the result is 0. In order to avoid concluding that there is no variation within each of these groups, a familiar strategy is used: the differences from the mean are squared (shown in the final column) before they are added together. Since the numbers in the final column sum to 1,094, the within group sum of squares for this sample is 1,094.

The calculation of the between group sum of squares relies on a similar logic. Although Figure 8.1 illustrates the between group variation as the distance between the group means, in practice, this distance is established by finding how far each group mean is from the overall mean. The between group sum of squares is calculated by finding the distance between each group mean and the overall mean, and then squaring it. But, because it relies on group means, there is one additional step in the calculation: before the squared distances from the overall mean are summed, they are multiplied by the number of cases in each group, to adjust for unequal group sizes:

$$SS_{between\ group} = \sum n_{group} \left(\overline{x}_{group} - \overline{x}_{total} \right)^2$$

**between group
sum of squares**

In the formula for the between group sum of squares, the subscript "total" is added to the usual symbol for the overall sample mean ($\overline{x}$), to distinguish it from the group means. Just like the within group sum of squares, calculating the between group sum of squares is easiest using a table, such as the one in Figure 8.4. Since there are three groups, the table has three rows: one for the employed group, one for the self-employed group, and one for the not employed group. The group means appear in the within group sum of squares table. The overall mean is calculated like any other sample mean: add together all of the mental-health scores, and divide by the number of cases (ignoring group membership). Then, find the difference between each group mean and the overall mean, and square it. The n_{group} symbol instructs you to multiply the result for each group by the number of cases in that group, before summing the rows. So, for the employed group, which includes four people, the squared deviation from the mean (26.0) is multiplied by 4 to get 104.0 in the final column. Similarly, for the self-employed group, which includes four people, the squared deviation from the mean (37.2) is multiplied by 4 to get 148.8 in the final column. For the not employed group, which includes three people, the squared deviation from the mean (222.0) is multiplied by 3 to get 666.0 in the final column. Since the numbers in the final column sum to 918.8, the between group sum of squares for this sample is equal to 918.8.

One way to find the total sum of squares is to simply add together the between group sum of squares and the within group sum of squares. But you can also independently find the total sum of squares using the table method; this is a useful way to check your calculations. The calculation of the total sum of squares is almost

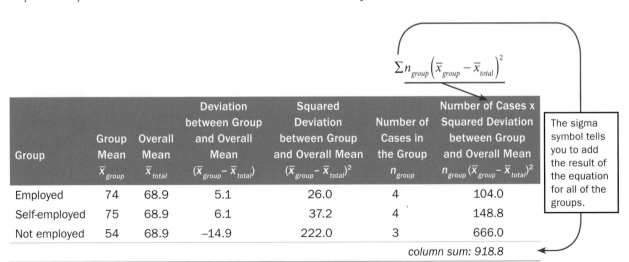

$$\sum n_{group} \left(\overline{x}_{group} - \overline{x}_{total} \right)^2$$

Group	Group Mean $\overline{x}_{group}$	Overall Mean $\overline{x}_{total}$	Deviation between Group and Overall Mean $(\overline{x}_{group} - \overline{x}_{total})$	Squared Deviation between Group and Overall Mean $(\overline{x}_{group} - \overline{x}_{total})^2$	Number of Cases in the Group n_{group}	Number of Cases x Squared Deviation between Group and Overall Mean $n_{group}(\overline{x}_{group} - \overline{x}_{total})^2$	
Employed	74	68.9	5.1	26.0	4	104.0	The sigma symbol tells you to add the result of the equation for all of the groups.
Self-employed	75	68.9	6.1	37.2	4	148.8	
Not employed	54	68.9	–14.9	222.0	3	666.0	
						column sum: 918.8	

Figure 8.4   Translating the Between Group Sum of Squares Formula into a Table (Hypothetical Data)

Source: Author generated; Calculated using data from Statistics Canada, 2014.

identical to the calculation of the within group sum of squares, with one important change: instead of finding the deviation of each case from its group mean, find the deviation of each case from the overall mean. The only difference between the formula for the within group sum of squares and the formula for the total sum of squares is the subscript that designates which mean is subtracted from the value for each case:

total sum of squares (version 2)

$$SS_{total} = \Sigma \left(x_i - \bar{x}_{total} \right)^2$$

To calculate the total sum of squares, use a table like the one in Figure 8.3, but replace the fourth column with the overall mean. The table in Figure 8.5 incorporates this change and shows how it affects the remaining calculations. Since the results in the final column sum to 2,012.8, the total sum of squares for this sample is equal to 2,012.8. This gives us confidence in our previous calculations, since the within group sum of squares (1,094) and the between group sum of squares (918.8) also sum to 2,012.8.

Person (Case)	Employment Status (Group)	Mental-Health Score x_i	Overall Mean $\bar{x}_{total}$	Deviation from Overall Mean $(x_i - \bar{x}_{total})$	Squared Deviation from Overall Mean $(x_i - \bar{x}_{total})^2$
Manroop	Employed	80	68.9	11.1	123.2
Stephanie	Employed	56	68.9	−12.9	166.4
Katya	Employed	78	68.9	9.1	82.8
Ramez	Employed	82	68.9	13.1	171.6
Runa	Self-employed	58	68.9	−10.9	118.8
Sharon	Self-employed	84	68.9	15.1	228.0
Novera	Self-employed	76	68.9	7.1	50.4
Mac	Self-employed	82	68.9	13.1	171.6
Jodie	Not employed	57	68.9	−11.9	141.6
Kristen	Not employed	42	68.9	−26.9	723.6
Jullian	Not employed	63	68.9	−5.9	34.8
					column sum: 2,012.8

The sigma symbol tells you to add the result of the equation for all of the cases.

Figure 8.5   Translating the Total Sum of Squares Formula into a Table

Calculating an F-Statistic

At this point, you know how to find the between group sum of squares and the within group sum of squares using sample data. You might be tempted decide whether between group variation is bigger or smaller than the within group variation by simply comparing these two numbers. In the hypothetical example, the between group sum of squares (918.8) is smaller than the within group sum of squares (1,094). But, on its own, this information isn't that useful because it doesn't account for the number of groups being compared or for the sample size. And, much like in a t-test, we are left with the question: In the sample, how much bigger does the between group variation need to be than the within group variation in order for a researcher to assert that there is likely a difference between the group means in the population? Instead of a t-statistic and a t-distribution, one-way ANOVA tests use an F-statistic and an **F-distribution** to make this assessment.

F-distribution A probability distribution used to determine the likelihood of randomly selecting a sample with the observed ratio of between group variation to within group variation (or a larger ratio), if the group means are equal in the population.

The F-statistic accounts for the number of groups being compared and for the sample size by dividing each sum of squares by its degrees of freedom. The degrees of freedom of the between group sum of squares accounts for the number of groups being compared. The between group degrees of freedom is equal to the number of groups minus 1. You subtract 1 from the number of groups because if you know the overall mean and the number of cases in each group, the mean of the final group can always be calculated from the means of the other groups; it is no longer a "free" parameter.

$$df_{between\ group} = (number\ of\ groups - 1)$$

between group degrees of freedom (df_1)

The between group degrees of freedom is sometimes denoted as df_1. In the hypothetical example in this chapter there are three groups, and so the between group degrees of freedom is 2 ($3 - 1 = 2$).

The degrees of freedom of the within group sum of squares accounts for the sample size. Like other statistics that rely on means, the number of degrees of freedom is the number of cases minus 1, for each group. You subtract 1 from the number of cases because if you know a group mean, the score of the final case can always be calculated from the scores of the other cases; it is no longer a "free" parameter. (See the description of "degrees of freedom" in Chapter 7.) So, the within group degrees of freedom is equal to the number of cases in the first group minus 1, plus the number of cases in the second group minus 1, plus the number of cases in the third group minus 1, and so on for all of the groups:

$$df_{within\ group} = (n_{group1} - 1) + (n_{group2} - 1) + (n_{group3} - 1) + \ldots$$

within group degrees of freedom (df_2)

The within group degrees of freedom is sometimes denoted as df_2. In the hypothetical example in this chapter, the employed group has four cases, the self-employed group has four cases, and the not employed group has three cases. Thus, the within group degrees of freedom is $4 - 1$, plus $4 - 1$, plus $3 - 1$ and, thus, is equal to 8 ($3 + 3 + 2 = 8$).

The F-statistic shows whether the between group variation is larger or smaller than the within group variation, after accounting for the sample size and the number of groups. As a result, it is sometimes called an F-ratio. The numerator of the F-statistic captures information about the between group variation, and the denominator captures information about the within group variation. The between group variation is captured in the numerator by dividing the between group sum of squares by the between group degrees of freedom. The within group variation is captured in the denominator by dividing the within group sum of squares by the within group degrees of freedom. It might be helpful to think about the numerator of the fraction in the F-statistic equation as showing the amount of "explained variation," or the amount of variation attributable to group membership, and the denominator as showing the amount of "unexplained variation," or the amount of variation not attributable to group membership. The formula for an F-statistic or F-ratio is:

F-statistic (or F-ratio)

$$F = \frac{SS_{between\ group} / df_{between\ group}}{SS_{within\ group} / df_{within\ group}}$$

Let's calculate the F-statistic for our hypothetical example by substituting the between group sum of squares, the within group sum of squares, and the degrees of freedom into the equation:

$$F = \frac{SS_{between\ group} / df_{between\ group}}{SS_{within\ group} / df_{within\ group}}$$

$$F = \frac{918.8/2}{1094/8}$$

$$= \frac{459.4}{136.8}$$

$$= 3.36$$

In Chapter 7, you learned how to interpret a t-statistic in relation to a t-distribution, which changes shape depending on the size of the sample (or degrees of freedom). Similarly, the F-statistic is interpreted in relation to an F-distribution, which also changes shape depending on the number of groups being compared and the size of the sample (captured in the between group and within group degrees of freedom). Figure 8.6 shows how the F-distribution changes shape with different degrees of freedom. Unlike the t-distribution, the F-distribution is asymmetrical; that

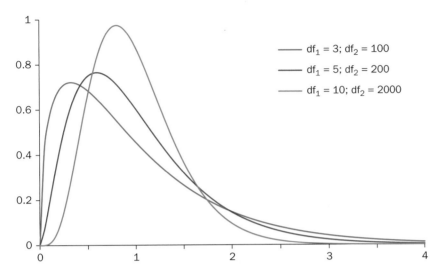

Figure 8.6   The F-Distribution, with Various Degrees of Freedom

is, it is right-skewed and has only one tail. The distribution begins at 0; it is not possible to obtain a negative F-statistic, since it's not possible to have negative variation.

The F-distribution is a probability distribution. The F-statistic produced by a randomly selected sample identifies a point on the horizontal axis of the F-distribution. The amount of the area under the curve that is beyond the value of the F-statistic on the horizontal axis shows the probability of randomly selecting a sample with the observed ratio of between group variation to within group variation (or a larger ratio), if the group means are equal in the population. Like all tests of statistical significance, this probability is represented as a p-value. The p-value associated with each F-statistic shows the probability of selecting a sample with the observed relationship (or a relationship of greater magnitude), if the null hypothesis is true in the population.

Figure 8.7 shows an F-distribution corresponding to the degrees of freedom in the hypothetical example ($df_1 = 2$, $df_2 = 8$). Nine per cent of the area under the curve is beyond the calculated F-statistic of 3.36. Thus, the probability of randomly selecting a sample that produces an F-statistic of this size (or larger) from a population in which the group means are equal is 0.09. In other words, even if there is no relationship between employment status and mental-health scores in the population, there is still a 9 per cent chance of randomly selecting a sample (of 11 cases) with these results. Since the p-value is larger than 0.05 (or 5 per cent), we fail to reject the null hypothesis; for our hypothetical example, then, we cannot be confident that employment status is related to people's mental-health scores in the population. In other words, it is entirely possible that the average mental-health scores are the same in each group in the (hypothetical) population. When the amount of between group variation is compared to the amount of within group variation in the sample, the ratio isn't large enough to assert that, in the population, there is likely a relationship between group membership and the dependent variable.

Researchers typically use online or interactive calculators, or statistical software to determine the proportion of the area under the curve that is beyond a

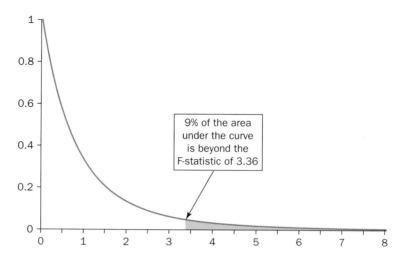

Figure 8.7 **Finding the Probability of Randomly Selecting a Sample with an F-Statistic of 3.36 or Higher Using an F-Distribution (df$_1$ = 2, df$_2$ = 8)**

specific point on the horizontal axis of an F-distribution. (One such calculator is included in a spreadsheet tool available from the Student Resources area of the companion website for this book.) Alternatively, researchers can use online calculators or printed tables to determine the critical value of the F-statistic for a specific F-distribution. Recall that the critical value of each test statistic is the cut-off point where the probability of making a type I error matches the alpha value. For an alpha value of 0.05, the critical value of the F-statistic is the point on the horizontal axis of an F-distribution that has only 5 per cent of the area under the curve beyond it.

For the F-distribution corresponding to the hypothetical example, using an alpha value of 0.05, the critical value is 4.46. (See Figure 8.8.) In other words, to obtain a p-value that is less than 0.05, the F-statistic must be higher than 4.46. Since the F-statistic for the hypothetical example (3.36) is lower than the critical value (4.46), we cannot assert that a relationship between employment status and mental health is likely to exist in the population.

Now that you know how to calculate and interpret one-way ANOVA tests, let's return to the real-world CCHS example of employment status and mental-health scores. Table 8.1 shows the means for each of the three groups: employees, the self-employed, and people who were not employed. The differences between the group means are relatively small. Based on this sample data, are there likely to be differences between the group means in the larger population?

An analysis using SPSS shows that for this real-world example, the between group sum of squares is 27,107 (with 2 degrees of freedom) and the within group sum of squares is 2,242,076 (with 19,199 degrees of freedom). The resulting F-statistic is 116. Using an alpha value of 0.05, the critical value of the F-statistic for an F-distribution corresponding to these degrees of freedom is 3. Since the F-statistic of 116 is higher than 3, we can assert that there is likely a relationship between employment status and mental-health scores in the population of working-aged people living in Canada; that is, at least one of the groups has an average mental-health score that is different from the others.

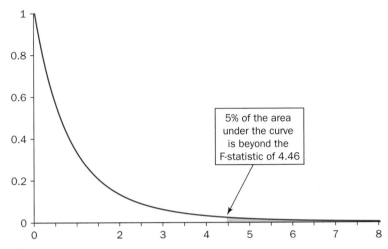

Figure 8.8   Critical Value of the F-Statistic for an F-Distribution (df$_1$ = 2, df$_2$ = 8; alpha value = 0.05)

Step-by-Step: One-Way ANOVA Tests

Step 1: Identify the research hypothesis and the null hypothesis.

Between Group Sum of Squares

$$SS_{between\ group} = \sum n_{group} \left(\bar{x}_{group} - \bar{x}_{total} \right)^2$$

Step 2: Count the total number of cases in each group (n_{group1}, n_{group2}, n_{group3}, . . .) Do not include cases that are missing information for the variable x.

Step 3: Find the mean of the variable for the cases in each group ($\bar{x}_{group1}$, $\bar{x}_{group2}$, $\bar{x}_{group3}$, . . .).

Step 4: Find the overall mean of the variable ($\bar{x}_{total}$). Ignore the groups when calculating the overall mean (but each case used in the calculation should belong to one of the groups).

Step 5: For *each group separately*, subtract the overall mean (from Step 4) from the group mean (from Step 3). Complete this calculation once for each group.

Step 6: For *each group separately*, square the result of Step 5. Complete this calculation once for each group.

Step 7: For *each group separately*, multiply the result of Step 6 by the number of cases in the group (from Step 2). Complete this calculation once for each group.

Step 8: Add together the results of Step 7 for each group to find the between group sum of squares.

Continued

Within Group Sum of Squares

$$SS_{within\ group} = \Sigma\left(x_i - \overline{x}_{group}\right)^2$$

Step 9: Identify which group each case belongs to.

Step 10: For *each case individually*, subtract the mean of its group (from Step 3) from the value of the case (x_i) to find the deviation from the group mean. Complete this calculation once for each case.

Step 11: For *each case individually*, square the deviation from the group mean (the result of Step 10). Complete this calculation once for each case.

Step 12: Add together all of the squared deviations from the group means (the results of Step 11) to find the within group sum of squares.

Total Sum of Squares

Tip: Only the between group sum of squares and the within group sum of squares are needed to calculate the F-statistic. Calculating the total sum of squares provides a way to check your work: the total sum of squares should be equal to the between group sum of squares plus the within group sum of squares.

$$SS_{total} = \Sigma\left(x_i - \overline{x}_{total}\right)^2$$

Step 13: For *each case individually*, subtract the overall mean (from Step 4) from the value of the case (x_i) to find the deviation from the overall mean. Complete this calculation once for each case.

Step 14: For *each case individually*, square the deviation from the overall mean (the result of Step 13). Complete this calculation once for each case.

Step 15: Add together all of the squared deviations from the overall mean (the results of Step 14) to find the total sum of squares.

Degrees of Freedom

$$df_{between\ group} = \left(number\ of\ groups - 1\right)$$

$$df_{within\ group} = \left(n_{group1} - 1\right) + \left(n_{group2} - 1\right) + \left(n_{group3} - 1\right) + \ldots$$

Step 16: Subtract 1 from the number of groups to find the between group degrees of freedom.

Step 17: For *each group separately*, subtract 1 from the number of cases in the group (from Step 2). Complete this calculation once for each group.

Step 18: Add together the results of Step 17 for each group to find the within group degrees of freedom.

F-Statistic

$$F = \frac{SS_{between\ group}\ /\ df_{between\ group}}{SS_{within\ group}\ /\ df_{within\ group}}$$

Step 19: Divide the between group sum of squares (from Step 8) by the between group degrees of freedom (from Step 16) to find the numerator of the F-statistic equation.

Step 20: Divide the within group sum of squares (from Step 12) by the within group degrees of freedom (from Step 18) to find the denominator of the F-statistic equation.

Step 21: Divide the result of Step 19 by the result of Step 20 to find the F-statistic.

Statistical Significance

Step 22: Determine the alpha value you want to use for the one-way ANOVA test. The most common alpha value used in the social sciences is 0.05.

Step 23: Determine the critical value of the F-statistic, using the alpha value you selected in Step 22, for an F-distribution with the degrees of freedom you found in Steps 16 and 18. This can be done using a printed table of the critical values of the F-distribution, an online calculator, or the spreadsheet tool available from the Student Resources area of the companion website for this book.

Step 24: Compare the F-statistic (from Step 21) to the critical value (from Step 23). If the F-statistic is higher than the critical value, the result is statistically significant: you can reject the null hypothesis and state that at least one of the group means is likely to be different in the population. If the F-statistic is lower than the critical value, the result is not statistically significant: you fail to reject the null hypothesis and state that you are not confident that the group means are different in the population.

Post-Hoc Tests

When a one-way ANOVA test produces a statistically significant result, it indicates that *at least one* group mean is likely to be different from another group mean in the population. It does not indicate that every group mean will be significantly different from every other group mean. Because one-way ANOVA tests are calculated by pooling together all of the within group variation and all of the between group variation, they don't provide any information about *which* group means are likely to be different than others in the population. In order to learn more, researchers use **post-hoc tests**. *Post hoc* is Latin for "after this." Post-hoc tests are performed after a statistically significant relationship has been established, and they typically provide more details about the relationship that was established.

post-hoc tests Tests conducted after a statistically significant relationship has been established.

There are many different types of post-hoc tests that researchers can use in conjunction with a one-way ANOVA test. One of the most common is called the "least significant differences" (LSD) test. This test is popular because it is the equivalent of running a t-test of independent means between each possible pair of groups. For instance, since the categorical variable for employment status divides the sample into three groups, three pairs of relationships are tested: employees are compared to the self-employed, employees are compared to people who are not employed, and the self-employed are compared to people who are not employed. (These three comparisons correspond to the three rows in Table 8.2.)

The results of the one-way ANOVA test using CCHS data shows that there is likely a relationship between employment status and mental-health scores for people living in Canada. Let's use an LSD post-hoc test to determine which groups have an average mental-health score that is likely to be different from the others. Table 8.4 shows the results of the LSD post-hoc test in a matrix format. The values in each cell show the size of the difference between group means. For each cell, the mean mental-health score for the group in the column is subtracted from the mean mental-health score of the group in the row, to find the mean difference. As a result, a positive number indicates that the group in the row has a mean that is *higher* than the group in the column. A negative number indicates that the group in the row has a mean that is *lower* than the group in the column. So, on average, employees have a mental-health score that is 1.5 points lower than the mental-health score for self-employed people. And, on average, employees have a mental-health score that is 2.3 points higher than people who are not in the labour force. Finally, people who are self-employed have a mental-health score that is, on average, 3.8 points higher than people who are not in the labour force. (These mean differences correspond with those shown in Table 8.2.)

Technically, there is no need to list the mean differences below the diagonal in the matrix (printed in grey), since they are the same as those above the diagonal, only with the opposite sign. Since, on average, employees have a mental-health score that is 1.5 points lower than that of the self-employed, the reverse must also be true: on average, the self-employed have a mental-health score that is 1.5 points higher than that of employees.

The asterisk (*) beside a mean difference in the post-hoc table shows that a t-test of independent means for that pair of groups is statistically significant, using an alpha value of 0.05. In other words, a t-test of independent means comparing the average mental-health score for employees and the average mental-health score for the self-employed (omitting people who are not employed) produces a p-value that is less than 0.05. Since all of the mean differences are statistically significant, it is likely that all three of the group means are different from each other in the population.

I describe LSD post-hoc tests here because they are easy to understand—and because you just learned about t-tests in Chapter 7. One limitation of LSD post-hoc tests, however, is that they use the version of the t-test of independent means that assumes that the group variances are equal in the population. There are many other

Table 8.4   **The Results of Least Significant Differences (LSD) Post-Hoc Tests**

	Employee	Self-employed	Not in the labour force
Employee	–	−1.5*	2.3*
Self-employed	1.5*	–	3.8*
Not in the labour force	−2.3*	−3.8*	–

*Indicates that the mean difference is significant at the p < 0.05 level.
Source: Author generated; Calculated using data from Statistics Canada, 2014.

post-hoc tests that can be used with a one-way ANOVA test, but it is beyond the scope of this book to describe them all. Some post-hoc tests do not assume that the group variances are equal in the population. Other tests—such as the Bonferroni post-hoc test—make a correction to compensate for simultaneously conducting multiple tests of statistical significance. Recall that an alpha value of 0.05 is associated with a 5 per cent chance of making a type I error (when no relationship actually exists in the population, but a researcher claims that it likely does exist based on analysis of sample data). This is equivalent to a 1 in 20 chance of making a type I error. As a result, when researchers conduct multiple significance tests, one may produce a p-value less than 0.05 just by chance, even though no relationship exists in the population. For example, post-hoc tests for a one-way ANOVA test with six groups results in 21 unique pairs of groups—and a t-test of independent means is performed on each pair. Since there is a 1 in 20 chance of making a type I error, 1 of the 21 tests is expected to produce a p-value less than 0.05 just by chance, and not because a relationship actually exists in the population. And a researcher has no way of knowing *which* of the 21 tests produced a significant result just by chance. To compensate, several post-hoc tests adjust the alpha values to account for the number of significance tests being conducted at the same time.

One-Way ANOVA Tests in Action

So far, we have established that employment status is likely related to the mental health of working-aged people living in Canada, although there is only a small effect. Let's now expand our inquiry to find out whether household income is related to mental-health scores, for all people living in Canada. Since there is more variation in household income, there are more groups (and pairs of groups) to consider. In this analysis, I use a relative measure of household income—income quintile group—which is calculated separately for each province in order to better account for regional income disparities. People in the highest income quintile group are people who live in households with incomes that are in the top 20 per cent, within their province. People in the lowest income quintile group are people who live in households with incomes that are in the lowest 20 per cent, within their province. This measure captures people's relative economic advantage or disadvantage by comparing how their household income compares to other households in the same geographic area.

The independent variable in this example is a person's household income quintile group, and, once again, the dependent variable is a person's positive mental-health score. The research hypothesis is that relative household income is related to mental-health scores in the population. The null hypothesis is that there is no relationship between relative household income and mental-health scores in the population. Or, alternatively, the null hypothesis is that each household income quintile group has the same average mental-health score in the population.

A means comparison shows a clear pattern in the relationship between mental-health scores and household income quintile group. Table 8.5 shows that people with household incomes in the lowest quintile group have the lowest average

Table 8.5   The Relationship between Household Income Quintile Group and Mental-Health Scores

Group	Frequency	Percentage	Mental-Health Score Mean	Std. Deviation
Lowest income quintile group	4,419	19.3	52.2	12.2
Income quintile group 2	4,550	19.9	53.9	11.2
Income quintile group 3	4,629	20.3	54.4	10.8
Income quintile group 4	4,705	20.6	55.0	10.1
Highest income quintile group	4,547	19.9	55.8	9.5
Total/Overall	22,850	100.0	54.3	10.8

Source: Author generated; Calculated using data from Statistics Canada, 2014.

mental-health scores. Average mental-health scores are progressively higher for each higher quintile group, and people with household incomes in the highest quintile group have the highest average mental-health scores. Despite this clear pattern, the differences between means are relatively small. The largest mean difference, between the highest income quintile group and the lowest, is only 3.6 points (55.8 compared to 52.2). This mean difference has a Cohen's d of 0.33 (3.6 divided by the pooled standard deviation of 10.8).

A one-way ANOVA test is used to assess whether there is likely a relationship between household income quintile group and mental-health scores in the larger population. The results show that the F-statistic is 70.364, which is associated with a p-value less than 0.05. It would be unlikely to randomly select a sample with this relationship if the null hypothesis were true in the population. Thus, the null hypothesis can be rejected, and we can conclude that there is likely a relationship between household income quintile group and mental-health scores in the larger population.

Post-hoc tests show which group means are significantly different from the others. In the slightly larger matrix of post-hoc results in Table 8.6, I have not shown the mean differences below the diagonal, for clarity. These post-hoc results show that there is a statistically significant difference between each pair of household income groups, using an alpha value of 0.05.

Plotting the mean differences on an error-bar graph makes these results easier to understand. (See Figure 8.9.) Error-bar graphs show the mean of each group and its 95 per cent confidence interval. The graph helps to contextualize the results of the one-way ANOVA test and post-hoc tests by illustrating that the average mental-health score for people in households in the lowest income quintile group is likely to be much lower than the average mental-health score for people in households in the second income quintile group. (I have adjusted the scale of the y-axis to show the differences more clearly.) The graph also shows that the average mental-health score for people in households in the highest income quintile group is likely to be much higher than the average mental-health score for people in households in the fourth income quintile group. Overlaps between the confidence intervals for the three middle household income quintile groups make it harder to be confident that these

Table 8.6   **The Results of Post-Hoc Tests for the Relationship between Household Income Quintile Group and Mental-Health Scores, Using the Least Significant Differences Method**

	Lowest quintile group	2nd quintile group	3rd quintile group	4th quintile group	Highest quintile group
Lowest quintile group	–	–1.7*	–2.2*	–2.8*	–3.6*
2nd quintile group		–	–0.5*	–1.1*	–1.9*
3rd quintile group			–	–0.6*	–1.4*
4th quintile group				–	–0.8*
Highest quintile group					–

*Indicates that the mean difference is significant at the $p < 0.05$ level.
Source: Author generated; Calculated using data from Statistics Canada, 2014.

groups have different average mental-health scores in the population, despite the statistically significant post-hoc tests. The size of the differences between these three group means (0.5 and 0.6 points) also makes it harder to be confident that there is a practical or meaningful difference in mental-health scores.

Overall, these results suggest that the relationship between household income and mental health is particularly pronounced for people with the lowest relative household incomes and for people with the highest relative household incomes.

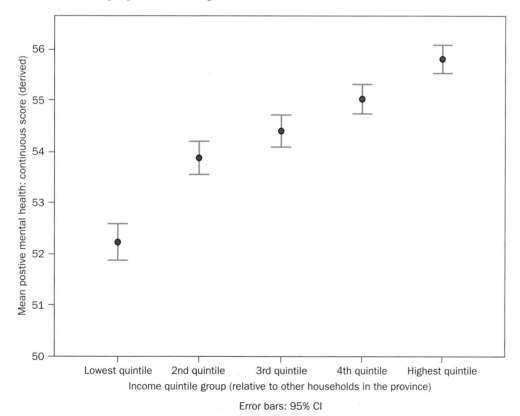

Figure 8.9   **Error-Bar Graph of the Relationship between Mental-Health Scores and Household Income Quintile Group**

Source: Author generated; Calculated using data from Statistics Canada, 2014.

The direction of the relationship matches our expectations: those with lower household incomes have lower positive mental-health scores, and those with higher household incomes have higher positive mental-health scores, on average. These results provide additional support for the idea that economic inequality is a social determinant of mental health (Manseau 2014).

How Does It Look in SPSS?

One-Way ANOVA

The One-Way ANOVA procedure, with the "LSD" test selected in the Post-Hoc option, produces results that look like those in Image 8.1.

ANOVA

Positive mental health: continuous score (Derived) (A)

	(B) Sum of Squares	(C) df	(D) Mean Square	(E) F	(F) Sig.
Between Groups	32714.169	4	8178.542	70.364	.000
Within Groups	2655447.716	22846	116.233		
Total	2688161.885	22850			

Post Hoc Tests

Multiple Comparisons

Dependent Variable: Positive mental health: continuous score (Derived)

(G) LSD

(H) (I) Income quintile group (relative to other households in the province)	(J) Income quintile group (relative to other households in the province)	(J) Mean Difference (I-J)	(K) Std. Error	(L) Sig.	(M) 95% Confidence Interval Lower Bound	Upper Bound
Lowest quintile group	2nd quintile group	-1.650*	.228	.000	-2.09	-1.20
	3rd quintile group	-2.174*	.227	.000	-2.62	-1.73
	4th quintile group	-2.802*	.226	.000	-3.24	-2.36
	Highest quintile group	-3.582*	.228	.000	-4.03	-3.14
2nd quintile group	Lowest quintile group	1.650*	.228	.000	1.20	2.09
	3rd quintile group	-.526*	.225	.019	-.97	-.09
	4th quintile group	-1.154*	.224	.000	-1.59	-.71
	Highest quintile group	-1.934*	.226	.000	-2.38	-1.49
3rd quintile group	Lowest quintile group	2.174*	.227	.000	1.73	2.62
	2nd quintile group	.526*	.225	.019	.09	.97
	4th quintile group	-.627*	.223	.005	-1.06	-.19
	Highest quintile group	-1.408*	.225	.000	-1.85	-.97
4th quintile group	Lowest quintile group	2.802*	.226	.000	2.36	3.24
	2nd quintile group	1.154*	.224	.000	.71	1.59
	3rd quintile group	.627*	.223	.005	.19	1.06
	Highest quintile group	-.781*	.224	.000	-1.22	-.34
Highest quintile group	Lowest quintile group	3.582*	.228	.000	3.14	4.03
	2nd quintile group	1.934*	.226	.000	1.49	2.38
	3rd quintile group	1.408*	.225	.000	.97	1.85
	4th quintile group	.781*	.224	.000	.34	1.22

*. The mean difference is significant at the 0.05 level.

Image 8.1   **An SPSS One-Way ANOVA Test**

A. The label of the ratio-level dependent variable appears at the top of the ANOVA results.

B. The "Sum of Squares" column lists the between group sum of squares, the within group sum of squares, and the total sum of squares in three rows. The total sum of squares is the sum of the between group sum of squares and the within group sum of squares.

C. The "df" column lists the between group degrees of freedom, the within group degrees of freedom, and the total degrees of freedom in three rows. The total degrees of freedom is the sum of the between group degrees of freedom and the within group degrees of freedom.

D. The "Mean Square" column shows the sum of squares divided by their respective degrees of freedom, for the between group row and the within group row. The number in the between group row ($32{,}714.169 \div 4 = 8{,}178.542$) becomes the numerator in the F-statistic equation. The number in the within group row ($2{,}655{,}447.716 \div 22{,}846 = 116.233$) becomes the denominator in the F-statistic equation.

E. This column shows the F-statistic: 8178.542 divided by 116.233 is equal to 70.364. (The results produced by SPSS vary slightly from those produced by hand calculations because it retains more decimal places.)

F. The "Sig." column shows the p-value associated with this F-statistic. It shows the likelihood of randomly selecting a sample with the observed ratio of between group variation to within group variation (or a larger ratio), if the group means are equal in the population. This p-value indicates that there is less than a 0.1 per cent chance of selecting this sample from a population in which there is no difference between the group means ($p < 0.001$).

G. This label shows that the post-hoc tests are LSD, or "least significant differences," tests.

H. The label at the top of the column shows the variable used to define the groups.

I. A row of results is printed for each pair of groups. The first row shows how the lowest quintile group compares to the second quintile group, the second row shows how the lowest quintile group compares to the third quintile group, and so on. Notice that there are two rows for each pairing: the first row shows how the lowest quintile group compares to the second quintile group, and the fifth row shows how the second quintile group compares to the lowest quintile group. The results of the first and the fifth row are identical, only the positive/negative signs are reversed. (The sign of the difference doesn't matter and can be ignored.) The results in the columns that I have labelled J to M mirror the results produced by an independent-samples t-test for each pair of groups. (Again, the results produced by SPSS vary slightly from those shown in Table 8.6 because it retains more decimal places.)

J. The "Mean Difference" column shows the size of the difference between the means for each pair of groups (listed in the rows). The first row shows that there is a 1.7 point difference between the average mental-health score of

Continued

people in the lowest income quintile group and people in the second income quintile group (52.2 − 53.9 = −1.7). The largest difference in means is between the lowest income quintile group and the highest income quintile group (3.6 points). As specified by the footnote, an asterisk (*) indicates that the difference between the two group means is statistically significant at the $p < 0.05$ level (using a t-test of independent means).

K.   The "Std. Error" column shows an estimate of the standard error; it's an estimate of the standard deviation of the sampling distribution of mean differences for each pairing.

L.   The "Sig." column shows the p-value associated with the results of a t-test of independent means between each pair of groups (listed in the rows). It shows the likelihood of randomly selecting a sample with the observed difference (or greater) between the two group means, if there were actually no difference between the two group means in the population. These results suggest that it would be unlikely to select a sample with these differences between group means if there is no relationship between income quintile group and mental-health scores in the population (since all of the p-values are less than 0.05).

M.   The "95% Confidence Interval" columns show the lower and upper bounds of the mean difference between each pair of groups that is likely to exist in the population. Since the upper and lower bounds of the confidence interval do not cross over 0 for any of the pairings, the difference between the mean of each pair of groups is unlikely to be 0 in the larger population.

Other Types of ANOVA Tests

In addition to one-way ANOVA tests, there are several other statistical significance tests in the ANOVA family. Two-way ANOVA tests are used to analyze how two categorical independent variables are simultaneously related to a ratio-level dependent variable. Two-way ANOVA tests account for how the two independent variables might combine or interact to influence the dependent variable. Multiple ANalysis Of VAriance (MANOVA) tests are used to analyze how a categorical independent variable is simultaneously related to two or more ratio-level dependent variables. Finally, repeated measures ANOVA tests (also called within-subject ANOVA tests) are used when data are paired or linked in some way. A repeated measures ANOVA test is the conceptual equivalent of a paired t-test, or a dependent samples t-test, and is typically used in experiments with pre-test and post-test measurements. If you are analyzing data in one of these circumstances, be sure to do some additional research about how to conduct and interpret these other types of ANOVA tests.

When to Use T-Tests and When to Use ANOVA Tests

Both t-tests of independent means and one-way ANOVA tests rely on several assumptions. First, both tests assume that the data are from a randomly selected, unbiased sample of a population. When this first assumption is violated, it does not make any sense to use tests of statistical significance to assess the reliability of a relationship. Second, both tests assume that the dependent variable is approximately normally distributed, with limited kurtosis and skew. Both tests are relatively forgiving when this assumption is violated, however. ANOVA tests also rely on the assumption that each group defined by the independent variable has roughly equal variances in the population, but, again, it is relatively forgiving when this assumption is violated. Unequal population variances become a substantial concern only when the number of cases in each group in the sample is very unbalanced. Finally, both t-tests of independent means and one-way ANOVA tests assume that the groups are independent of each other and that cases are not linked or paired in any way. If the sample has linked cases, use a paired t-test or a repeated measures ANOVA test instead.

T-tests and ANOVA tests both rely on comparing group means and the amount of variation around those means. In Chapter 4, you learned that the mean is sometimes a poor measure of the centre of a distribution; similarly, you learned that the standard deviation (or variance) is sometimes a poor measure of the spread of a distribution. If the mean and standard deviation are poor measures of the centre and spread of a ratio-level dependent variable, neither t-tests nor ANOVA tests should be used. Either select a different statistical test, or modify the variable so that the mean and standard deviation provide good measures of the centre and spread.

T-tests of independent means and one-way ANOVA tests are both used to assess the reliability of a relationship between a categorical independent variable and a ratio-level dependent variable. In general, t-tests are used with small sample data, especially when there are fewer than 100 cases in each group. ANOVA tests are used with larger samples and when a researcher wants to make comparisons between more than two groups. Since ANOVA tests are more flexible, many researchers with larger samples always use them to compare group means, regardless of the number of groups being compared.

Best Practices in Presenting Results

Writing about the Results of One-Way ANOVA Tests

Writing about the results of one-way ANOVA tests is similar to writing about other tests of statistical significance. The test results are reported in parentheses, just as you would do for a citation, following a claim about the likely presence or absence of a relationship in the population. The citation for a one-way ANOVA test should include the F-statistic, both degrees of freedom, and the associated p-value. By convention, the between group degrees of freedom is listed first, followed by the within group degrees of freedom (often labelled df_1 and df_2). For example, a researcher reporting the results of the one-way ANOVA test for the relationship between household income and mental-health scores might say this: "A person's household income quintile group, relative to others living the same province, is related to average mental-health scores ($F = 70.364$, $df_1 = 4$, $df_2 = 22846$, $p < 0.001$)." Following this assertion, the researcher could go on to report the direction of the relationship and the magnitude of the differences between the groups.

What You Have Learned

In this chapter, you learned how to assess the reliability of a relationship between an independent categorical variable with more than two attributes or groups and a ratio-level dependent variable. You were introduced to the logic of ANalysis Of VAriance (ANOVA) tests, specifically one-way ANOVA tests. You learned to calculate the total sum of squares, the between group sum of squares, and the within group sum of squares, as well as their associated degrees of freedom, and used these elements to calculate an F-statistic. The F-statistic, along with an F-distribution, is used to assess the likelihood of randomly selecting a sample with the observed ratio of between group variation to within group variation (or a larger ratio), if there is no relationship between group membership and the dependent variable in the population. In addition to one-way ANOVA tests, you also learned how to use post-hoc tests to identify which groups are likely to have significantly different means from each other.

This chapter extended the research focus introduced in Chapter 7, which investigated the relationship between economic advantage or disadvantage and mental health.

The analysis in Chapter 7 established that people who had trouble meeting their basic expenses were likely to have lower levels of positive mental health. The analysis in this chapter shows that people in households in the lowest income quintile group have lower positive mental-health scores, on average, than even those in the second-lowest household income group. In contrast, people in households in the highest income quintile group have higher positive mental-health scores, on average, than people in the second-highest household income group. Employment status also appears to be related to mental-health scores among working-aged people: those who are not employed have lower mental-health scores, on average, than both employees and the self-employed. In general, people living in low income households and people who are not employed appear to have lower mental-health scores than people living in higher income households and people who are employed. People who are economically disadvantaged appear to have lower levels of mental health, on average, which can result in less resilience and less ability to successfully negotiate everyday interactions.

Check Your Understanding

Check your understanding by answering the following questions:

1. How do one-way ANOVA tests use variation to assess differences between group means?
2. How are the sum of squares used in one-way ANOVA tests related to the calculation of the standard deviation and/or the variance?
3. What is the relationship between the total sum of squares, the between group sum of squares, and the within group sum of squares in a sample?

4. How is the process of comparing a t-statistic to a t-distribution similar to the process of comparing an F-statistic to an F-distribution? What is different?
5. Why do researchers use post-hoc tests? What information do post-hoc tests provide?
6. In what situations would you use a t-test of independent means? In what situations would you use a one-way ANOVA test?

Practice What You Have Learned

Check to see if you can apply the key concepts in this chapter by answering the following questions. Keep two decimal places in any calculations.

1. Your school's student-life office is investigating how social integration is related to student success. Researchers in the office survey a random sample of

students at the school about how connected they are to other students. One survey question asks students how many friends they have at school. The answers from first-year students are as follows:

2 friends	1 friend	5 friends	2 friends	5 friends
2 friends	4 friends	6 friends	4 friends	1 friend
4 friends	0 friends	3 friends		

The answers from second-year students are as follows:

6 friends	5 friends	2 friends	4 friends	6 friends
5 friends	3 friends	4 friends	2 friends	6 friends
4 friends	1 friend			

The answers from upper-year students are as follows:

12 friends	9 friends	7 friends	3 friends	8 friends
16 friends	5 friends	8 friends	6 friends	9 friends
8 friends	10 friends	9 friends	6 friends	4 friends

   a.   Find the mean number of friends among students in the sample overall, regardless of their year of study.

   b.   Find the mean number of friends among students in each year of study.

   c.   Describe the magnitude of the relationship between students' year of study and their number of friends, that is, the size of the difference between the means in the sample.

2.   It can be particularly difficult for students to make friends at school during their first year. A researcher in the student-life office wants to know how the average number of friends among first-year students compares to the average number of friends among second-year students.

   a.   Determine which variable should be treated as independent and which should be treated as dependent.

   b.   State a non-directional hypothesis that the researcher can test.

   c.   State the null hypothesis associated with the research hypothesis you identified in (b).

3.   In spite of the small sample size, the researcher decides to use a one-way ANOVA test to assess whether, in the population, there is a relationship between the number of friends that students have and whether they are in their first or second year. This analysis uses only the first-year and second-year students' answers.

   a.   Find the mean number of friends among all of the first- and second-year students in the sample (exclude the upper-year students).

   b.   Calculate the total sum of squares, the within group sum of squares, and the between group sum of squares.

4.   For the one-way ANOVA test described in question 3:

   a.   Calculate the between group degrees of freedom and the within group degrees of freedom.

   b.   Calculate the F-statistic.

   c.   Using an alpha value of 0.05, the critical value of an F-statistic with these degrees of freedom is 4.28. Based on your answer to (b), state whether you reject or fail to reject the null hypothesis. Is there likely to be a relationship between students' year of study and their number of friends in the school population?

5.   The researcher from the student life office decides to expand their analyses to include students from all years. They use a one-way ANOVA test to assess whether, in the population, there is a relationship between the number of friends people have and whether they are first-year, second-year, or upper-year students. Calculate the total sum of squares, the within group sum of squares, and the between group sum of squares for this test.

6.   For the one-way ANOVA test described in question 5:

   a.   Calculate the between group degrees of freedom and the within group degrees of freedom.

   b.   Calculate the F-statistic.

   c.   Using an alpha value of 0.05, the critical value of an F-statistic with these degrees of freedom is 3.25. Based on your answer to (b), state whether you reject or fail to reject the null hypothesis. Is there likely to be a relationship between students' year of study and their number of friends in the school population?

7. A course research project asks you to investigate the relationship between precarious work and mental health. The 2012 CCHS asks respondents to indicate whether they "Agree," "Neither agree nor disagree," or "Disagree" with this statement: "Your job security was good." You think people's answers to this question might be related to their positive mental-health score.

    a. Determine which variable should be treated as independent and which should be treated as dependent.
    b. State a non-directional research hypothesis that you can test.
    c. State the null hypothesis for the research hypothesis you identified in (b).

8. Overall, a random sample of 1,636 workers who responded to the 2012 CCHS have an average positive mental-health score of 54.5 (s.d. = 10.3). Among people who agreed that they have good job security, the average is 55.2 (s.d. = 9.9, n = 1,286). Among people who neither agreed nor disagreed that they have good job security, the average is 52.5 (s.d. = 11.1, n = 149), and among people who disagreed that they have good job security, the average is 51.9 (s.d. = 11.3, n = 201).

    a. Describe the magnitude of the relationship between positive mental-health scores and perceptions of job security in the sample, that is, the size of the differences between the three means.
    b. Calculate Cohen's d for the difference between the average mental-health score of the group who agree they have good job security and the group who disagree that they have good job security. Would you describe the effect as small, medium, or large?

9. For the relationship between positive mental-health scores and perceptions of job security described in question 8, the total sum of squares is 172,658 and the between group sum of squares is 2,504.

    a. Find the within group sum of squares.
    b. Calculate the between group degrees of freedom and the within group degrees of freedom.
    c. Calculate the F-statistic.
    d. Using an alpha value of 0.05, the critical value of an F-statistic with these degrees of freedom is 3.00. Based on your answer to (c), state whether you reject or fail to reject the null hypothesis. Is there likely to be a relationship between perceptions of job security and positive mental-health scores among workers in the Canadian population?

Table 8.7 The Results of Post-Hoc Tests for the Relationship between Perceived Job Security and Positive Mental-Health Scores, Using the Least Significant Differences Method

	Good job security: Agree	Good job security: Neither	Good job security: Disagree
Good job security: Agree	–	2.7*	3.2*
Good job security: Neither		–	0.5
Good job security: Disagree			–

*Indicates that the mean difference is significant at the p < 0.05 level.

10. Table 8.7 shows the results of post-hoc tests for the relationship between perceived job security and positive mental-health scores. In the population of workers in Canada, which groups are likely to have significantly different mental-health scores, on average, than which other groups? How might you explain these results?

11. Table 8.8, excerpted from a Statistics Canada publication, uses data from the mental-health component of the 2012 CCHS to estimate the number and percentage of people who have "complete mental health," that is, people who do not have any mental and/or substance use disorders and whose answers to questions on the positive mental-health scale indicate that they are flourishing.

    a. Is there likely a difference between the percentage of men and the percentage of women with complete mental health in the Canadian population? Explain how you determined your answer.
    b. Is there likely a difference between the percentage of people with and the percentage of people without a post-secondary education who have complete mental health in the Canadian population? List two ways that you can make this assessment using the information in this table.

Table 8.8   Prevalence of Complete Mental Health, by Selected Characteristics, Household Population Aged 15 or Older, Canada Excluding Territories, 2012

Characteristics	Complete Mental Health			
		Prevalence		
	Estimated Number '000	%	95% Confidence Interval	
			From	To
Sex				
Male[†]	9,668	71.9	70.6	73.3
Female	10,123	73.1	71.8	74.3
Post-secondary education				
Yes[†]	14,514	73.7	72.5	74.7
No	3,824	69.0**	67.0	70.9
Employment status				
Has job[†]	12,941	73.5	72.3	74.7
Permanently unable to work	315	47.0**	41.2	53.0
Does not have job	5,232	71.3*	69.7	72.9

†Reference group
*Significantly different from reference group (p < 0.05)
**Significantly different from reference group (p < 0.01)
Source: 2012 Canadian Community Health Survey—Mental Health.

Source: Excerpt from Gilmour 2014, 7.

12. Looking at the information in Table 8.8:

   a. Determine whether, in the Canadian population, there is likely a relationship between people's employment status and whether or not they have complete mental health. If there is a relationship, which groups are you confident there is a difference between?

   b. Explain what the * and the ** symbols in this table indicate.

13. The graph in Figure 8.10 shows the results of the Ontario Student Drug Use and Health Survey, conducted by the Centre for Addiction and Mental Health. Students enrolled in grades 7 to 12 across Ontario are randomly sampled and asked to complete anonymous questionnaires. The vertical axis of the graph shows the percentage of students who had moderate-to-serious psychological distress (symptoms of depression and anxiety) in the past month. The horizontal axis of the graph indicates that the results are shown for students overall, for males (M) and females (F), for students in grades 7 to 12, and for students in four regions: Toronto (TO), northern Ontario (N), western Ontario (W), and eastern Ontario (E).

   a. Is there likely a difference between the percentage of males and females who have moderate-to-serious psychological distress in the population of Ontario grade 7 to 12 students? Explain how you arrived at your answer.

   b. Is there a statistically significant difference between the percentage of males and females who have moderate-to-serious psychological distress? Explain how you know.

14. Using the information in the graph in Figure 8.10:

   a. Determine whether there is likely a difference between the percentage of students in each grade who have moderate-to-serious psychological distress in the population of Ontario grade 7 to 12 students. In which grades is there likely to be a higher (or lower) percentage of students in psychological distress? Explain how you arrived at your answer.

   b. Is there a statistically significant difference between the percentage of students in each grade who have moderate-to-serious psychological distress? Explain how you know.

15. Using the information in the graph in Figure 8.10:

   a. Determine whether there is likely a difference between the percentage of students in each region who have moderate-to-serious psychological distress in the population of Ontario grade 7 to 12 students. In which regions is there likely to be a higher (or lower) percentage of students in psychological distress? Explain how you arrived at your answer.

   b. Is there a statistically significant difference between the percentage of students in each region who have moderate-to-serious psychological distress? Explain how you know.

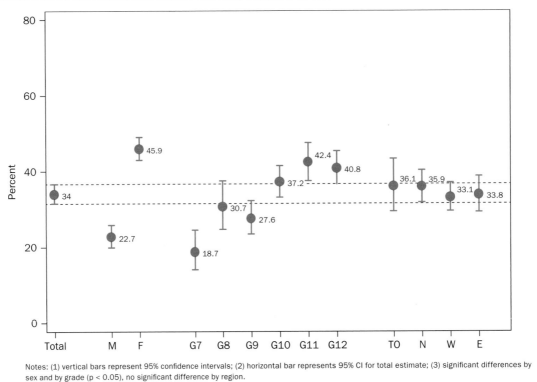

Notes: (1) vertical bars represent 95% confidence intervals; (2) horizontal bar represents 95% CI for total estimate; (3) significant differences by sex and by grade (p < 0.05), no significant difference by region.

Figure 8.10   Percentage of Ontario Grade 7–12 Students Indicating Moderate-to-Serious Psychological Distress in the Past Month by Sex, Grade, and Region, 2015 OSDUHS

Source: Boak et al. 2016, 17.

Practice Using Statistical Software (IBM SPSS)

Answer these questions using IBM SPSS and the GSS27.sav or the GSS27_student.sav dataset available from the Student Resources area of the companion website for this book. Weight the data using the "Standardized person weight" [STD_WGHT] variable you created following the instructions in Chapter 5. Report two decimal places in your answers, unless fewer are printed by IBM SPSS. It is imperative that you save the dataset to keep any new variables that you create.

1. Select the "Descriptive" checkbox in the One-Way ANOVA procedure Options to find the mean and 95 per cent confidence interval for the mean of "Number of close friends" [SCF_100C], divided by "Education - Highest degree (4 categories)" [DH1GED], and to produce a one-way ANOVA test assessing the relationship between these two variables.

   a. Describe the magnitude of the relationship, that is, the size of the difference between the means in the sample.

   b. How do the 95 per cent confidence intervals for the mean number of close friends among each of the four groups compare to each other? In the population, are people with higher levels of education likely to have more (or fewer) close friends, on average, than people with lower levels of education?

2. Consider the one-way ANOVA test produced in question 1:

   a. State a non-directional research hypothesis for this relationship.

   b. State the null hypothesis associated with the non-directional research hypothesis you identified in (a).

c. Interpret the F-statistic and its associated p-value in relation to the null hypothesis you identified in (b). In the population, is there likely to be a relationship between people's highest level of education and their number of close friends?

3. Use the Post-Hoc option in the One-Way ANOVA procedure to produce LSD post-hoc tests, with a 0.05 significance level, for the relationship between "Number of close friends" [SCF_100C] and "Education - Highest degree (4 categories)" [DH1GED].

    a. Which educational group(s) have a significantly different number of close friends, on average, than others in the population?

    b. Does your answer to (a) correspond with your answer to question 1(b)? Describe any similarities and differences between your conclusions.

4. Use the Chart Builder tool to create an error-bar graph showing the mean and 95 per cent confidence interval for the mean of "Number of close friends" [SCF_100C] divided by "Education - Highest degree (4 categories)" [DH1GED]. Explain what the graph shows and how it reflects the results of question 3(a).

5. The variable "Number of paid hours worked per week - All jobs" [WKWEHRC] shows how many paid hours people work at all of their jobs, among people who are working for pay. (People who are not working for pay are assigned the attribute "Valid skip.") Select the "Descriptive" checkbox in the One-Way ANOVA procedure Options to find the mean and 95 per cent confidence interval for the mean of this variable, divided by "Education - Highest degree (4 categories)" [DH1GED], and to produce a one-way ANOVA test assessing the relationship between these two variables.

    a. In the overall sample, what is the average number of hours people spend doing paid work? In the overall population, what is the average number of hours people spend doing paid work likely to be?

    b. Describe the magnitude of the relationship between the two variables, that is, the size of the difference between the means in the sample.

c. How do the 95 per cent confidence intervals for the mean number of hours spent doing paid work among each of the four groups compare to each other? In the population, are people with higher levels of education likely to spend more or less time doing paid work, on average, than people with lower levels of education?

6. Consider the ANOVA table of results in the output from question 5:

    a. State a non-directional research hypothesis for this relationship.

    b. State the null hypothesis associated with the non-directional research hypothesis you identified in (a).

    c. Interpret the F-statistic and its associated p-value in relation to the null hypothesis you identified in (b). In the population, is there likely to be a relationship between people's highest level of education and the amount of time they spend doing paid work?

7. Use the Post-Hoc option in the One-Way ANOVA procedure to produce LSD post-hoc tests, with a 0.05 significance level, for the relationship between "Number of paid hours worked per week - All jobs" [WKWEHRC] and "Education - Highest degree (4 categories)" [DH1GED].

    a. Which educational group(s) spend a significantly different number of hours doing paid work, on average, than others in the population?

    b. Does your answer to (a) correspond with your answer to question 5(c)? Describe any similarities and differences between your conclusions.

8. Use the Chart Builder tool to create an error-bar graph showing the mean and 95 per cent confidence interval for the mean of "Number of paid hours worked per week - All jobs" [WKWEHRC] divided by "Education - Highest degree (4 categories)" [DH1GED]. Explain what the graph shows and how it reflects the results of question 7(a).

Key Formulas

Total sum of squares (version 1)

$$SS_{total} = SS_{between\ group} + SS_{within\ group}$$

Total sum of squares (version 2)

$$SS_{total} = \Sigma\left(x_i - \overline{x}_{total}\right)^2$$

Between group sum of squares

$$SS_{between\ group} = \Sigma n_{group}\left(\overline{x}_{group} - \overline{x}_{total}\right)^2$$

Within group sum of squares

$$SS_{within\ group} = \Sigma\left(x_i - \overline{x}_{group}\right)^2$$

Between group degrees of freedom

$$df_{between\ group} = \left(number\ of\ groups - 1\right)$$

Within group degrees of freedom

$$df_{within\ group} = \left(n_{group1} - 1\right) + \left(n_{group2} - 1\right) + \left(n_{group3} - 1\right) + \ldots$$

F-statistic (or F-ratio)

$$F = \frac{SS_{between\ group}\ /\ df_{between\ group}}{SS_{within\ group}\ /\ df_{within\ group}}$$

References

Boak, Angela, Hayley A. Hamilton, Edward M. Adlaf, Joanna L. Henderson, and Robert E. Mann. 2016. *The Mental Health and Well-Being of Ontario Students, 1991–2015 OSDUHS Highlights*. Toronto: Centre for Addiction and Mental Health. http://www.camh.ca/en/research/news_and_publications/ontario-student-drug-use-and-health-survey/Documents/2015%20OSDUHS%20Documents/2015OSDUHS_Highlights_MentalHealthReport.pdf.

Gilmour, Heather. 2014. "Positive Mental Health and Mental Illness." Catalogue no. 82–003-X. Ottawa: Statistics Canada. http://www.statcan.gc.ca/pub/82–003-x/2014009/article/14086-eng.pdf.

Llena-Nozal, Ana. 2009. "The Effect of Work Status and Working Conditions on Mental Health in Four OECD Countries." *National Institute Economic Review* 209 (July): 72–87.

Manseau, Marc W. 2014. "Economic Inequality and Poverty as Social Determinants of Mental Health." *Psychiatric Annals* 44 (1): 32–38. doi:10.3928/00485713–20140108–06.

Paul, Karsten I., and Klaus Moser. 2009. "Unemployment Impairs Mental Health: Meta-Analyses." *Journal of Vocational Behavior* 74 (3): 264–82. doi:10.1016/j.jvb.2009.01.001.

Statistics Canada. 2014. "Canadian Community Health Survey, 2012: Mental Health Component." *Public Use Microdata File*. Ottawa, ON: Statistics Canada.

———. 2015. *Labour Force Survey*. Public Use Microdata File.

Assessing Relationships between Categorical Variables

Learning Objectives

In this chapter, you will learn:

- What proportionate reduction in error measures show

- How to calculate lambda and gamma measures

- The logic of chi-square tests

- How to calculate and interpret a chi-square test of independence

- How the elaboration model is used to investigate more complex relationships

- How to present cross-tabulation results

Introduction

In chapters 7 and 8, you learned how to assess a relationship between a categorical independent variable and a ratio-level dependent variable. In this chapter, you will learn how to assess relationships between two categorical variables. The first half of the chapter introduces proportionate reduction in error measures, which are used to assess the magnitude of a relationship between two categorical variables. The second half of the chapter describes how to calculate and interpret a chi-square test of independence, which is used to assess the reliability of a relationship between two categorical variables. The chapter concludes by showing how elaboration models are used to conduct multivariate analyses.

The research focus of this chapter is the relationship between racialization, contact with police, and perceptions of police. In the United States, a series of high-profile incidents have highlighted questionable police conduct towards Black people, including the shooting of unarmed suspects and deaths in police custody. Public outrage about these incidents coalesced around the "Black Lives Matter" movement, which builds on a long history of organizing by Black and racialized communities (Petersen-Smith 2015). Although the politics of race might sometimes appear less stark in Canada than in the US, racial discrimination—including discrimination against Aboriginal peoples—is also entrenched within the Canadian criminal justice system (Chan and Mirchandani 2002, Comack 2012). Most recently, debates about the use of "carding" by police forces have gained

Coast-to-Coast/iStockphoto

Photo 9.1   **The Black Lives Matter movement has staged ongoing protests against police brutality and the use of "carding."**

public attention. In some Canadian cities, police officers routinely stop and question racialized people—especially racialized young men—to collect identifying information from them as they go about their daily activities (Rankin and Winsa 2012). Community leaders have spoken out about how these policing practices erode people's trust in police and in the justice system more generally (Cole 2015).

Canada's 2014 General Social Survey (GSS) on Victimization asks respondents whether they think their "local police force does a good job, an average job, or a poor job of treating people fairly?" (See the "Spotlight on Data" box for more information about this survey.) Since many rural communities do not have local police forces (and rely instead on the Royal Canadian Mounted Police [RCMP] or a provincial service), the analyses in this chapter only include people living in urban centres. Overall, more than two-thirds of people (68 per cent) living in Canadian urban centres say that their local police force does a good job of treating people fairly. Another 26 per cent say they do an average job, and only 7 per cent say their local police do a poor job of treating people fairly. In general, this is good news: most people think the police treat people fairly. In addition to asking about people's perceptions of police fairness, the GSS collects information about whether people have come into contact with police during the previous 12 months. The survey asks about police contact in a variety of contexts: in a public information session, because of problems with respondents' own mental health or alcohol/drug use or family members' mental health or alcohol/drug use, for a traffic violation, as a witness to a crime, or when arrested by police. Almost one-third of people living in urban centres (30 per cent) report having some contact with police in the previous year, reinforcing the notion that local police are an integral part of Canadian communities. In this chapter, statistical analysis is used to investigate the following:

- Are racialized people more or less likely than non-racialized people to have contact with police?
- Are people's perceptions of police fairness related to their age?
- Are people's perceptions of police fairness related to whether they have had contact with police?
- How does the relationship between contact with police and perceptions of police fairness change when racialization is taken into account?

It is particularly difficult to measure "race" or racialization, since race is a social construct that varies over time and in different contexts. Statistics Canada typically identifies whether a person is a "visible minority," using the same definition as the federal government's Employment Equity Act: "persons, other than aboriginal peoples, who are non-Caucasian in race or non-white in colour" (Government of Canada 1995). In the General Social Survey, the question used to determine "visible minority status" starts by saying "People in Canada come from many racial or cultural groups. You may belong to more than one group on the following list." The GSS then asks, "Are you?" and provides a list to select from: "White, Chinese, South Asian, Black, Filipino, Latin American, Southeast Asian, Arab, West Asian,

Spotlight on Data

The General Social Survey on Victimization

The "Spotlight on Data" box in Chapter 6 introduced Statistics Canada's General Social Survey (GSS). The GSS collects information about a different topic each year, rotating through five main topics. This chapter uses information from the 2014 GSS on Victimization. The goal of this survey is to understand how people living in Canada perceive crime and the justice system as well as to collect information about their experiences of victimization (Statistics Canada 2016).

The population for the 2014 GSS is non-institutionalized residents aged 15 and older living in Canada's provinces and territories. The analyses in this chapter focus on people living in urban areas; thus, only people living in nine provinces are included. (PEI is excluded.) In the 2014 GSS, people living in Canada's provinces were selected from Statistics Canada's new telephone sampling frame, which combines information about land-line and cellular telephone numbers for each household from the Census and other administrative sources (Statistics Canada 2016). Most respondents completed the survey over the telephone (although some data were collected over the Internet in a pilot project). The response rate for people living in Canada's provinces was 53 per cent.

Korean, Japanese or another group." People can indicate that they belong to more than one group. Statistics Canada then groups people's responses: those who say they are only Caucasian and those who say they are only Aboriginal, as well as those who claim White/Latin American or White/Arab–West Asian multiple responses are classified as "not a visible minority." People who choose any other racial or ethnic affiliation, including those with other multiple responses, are classified as "visible minorities." Using this schema, approximately 17 per cent of all people living in Canada are considered visible minorities, and 20 per cent of people living in urban centres are considered visible minorities.

As a social scientist, you might be critical of this approach to measuring "race" and racialization. The question's reference to "cultural groups" and the failure to clearly distinguish between "race," ethnicity, and nationality in the list of response options makes it hard to know what is actually being captured. This conflation, as well as the exclusion of Aboriginal people from the category, certainly makes "visible minority status" an imperfect measure of the complexities of "race" and racialization in Canada. Even the label "visible minority" is problematic since it suggests that people outside the category are invisible (although critical race theorists might agree that "invisible majority" is an appropriate label). Despite these concerns, however, Statistics Canada's data about visible minority people is the most generalizable information available about the experiences of racialized people living in Canada. There are very limited data available about "race" and racism in the Canadian criminal justice system. Canada's official crime statistics do not include information about the race of the victim or the persons charged, beyond Aboriginal status. And, since 2003, many Canadian police services have either failed to report Aboriginal status information or formally refuse to do so (Millar and Owusu-Bempah 2011). Statistics Canada's publicly available victim surveys either omit information on race and ethnicity entirely, or collect only information about "visible minority" status.

Assessing the Magnitude of Relationships between Categorical Variables

The most common strategy that researchers use to describe the size of a relationship between two categorical variables is to show how the distribution of the dependent variable changes depending on group membership. Typically, this involves reporting the percentage of people in each group with each attribute (like in a cross-tabulation). For example, Table 9.1 is a cross-tabulation between visible minority status and contact with police, showing the counts (frequencies) and the column percentages in each cell. Visible minority status is treated as the independent variable, and contact with police is treated as the dependent variable. The results suggest that, in urban centres, people who are visible minorities are actually less likely to have contact with police than those who are not visible minorities. Thirty-two (32) per cent of people who are not visible minorities had contact with police in the past 12 months, compared to 26 per cent of people who are visible minorities. About the same proportion of people who are visible

Table 9.1   A Cross-Tabulation Showing the Relationship between Visible Minority Status and Contact with Police

Contact with Police in Last 12 Months (any reason)		Visible Minority		Total
		No	Yes	
No	Count	14,904	4,016	18,920
	Column %	68.5%	73.7%	69.6%
Yes	Count	6,849	1,430	8,279
	Column %	31.5%	26.3%	30.4%
Total	Count	21,753	5,446	27,199
	Column %	100.0%	100.0%	100.0%

Source: Author generated; Calculated using data from Statistics Canada, 2016.

minorities and those who are not report having contact with police for a negative reason (such as being stopped for a traffic violation). People who are visible minorities are slightly more likely to report that they had contact with police as a witness to a crime or for "other" reasons, which include being arrested or having mental-health or drug/alcohol problems.

Proportionate Reduction in Error Measures

One way to describe the magnitude of a relationship between two categorical variables is using **proportionate reduction in error (PRE) measures**. PRE measures show how much the error in predicting the attributes of the dependent variable can be reduced if the attributes of the independent variable are known. In other words, they show how much better predictions about the dependent variable are if group membership (defined by the independent variable) is taken into account. In this section, you will learn to calculate and interpret two commonly used proportionate reduction in error measures: lambda (for nominal variables) and gamma (for ordinal variables).

Lambda is used to measure the magnitude of an association between two nominal-level variables or between a nominal-level and an ordinal-level variable. Lambda (λ) is a letter in the Greek alphabet. Lambda is an asymmetric measure of association, which means it produces different results depending on which variable is treated as independent and which is treated as dependent. Lambda relies on using the most common attribute (the mode) to predict the attribute of the dependent variable for every case, first ignoring group membership and then taking group membership into account. It shows how much better the predictions are when group membership is taken into account, compared to when group membership is ignored.

The calculation of lambda includes two components: E_1 and E_2. E_1 is the number of prediction errors made if group membership is not taken into account. To calculate E_1, find the total number of cases and subtract the number of cases with the most common attribute overall. E_1 shows the number of incorrect guesses if the

proportionate reduction in error (PRE) measures Show how much the error in predicting the attributes of the dependent variable can be reduced if the attributes of the independent variable are known.

lambda (λ) A PRE measure of the magnitude of an association between two nominal-level variables or between a nominal-level and an ordinal-level variable.

most common attribute is predicted for every case. E_2 is the number of prediction errors made when group membership *is* taken into account. To calculate E_2, find the total number of cases in each group and subtract the number of cases with the most common attribute within that group; then, add together the results for all of the groups. E_2 shows the number of incorrect guesses if the most common attribute in each group is predicted for every case in the group.

lambda

$$\lambda = \frac{E_1 - E_2}{E_1}$$

$$E_1 = n_{overall} - modal\ frequency_{overall}$$

$$E_2 = \Sigma\left(n_{group} - modal\ frequency_{group}\right)$$

Let's use a hypothetical example to illustrate. Table 9.2 shows a cross-tabulation between group membership and contact with police. There are 200 people overall. The most common answer overall is "No contact with police." If this answer is predicted for everyone, the prediction is correct for 110 cases and incorrect for 90 cases. Thus E_1 is equal to 90 (calculated as 200 − 110). Now, let's see if knowing which group a case belongs to helps to improve the predictions. In Group 1, the most common answer is "No contact with police." If this answer is predicted for everyone in Group 1, the prediction is correct for 80 cases and incorrect for 20 cases. In Group 2, the most common answer is "Had contact with police." If this answer is predicted for everyone in Group 2, the prediction is correct for 70 cases and incorrect for 30 cases. Thus, E_2 is equal to 50 (calculated as [100 − 80] + [100 − 70]).

These values of E_1 and E_2 are substituted into the lambda formula:

$$\lambda = \frac{E_1 - E_2}{E_1}$$

$$\lambda = \frac{90 - 50}{90}$$

$$= \frac{40}{90}$$

$$= 0.44$$

Table 9.2 A Cross-Tabulation between Group Membership and Contact with Police (Hypothetical Data)

Contact with Police	Group 1	Group 2	Total
No contact with police	80	30	110
Had contact with police	20	70	90
Total	100	100	200

Lambda ranges from 0 to 1. A lambda of 0 indicates that knowing about group membership doesn't result in any improvement in prediction, and a lambda of 1 indicates that knowing about group membership results in perfectly predicting the dependent variable. This lambda of 0.44 indicates that the error in predicting whether or not people have had contact with police can be reduced by 44 per cent if you know whether they belong to Group 1 or Group 2.

Now let's calculate lambda for the relationship between visible minority status and contact with police, using the GSS data in Table 9.1. There are 27,199 cases overall and the most common answer is "No contact with police," selected by 18,920 people. Thus, E_1 is 8,279 (27,199 − 18,920). For the 21,753 people who are not visible minorities, the most common answer is "No contact with police," selected by 14,904 people. For the 5,446 people who are visible minorities, the most common answer is also "No contact with police," selected by 4,016 people. Thus, E_2 is also 8,279 ([21,753 − 14,904] + [5,446 − 4,016]). When E_2 is subtracted from E_1 to find the numerator of the lambda equation, the result is 0. Since 0 divided by any quantity is 0, lambda is 0 for this relationship. In this example, the most common answer in each group is the same, and so knowing whether or not people are visible minorities does not help to predict whether or not they had contact with police.

Step-by-Step: Lambda

$$\lambda = \frac{E_1 - E_2}{E_1}$$

$$E_1 = n_{overall} - modal\ frequency_{overall}$$

$$E_2 = \Sigma \left(n_{group} - modal\ frequency_{group} \right)$$

Step 1: Count the total number of cases in each group defined by the independent variable (n_{group1}, n_{group2}, n_{group3}, . . .). Do not include cases that are missing information for the dependent variable.

Step 2: Add together the results of Step 1 for each group to find the total number of cases ($n_{overall}$).

Step 3: Identify the most common attribute of the dependent variable (the mode). Determine how many cases have this attribute to find the overall modal frequency ($modal\ frequency_{overall}$).

Step 4: Subtract the overall modal frequency (from Step 3) from the total number of cases (from Step 2) to find E_1.

Step 5: *Within each group*, identify the most common attribute of the dependent variable (the mode). Determine how many cases in the group have this attribute to find the modal frequency of the group. Do this for each group ($modal\ frequency_{group1}$, $modal\ frequency_{group2}$, $modal\ frequency_{group3}$, . . .).

Continued

Step 6: For *each group separately*, subtract the modal frequency of the group (from Step 5) from the total number of cases in the group (from Step 1). Complete this calculation once for each group.

Step 7: Add together the results of Step 6 for each group to find E_2.

Step 8: Subtract E_2 (from Step 7) from E_1 (from Step 4) to find the numerator of the lambda equation.

Step 9: Divide the result of Step 8 by E_1 (from Step 4) to find lambda.

How Does It Look in SPSS?

Lambda

The Crosstabs procedure, with the "Lambda" checkbox selected in the Statistics option, produces results that look like those in Image 9.1. The labels of the two variables in the cross-tabulation appear at the top of the table. Above each cross-tabulation, SPSS prints a "Case Processing Summary" (not shown) that lists the number and percent of valid cases, missing cases, and total cases.

Contact with police - any reason * Visible minority status of the respondent Crosstabulation

Count

			(D) Visible minority status of the respondent		
			(E) Not a visible minority	Visible minority	(F) Total
Contact with police - (B) No any reason (A)			14904	4016	18920
	Yes		6849	1430	8279
Total	(C)		21753	5446	27199

(G) **Directional Measures**

			Value	Asymptotic Standard Error[a]	Approximate T	Approximate Significance
Nominal by Nominal (H)	Lambda (I)	Symmetric	.000	.000	.[b]	.[b]
		Contact with police - any (J) reason Dependent	.000	.000	.[b]	.[b]
		Visible minority status of (K) the respondent Dependent	.000	.000	.[b]	.[b]
	Goodman and Kruskal tau (L)	Contact with police - any reason Dependent	.002	.001		.000[c]
		Visible minority status of the respondent Dependent	.002	.001		.000[c]

a. Not assuming the null hypothesis.

b. Cannot be computed because the asymptotic standard error equals zero.

c. Based on chi-square approximation

Image 9.1 **An SPSS Lambda Measure**

A. The label of the row variable appears on the far left. The "Contact with police" variable is placed in the rows because I am treating it as the dependent variable in this analysis.

B. The attributes of the row variable are listed in a vertical column. Only valid attributes are shown in an SPSS cross-tabulation; rows for attributes designated as missing are not printed.

C. The "Total" row shows the overall totals for the column variable. Overall, there were 27,199 people in the sample: 21,753 of them were not a visible minority, and 5,446 of them were.

D. The label of the column variable appears above the columns. The "Visible minority status" variable is placed in the columns because I am treating it as the independent variable in this analysis.

E. The attributes of the column variable are listed at the top of the columns. The first column shows information for first group: among the 21,753 people who are not visible minorities, 14,904 had no contact with police, and 6,849 had contact. The second column shows information for the second group: among the 5,446 people who are visible minorities, 4,016 had no contact with police and 1,430 had contact.

F. The "Total" column shows the overall totals for the row variable. Overall, there were 27,199 people in the sample: 18,920 of them had no contact with police and 8,279 had contact.

G. This title indicates that these statistics are "Directional" or asymmetric measures. In other words, the results change depending on which variable is treated as independent and which is treated as dependent.

H. This label indicates that these measures are used to assess the relationship between two nominal-level variables. Lambda can also be used to assess the relationship between one nominal-level variable and one ordinal-level variable.

I. These three rows report lambda measures. Since lambda is an asymmetric measure, the first row can be ignored. The second row shows lambda when "Contact with police" is treated as the dependent variable; the third row shows lambda when "Visible minority status" is treated as the dependent variable.

J. This result corresponds to the lambda calculated in this chapter. Since "Contact with police" is treated as the dependent variable in this analysis, it is the statistic that should be reported. A lambda of 0 indicates that knowing whether or not someone is a visible minority does not help us to reduce our error in predicting whether people have had contact with police. The other numbers in this row can be ignored.

K. This is the lambda that would be obtained if "Visible minority status" were used as the dependent variable. Since this does not match the approach used in this analysis, it should not be reported. The other numbers in the row can also be ignored.

L. Goodman and Kruskal's tau statistics are printed by default when lambda is requested. You can ignore these rows.

gamma (γ) A PRE measure of the magnitude and direction of an association between two ordinal-level variables.

Gamma is used to measure the magnitude of a relationship between two ordinal-level variables. Gamma (γ) is also a letter in the Greek alphabet. Compared to lambda, gamma has two main advantages: first, it uses more information than lambda because it accounts for the order or ranking of attributes, and not just the most common attribute; second, it also provides information about the direction of the relationship. Unlike lambda, gamma is a symmetric measure of association, which means that it doesn't matter which variable is treated as independent and which is treated as dependent; the result is the same either way.

Gamma relies on counting *pairs* of cases so that every case in the sample is compared to every other case in the sample. The formula used to find the total number of pairs of cases in a sample is:

$$number\ of\ pairs = \frac{n(n-1)}{2}$$

The first *n* in the formula represents the number of cases in the dataset. Each individual case can be paired with every other case in the dataset, *except* itself, which is why *n* is multiplied by *n* − 1. This approach double-counts each pair, though, because it includes pairs with the same two cases in a different order. In other words, it treats the pairing of case 1 and case 2 and the pairing of case 2 and case 1 as two different pairs of cases. Since the order of the cases within a pair doesn't matter, *n*(*n* − 1) is divided by two in order to find the number of unique pairs.

concordant pairs of cases Formed when one case ranks higher than another case on both the independent variable and the dependent variable (or lower than another case on both).

discordant pairs of cases Formed when one case ranks higher than another case on the independent variable and lower than it on the dependent variable, or vice versa.

Gamma is calculated by finding the number of **concordant pairs of cases** and the number of **discordant pairs of cases**. The word *concordant* refers to something that is consistent or in agreement. A concordant pair of cases occurs when one case ranks higher than another case on both the independent variable and the dependent variable. A concordant pair of cases also occurs when a case ranks lower than another case on both the independent variable and the dependent variable. The word *discordant* refers to something that is inconsistent or not in agreement. A discordant pair of cases occurs when one case ranks higher than another case on the independent variable and lower than it on the dependent variable, or vice versa. The gamma calculation ignores pairs where a case is ranked the same as another case on either the independent variable or the dependent variable. These are called "tied" cases. As a result, gamma tends to underestimate the strength of an association if there are many tied cases.

It's easiest to learn how to count concordant and discordant pairs of cases using some illustrations. Imagine a cross-tabulation where the independent variable is listed from lowest to highest in the columns, left to right, and the dependent variable is listed from lowest to highest in the rows, top to bottom. For simplicity, let's start with a two-by-two cross-tabulation (see Figure 9.1), and begin with the top left cell (cell A). To find cases that form a concordant pair with those in cell A, we need to find cases ranked above them on *both* the independent variable and the dependent variable. The cases in cell B are ranked above those in cell A on the independent variable but not on the dependent variable. The cases in cell C are ranked above those in cell A on the dependent variable but not on the independent variable. Only

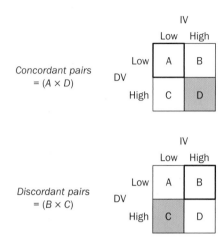

Figure 9.1   Concordant and Discordant Pairs of Cases in a 2 x 2 Cross-Tabulation

the cases in cell D are ranked above those in cell A on both the independent variable and the dependent variable. So, the number of concordant pairs of cases is the number of cases in cell A multiplied by the number of cases in cell D (A x D). For the cases in the three remaining cells in the cross-tabulation (cells B, C, and D), there are no cases ranked above them on both the independent variable and the dependent variable.

A similar strategy is used to find discordant pairs of cases. It's easiest to start with the cell ranked highest on the independent variable and lowest on the dependent variable in the top right (cell B). To find cases that form a discordant pair with the cases in cell B, we need to find cases that are ranked below them on the independent variable but above them on the dependent variable. Only the cases in cell C meet these criteria. For the cases in the remaining cells (cell A, cell C, and cell D), there are no cases ranked below them on the independent variable but above them on the dependent variable. So, the number of discordant pairs is the number of cases in cell B multiplied by the number of cases in cell C (B x C).

When the independent or dependent variable has more than two attributes, the same logic applies, but there are more pairs of cases to account for. Figure 9.2 shows how to find concordant and discordant pairs in a two-by-three cross-tabulation. Once again, the independent variable is listed from lowest to highest in the columns, left to right, and the dependent variable is listed from lowest to highest in the rows, top to bottom. Cases that form a concordant pair with those in cell A are found in cells D and F. So the number of cases in cells D and F are added together, and multiplied by the number of cases in cell A. But, in a two-by-three cross-tabulation, the cases in cell A are no longer the only ones with cases ranked above them on both the independent variable and the dependent variable. The cases in cell C also have cases that are ranked above them on both variables: those in cell F. So, the number of cases in cell C and cell F are multiplied together and added to the number of concordant pairs.

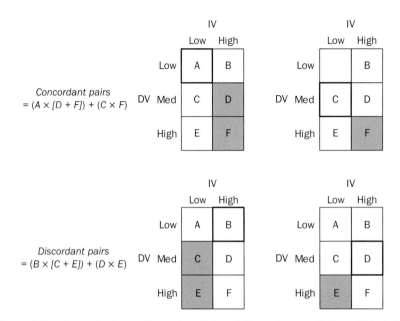

Figure 9.2   **Concordant and Discordant Pairs of Cases in a 2 x 3 Cross-Tabulation**

A similar process is used to find the number of discordant pairs of cases in larger cross-tabulations. Start with the cases with the highest attribute of the independent variable and the lowest attribute of the dependent variable, in cell B. Then, find the number of cases in the cells ranked lower than those in cell B on the independent variable and higher than them on the dependent variable, which are the cases in cells C and E. The sum of the cases in cells C and E is multiplied by the number of cases in cell B, to calculate the number of discordant pairs. Once again, though, the cases in cell B are not the only ones with cases ranked lower than them on the independent variable and higher than them on the dependent variable. Cell D also meets this criteria, since the cases in cell E are ranked lower on the independent variable and higher than them on the dependent variable. So, the number of cases in cell D and cell E are multiplied together and added to the number of discordant pairs.

Once you have tallied the number of concordant and discordant pairs of cases, it's easy to calculate gamma. The numerator of the gamma formula captures whether there are more concordant pairs than discordant pairs; if so, gamma will be positive. If there are more discordant pairs than concordant pairs, gamma will be negative. The denominator of the formula effectively standardizes the result, by dividing it by the total number of non-tied pairs of cases. The difference between the number of concordant and discordant pairs is shown as a proportion of the total number of (non-tied) pairs of cases in the sample.

Gamma ranges from −1 to +1. The sign of gamma indicates whether the re-lationship is positive or negative. A positive gamma indicates that there are more

gamma

$$\gamma = \frac{Concordant\ pairs - Discordant\ pairs}{Concordant\ pairs + Discordant\ pairs}$$

Table 9.3 A Cross-Tabulation Showing the Relationship between Age Group and Perceptions of Police Fairness

Do you think your local police force does a good job, an average job or a poor job of treating people fairly?		Age Group			
		Young (15 to 34)	Middle-Aged (35 to 54)	Older (55 and up)	Total
A poor job	Count	Ⓐ 811	Ⓑ 543	Ⓒ 338	1,692
	Column %	9.5%	6.3%	4.2%	6.7%
An average job	Count	Ⓓ 2,463	Ⓔ 2,173	Ⓕ 1,838	6,474
	Column %	29.0%	25.3%	22.8%	25.7%
A good job	Count	Ⓖ 5,223	Ⓗ 5,889	Ⓘ 5,868	16,980
	Column %	61.5%	68.4%	72.9%	67.5%
Total	Count	8,497	8,605	8,044	25,146
	Column %	100.0%	100.0%	100.0%	100.0%

Source: Author generated; Calculated using data from Statistics Canada, 2016.

positive relationships in the data: people with higher attributes of the independent variable have higher attributes of the dependent variable (or people with lower attributes of the independent variable have lower attributes of the dependent variable). A negative gamma indicates that there are more negative relationships in the data: people with higher attributes of the independent variable have lower attributes of the dependent variable (or vice versa). The size of gamma provides information about the magnitude of the relationship between the two variables. Like lambda, it shows how much the error in predicting the attributes of the dependent variable can be reduced if the attributes of the independent variable are known.

Now that you understand the basics, let's use gamma to assess whether people's perceptions of police fairness are related to their age group. Recall that the GSS asks people to indicate whether their local police did a poor job, an average job, or a good job of treating people fairly. Age is divided into three groups that loosely represent life stages: young people (15 to 34 years old), middle-aged people (35 to 54 years old), and older people (55 years old and up). Table 9.3 shows a clear pattern in the relationship between people's age group and their perceptions of police fairness. Young people are more likely to report that the local police do a poor job of treating people fairly, with 10 per cent saying that police do a poor job, compared to 6 per cent of middle-aged people and only 4 per cent of older people who gave this response. Conversely, almost three-quarters (73 per cent) of older people say that the local police do a good job of treating people fairly, compared to only 68 per cent of middle-aged people, and 62 per cent of young people.

The gamma calculation uses the number of cases in each cell (not the percentages). To make it easy to follow, I've labelled the cells in the table with letters. Although the numbers become quite large, the math remains simple. The number of concordant pairs of cases is:

$$Concordant\ pairs = \Big(A \times \big[E+F+H+I\big]\Big) + \Big(B \times \big[F+I\big]\Big) + \Big(D \times \big[H+I\big]\Big) + \Big(E \times I\Big)$$

$$= \Big(811 \times \big[2173+1838+5889+5868\big]\Big) + \Big(543 \times \big[1838+5868\big]\Big)$$

$$+ \Big(2463 \times \big[5889+5868\big]\Big) + \Big(2173 \times 5868\Big)$$

$$= \Big(811 \times 15768\Big) + \Big(543 \times 7706\Big) + \Big(2463 \times 11757\Big) + \Big(2173 \times 5868\Big)$$

$$= 12787848 + 4184358 + 28957491 + 12751164$$

$$= 58680861$$

Similarly, the number of discordant pairs of cases is:

$$Discordant\ pairs = \Big(C \times \big[D+E+G+H\big]\Big) + \Big(B \times \big[D+G\big]\Big) + \Big(F \times \big[G+H\big]\Big) + \Big(E \times G\Big)$$

$$= \Big(338 \times \big[2463+2173+5223+5889\big]\Big) + \Big(543 \times \big[2463+5223\big]\Big)$$

$$+ \Big(1838 \times \big[5223+5889\big]\Big) + \Big(2173 \times 5223\Big)$$

$$= \Big(338 \times 15748\Big) + \Big(543 \times 7686\Big) + \Big(1838 \times 11112\Big) + \Big(2173 \times 5223\Big)$$

$$= 5322824 + 4173498 + 20423856 + 11349579$$

$$= 41269757$$

The number of concordant and discordant pairs is substituted into the gamma formula:

$$\gamma = \frac{Concordant\ pairs - Discordant\ pairs}{Concordant\ pairs + Discordant\ pairs}$$

$$\gamma = \frac{58680861 - 41269757}{58680861 + 41269757}$$

$$= \frac{17411104}{99950618}$$

$$= 0.174$$

Since the number of concordant pairs is larger than the number of discordant pairs, there is a positive relationship between the two variables—that is, lower attributes of one variable are associated with lower attributes of the other variable, and higher attributes of one variable are associated with higher attributes of the other variable. So, in general, younger people give lower ratings of police fairness, and older people give higher ratings of police fairness. Because gamma is a proportional reduction in error (PRE) measure, a gamma of 0.174 shows that we can reduce our error in predicting ratings of police fairness by 17 per cent if we know a person's age group.

When reporting gamma measures calculated by statistical software, be aware that positive and negative relationships are determined using the values (numbers) on a variable, not the attributes themselves. To obtain correct results, each ordinal-level variable must have low values (numbers) assigned to lower attributes, and high values (numbers) assigned to higher attributes. In the original GSS variable capturing perceptions of police fairness, "Good" was assigned the value "1," "Average" was assigned the value "2," and "Poor" was assigned the value "3." Since the software does not read English, it does not know that a "Good" rating is conceptually higher than an "Average" rating; it assumes that higher values indicate higher ratings. If necessary, recode variables before using statistical software to produce gamma measures.

Although gamma is easy to understand, some researchers are critical of the fact that it ignores tied pairs. To find out how many tied pairs are being ignored, find the total number of pairs in a sample, and subtract the number of concordant and discordant pairs. In this sample, the total number of pairs is:

$$number\ of\ pairs = \frac{n(n-1)}{2}$$
$$number\ of\ pairs = \frac{25146(25146-1)}{2}$$
$$= \frac{632296170}{2}$$
$$= 316148085$$

The total number of concordant and discordant pairs used in the calculation of gamma is shown in the denominator of the gamma equation: 99,950,618. So, in this example, there are 216,197,467 tied pairs (316,148,085 − 99,950,618). In other words, more than two-thirds (68 per cent) of the pairs of cases in this example are tied and thus ignored in the gamma calculation. When there are many tied pairs, some researchers report Kendall's tau (τ), which is a measure of association that is similar to gamma but which also accounts for tied pairs. There are two common versions of tau: tau-b (τ_b) is used with cross-tabulations that have an equal number of rows and columns; tau-c (τ_c) is used with cross-tabulations that have an unequal number of rows and columns. Like gamma, tau ranges from −1 to +1, and the sign indicates whether the relationship is positive or negative. The size of tau indicates the strength of the relationship, but, unlike gamma, tau is not a proportional reduction in error measure, so it cannot be interpreted in terms of reducing prediction errors. Instead, it is interpreted in a way similar to Pearson's correlation coefficient, which is described Chapter 10. In general, though, a higher tau (regardless of the sign) indicates a stronger relationship between the variables.

Overall, proportionate reduction in error statistics provide a useful way to assess the magnitude of a relationship between two categorical variables. Lambda and gamma each provide a one-number summary of the relative strength of the relationship. Gamma also provides information about the direction of a relationship, since it is used with ordinal-level variables.

Step-by-Step: Gamma

$$\gamma = \frac{Concordant\ pairs - Discordant\ pairs}{Concordant\ pairs + Discordant\ pairs}$$

Step 1: Prepare a cross-tabulation of the two variables you are assessing the relationship between so that the attributes of each variable are listed from lowest to highest, moving from top to bottom in the rows, and moving from left to right in the columns. Do not include a total row or a total column in the cross-tabulation (or cross them if out if they have been printed). Do not include attributes that represent missing information in the cross-tabulation.

Step 2: For *each cell* in the cross-tabulation, identify all of the cells that are located both below *and* to the right of the cell. (See Figure 9.2 for an example.) Some cells will not have any other cells located both below and to the right of them.

Step 3: For *each cell separately*, add together the number of cases in all of the cells that are located both below and to the right of them (the cells you identified in Step 2). Complete this calculation once for each cell that has other cells located both below and to the right of them.

Step 4: For *each cell separately*, multiply the number of cases in the cell by the sum of the cases in the cells located below and to the right of them (from Step 3). Complete this calculation once for each cell that has other cells located both below and to the right of them.

Step 5: Add together the results of Step 4 for each cell to find the number of concordant pairs.

Step 6: For *each cell* in the cross-tabulation, identify all of the cells that are located both below *and* to the left of the cell. (See Figure 9.2 for an example.) Some cells will not have any other cells located both below and to the left of them.

Step 7: For *each cell separately*, add together the number of cases in all of the cells that are located both below and to the left of them (the cells you identified in Step 6). Complete this calculation once for each cell that has other cells located both below and to the left of them.

Step 8: For *each cell separately*, multiply the number of cases in the cell by the sum of the cases in the cells located below and to the left of them (from Step 7). Complete this calculation once for each cell that has other cells located both below and to the left of them.

Step 9: Add together the results of Step 8 for each cell to find the number of discordant pairs.

Step 10: Subtract the number of discordant pairs (from Step 9) from the number of concordant pairs (from Step 5) to find the numerator of the gamma equation.

Step 11: Add together the number of discordant pairs (from Step 9) and the number of concordant pairs (from Step 5) to find the denominator of the gamma equation.

Step 12: Divide the result of Step 10 by the result of Step 11 to find gamma.

How Does It Look in SPSS?

Gamma

The Crosstabs procedure, with the "Gamma" checkbox selected in the Statistics option, produces results that look like those in Image 9.2. The labels of the two variables in the cross-tabulation appear at the top of the table. Above each cross-tabulation, SPSS prints a "Case Processing Summary" (not shown) that lists the number and percent of valid cases, missing cases, and total cases.

Perception (local police) - Treating people fairly * Age Group Crosstabulation

Count

(A)		Age group			Total
		Young (15-34)	Middle-aged (35-54)	Older (55 and up)	
Perception (local police) - Treating people fairly	Poor job	811	543	338	1692
	Average job	2463	2173	1838	6474
	Good job	5223	5889	5868	16980
Total		8497	8605	8044	25146

(B) **Symmetric Measures**

		Value	Asymptotic Standard Error[a]	Approximate T[b]	Approximate Significance
Ordinal by Ordinal (C) Gamma	(D)	.174	.010	17.000	.000
N of Valid Cases	(E)	25146			

a. Not assuming the null hypothesis.

b. Using the asymptotic standard error assuming the null hypothesis.

Image 9.2   **An SPSS Gamma Measure**

A. See A–F in the "How Does It Look in SPSS? Lambda" box for information on how to read a cross-tabulation.

B. This title indicates that these statistics are "Symmetric" measures. In other words, the results are the same, regardless of which variable is treated as independent and which is treated as dependent.

C. This label indicates that gamma is used to assess the relationship between two ordinal-level variables.

Continued

> D. This result corresponds to the gamma calculated in this chapter. A gamma of 0.174 indicates that knowing a person's age group can help us to reduce our error in predicting perceptions of police fairness by 17.4 per cent. The other numbers in this row can be ignored.
>
> E. This row shows how many (valid) cases were used to calculate gamma.

The Chi-Square Test of Independence

chi-square test of independence A non-parametric statistical significance test used to assess the reliability of a relationship between two categorical variables.

The most common way that researchers assess the reliability of a relationship between two categorical variables is by using a **chi-square test of independence**. Chi (χ) is a letter in the Greek alphabet that looks like an italicized letter *x*. *Chi* is pronounced like the word *eye*, with a hard *k* sound at the beginning; thus, it rhymes with *tie*. It's not pronounced *chai* like the milk tea or *chee* like the energy flow (which is also spelled *chi*). Like other tests of statistical significance, the chi-square test of independence provides information about the probability of randomly selecting a sample with the observed relationship (or one of greater magnitude) if the null hypothesis is true in the population.

parametric tests Tests that assume that the variables being tested have an underlying normal distribution in the population and that rely on estimating population parameters.

non-parametric tests Tests that do not rely on any assumptions about the underlying distribution of the variables being tested and that do not rely on estimating population parameters.

Up to this point, the tests of statistical significance that I have described—t-tests and ANOVA tests—have been **parametric tests**. Parametric tests assume that the variables being tested have an underlying normal distribution in the population. They are called parametric tests because they rely on estimating population parameters, such as means and variances. Chi-square tests of independence, however, are **non-parametric tests**. Non-parametric tests do not rely on any assumptions about the underlying distribution of the variables being tested, nor do they typically involve estimating population parameters. For instance, instead of using means, chi-square tests rely on comparing the frequencies that are observed in the sample, with the frequencies that are expected if the null hypothesis is true. The calculation of a chi-square statistic is also slightly different than the calculation of other test statistics because it relies on finding the contribution of every cell in a cross-tabulation, instead of the contribution of every case in a sample.

Table 9.4 is a cross-tabulation of the relationship between perceptions of police fairness and whether or not a person had contact with police, using GSS data. There appears to be a relationship: people who had contact with police are more likely to say that police do a "Poor" or "Average" job of treating people fairly, compared to people who had no contact with police. One in ten (11 per cent) of people who had contact with police say that they do a poor job of treating people fairly, compared to only 5 per cent of people who had no contact with police—a difference of 6 percentage points. Similarly, 70 per cent of people who had no contact with police say that they do a good job of treating people fairly, compared to only 61 per cent of people who had contact with police. But, before making claims about how contact with police affects people's perceptions of police fairness, we want to be confident that the relationship that appears in this sample is likely to exist in the larger population.

Table 9.4   A Cross-Tabulation Showing the Relationship between Contact with Police and Perceptions of Police Fairness

Do you think your local police force does a good job, an average job, or a poor job of treating people fairly?		Had Contact with Police		
		No	Yes	Total
A poor job	Count	821	871	1,692
	Column %	4.7%	11.1%	6.7%
An average job	Count	4,316	2,157	6,473
	Column %	24.9%	27.6%	25.7%
A good job	Count	12,184	4,791	16,975
	Column %	70.3%	61.3%	67.5%
Total	Count	17,321	7,819	25,140
	Column %	100.0%	100.0%	100.0%

Source: Author generated; Calculated using data from Statistics Canada, 2016.

The calculation of the chi-square test of independence relies on the idea of **expected frequencies** (or expected counts). The expected frequencies are the number of cases that researchers expect to be in each cell in a cross-tabulation if the null hypothesis is true; that is, if group membership (the independent variable) is not related to the dependent variable in the population. The expected frequencies are compared with the frequencies that are actually observed in each cell of a cross-tabulation produced from sample data (the **observed frequencies** or observed counts). The size of the difference between the expected frequencies and the observed frequencies is used to determine the likelihood of randomly selecting a sample with the observed relationship from a population in which group membership is not related to the dependent variable.

To illustrate observed and expected frequencies, let's return to the hypothetical example in Table 9.2. The dependent variable is "Contact with police." There are 200 people overall: 110 of them had no contact with police and 90 had contact. So, overall, 55 per cent of people had no contact with police, and 45 per cent had contact. The independent variable divides the sample into two groups with 100 people in each. Since the independent variable has two attributes and the dependent variable has two attributes, there are four cells in the cross-tabulation. An expected frequency can be calculated for each of the four cells:

- People in Group 1 who had no contact with police
- People in Group 1 who had contact with police
- People in Group 2 who had no contact with police
- People in Group 2 who had contact with police

If the null hypothesis is true and there is no relationship between group membership and contact with police, then the distribution of cases within each group is expected to be the same as the distribution within the sample overall. Since 55 per cent

expected frequencies The number of cases that researchers expect to be in each cell of a cross-tabulation if the null hypothesis is true.

observed frequencies The number of cases that are actually in each cell of a cross-tabulation.

Table 9.5   **A Cross-Tabulation Showing Expected Frequencies (Hypothetical Data)**

	Group 1	Group 2	Total
No contact with police	80 [Expected = 55]	30 [Expected = 55]	110 [55%]
Had contact with police	20 [Expected = 45]	70 [Expected = 45]	90 [45%]
Total	100	100	200

of the sample overall had no contact with police, 55 per cent of people in Group 1 are expected to have no contact with police, and 55 per cent of people in Group 2 are expected to have no contact with police. Since Group 1 has 100 people in it, the number of people expected to have no contact with police is 55 per cent of 100 people, or 55 people. Similarly, since Group 2 has 100 people in it, the number of people expected to have no contact with police is 55 per cent of 100 people, or 55 people. (See Table 9.5.) The same logic extends to the calculation of how many people are expected to have contact with police. If the null hypothesis is true and group membership is not related to contact with police, since 45 per cent of the sample overall had contact with police, 45 per cent of people in Group 1 are expected to have contact with police, and 45 per cent of people in Group 2 are expected to have contact with police.

The expected frequencies are easy to calculate for this example, but a formula is useful in more complicated situations. To calculate the expected frequency (f_e) for any cell in a cross-tabulation, find the total number of cases in the same row as the cell, multiply it by the total number of cases that are in the same column as the cell, and then divide the result by the total number of cases in the cross-tabulation overall. So for the "Group 1, no contact" cell in Table 9.5, the "Total" column shows there are 110 cases in the same row, and the "Total" row shows there are 100 cases in the same. To find the expected frequency for the "Group 1, no contact" cell, multiply 110 by 100, and then divide the result by 200 (the total number of cases). The same approach is used for each cell, substituting the different row and column totals into the equation. Try it yourself for one of the other cells in Table 9.5.

expected frequencies

$$f_e = \frac{(row\ total)(column\ total)}{overall\ total}$$

The chi-square test of independence compares the expected frequencies for each cell to the observed frequencies in each cell to determine the likelihood that the sample was randomly selected from a population in which there is no relationship between the two variables. The formula used to calculate a chi-square statistic is:

chi-square statistic

$$\chi^2 = \Sigma \frac{(f_o - f_e)^2}{f_e}$$

The formula for a chi-square statistic includes a sigma (Σ), which tells you to sum together the results of the calculation; but in this context, it refers to summing the results for each *cell* in the cross-tabulation, not the results for each case in the dataset. So, for the hypothetical example in Table 9.5, the calculation is completed for each of the four cells (shaded in light teal), and the results are added together (cells in the total row and total column aren't included). In the chi-square formula, the f_e symbol denotes the expected frequencies, and the f_o symbol denotes the observed frequencies, that is, the number of cases in each cell in the sample. The difference between the observed frequency and the expected frequency for each cell contributes to the size of the chi-square statistic: large differences between the observed and the expected frequencies result in a large chi-square statistic, whereas small differences between the observed and the expected frequencies result in a small chi-square statistic.

Like other formulas that include a sigma (Σ), the chi-square formula is easiest to implement using a table. Figure 9.3 shows how to calculate the chi-square statistic for the hypothetical example in Table 9.5. There is one row for each cell that lists the observed and expected frequencies. The difference between the observed and expected frequency for each cell is squared, and the result is divided by the expected frequency. The results in the final column are summed to obtain the chi-square statistic, which is interpreted in relation to a chi-square distribution.

Like the F-distribution, the chi-square distribution is a probability distribution that changes shape depending on the number of degrees of freedom. The degrees of freedom of a chi-square distribution account for the number of cells in the cross-tabulation. Many people find the concept of degrees of freedom, or free parameters, easiest to understand in the context of cross-tabulations. For chi-square tests of independence, the number of degrees of freedom is the number of rows in the cross-tabulation minus 1, multiplied by the number of columns in the

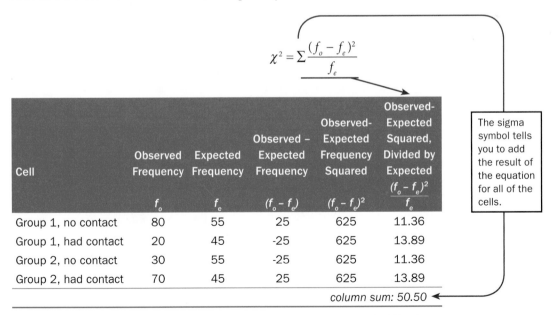

$$\chi^2 = \Sigma \frac{(f_o - f_e)^2}{f_e}$$

Cell	Observed Frequency f_o	Expected Frequency f_e	Observed – Expected Frequency $(f_o - f_e)$	Observed- Expected Frequency Squared $(f_o - f_e)^2$	Observed- Expected Squared, Divided by Expected $\frac{(f_o - f_e)^2}{f_e}$
Group 1, no contact	80	55	25	625	11.36
Group 1, had contact	20	45	-25	625	13.89
Group 2, no contact	30	55	-25	625	11.36
Group 2, had contact	70	45	25	625	13.89
					column sum: 50.50

The sigma symbol tells you to add the result of the equation for all of the cells.

Figure 9.3   Translating the Chi-Square Formula into a Table (Hypothetical Data)

cross-tabulation minus 1. This is because if you know the total number (or percentage) of cases, the number of cases in the last row and the last column are already determined; that is, they are fixed (not free) parameters. If you have played Sudoku games, this process will be especially familiar to you.

Figure 9.4 illustrates the idea of fixed and free parameters. For the two-by-two cross-tabulation in the left panel, as soon as one cell is filled in, the remainder of the cells can be filled in using simple subtraction. In other words, only one cell is "free" to vary; once the number of cases in one cell is established, the remaining cells are fixed. Thus, a two-by-two cross-tabulation has only one degree of freedom: two rows minus 1 is equal to 1 and two columns minus 1 is equal to 1, and 1 multiplied by 1 is equal to 1 ([2 − 1] x [2 − 1] = 1). A two-by-three cross-tabulation is shown in the middle panel of Figure 9.4. When only one cell is filled in, there isn't enough information to complete the rest of the table. But when two cells are filled in, the number of cases in the remaining cells are fixed. Thus, a two-by-three cross-tabulation has two degrees of freedom: three rows minus 1 is equal to 2, and two columns minus 1 is equal to 1, and 2 multiplied by 1 is equal to 2 ([3 − 1] x [2 − 1] = 2). The three-by-three cross-tabulation in the right panel has four degrees of freedom: three rows minus 1 is equal to 2, and three columns minus 1 is equal to 2, and 2 multiplied by 2 is equal to 4.

The left panel of Figure 9.5 shows how the shape of the chi-square distribution changes depending on the number of degrees of freedom. The chi-square statistic produced by a randomly selected sample identifies a point on the horizontal axis of the distribution. And, just like for other distributions, the amount of the area under the curve that is beyond that point on the horizontal axis is used to determine the probability of randomly selecting a sample that would produce a chi-square statistic of this size (or larger), if no relationship exists in the population. Like all tests of statistical significance, this probability is represented as a p-value.

The hypothetical two-by-two cross-tabulation shown in Table 9.5 has one degree of freedom. For an alpha value of 0.05, the critical value of the chi-square

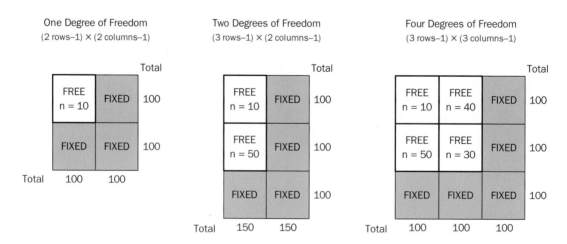

Figure 9.4 **How Free and Fixed Parameters Reflect "Degrees of Freedom" (Hypothetical Data)**

statistic is the point on the horizontal axis of a chi-square distribution that has only 5 per cent of the area under the curve beyond it. For a chi-square distribution with one degree of freedom, using an alpha value of 0.05, the critical value is 3.84. (See the right panel of Figure 9.5.) In other words, to obtain a p-value that is less than 0.05, the chi-square statistic must be higher than 3.84. Since the chi-square statistic for the hypothetical example (50.50) is higher than the critical value (3.84), there is less than a 5 per cent chance of randomly selecting a sample with the observed relationship (or a relationship of greater magnitude), if there is no relationship between group membership and the dependent variable in the population. As a result, we reject the null hypothesis and assert that there is likely a relationship between the two variables in the (hypothetical) population.

In Table 9.6 we return to the real-world GSS results from Table 9.4 that show the relationship between contact with police and perceptions of police fairness. In the sample, people who had contact with police had worse perceptions of police fairness than people who had no contact with police. We want to assess whether this relationship is likely to exist in the larger population. The research hypothesis is that there is a relationship between having contact with police and perceptions of police fairness in the population. The null hypothesis is that there is no relationship between having contact with police and perceptions of police fairness in the population. Let's begin by finding the expected frequencies for each cell in the cross-tabulation. For the top-left cell (people who had no contact with police and who think police do a poor job of treating people fairly), the expected frequency is calculated by multiplying the row total by the column total (1,692 x 17,321) and dividing the result by the grand total (25,140). The number of people expected in this cell if there is no relationship between having contact with police and perceptions of police fairness is 1,165.76. Notice that expected frequencies can include partial people (in other words, 0.76 of a person). Even though it's not possible to have a partial respondent or a partial case in the data, it's important to retain these decimals for mathematical accuracy. Try calculating the expected frequency for another cell yourself.

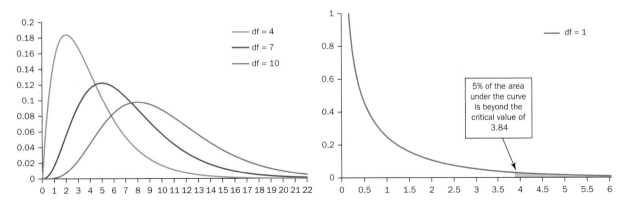

Figure 9.5 The Chi-Square Distribution for Various Degrees of Freedom, and the Critical Value of the Chi-Square Statistic for a Chi-Square Distribution with One Degree of Freedom (alpha value = 0.05)

Table 9.6   **Observed and Expected Frequencies (Counts) in a Cross-Tabulation**

Do you think your local police force does a good job, an average job, or a poor job of treating people fairly?		Had Contact with Police		
		No	Yes	Total
Poor job	Count (Observed Frequency)	821	871	1,692
	Expected Frequency	1,165.76	526.24	1,692.00
Average job	Count (Observed Frequency)	4,316	2,157	6,473
	Expected Frequency	4,459.78	2,013.22	6,473.00
	Count (Observed Frequency)	12,184	4,791	16,975
Good job	Expected Frequency	11,695.46	5,279.54	16,975.00
Total	Count (Observed Frequency)	17,321	7,819	25,140
	Expected Frequency	17,321.00	7,819.00	25,140.00

Source: Author generated; Calculated using data from Statistics Canada, 2016.

Since this cross-tabulation has six cells (excluding the total row and column), the table used to implement the chi-square formula will have six rows. When the calculations are completed for each cell and the results are summed, a chi-square statistic of 408.341 is obtained. (See Figure 9.6.)

The cross-tabulation in this example has two degrees of freedom ([three rows minus 1] multiplied by [two columns minus 1]). For a chi-square distribution with two degrees of freedom, using an alpha value of 0.05, the critical value is 5.99. Since the chi-square statistic of 408.34 is higher than 5.99, we are confident that there is less than a 5 per cent chance of randomly selecting a sample with this relationship, if there is no relationship between contact with police and perceptions of police fairness in the larger population. As a result, we can reject the null hypothesis and assert that a relationship likely exists in the population. It is very likely that, in the larger population, people who had contact with police have poorer perceptions of police fairness than people who had no contact with police. People who had no contact with police in the past 12 months, and are basing their evaluations on second-hand knowledge, media representations, or more dated contact, are more likely to say that police do a good job of treating people fairly.

For chi-square statistics to be accurate, every cell must have an expected frequency of 1 or more, and at least 80 per cent of cells must have an expected frequency of 5 or more. In Table 9.6 the lowest expected frequency is 526.24. If a cross-tabulation has expected frequencies lower than 1, or fewer than 80 per cent of cells with an expected frequency of 5 or more, there are two potential remedies. If the sample size is small, use an alternate test of statistical significance. If the sample size is large, but there are still some small expected frequencies, there are likely some small groups within the sample. Eliminating the small groups by merging them with others will reduce the number of cells with low expected frequencies. (See the "Hands-on Data Analysis Box" in Chapter 3 for a discussion of how to recode variables.)

Compared to many other statistical tests, the results of chi-square tests of independence are influenced strongly by sample size. In part, this is because the degrees

$$\chi^2 = \Sigma \frac{(f_o - f_e)^2}{f_e}$$

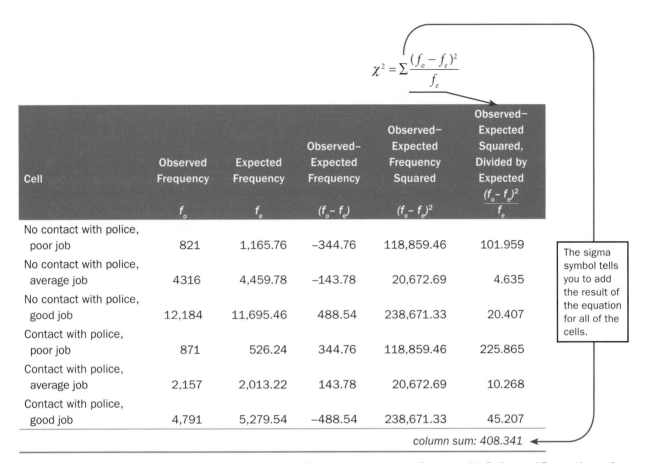

Cell	Observed Frequency f_o	Expected Frequency f_e	Observed–Expected Frequency $(f_o - f_e)$	Observed–Expected Frequency Squared $(f_o - f_e)^2$	Observed–Expected Squared, Divided by Expected $\frac{(f_o - f_e)^2}{f_e}$
No contact with police, poor job	821	1,165.76	−344.76	118,859.46	101.959
No contact with police, average job	4316	4,459.78	−143.78	20,672.69	4.635
No contact with police, good job	12,184	11,695.46	488.54	238,671.33	20.407
Contact with police, poor job	871	526.24	344.76	118,859.46	225.865
Contact with police, average job	2,157	2,013.22	143.78	20,672.69	10.268
Contact with police, good job	4,791	5,279.54	−488.54	238,671.33	45.207
				column sum:	408.341

The sigma symbol tells you to add the result of the equation for all of the cells.

Figure 9.6 Calculating the Chi-Square Statistic for the Relationship between Contact with Police and Perceptions of Police Fairness

Source: Author generated; Calculated using data from Statistics Canada, 2016.

of freedom account for the number of *cells,* not the number of *cases* used to calculate the chi-square statistic. As a result, large samples tend to produce statistically significant results, even when the magnitude of the relationship is small; conversely, small samples tend to produce results that are not statistically significant, even when the magnitude of the relationship is large. This tendency is illustrated in Figure 9.7. The cross-tabulation on the left has no expected frequencies lower than 5 and produces a chi-square statistic of 0.551, which corresponds with a p-value of 0.458. The cross-tabulation on the right shows an identical relationship, but the number of

	Group 1	Group 2	Total
Attribute 1	10	13	23
Attribute 2	12	10	22
Total	22	23	45

$\chi^2 = 0.551$; df = 1; p = 0.458

	Group 1	Group 2	Total
Attribute 1	100	130	230
Attribute 2	120	100	220
Total	220	230	450

$\chi^2 = 5.51$; df = 1; p = 0.019

Figure 9.7 How Sample Size Affects a Chi-Square Test of Independence (Hypothetical Data)

Step-by-Step: Chi-Square Test of Independence

Step 1: Identify the research hypothesis and the null hypothesis.

Expected Frequencies

$$f_e = \frac{(row\ total)(column\ total)}{overall\ total}$$

Step 2: Prepare a cross-tabulation of the two variables you are assessing the relationship between, with the independent variable in the columns and the dependent variable in the rows. Include a total row and a total column. Do not include attributes that represent missing information in the cross-tabulation.

Step 3: Count the total number of cases (the *overall total*). Do not include cases that are missing information for either variable.

For Steps 4 to 7, do not include the cells in the total row or the total column.

Step 4: For *each cell* in the cross-tabulation, find the number of cases that are located in the same row (the *row total*). Find this number in the total column, for the row that the cell is located in.

Step 5: For *each cell* in the cross-tabulation, find the number of cases that are located in the same column (the *column total*). Find this number in the total row, for the column that the cell is located in.

Step 6: For *each cell separately*, multiply the row total (from Step 4) by the column total (from Step 5).

Step 7: For *each cell separately*, divide the result of Step 6 by the overall total (from Step 3) to find the expected frequency for the cell (f_e).

Chi-Square Statistic

$$\chi^2 = \Sigma \frac{(f_o - f_e)^2}{f_e}$$

For Steps 8 to 11, do not include the cells in the total row or the total column.

Step 8: For *each cell* in the cross-tabulation, find the number of cases in the cell; this is the observed frequency for the cell (f_o).

Step 9: For *each cell separately*, subtract the expected frequency (from Step 7) from the observed frequency (from Step 8).

Step 10: For *each cell separately*, square the result of Step 9.

Step 11: For *each cell separately*, divide the result of Step 10 by the expected frequency (from Step 7).

Step 12: Add together the results of Step 11 for each cell to find the chi-square statistic.

Degrees of Freedom

$$df_{\chi^2} = (rows - 1)(columns - 1)$$

Step 13: Count the number of attributes in the dependent variable (*rows*). Do not include attributes with no cases or attributes that represent missing information. This count will match the number of rows in the cross-tabulation, excluding the total row.

Step 14: Count the number of attributes in the independent variable (*columns*). Do not include attributes with no cases or attributes that represent missing information. This count will match the number of columns in the cross-tabulation, excluding the total column.

Step 15: Subtract 1 from the number of rows (from Step 13).

Step 16: Subtract 1 from the number of columns (from Step 14).

Step 17: Multiply the result of Step 15 and Step 16 to find the degrees of freedom.

Statistical Significance

Step 18: Determine the alpha value you want to use for the chi-square test. The most common alpha value used in the social sciences is 0.05.

Step 19: Determine the critical value of the chi-square statistic, using the alpha value you selected in Step 18, for a chi-square distribution with the degrees of freedom you found in Step 17. This can be done using a printed table of the critical values of the chi-square distribution, an online calculator, or the spreadsheet tool available from the Student Resources area of the companion website for this book.

Step 20: Compare the chi-square statistic (from Step 12) to the critical value (from Step 19). If the chi-square statistic is higher than the critical value, the result is statistically significant: you can reject the null hypothesis and state that a relationship likely exists between the two variables in the population. If the chi-square statistic is lower than the critical value, the result is not statistically significant: you fail to reject the null hypothesis and state that you are not confident that a relationship exists between the two variables in the population.

people in each cell is multiplied by 10. Because of the larger sample size, it produces a chi-square statistic of 5.51 (exactly 10 times larger than the cross-tabulation on the left), which has a p-value of 0.019—below the usual alpha value of 0.05.

Because of this sensitivity to sample size, when you use chi-square tests of independence be sure to always describe the size and direction of the relationship, as well as the statistical significance. One way to do this is by using chi-square-based measures of association, which standardize the chi-square statistic and take the sample size into account.

Chi-Square-Based Measures of Association

Earlier in this chapter, you learned about proportionate reduction in error (PRE) measures of association—lambda and gamma—which are used to assess the magnitude of the relationship between two categorical variables. The chi-square statistic is used primarily to determine the reliability of a relationship. But the chi-square statistic is also used to calculate two measures of association for categorical variables: phi and Cramér's V. **Phi** (ϕ) is used to assess the magnitude of a relationship between two categorical variables that have only two attributes each (two-by-two cross-tabulations), whereas **Cramér's V (V)** is used to assess the magnitude of a relationship between two categorical variables where one or both variables have more than two attributes. Conceptually, phi and Cramér's V can be thought of as measures of effect size, like Cohen's d.

> **phi (ϕ)** A chi-square-based measure of association used to assess the magnitude of a relationship between two categorical variables that have only two attributes each.

> **Cramér's V (V)** A chi-square-based measure of association used to assess the magnitude of a relationship between two categorical variables, where one or both variables have more than two attributes.

Phi and Cramér's V usually range from 0 to 1, where 0 indicates that there is no relationship, and 1 indicates that there is a perfect relationship between the two variables. Much like for Cohen's d, there is no absolute interpretation of phi and Cramér's V; instead, researchers use them to classify the effect of the independent variable on the dependent variable as small, medium, or large (or a relationship as weak, moderate, or strong). Following Cohen (1988), phi and Cramér's V can be interpreted as follows (notice that these thresholds are substantially lower than those for Cohen's d):

- 0.1 is a small effect (or a weak relationship).
- 0.3 is a medium effect (or a moderate relationship).
- 0.5 is a large effect (or a strong relationship).

Phi is calculated as follows, where χ^2 is the chi-square statistic for the relationship:

phi
$$\phi = \sqrt{\frac{\chi^2}{n}}$$

The two-by-two cross-tabulation in Table 9.1 shows the relationship between visible minority status and contact with police. There are 27,199 cases and the relationship produces a chi-square statistic of 56.213 ($p < 0.001$). As a result, we are relatively confident that there is a relationship between visible minority status and contact with police in the larger population. The phi for this relationship is:

$$\phi = \sqrt{\frac{\chi^2}{n}}$$

$$\phi = \sqrt{\frac{56.213}{27199}}$$

$$= \sqrt{0.0021}$$

$$= 0.046$$

This phi of 0.046, which is less than 0.1, indicates that the magnitude of the relationship between visible minority status and having contact with police is quite small.

The calculation of Cramér's V is similar to phi but includes an adjustment for the number of variable attributes. The formula is:

$$V = \sqrt{\frac{\chi^2}{n(\min[rows-1],[columns-1])}}$$

Cramér's V

In the denominator of the fraction in the Cramér's V formula, the number of cases is multiplied by *either* the number of rows minus 1, or the number of columns minus 1—whichever is lower (the minimum).

The three-by-two cross-tabulation in Table 9.6 shows the relationship between contact with police and perceptions of police fairness. There are 25,140 cases, and the relationship produces a chi-square statistic of 408.341. The number of rows minus 1 is equal to 2, and the number of columns minus 1 is equal to 1. The lower of these two values is 1, which is used in the calculation of Cramér's V as follows:

$$V = \sqrt{\frac{\chi^2}{n(\min[rows-1],[columns-1])}}$$
$$= \sqrt{\frac{408.341}{25140(1)}}$$
$$= \sqrt{0.0162}$$
$$= 0.127$$

The Cramér's V of 0.127 indicates that the magnitude of the relationship between contact with police and perceptions of police fairness is slightly larger than the magnitude of the relationship between visible minority status and contact with police. Overall, though, the size (or magnitude) of both relationships is relatively small.

Step-by-Step: Phi

$$\phi = \sqrt{\frac{\chi^2}{n}}$$

Step 1: Count the total number of cases (n). Do not include cases that are missing information for either variable.

Step 2: Find the chi-square statistic for the relationship between the two variables (χ^2).

Step 3: Divide the chi-square statistic (from Step 2) by the total number of cases (from Step 1).

Step 4: Find the square root of the result of Step 3 to find phi.

Step-by-Step: Cramér's V

$$V = \sqrt{\frac{\chi^2}{n(\min[rows-1],[columns-1])}}$$

Step 1: Count the total number of cases (*n*). Do not include cases that are missing information for either variable.

Step 2: Find the chi-square statistic for the relationship between the two variables (χ^2).

Step 3: Count the number of attributes in the dependent variable (*rows*). Do not include attributes with no cases or attributes that represent missing information. This count will match the number of rows in the cross-tabulation, excluding the total row.

Step 4: Count the number of attributes in the independent variable (*columns*). Do not include attributes with no cases or attributes that represent missing information. This count will match the number of columns in the cross-tabulation, excluding the total column.

Step 5: Compare the count from Step 3 and Step 4 and determine which is lower. This is the minimum count.

Step 6: Subtract 1 from the minimum count (from Step 5).

Step 7: Multiply the result of Step 6 by the total number of cases (from Step 1) to find the denominator of the fraction in the Cramér's V equation.

Step 8: Divide the chi-square statistic (from Step 2) by the result of Step 7.

Step 9: Find the square root of the result of Step 8 to find Cramér's V.

How Does It Look in SPSS?

Chi-Square, Phi, and Cramér's V

The Crosstabs procedure, with the "Observed" and "Expected" checkboxes selected in the Cells option, and the "Phi and Cramer's V" checkbox selected in the Statistics option, produces results that look like those in Image 9.3. The labels of the two variables in the cross-tabulation appear at the top of the table. Above each cross-tabulation, SPSS prints a "Case Processing Summary" (not shown) that lists the number and percent of valid cases, missing cases, and total cases.

A. See A–F in the "How Does It Look in SPSS? Lambda" box for information on how to read a cross-tabulation.

B. The "Expected Count" row shows the number of cases expected in each cell if there is no relationship between the two variables in the population. In this sample, there are 821 people who had no contact with police and who think

Perception (local police) - Treating people fairly * Contact with police - any reason Crosstabulation

(A)			Contact with police - any reason		Total
			No	Yes	
Perception (local police) - Treating people fairly	Poor job	Count	821	871	1692
	(B)	Expected Count	1165.8	526.2	1692.0
	Average job	Count	4316	2157	6473
		Expected Count	4459.8	2013.2	6473.0
	Good job	Count	12184	4791	16975
		Expected Count	11695.5	5279.5	16975.0
Total		Count	17321	7819	25140
		Expected Count	17321.0	7819.0	25140.0

(C) **Chi-Square Tests**

		Value	df	Asymptotic Significance (2-sided)
Pearson Chi-Square	(D)	408.334[a]	2	.000
Likelihood Ratio		383.583	2	.000
Linear-by-Linear Association	(E)	345.586	1	.000
N of Valid Cases	(F)	25140		

(G) a. 0 cells (0.0%) have expected count less than 5. The minimum expected count is 526.24.

(H) **Symmetric Measures**

			Value	Approximate Significance
Nominal by Nominal (I)	Phi	(J)	.127	.000
	Cramer's V	(K)	.127	.000
N of Valid Cases		(L)	25140	

Image 9.3 **An SPSS Chi-Square Test of Independence and Associated Measures**

that the police do a poor job of treating people fairly. If there is no relationship between contact with police and perceptions of police fairness, 1,165.8 people in the sample are expected to have no contact with police and think the police do a poor job of treating people fairly. The expected count will always match the actual count in the total row and the total column.

C. This title indicates that this table shows the "Chi-Square Test" results.

D. The results in the "Pearson Chi-Square" row correspond to the calculations in this chapter. The "Value" column shows that the chi-square statistic is

Continued

408.334 (calculated as 408.341 in Figure 9.6; the two results vary slightly because SPSS retains more decimal places). The "df" column shows the degrees of freedom, which are used to determine the shape of the chi-square distribution that the chi-square statistic is evaluated against. The "Asymptotic Significance (2-sided)" column shows the p-value associated with this chi-square statistic (with these degrees of freedom). That is, it shows the likelihood of randomly selecting a sample with the observed relationship (or a larger relationship), if no relationship exists in the population. This p-value indicates that there is less than a 0.1 per cent chance of selecting this sample from a population in which there is no relationship between contact with police and perceptions of police fairness ($p < 0.001$).

E. The "Likelihood Ratio" and "Linear-by-Linear Association" rows can be ignored.

F. This row shows how many (valid) cases were used to calculate the chi-square statistic.

G. The footnote reports the percentage of cells in the cross-tabulation with an expected count less than 5, and the lowest expected count. If more than 20 per cent of cells have an expected count less than 5, or the lowest expected count is less than 1, the chi-square statistic should not be used to test statistical significance.

H. This title indicates that these are "Symmetric" measures. In other words, the results are the same, regardless of which variable is treated as independent and which is treated as dependent.

I. This label indicates that phi and Cramér's V are used to assess the relationship between two nominal-level variables. They can also be used to assess the relationship between a nominal-level variable and an ordinal-level variable, or between two ordinal-level variables.

J. Phi and Cramér's V are both printed, regardless of how many rows and columns are in the cross-tabulation. Phi should only be reported for two-by-two cross-tabulations. The other number in the row can be ignored.

K. This result corresponds to the Cramér's V calculated in this chapter. A Cramér's V of 0.127 indicates that the magnitude of the relationship between contact with police and perceptions of police fairness is relatively small. Cramér's V should only be reported for cross-tabulations larger than two-by-two. The other number in the row can be ignored.

L. This row shows how many (valid) cases were used to calculate phi and Cramér's V.

Extending Cross-Tabulations: The Elaboration Model

elaboration model An analytic approach based on assessing a relationship between two variables, and then investigating how the relationship changes after controlling for a third variable.

The **elaboration model** gives researchers the conceptual tools they need to move toward multivariate analysis and more complex statistical modelling. This analytic approach relies on first assessing the relationship between two variables, and then investigating how that relationship changes after controlling for a third variable. The elaboration model was developed by American sociologist Paul Lazarsfeld, who identified four scenarios that can occur: replication, specification, explanation, and interpretation.

The elaboration model can be thought of as a conceptual equivalent to the experimental method. In the social sciences, researchers often cannot use experimental research designs that randomly assign cases to groups—typically, a control group and one or more treatment groups. But quantitative researchers analyzing secondary data can divide the cases into meaningful groups based on their characteristics and then investigate how relationships differ between the two groups. In other words, they can account for (or control for) group membership in their analyses. Even though the elaboration model is conceptually similar to an experimental research design, because the groups in an elaboration model are not based on random assignment researchers cannot use the results to make causal claims.

The first step in developing an elaboration model is to investigate the original relationship between two variables in a sample overall. The original relationship is sometimes referred to as the **zero-order relationship**. Then, the sample is divided into subgroups based on the attributes of a third variable. (See the "Hands-on Data Analysis" box to learn how to do this.) The researcher then assesses the relationship between the same two variables for each subgroup within the sample separately. The relationships within each subgroup are sometimes called **partial relationships**.

zero-order relationship The original relationship between two variables in a sample overall.

partial relationship The relationship between two variables in a subgroup of the sample, or after controlling for a third variable.

Replication and Specification

To illustrate the usefulness of the elaboration model, let's consider four possible scenarios. The first scenario is **replication**. When replication occurs, the relationship between the two variables in each subgroup is roughly the same as in the sample overall. In other words, controlling for the third variable doesn't substantially influence the relationship between the independent variable and the dependent variable.

The second scenario is **specification**. When specification occurs, the relationship between the two variables in one or more subgroups is stronger than the relationship in the sample overall and is weaker (or disappears entirely) in other subgroups. In this situation, controlling for the third variable shows how the relationship between the independent variable and the dependent variable is influenced (or moderated) by subgroup membership.

replication When the relationship between two variables in each subgroup of the sample is roughly the same as the relationship between the two variables in the sample overall.

specification When the relationship between two variables in one or more subgroups of the sample is stronger than the relationship in the sample overall and is weaker in other subgroups.

Explanation and Interpretation

The third and fourth scenarios are explanation and interpretation. In both cases, the relationship between two variables in each subgroup of the sample is weaker than the relationship in the sample overall. The distinction between explanation and interpretation depends on whether the third, controlling variable is conceptualized

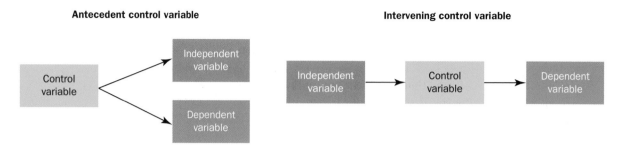

Figure 9.8   A Conceptual Illustration of Antecedent and Intervening Control Variables

antecedent variable A variable that occurs before both the independent variable and the dependent variable in time and, thus, may influence both of them.

intervening variable A variable that occurs between the independent variable and dependent variable in time; it may be influenced by the independent variable and may also influence the dependent variable.

explanation When the relationship between two variables in each subgroup of the sample is weaker than the relationship in the sample overall, as a result of controlling for an antecedent variable.

interpretation When the relationship between two variables in each subgroup of the sample is weaker than the relationship in the sample overall, as a result of controlling for an intervening variable.

suppression When the relationship between two variables in each subgroup of the sample is stronger than the relationship between the two variables in the sample overall.

as an **antecedent variable** or an **intervening variable**. An antecedent, or contextual, variable occurs before both the independent variable and the dependent variable in time and, thus, may influence both of them. An intervening variable occurs between the independent variable and dependent variable in time; it may be influenced by the independent variable and may also influence the dependent variable. Figure 9.8 provides a conceptual illustration of the difference between antecedent and intervening variables.

If, compared to the sample overall, the relationship between two variables becomes weaker (or disappears entirely) in each of the subgroups defined by an *antecedent variable*, the result is **explanation**. When this occurs, it suggests that the original relationship was not genuine and was only an artifact of the relationship that the control variable had with both the independent variable and the dependent variable. In other words, the relationship between the independent variable and the dependent variable is explained entirely by subgroup membership. Once subgroup membership is taken into account, the original relationship between the two variables becomes weaker or disappears.

If, compared to the sample overall, the relationship between two variables becomes weaker (or disappears entirely) in each of the subgroups defined by an *intervening variable*, the result is **interpretation**. When this occurs, the researcher knows that the independent variable is not directly affecting the dependent variable; rather, it is working through another mechanism (the control variable). This knowledge can help the researcher to develop a more nuanced understanding of how the independent and dependent variables are related.

Suppression and Distortion

Other researchers have refined Lazarsfeld's original elaboration model by setting out two additional scenarios that might occur. The first additional scenario is **suppression**, which occurs when *all* of the partial relationships are stronger than the zero-order relationship. In other words, a relationship between two variables appears only when a third variable is controlled for. This outcome typically occurs when the third variable has a positive relationship with one of the original variables and a negative relationship with the other original variable; they effectively cancel each other out, making it appear as though there is no relationship between the two original variables.

The second additional scenario is **distortion**, which occurs when *all* of the partial relationships are in the opposite direction of the zero-order relationship. In other words, when a third variable is controlled for, the direction of the relationship changes: a positive relationship becomes negative (or vice versa), or a group that was more likely to do something becomes less likely to do something (in all of the partial relationships).

In social science data analyses, the most common outcomes of the elaboration model are replication and specification. The other four scenarios (explanation, interpretation, suppression, and distortion) tend to occur more rarely.

distortion When the relationship between two variables in each subgroup of the sample is in the opposite direction of the relationship in the sample overall.

The Elaboration Model in Action

Let's use an elaboration model to investigate how the relationship between contact with police and perceptions of police fairness changes when racialization is taken into account. The original—or zero-order—relationship between contact with police and perceptions of police fairness is shown in Table 9.4. Among people who had contact with police, 11 per cent said that police did a poor job of treating people fairly, compared to only 5 per cent among people who had no contact with police, for a 6 percentage point difference. Table 9.7 shows what happens when you take into account—or control for—whether someone is a visible minority or not. The top cross-tabulation in Table 9.7 shows the relationship between contact with police and perceptions of police fairness among people who are not visible minorities. Within this subgroup, 11 per cent of people who had contact with police say they do a poor job of treating people fairly, compared to 5 per cent among people who had no contact with police, for a 6 percentage point difference. In other words, the partial relationship for people who are not visible minorities is the same as in the sample overall. (Both have a 6 percentage point difference.) The bottom cross-tabulation in Table 9.7 shows that among visible minorities, 13 per cent of people who had contact with police say they do a poor job of treating people fairly, compared to 5 per cent among people who had no contact with police, for an 8 percentage point difference. In other words, the partial relationship for people who are visible minorities is slightly stronger than the original relationship. This result provides an example—albeit a relatively weak one—of specification: the partial relationship is stronger than the original relationship for one subgroup.

This analysis shows that the relationship between contact with police and perceptions of police fairness is slightly more pronounced for people who are visible minorities. Compared to non-racialized people who had contact with police, racialized people who had contact with police are more likely to say that police do a poor job of treating people fairly. Overall, the use of elaboration models—investigating a relationship between two variables in a sample overall, then dividing the sample into subgroups and investigating the relationship for each subgroup separately—can give researchers substantial insight into the social processes and structures that underlie relationships between variables. Undertaking the type of exploratory data analysis that the elaboration model promotes will also help you to develop multivariate analyses.

Table 9.7   **An Example of the Elaboration Model**

Cross-tabulation for people who are not visible minorities:

| Do you think your local police force does a good job, an average job, or a poor job of treating people fairly? | Had Contact with Police | | |
	No (n = 13,427)	Yes (n = 6,402)	Total (n = 19,829)
A poor job	4.5%	10.6%	6.5%
An average job	24.4%	27.4%	25.4%
A good job	71.0%	62.0%	68.1%
Total	100.0%	100.0%	100.0%

Cross-tabulation for people who are visible minorities:

| Do you think your local police force does a good job, an average job, or a poor job of treating people fairly? | Had Contact with Police | | |
	No (n = 3,571)	Yes (n = 1,304)	Total (n = 4,875)
A poor job	4.6%	12.6%	6.7%
An average job	26.5%	28.8%	27.1%
A good job	68.9%	58.7%	66.1%
Total	100.0%	100.0%	100.0%

Source: Author generated; Calculated using data from Statistics Canada, 2016.

Hands-on Data Analysis

Splitting a Dataset

There are several ways to put the elaboration model into practice using statistical software. Some statistical software allows you to add a third variable or a control variable to a cross-tabulation. But the easiest way to investigate how a third variable influences more than one bivariate relationship is to virtually divide the dataset into groups in order to conduct parallel analyses.

Most statistical software programs have a Split File tool that allows you to specify a variable to use to divide a dataset into subgroups. Only categorical variables with relatively few attributes should be used to split a dataset. If necessary, recode the grouping variable before splitting the dataset so that there are a manageable number of results to interpret. Once a dataset is virtually divided into subgroups, the results of every procedure—frequency tables, graphs, means comparisons, cross-tabulations—are produced for each subgroup separately, until you deactivate the Split File tool. Depending on the options you select when you split the dataset, the results for each subgroup are either presented in separate tables or combined together in a single, large table.

Best Practices in Presenting Results

Writing about Cross-Tabulation Results

Writing about differences between groups is relatively straightforward but can take some practice. There are three general guidelines to follow in order to effectively present cross-tabulation results.

First, always report the percentage of people in each group with an attribute, not the frequency. You must also clearly identify the group that you are reporting percentages for:

> *Incorrect*: 871 people who had contact with police say that the local police do a poor job of treating people fairly. [Although this is technically accurate, it provides no sense of the context or proportion.]
> *Correct*: Among people who had contact with police, 11 per cent report that the local police do a poor job of treating people fairly.
> *Incorrect*: Eleven (11) per cent say that the local police do a poor job of treating people fairly.
> *Correct*: Eleven (11) per cent of people who had contact with police say that the local police do a poor job of treating people fairly.

Second, be sure to indicate which groups you are comparing. If you say that members of a group are more or less likely to have some characteristic, you must specify who you are comparing them to. For example:

> *Incorrect:* People who had contact with police are more likely to say that the local police do a poor job of treating people fairly.
> *Correct:* People who had contact with police are more likely than people who had no contact with police to say that the local police do a poor job of treating people fairly.

In this example, it might seem as though it's common sense to compare people who had contact with police to people who had no contact with police, but for many variables it's not always obvious who the comparison group is.

Third, and finally, it's crucial that you report comparable statistics for each group. So, if you report the percentage of people in one group who gave a "poor" rating, also report the percentage of people in the other group or groups who gave a "poor" rating. Then, if it is useful to your argument, you can report the percentage of people in each of the groups who gave a "good" rating. It's bad practice to report the percentage of people in one group who gave a "poor" rating, compared to the percentage of people in another group who gave a "good" rating because the reader has no way of knowing how the proportion of people who gave a "poor" rating in the first group compares to the other group.

Incorrect: Eleven (11) per cent of people who had contact with police say that the local police do a poor job of treating people fairly, compared to 70 per cent of people who had no contact with police who say that the local police do a good job of treating people fairly.

Correct: Eleven (11) per cent of people who had contact with police say that the local police do a poor job of treating people fairly, compared to 5 per cent of people who had no contact with police who say the same.

Correct: Sixty-one (61) per cent of people who had contact with police say that the local police do a good job of treating people fairly, compared to 70 per cent of people who had no contact with police who say the same.

Even better: People who had contact with police are more likely to say that the local police do a poor job of treating people fairly: 11 per cent of people who had contact with police report that they do a poor job, compared to only 5 per cent of people who had no contact with police.

Reporting the results of chi-square tests of independence is similar to reporting the results of other statistical significance tests (described in Chapter 7). The test statistic—the chi-square statistic—and the associated degrees of freedom and p-value are reported after a statement of the result, similar to a literature citation. For instance, a researcher reporting the results of the analysis in this chapter might write: "Compared to people who had no contact with police, people who had contact with police are significantly more likely to say that the local police do a poor job of treating people fairly ($\chi^2 = 408.341$, df = 2, p < 0.001)."

Showing Multiple Bivariate Relationships in a Single Table

In Chapter 2, you learned how to combine the results of several frequency tables into a single large table. A similar strategy can be used to combine the results of several cross-tabulations and means comparisons into a single large table, as long as there is one variable that is common to all of the bivariate relationships (usually an independent variable). Combining results into a single table saves space and makes it easier to see patterns in the data.

Table 9.8 shows the results of two cross-tabulations and a means comparison in a single table. The variable that is common to all of the relationships—visible minority status—is shown in the columns, and the remaining variables are in the rows. For categorical variables (like contact with police and perceptions of police fairness), the rows list column percentages. For ratio-level variables (like age), the rows list means and standard deviations.

In tables that show multiple bivariate relationships, statistical significance is usually indicated by an asterisk (*) following the name of the variable in the row, and a footnote is used to indicate the alpha value. Sometimes researchers use different numbers of asterisks to denote relationships that are significant at different alpha values: one asterisk (*) often indicates an alpha value of 0.05, two asterisks (**) often indicate an alpha value of 0.01, and three asterisks (***) often indicate an alpha value of 0.001. The type of significance test that was used usually isn't specified.

Table 9.8   **A Table That Shows Three Bivariate Relationships**

	Visible Minority Status	
	Not a Visible Minority (n = 21,753)	Visible Minority (n = 5,446)
Had contact with police in past year*		
No	68.5%	73.7%
Yes	31.5%	26.3%
Does your local police force do a good, average, or poor job of treating people fairly?*		
Poor job	6.5%	6.7%
Average job	25.4%	27.1%
Good job	68.1%	66.1%
Age (years)*		
Mean	47.3	39.0
Standard deviation	18.5	15.5

*Indicates a statistically significant difference at the $p < 0.05$ level.
Source: Author generated; Calculated using data from Statistics Canada, 2016.

It's also good practice to list the sample size (n) for each column. This allows readers to identify any relationships that rely on a small number of cases. Sometimes each bivariate relationship has a slightly different number of cases because of missing information for some cases. As long as the number of missing cases in each relationship is relatively small, it's common to report the highest number of cases in each column. Sometimes researchers include a footnote indicating that missing data are excluded for each variable.

What You Have Learned

This chapter introduced strategies for assessing the magnitude and reliability of a relationship between two categorical variables. Two proportionate reduction in error (PRE) measures were described: lambda for nominal variables and gamma for ordinal variables. You learned how to calculate and interpret one of the most commonly used statistical significance tests: the chi-square test of independence. You were also introduced to two chi-square-based measures of association: phi and Cramér's V. And, as a first step toward multivariate analyses, you learned how the elaboration model helps researchers to see patterns in data and to develop a more nuanced understanding of social processes.

The research focus of this chapter was the relationship between racialization, contact with police, and perceptions of police fairness among people living in Canadian urban centres. Data from Canada's 2014 General Social Survey show that people who are visible minorities are slightly less likely to have contact with police than people who are not visible minorities. This result is somewhat surprising, although people who are visible minorities were more likely to report contact with police in the context of being a victim of a crime or for other reasons, such as arrest, mental health, and drug/alcohol problems. Compared to people who had no contact with police, people who had contact with police were more likely to say that the local police do a poor job of treating people fairly. This relationship is even more pronounced among people who are visible minorities. The relatively small magnitude of these relationships, however, suggests that although there may be a relationship between racialization and policing practices, this relationship may not be consistent among all of the groups whom Statistics Canada classifies as visible minorities. A more nuanced understanding and measurement of racialization is needed to determine how Black people's perceptions of police and experiences of policing differ from those of non-racialized people and people who are racialized in other ways.

Check Your Understanding

Check to see if you understand the key concepts in this chapter by answering the following questions:

1. What do proportional reduction in error measures show?
2. What is the difference between lambda and gamma?
3. What do expected frequencies represent?
4. What do the results of a chi-square test of independence show?
5. What are the similarities and differences between a chi-square test of independence and a one-way ANOVA test?
6. What are the benefits of using the elaboration model?
7. What is the difference between the six possible elaboration model scenarios: replication, specification, explanation, interpretation, suppression, and distortion?
8. How do you construct a table that displays more than one bivariate relationship?

Practice What You Have Learned

Check to see if you can apply the key concepts in this chapter by answering the following questions. Keep two decimal places in any calculations.

1. As part of a class project, you and your peers survey a random sample of 200 people living in your community. You are interested in finding out whether or not people who have immigrated to Canada are more or less likely to trust police than people who were born in Canada. Among those who were born in Canada, 99 people said that, in general, police can be trusted; 51 people said that, in general, police cannot be trusted. Among those who were not born in Canada, 32 people said that, in general, police can be trusted, and 18 people said that, in general, police cannot be trusted.

    a. Determine which variable should be treated as independent and which should be treated as dependent.
    b. State a non-directional research hypothesis that you can test.

    c. State the null hypothesis associated with the research hypothesis you identified in (b).

2. Using the information from question 1:

    a. Construct a cross-tabulation that include counts and column percentages.
    b. Describe the magnitude of the relationship between the two variables in the sample, that is, the differences between the column percentages.

3. Using the cross-tabulation you constructed in question 2, calculate lambda. Explain what the result shows. How much can you reduce the error in predicting people's trust in police if you know whether or not they were born in Canada?

4. Using the cross-tabulation you constructed in question 2, and keeping four decimal places in the calculation:

    a. Determine the expected frequencies in each cell.
    b. Calculate the chi-square statistic.

c. Calculate the degrees of freedom of the chi-square statistic.

d. Using an alpha value of 0.05, the critical value of a chi-square statistic with these degrees of freedom is 3.84. Based on your answer to (b), state whether you reject or fail to reject the null hypothesis. In the population, is there likely to be a relationship between whether or not people were born in Canada and their trust in police?

5. Using the chi-square statistic from question 4(b), and keeping four decimal places in the calculation, find phi. Explain what phi shows. Would you describe the effect as small, medium, or large?

6. In preparation for an upcoming campaign, the Canadian Civil Liberties Association (www.ccla.org) surveys a random sample of Canadians to investigate how people's perceptions of police are related to various characteristics. They circulate the cross-tabulation in Table 9.9, which shows how people's perceptions of police are related whether they live in a large urban area, a suburban area, or a rural area.

a. Determine which variable should be treated as independent and which should be treated as dependent.

b. State a non-directional research hypothesis that you can test.

c. State the null hypothesis associated with the research hypothesis you identified in (b).

7. Using the information in Table 9.9:

a. Make a version of this cross-tabulation that shows the column percentages.

b. Describe the magnitude of the relationship between the two variables in the sample, that is, the differences between the column percentages.

8. Using the information in Table 9.9, calculate gamma. Explain what the result shows. How much can you reduce the error in predicting people's perceptions of police if you know whether they live in an urban area, a suburban area, or a rural area?

9. Calculate the total number of pairs of cases in Table 9.9. Determine the percentage of pairs that are used in the gamma calculation and the percentage of tied pairs, which are excluded from the gamma calculation.

10. Using the information in Table 9.9:

a. Determine the expected frequencies in each cell.

b. Calculate the chi-square statistic.

c. Calculate the degrees of freedom of the chi-square statistic.

d. Using an alpha value of 0.05, the critical value of a chi-square statistic with these degrees of freedom is 9.49. Based on your answer to (b), state whether you reject or fail to reject the null hypothesis. In the population, is there likely to be a relationship between whether people live in an urban area, a suburban area, or a rural area and their perceptions of police?

11. Using the chi-square statistic from question 10(b), calculate Cramér's V. Explain what Cramér's V shows. Would you describe the effect as small, medium, or large?

12. You find a published research report that shows two cross-tabulations of the relationship between gender and perceptions of how well police enforce laws: one for people who are visible minorities and one for people who are not visible minorities. You want to use the elaboration model to determine whether or not the relationship between gender and perceptions of how well police enforce laws is influenced by whether or not people are visible minorities. Find the partial relationships by calculating column percentages for the two cross-tabulations in Table 9.10.

Table 9.9 A Cross-Tabulation Showing the Relationship between Perceptions of Police and Place of Residence (Hypothetical Data)

Overall, would you say that police in your community are doing:	Place of Residence			
	Urban Area	Suburban Area	Rural Area	Total
A poor job	59	24	2	85
A good job	250	121	12	383
An excellent job	187	230	120	537
Total	496	375	134	1,005

Table 9.10   Two Cross-Tabulations Showing the Relationship between Gender and Perceptions of How Well Police Enforce Laws, One for People Who Are Not Visible Minorities and One for Those Who Are (Hypothetical Data)

Cross-tabulation for people who are not visible minorities:

Do you think your local police force does a good job, an average job, or a poor job of enforcing the laws?	Gender		Total
	Men	Women	
A poor or average job	126	119	245
A good job	173	211	384
Total	299	330	629

Cross-tabulation for people who are visible minorities:

Do you think your local police force does a good job, an average job, or a poor job of enforcing the laws?	Gender		Total
	Men	Women	
A poor or average job	63	54	117
A good job	90	211	301
Total	153	265	418

13. Merge the two cross-tabulations in Table 9.10 into a single cross-tabulation that shows the relationship between gender and perceptions of how well police enforce laws, regardless of whether or not someone is a visible minority.

a. Calculate column percentages for the merged cross-tabulation to find the zero-order relationship.

b. Compare the zero-order relationship to the partial relationships you found in question 12. Determine whether this an example of replication, specification, explanation, interpretation, suppression, or distortion, and explain why.

c. Describe how the relationship between gender and perceptions of how well police enforce laws changes when visible minority status is taken into account.

14. Owusu-Bempah (2014) investigated how people's perceptions of police bias are related to their racial identification. He asked a random sample of participants from the Greater Toronto Area whether or

Table 9.11   Respondents Who Perceive Police Bias, by Race

Police treat Black people worse than White people:	Racial Identification			Total
	Black	Chinese	White	
Yes	379	244	282	905
No	135	260	223	618
Total	514	504	505	1,523

Source: Derived from Owusu-Bempah 2014, 77.

not they thought police treat Black people differently than White people and, if so, whether they treat Black people better or worse than they treat White people. The results in Table 9.11 are derived from his study.

a. Determine which variable should be treated as independent and which should be treated as dependent.

b. State a non-directional research hypothesis that you can test.

c. State the null hypothesis associated with the research hypothesis you identified in (b).

15. Using the information in Table 9.11, calculate lambda. Explain what the result shows. How much can you reduce the error in predicting people's perceptions of police bias against Black people if you know their racial identification?

16. Using the information in Table 9.11:

a. Determine the expected frequencies in each cell.

b. Calculate the chi-square statistic.

c. Calculate the degrees of freedom of the chi-square statistic.

d. Using an alpha value of 0.05, the critical value of a chi-square statistic with these degrees of freedom is 5.99. Based on your answer to (b), state whether you reject or fail to reject the null hypothesis. In the Greater Toronto population, is there likely to be a relationship between people's racial identification and their perceptions of police bias?

17. The graph in Figure 9.9 is excerpted from a Statistics Canada publication describing perceptions of police in Canada's three territories. It shows how

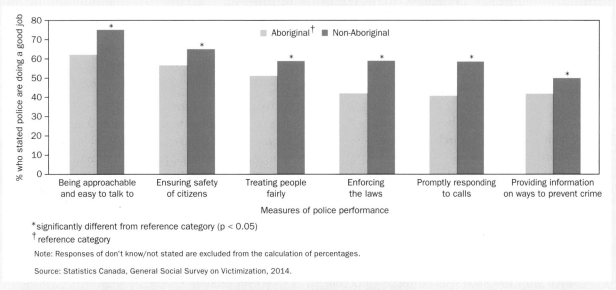

*significantly different from reference category (p < 0.05)
†reference category

Note: Responses of don't know/not stated are excluded from the calculation of percentages.

Source: Statistics Canada, General Social Survey on Victimization, 2014.

Figure 9.9 **Perceptions of Police Performance, by Aboriginal Identity, 2014**

Source: Cotter 2016, 5.

perceptions of various aspects of police performance are related to Aboriginal identity.

a. This graph displays the results of six different cross-tabulations. What are they?

b. What do the * symbols indicate?

c. Describe the general pattern that is evident in this chart. Is this same pattern likely to occur in the population of people living in Canada's three territories?

Practice Using Statistical Software (IBM SPSS)

Answer these questions using IBM SPSS and the GSS27.sav or the GSS27_student.sav dataset available from the Student Resources area of the companion website for this book. Weight the data using the "Standardized person weight" [STD_WGHT] variable you created following the instructions in Chapter 5. Report two decimal places in your answers, unless fewer are printed by IBM SPSS. It is imperative that you save the dataset to keep any new variables that you create.

1. The variable "Donated money or goods - 12 months" [VCG_340] shows people's answers to this question: "In the past 12 months, did you donate money or goods to any organization or charity?" Use the Crosstabs procedure to produce a cross-tabulation showing the relationship between this variable (in the rows) and "Sex of respondent" [SEX] (in the columns). Use the Cells option in the Crosstabs procedure to select the "Observed" and "Expected" count checkboxes, as well as the "Column" percentage checkbox. Describe the magnitude of the relationship between the two variables, that is, the differences between the column percentages in the sample.

2. Use the Statistics option in the Crosstabs procedure to find "Lambda" for the relationship between "Sex of respondent" [SEX] and "Donated money or goods - 12 months" [VCG_340].

   a. Determine which variable should be treated as independent and which should be treated as dependent.

   b. Explain what lambda shows.

3. Use the Statistics option in the Crosstabs procedure to produce the "Chi-square" statistic for the relationship between "Sex of respondent" [SEX] and "Donated money or goods - 12 months" [VCG_340].

    a. State a non-directional research hypothesis for this relationship.

    b. State the null hypothesis associated with the non-directional research hypothesis you identified in (a).

    c. Interpret the chi-square statistic and its associated p-value in relation to the null hypothesis you identified in (b). In the population, is there likely to be a relationship between people's sex/gender and whether or not they donated money or goods in the past year?

4. Use the Statistics option in the Crosstabs procedure to find "Phi and Cramer's V" for the relationship between "Sex of respondent" [SEX] and "Donated money or goods - 12 months" [VCG_340]. Explain what phi shows. (Cramér's V is not meaningful for this relationship.) Would you describe the effect as small, medium, or large?

5. To use the elaboration model, select "Organize output by groups" in the Split File tool and divide the cases by "Visible minority status of the respondent" [VISMIN]. Then, use the Crosstabs procedure to produce two cross-tabulations showing the relationship between "Sex of respondent" [SEX] (in the columns) and "Donated money or goods - 12 months" [VCG_340] (in the rows): one for people who are visible minorities and one for people who are not visible minorities. (Additional crosstabs will be produced for people who are missing information about their visible minority status; they can be ignored.) Use the Cells option in the Crosstabs procedure to display "Column" percentages.

    a. Describe how the relationship between sex/gender and making a donation changes when visible minority status is taken into account.

    b. Compare the partial relationships in the two cross-tabulations to the zero-order relationship in the cross-tabulation from question 1. Determine whether this an example of replication, specification, explanation, interpretation, suppression, or distortion, and explain why.

    After completing this question, use the Split File tool to return to using all the cases in a single group.

6. The variable "Interest in politics" [REP_05] shows people's answers to this question: "Generally speaking, how interested are you in politics (e.g. international, national, provincial or municipal)?" The variable "Federal election - Vote in next election" [VBR_25] shows people's answers to this question: "How likely is it that you will vote in the next federal election?" In the *Variable View*, designate the attribute "Undecided" (5) as missing for the latter variable [VBR_25]. Then, use the Crosstabs procedure to produce a cross-tabulation showing the relationship between "Interest in politics" [REP_05] (in the columns) and "Federal election - Vote in next election" [VBR_25] (in the rows). Use the Cells option in the Crosstabs procedure to display the "Observed" and "Expected" counts, as well as "Column" percentages. Describe the magnitude of the relationship between the two variables, that is, the differences between the column percentages in the sample.

7. Use the Statistics option in the Crosstabs procedure to find "Gamma" for the relationship between "Interest in politics" [REP_05] and "Federal election - Vote in next election" [VBR_25]. Explain what gamma shows.

8. Use the Statistics option in the Crosstabs procedure to find the "Chi-square" statistic for the relationship between "Interest in politics" [REP_05] and "Federal election - Vote in next election" [VBR_25].

    a. State a non-directional research hypothesis for this relationship.

    b. State the null hypothesis associated with the non-directional research hypothesis you identified in (a).

    c. Interpret the chi-square statistic and its associated p-value in relation to the null hypothesis you identified in (b). In the population, is there likely to be a relationship between people's interest in politics and whether they intend to vote in the next federal election?

9. Use the Statistics option in the Crosstabs procedure to find "Phi and Cramer's V" for the relationship between "Sex of respondent" [SEX] and "Donated money or goods - 12 months" [VCG_340]. Explain what Cramér's V shows. (Phi is not meaningful for this relationship.) Would you describe the effect as small, medium, or large?

Key Formulas

Lambda

$$\lambda = \frac{E_1 - E_2}{E_1}$$

$$E_1 = n_{overall} - modal\ frequency_{overall}$$

$$E_2 = \Sigma\left(n_{group} - modal\ frequency_{group}\right)$$

Gamma

$$\gamma = \frac{Concordant\ pairs - Discordant\ pairs}{Concordant\ pairs + Discordant\ pairs}$$

Chi-square statistic

$$\chi^2 = \Sigma \frac{(f_o - f_e)^2}{f_e}$$

Expected frequencies

$$f_e = \frac{(row\ total)(column\ total)}{overall\ total}$$

Degrees of freedom of the chi-square statistic

$$df_{\chi^2} = (rows - 1)(columns - 1)$$

Phi

$$\phi = \sqrt{\frac{\chi^2}{n}}$$

Cramér's V

$$V = \sqrt{\frac{\chi^2}{n(\min[rows - 1], [columns - 1])}}$$

References

Canada. 1995. Employment Equity Act. Statutes of Canada 1995.

Chan, Wendy, and Kiran Mirchandani, eds. 2002. *Crimes of Colour: Racialization and the Criminal Justice System in Canada.* Peterborough, ON: Broadview Press.

Cohen, Jacob. 1988. *Statistical Power Analysis for the Behavioral Sciences.* 2nd ed. Hillsdale, NJ: L. Erlbaum Associates.

Cole, Desmond. 2015. "The Skin I'm In: I've Been Interrogated by Police More than 50 Times—All Because I'm Black." *Toronto Life,*

April 21. http://torontolife.com/city/life/skin-im-ive-interrogated-police-50-times-im-black/.

Comack, Elizabeth. 2012. *Racialized Policing: Aboriginal People's Encounters with Police*. Halifax: Fernwood.

Cotter, Adam. 2016. "Perceptions of Police Performance in the Territories, 2014." Catalogue no. 89–652-X2016005. Ottawa: Statistics Canada. http://www.statcan.gc.ca/pub/89–652-x/89–652-x2016005-eng.pdf.

Millar, Paul, and Akwasi Owusu-Bempah. 2011. "Whitewashing Criminal Justice in Canada: Preventing Research through Data Suppression." *Canadian Journal of Law and Society* 26 (03): 653–61. doi:10.3138/cjls.26.3.653.

Owusu-Bempah, Akwasi. 2014. "Black Males' Perceptions of and Experiences with the Police in Toronto." Doctoral Dissertation, Centre for Criminology & Sociolegal Studies, University of Toronto. https://tspace.library.utoronto.ca/bitstream/1807/68227/1/Owusu-Bempah_Akwasi_201411_PhD_thesis.pdf.

Petersen-Smith, Khury. 2015. "Black Lives Matter: A New Movement Takes Shape." *International Socialist Review*, Spring. http://isreview.org/issue/96/black-lives-matter.

Rankin, Jim, and Patty Winsa. 2012. "Known to Police: Toronto Police Stop and Document Black and Brown People Far More than Whites." *The Toronto Star*, March 9. http://www.thestar.com/news/insight/2012/03/09/known_to_police_toronto_police_stop_and_document_black_and_brown_people_far_more_than_whites.html.

Statistics Canada. 2016. "General Social Survey—Victimization (GSS) 2014." http://www23.statcan.gc.ca/imdb/p2SV.pl?Function=getSurvey&SDDS=4504.

Assessing Relationships between Ratio-Level Variables

10

Learning Objectives

In this chapter, you will learn:

- Some strategies for describing linear relationships

- How to graph a relationship between two ratio-level variables

- How to calculate and interpret Pearson's correlation coefficient

- How to read correlation matrices

- How to calculate and interpret Spearman's rank-order correlation coefficient

- The differences between Pearson's and Spearman's correlation coefficients

- How partial correlations are used to investigate more complex relationships

- How to write about correlations

Introduction

This chapter introduces strategies for assessing the magnitude and reliability of relationships between two ratio-level variables. I begin by introducing scatter-plot graphs and illustrating three ways to characterize relationships based on their appearance. Then, I describe how to calculate and interpret two commonly used correlation coefficients and the tests of statistical significance associated with each. The chapter concludes by illustrating how the elaboration model is used with correlations. Taken together, the material in this chapter provides a conceptual background for understanding linear regression, which is introduced in Chapter 11.

In this chapter, statistical analysis is used to investigate the employment earnings of recent immigrants to Canada. Canada needs immigrants to maintain and grow its population and labour force since fertility levels are not high enough to compensate for mortality (Statistics Canada 2012). Fortunately, Canada's reputation for safety and multiculturalism make it a preferred destination for many migrants. In 2015, Canada was the eighth-highest immigrant-receiving country in the world (Migration Policy Institute 2015). The largest numbers of immigrants to Canada come from China, India, the Philippines, Pakistan, and Iran (United Nations 2015).

Stacey Newman/iStockphoto

Photo 10.1   New Canadians are sworn in during a citizenship ceremony. Immigrants to Canada often experience disadvantages in the labour market.

The majority of migrants to Canada (63 per cent) enter via the economic immigrant class, as opposed to the family class or as refugees (Citizenship and Immigration Canada 2015). Within the economic class, the largest groups of immigrants are skilled workers and their families, as well as provincial/territorial nominees, many of whom are also skilled or semi-skilled workers. To immigrate to Canada as a skilled worker, a person must meet a series of criteria related to education, work experience, and language skills (age and existing ties to Canada are also considered). These criteria are intended to ensure that immigrants are able to integrate into the Canadian labour force. Over the past 25 years, however, the labour market outcomes for recent immigrants to Canada have deteriorated (Picot and Sweetman 2012). Each successive cohort of immigrants has had substantially lower employment earnings than the last, compared to the Canadian-born (Picot and Sweetman 2012). Despite being skilled workers, many immigrants report that their educational credentials and experience are not recognized by Canadian employers, that they are unable to find employment in their field, and that they must turn to lower paying "survival jobs" to support themselves and their families (Dean and Wilson 2009, Guo 2009).

This chapter uses data from the 2011 National Household Survey (NHS) to investigate the relationship between the length of time since immigration and employment earnings. (See the "Spotlight on Data" box for more information.) The goal is to understand whether recent immigrants' earnings increase as they become more settled in Canada. In this chapter, statistical analysis is used to find out the following:

- What is the relationship between the length of time since immigration and annual employment income for recent immigrants?
- How does visible minority status influence the relationship between the length of time since immigration and annual employment income for recent immigrants?

In this analysis, the length of time since immigration is measured as the number of years since becoming a permanent resident of Canada (i.e., a landed immigrant); information about time spent in Canada prior to becoming a landed immigrant is not available. Only people who immigrated to Canada between 2000 and 2010 are included; these people are referred to as "recent immigrants." The "Hands-on Data Analysis" box in this chapter describes how to filter cases so that only some types of people are included in statistical analyses.

In 2011, more than two-thirds (68 per cent) of recent immigrants were in the labour force. Among those employed, almost all were either employees of a company (92 per cent) or self-employed without any paid help (5 per cent). People who immigrated to Canada more recently were more likely to be employees and less likely to be self-employed than people who immigrated to Canada less recently. The small proportion of recent immigrants who started their own businesses and who employ others (only 3 per cent of those employed) are excluded from this analysis because their employment earnings depend on other factors, such as labour and operational costs and on what they elect to withdraw from their business as earnings. In addition, the 4 per cent of recent immigrants who report an employment income greater than $100,000 in 2010 are excluded from this analysis; these people do not fit the typical pattern of recent immigrants' earnings. Recent immigrants who reported no employment income or a loss in 2010 are also excluded. Among the recent immigrants included in this analysis, the average employment income in 2010 was $31,081 (s.d. = $23,862). Incomes ranged from $1,000 to $100,000. The median annual employment income was $26,000, and the middle 50 per cent of earners had incomes ranging from $12,000 to $45,000. The distribution is slightly right-skewed, with many cases clustered around the median income and fewer high-income earners.

Spotlight on Data

The National Household Survey

The National Household Survey (NHS) is the voluntary survey that replaced the long-form census in 2011. It collects information about the same topics as the

Continued

long-form census, including education, citizenship and immigration, ethnicity, mobility, labour-market activities, and income. The NHS data are used by all levels of government as well as community agencies, non-profit organizations, and businesses to support planning for programs and services (Statistics Canada 2013a).

The NHS population is the same as the census: all people who usually reside in Canada, including people living in the provinces, the territories, and on Aboriginal reserves. (See Chapter 5 for a discussion of census-taking on reserves.) People who live in institutions or collective dwellings (such as student residences, work camps, or hotels/motels) and full-time members of the Canadian Armed Forces stationed overseas are excluded (Statistics Canada 2013a). The NHS was distributed to a random sample of approximately one-third of households. Enumerators visited each selected household to drop off survey information and to provide instructions for completing the survey online. Mail questionnaires were used to follow up with respondents who did not complete the survey online. In remote areas and on Aboriginal reserves, all households were asked to complete the NHS, and data were collected in face-to-face interviews. Overall, 69 per cent of people selected to participate in the NHS actually did so. Data are weighted to account for the probability of selection and non-response and are calibrated to be approximately equal to the census counts in each area for age, sex, marital/common-law status, dwelling structure, household size, family structure, and language.

The large sample size of the NHS makes it a valuable source of information about recent immigrants. Some challenges, however, are associated with analyzing NHS employment and income data. The NHS collected information about employment and income for two different time periods. Information about people's labour force and employment status, occupation, industry, and place of work was collected for the survey reference week (1 to 7 May, 2011). But information about people's employment income, the number of weeks they worked, and whether they mainly worked full-time or part-time was collected for the 2010 calendar year. Thus, for people who changed jobs during the first four months of 2011, a mismatch exists in the data (and it is not possible to identify the affected cases). This is of particular concern when analyzing recent immigrants' employment: in one cohort of immigrants, less than half were employed continuously during their first four years in Canada (Preston et al. 2011).

The NHS also groups together immigrants who arrived in Canada in 2010 with those who arrived in the first four months of 2011. In order to limit the analyses to people living in Canada for at least part of 2010, immigrants who reported arriving in Canada in 2010/11 but who neither paid Canadian income taxes nor received any government transfer payments in 2010 are excluded. Although paying income tax and/or receiving government payments is not a perfect proxy for working in Canada, it indicates that the respondent was participating in the Canadian economy in some way.

Despite these limitations, the NHS is one of the best publicly available sources of information about the annual employment earnings of recent immigrants. Although the Labour Force Survey (described in Chapter 4) collects information about immigrant status, it only captures weekly—not yearly—earnings. Data from Statistics Canada's other major income survey, the Canadian Income Survey (CIS), is only available in Research Data Centres. (See the "Spotlight on Data" box in Chapter 1 for more information about Research Data Centres.)

Describing Relationships between Ratio-Level Variables

In this section, I describe three ways to characterize a relationship between two ratio-level variables:

1. Linear or non-linear
2. Homoscedastic or heteroscedastic
3. Monotonic or non-monotonic

The easiest way to begin is by examining a **scatterplot**, which is a graph that shows the relationship between variables by representing each case as a dot on the graph. The independent variable is typically plotted on the horizontal axis (the x-axis), and the dependent variable is plotted on the vertical axis (the y-axis). Figure 10.1 shows the relationship between years since immigration and annual employment income for 10 hypothetical cases. In this analysis, I treat years since immigration as the independent variable (shown on the horizontal axis) and annual employment income as the dependent variable (shown on the vertical axis). The dot farthest to the left represents a single person, who immigrated less than one year ago (recorded as 0 years) and who earns $20,000 in annual employment income. The dot farthest to the right represents another person, who immigrated nine years ago and who earns $39,000 in annual employment income. Scatterplots often reveal patterns in the relationship between two variables (or their absence). Figure 10.1 suggests that there is a positive relationship between these two variables: the more time since a person immigrated, the more employment income that person earns.

One of the simplest ways to describe a relationship between two ratio-level variables is as either linear or non-linear. A **linear relationship** refers to a "straight-line" relationship between two variables. A **non-linear relationship** refers to a relationship with any other type of pattern besides a straight line. For instance, the scatterplot in the left panel of Figure 10.2 shows a clear relationship between two variables, but it is not characterized by a straight line. Instead, it is characterized by a curve, or an arc, and thus it is a **curvilinear relationship**.

scatterplot A graph that shows the relationship between two variables by representing each case as a dot on the graph.

linear relationship A "straight-line" relationship between two ratio-level variables.

non-linear relationship A relationship between two ratio-level variables characterized by any other type of pattern besides a straight line.

curvilinear relationship A non-linear relationship characterized by the shape of a curve or an arc.

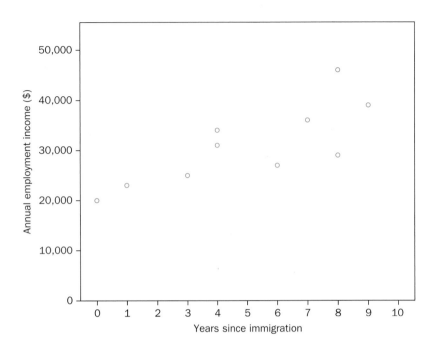

Figure 10.1   A Scatterplot of 10 Hypothetical Cases

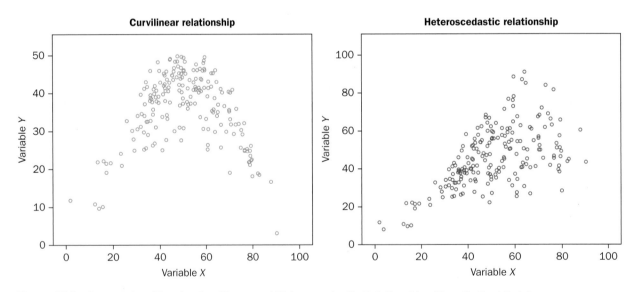

Figure 10.2   Scatterplots Showing Curvilinear and Heteroscedastic Relationships (Hypothetical Data)

Researchers also characterize relationships as being either **homoscedastic** or **heteroscedastic**. These words are easier to understand when you divide them into their two parts: the base, *scedastic*, and the prefix, *homo-* or *hetero-*. *Scedastic* comes from the Greek word for "dispersion" or "spread." The prefix *homo-* refers to sameness, while the prefix *hetero-* refers to difference (just like *homosexual* refers to same-sex relationships and *heterosexual* refers to different-sex relationships). So a homoscedastic relationship is one where the dispersion or spread of the dependent variable is roughly the same or consistent across all of the values on the independent variable. In contrast, a heteroscedastic relationship is one where the dispersion or spread of the dependent variable is different or changes across the values on the independent variable. Heteroscedastic relationships are recognizable by the "trumpet" pattern in the dots, as shown in the scatterplot in the right panel of Figure 10.2. In this example, cases with low values on the independent variable (x) have values on the dependent variable (y) that are clustered fairly close together. In contrast, cases with high values on the independent variable (x) have values on the dependent variable (y) that are quite spread out.

> **homoscedastic relationship** A relationship in which the dispersion of the dependent variable is the same across all the values on the independent variable.
>
> **heteroscedastic relationship** A relationship in which the dispersion of the dependent variable is different across different values on the independent variable.

Finally, characterizing relationships as either **monotonic** or **non-monotonic** can be useful. In monotonic relationships, an increase in one variable is consistently associated with an increase in a second variable or is consistently associated with a decrease in a second variable. By definition, all linear relationships are monotonic. But, unlike linear relationships, in monotonic relationships the size of the increase in the dependent variable does not need to be consistent across all of the values on the independent variable, and so some curvilinear relationships are also monotonic. As long as the trend is consistently in the same direction, a relationship is considered monotonic. The two top panels of Figure 10.3 show monotonic relationships. In the scatterplot in the left panel, the overall trend is for variable Y to increase as variable X increases. In the scatterplot in the right panel, the overall trend is for variable Y to decrease as variable X increases, even though it levels off a bit in the middle. Since the trend continues in the same direction, both of these relationships are considered monotonic. In contrast, a non-monotonic relationship is one where the direction of the relationship between two variables is not consistent across all of the values on a variable. The two bottom panels of Figure 10.3 show non-monotonic relationships. In the scatterplot in the left panel, variable Y first increases, then decreases, then increases again as variable X increases. In the curvilinear relationship shown in the scatterplot in the right panel, variable Y first increases and then decreases as variable X increases. Since the trend of the relationship changes direction, both of these relationships are considered non-monotonic.

> **monotonic relationship** A relationship in which an increase in one variable is consistently associated with an increase in a second variable or consistently associated with a decrease in a second variable.
>
> **non-monotonic relationship** A relationship in which the direction of the relationship between two variables is not consistent across all of the values on a variable.

The statistical techniques described in this chapter, as well as in chapters 11 through 13, should only be used with relatively homoscedastic, linear relationships. The results of correlation analyses can be misleading if a relationship is non-linear or heteroscedastic. More advanced statistical techniques can be used to analyze curvilinear and non-linear relationships.

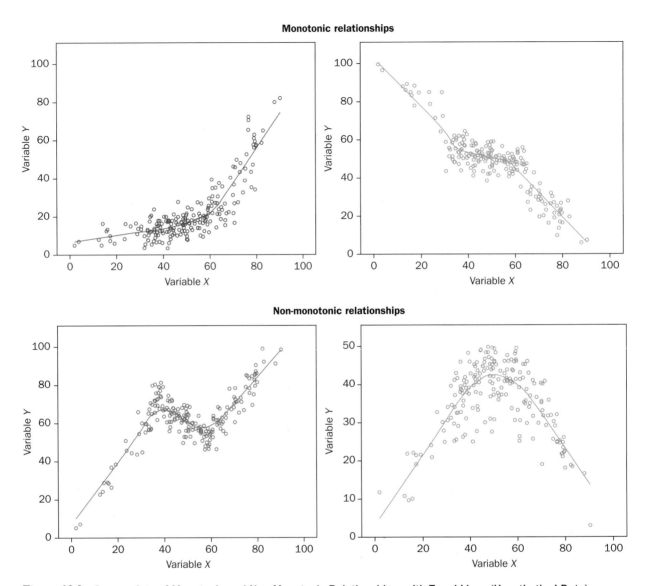

Figure 10.3 Scatterplots of Monotonic and Non-Monotonic Relationships, with Trend Lines (Hypothetical Data)

Hands-on Data Analysis

Selecting or Filtering Cases

Often, researchers don't want to use all of the cases in a dataset in their analyses. For example, in this chapter I am only interested in people who are recent immigrants to Canada. Most statistical software programs allow you to select cases to use in an analysis and to filter out the cases you do not want to use. You may have the option of temporarily filtering out cases or permanently deleting them from the dataset. I strongly recommend that you always temporarily filter cases to ensure that you don't accidentally delete any cases that you may need later.

One way to select cases is by using a combination of variable attributes to determine who will be included and who will be excluded. Begin by generating a frequency distribution for the variable(s) you want to base your selections on. Then, identify the values assigned to the attributes of the cases you want to include in your analysis. For example, the NHS dataset includes a variable for "Immigrant status," and the value "2" is used to denote "Immigrants." Then, use the Select cases or Filter cases tool in your statistical software to use only those cases with the attributes you are interested in. After selecting cases, generate a frequency distribution of the variable(s) you based your selections on to ensure that the correct cases are included and excluded.

Another way to select cases is by taking a random sample from a dataset. Most statistical software programs allow you to specify a number or percentage of cases to randomly select. This approach is used to improve data visualization, especially when there are many cases. For example, when the full NHS sample of recent immigrants is displayed in a scatterplot, discerning a pattern is impossible. (See the left panel of Figure 10.4.) There are so many cases that the dots are printed on top of one another, and it's hard to tell how many dots are clustered in any given area of the graph. The right panel of Figure 10.4 shows a scatterplot of years since immigration and of employment income in 2010 for a randomly selected 0.5 per cent of recent immigrants in the NHS sample. Using fewer cases makes the pattern easier to see.

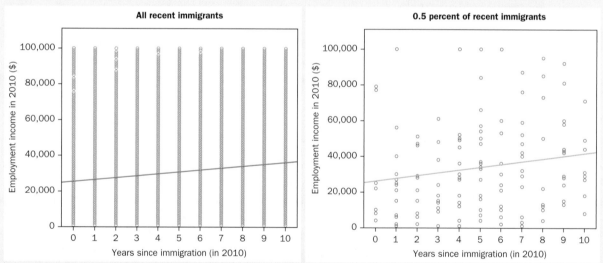

Figure 10.4 Scatterplots that Display All Recent Immigrants in the NHS Sample and a Random Selection of 0.5 Per Cent of Recent Immigrants from the NHS Sample

Source: Author generated; Calculated using data from Statistics Canada, 2013b.

Pearson's Correlation Coefficient

The magnitude of a relationship between two ratio-level variables is often described using a measure of association called **Pearson's correlation coefficient** (which is also sometimes called Pearson's product-moment correlation). It is denoted with a lower-case letter r and provides information about the strength and direction of the linear relationship between two variables. Pearson's correlation coefficient

Pearson's correlation coefficient (r)
A measure of association that provides information about the strength and direction of the linear relationship between two ratio-level variables.

ranges from –1 to +1. Like gamma (which you learned about in Chapter 9), Pearson's correlation coefficient is a symmetric measure, which means that it doesn't matter which variable is treated as independent and which is treated as dependent; the result is the same either way.

The sign of a correlation coefficient—whether it is a positive or a negative number—indicates whether a relationship is positive or negative. The size of a correlation coefficient, regardless of the sign, indicates the strength of a relationship. A coefficient of +/–1 indicates that there is a perfect positive or a perfect negative relationship between the two variables, and a coefficient of 0 indicates that there is no relationship between the two variables. But there is no universally agreed-upon interpretation of what characterizes a weak, moderate, or strong relationship. One common set of guidelines is as follows:

- +/−0.3 is the threshold for a weak relationship.
- +/−0.5 is the threshold for a moderate relationship.
- +/−0.7 is the threshold for a strong relationship.

Some researchers may use lower thresholds than these, however. Figure 10.5 shows scatterplots with a variety of correlation coefficients. In a perfect correlation, the line is perfectly diagonal and each case is located directly on the line. As the correlations become smaller, the line tilts toward horizontal, and the cases are spread farther from the line. When there is zero correlation, the line is perfectly horizontal and the cases are scattered randomly around it with no apparent relationship.

Calculating Pearson's Correlation Coefficient

covariance A measure of how two variables change (or vary) in relation to each other.

Pearson's correlation coefficient relies on the idea of **covariance**, which is a measure of how two variables change (or vary) in relation to each other. In other words, it shows whether change on one variable tends to be associated with change on a second variable. One way to capture the covariance between two variables is to find the mean of each variable, and then determine how far an individual case is above or below the mean of the first variable, and how far that same case is above or below the mean of the second variable. The deviations from the means are multiplied together, and then the results are added up for all of the cases. This is called the "sum of products" (abbreviated SP) because it is the sum of the products of the deviations from the means.

sum of products

$$SP = \Sigma\left(x_i - \bar{x}\right)\left(y_i - \bar{y}\right)$$

Technically, the sum of products could be divided by the number of cases to find the average product of the deviations from the mean. But, in practice, the sum

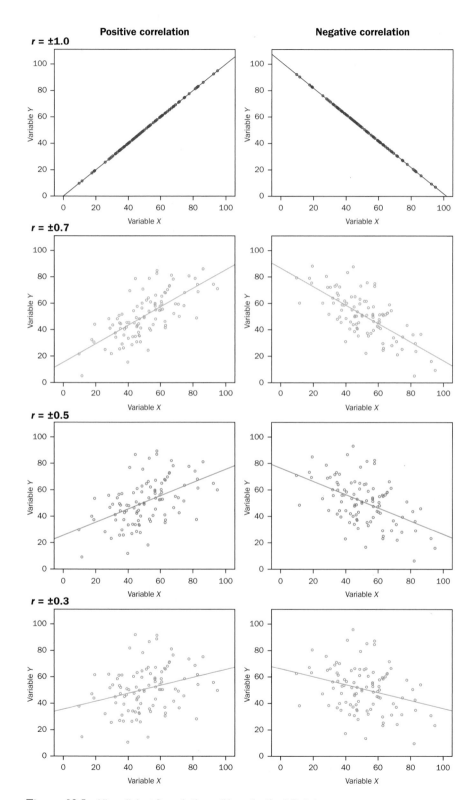

Figure 10.5   Visualizing Correlations (Hypothetical Data)

of products is divided by the number of cases minus 1, to correct for the underesti-mation that occurs when sample data are used to estimate variation in a population (as described in Chapter 5). So, the formula for the covariance is:

covariance

$$cov(x,y) = \frac{\Sigma(x_i - \bar{x})(y_i - \bar{y})}{n-1}$$

The first step in calculating Pearson's correlation coefficient is to find the sum of products. The sum of products is conceptually related to the sum of squares, which you learned about in Chapter 8. The key difference is that the sum of squares captures variation around the mean of a single variable, whereas the sum of products captures covariation around the means of two variables in combination.

Like other formulas that incorporate a sigma (Σ) symbol, it's easiest to cal-culate the sum of products using a table. The table in Figure 10.6 includes the same 10 hypothetical cases as the scatterplot in Figure 10.2. Let's take a moment to consider how the calculation works in practice. For people with values that are below the mean on both variables (such as Arjun), both deviations from the mean are negative. When they are multiplied together, the result is positive, and thus the case makes a positive contribution to the sum of products. For people with values that are above the mean on both variables (such as Kwame), both devia-tions from the mean are positive. When they are multiplied together, the result is again positive, and thus the case makes a positive contribution to the sum of products. But, for people with a value above the mean on one variable and a value below the mean on the other variable (such as Priya or Esengul), one of the devi-ations from the mean is positive and one of the deviations from the mean is nega-tive. When they are multiplied together, the result is negative, and thus the case makes a negative contribution to the sum of products. For people with a value that is exactly the same as the mean on one of the variables (such as Zhang Li), the deviation from the mean is 0, and when the two deviations from the mean are multiplied together the result is also 0. When all of the products of the deviations from the mean are added together, the result is either a positive number or a nega-tive number, which shows whether there are, overall, more positive or negative contributions. If the sum of products is positive, there is a positive relationship between the two variables. If the sum of products is negative, there is a negative relationship between the two variables.

Pearson's correlation coefficient is a standardized version of the sum of prod-ucts. Standardizing the sum of products ensures that the result will always be be-tween −1 and +1. This is achieved by dividing the sum of products by a measure of the amount of variation within each variable on its own: the square root of the sum

$$\Sigma(x_i - \bar{x})(y_i - \bar{y})$$

Person (case)	Mean of Years since Immigration x_i	Mean of Years since Immigration $\bar{x}$	Annual Employment Income ($) y_i	Mean Employment Income $\bar{y}$	Deviation from Mean of Years since Immigration $(x_i - \bar{x})$	Deviation from Mean of Employment Income $(y_i - \bar{y})$	Product of Deviations from Means $(x_i - \bar{x})(y_i - \bar{y})$
Arjun	0	5	20,000	31,000	-5	-11,000	55,000
Damia	1	5	23,000	31,000	-4	-8,000	32,000
Dmitri	3	5	25,000	31,000	-2	-6,000	12,000
Priya	4	5	34,000	31,000	-1	3,000	-3,000
Zhang Li	4	5	31,000	31,000	-1	0	0
Yusuf	6	5	27,000	31,000	1	-4,000	-4,000
Jessica	7	5	36,000	31,000	2	5,000	10,000
Esengul	8	5	29,000	31,000	3	-2,000	-6,000
Kwame	8	5	46,000	31,000	3	15,000	45,000
Alejandro	9	5	39,000	31,000	4	8,000	32,000

column sum: 173,000

The sigma symbol tells you to add the result of the equation for all of the cases.

Figure 10.6   Translating the Sum of Products Formula to a Table (Hypothetical Data)

of squares for each variable (SS_x and SS_y). So, while the sum of products captures how much the two variables vary together, the sum of squares captures how much each variable varies independently. The sum of squares is calculated in the same way as the total sum of squares in a one-way ANOVA test. It is simply the sum of the squared deviations from the (overall) mean of a variable:

$$SS_x = \Sigma(x_i - \bar{x})^2$$

$$SS_y = \Sigma(y_i - \bar{y})^2$$

sum of squares

Once again, it is easiest to calculate the sum of squares using a table. But you're already partway there because you found the deviation of each case from the mean of each variable in the table for the sum of products. This time, for each variable separately, square the deviations from the mean and add them together, as shown in Figure 10.7.

Once you have found the sum of products and the sum of squares for each variable, it's easy to calculate Pearson's correlation coefficient. The following two

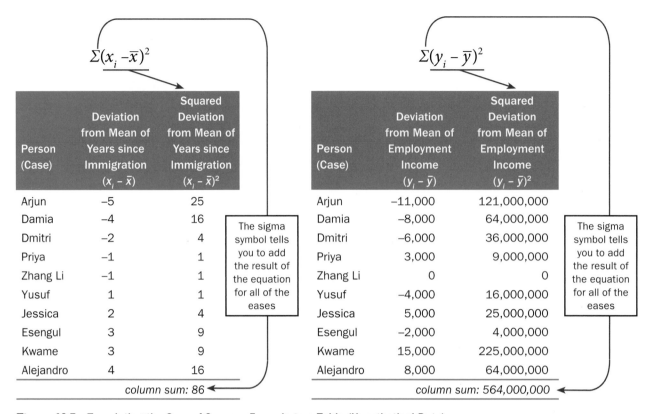

Figure 10.7   Translating the Sum of Squares Formula to a Table (Hypothetical Data)

formulas for Pearson's correlation coefficient are equivalent: one uses the notation for the sum of products and the sum of squares while the other substitutes in the formulas used to calculate each of the individual elements:

Pearson's r

$$r = \frac{SP}{\sqrt{SS_x}\sqrt{SS_y}} \quad or \quad r = \frac{\Sigma(x_i - \bar{x})(y_i - \bar{y})}{\sqrt{\Sigma(x_i - \bar{x})^2}\sqrt{\Sigma(y_i - \bar{y})^2}}$$

Conceptually, Pearson's correlation coefficient is equivalent to the covariance, divided by the product of the standard deviations of each variable. The sum of products is just a version of the covariance that does not adjust for the sample size since it's not divided by the number of cases minus 1. The square root of the sum of squares is just a version of the standard deviation that does not adjust for the sample size since it's not divided by the number of cases minus 1. Since the number of cases minus 1 is the same for both the covariance and the standard deviation, this adjustment is omitted from both the numerator and the denominator. Let's substitute the sum of products and the sums of squares we just calculated into the formula for Pearson's correlation coefficient:

$$r = \frac{SP}{\sqrt{SS_x}\sqrt{SS_y}} \quad or \quad r = \frac{\Sigma(x_i - \bar{x})(y_i - \bar{y})}{\sqrt{\Sigma(x_i - \bar{x})^2}\sqrt{\Sigma(y_i - \bar{y})^2}}$$

$$r = \frac{173000}{\sqrt{86}\sqrt{564000000}}$$

$$= \frac{173000}{(9.27)(23748.68)}$$

$$= \frac{173000}{220150.26}$$

$$= 0.786$$

This result indicates that there is a strong, positive relationship between the two variables in the hypothetical example since Pearson's correlation coefficient is positive and is greater than 0.7 (a common threshold for a strong relationship). One particularly nice feature of Pearson's correlation coefficient is that, when squared, it is also a proportionate reduction in error measure. You learned about proportionate reduction in error measures and how to interpret them in Chapter 9. For example, 0.786 squared is equal to 0.618. As a result, we can assert that the error in predicting recent immigrants' annual employment income would be reduced by 61.8 per cent if we knew how many years it has been since they immigrated. Or, the error in predicting the length of time since immigration would be reduced by 61.8 per cent if we knew people's annual employment income. You'll learn more about how to interpret the squared value of r in Chapter 11.

Pearson's correlation coefficient can also be interpreted in relation to the standard deviation of each variable. In this example, the correlation coefficient of 0.786 indicates that a one standard deviation increase in the independent variable (years since immigration) is associated with a 0.786 standard deviation increase in the dependent variable (employment income). For the 10 cases in the hypothetical example, the standard deviation of years since immigration is 3.09, and the standard deviation of annual employment income is $7,916.23. So, a 3.09 year (one standard deviation) increase in the amount of time since immigration is associated with a $6,222.16 increase in annual employment income (which is 0.786 of the standard deviation of $7,916.23). To make this result easier to understand, $6,222.16 can be divided by 3.09 in order to say that a one-year increase in the amount of time since immigration is associated with a $2,013.64 increase in annual employment income. (In fact, if this calculation is completed using all of the decimals in the correlation coefficient and the standard deviations, the result is $2,011.63, a number that will appear in the next chapter.) Researchers rarely provide this interpretation of Pearson's correlation coefficient, however, since these findings correspond with those produced by a linear regression, a more powerful technique that you'll learn about in Chapter 11.

Reading Correlation Matrices

You were first introduced to the matrix format in the discussion of post-hoc tests in Chapter 8. Most statistical software programs display correlations between two or more variables in a matrix format, as in Table 10.1. For correlations between two variables, the matrix will have two rows and two columns. But correlation matrices can show the relationships between more than two variables. When more than two variables are being assessed, the matrix will have the same number of rows and columns as there are variables. The diagonal cells in a correlation matrix (shaded light teal in Table 10.1) always show a correlation of 1. This is because each of these cells shows the correlation between a variable and itself, which is by definition a perfect correlation. So, the top-left shaded cell in the correlation matrix in Table 10.1 shows the correlation between the variable "Years since immigration" (in the columns) and the variable "Years since immigration" (in the rows). These perfect correlations are never reported since they are not a meaningful finding. The off-diagonal cells show the correlations between pairs of different variables. The same information is repeated above and below the diagonal. So, in Table 10.1, the information in the two white cells is identical. One cell shows the relationship between "Years since immigration" (in the columns) and "Annual employment income" (in the rows), and the other cell shows the relationship between "Annual employment income" (in the columns) and "Years since immigration" (in the rows). Since Pearson's correlation coefficient is a symmetric measure, only one set of these results should be reported; it doesn't matter which one. Because of this symmetry, researchers only need to look at *either* the cells above the diagonal of a correlation matrix or the cells below the diagonal of a correlation matrix but not both. In published correlation matrices, often only one set of results is printed (those above the diagonal or those below the diagonal), and the remaining cells are left blank.

The order of the variables is always the same in the rows and in the columns of a correlation matrix: the sequence of variables listed from top to bottom in the rows is identical to the sequence of variables listed from left to right in the columns. To save space and improve readability, published correlation matrices often only show labels for *either* the rows or the columns. If variable labels are shown for the rows, the columns are typically numbered 1, 2, 3, and so on, up to the maximum number of variables. (Alternatively, if variable labels are shown for

Table 10.1   **A Correlation Matrix (Hypothetical Data)**

	Years since Immigration	Annual Employment Income ($)
Years since Immigration	1.000	0.786
Annual Employment Income ($)	0.786	1.000

the columns, the rows are typically numbered.) To read these correlation tables, simply mentally replace the number "1" with the name of the first variable listed (in the rows/columns), replace the number "2" with the name of the second variable listed, and so on.

Interpreting T-Statistics for Pearson's Correlation Coefficient

The reliability of a correlation between two ratio-level variables is assessed using a statistical significance test that relies on a t-distribution. Like all tests of statistical significance, it shows the probability of randomly selecting a sample with the observed relationship (or one of greater magnitude) if no relationship exists in the population. For the hypothetical example, the research hypothesis is that there is a correlation between the number of years since immigration and annual employment income in the population of recent immigrants. The null hypothesis is that there is no correlation between the number of years since immigration and annual employment income in the population of recent immigrants. That is, the null hypothesis is that the Pearson's correlation coefficient is equal to 0 in the population. The t-statistic associated with Pearson's correlation coefficient is calculated using the following formula:

$$t = r\sqrt{\frac{n-2}{1-r^2}}$$

t-statistic for Pearson's r

So, for the hypothetical example, the t-statistic is:

$$t = \rho\sqrt{\frac{n-2}{1-\rho^2}}$$

$$t = 0.786\sqrt{\frac{10-2}{1-0.786^2}}$$

$$= 0.786\sqrt{\frac{8}{1-0.618}}$$

$$= 0.786\sqrt{\frac{8}{0.382}}$$

$$= 0.786\sqrt{20.942}$$

$$= 0.786(4.576)$$

$$= 3.597$$

This t-statistic of 3.597 is interpreted in relation to a t-distribution, with the degrees of freedom equivalent to the number of cases minus 2 (the same as for t-tests that assume that the groups have equal variances in the population). Since there are 10 cases in our hypothetical example, this t-statistic is interpreted in relation to a t-distribution with 8 degrees of freedom. The t-statistic identifies a point on the horizontal axis of the t-distribution, and the area under the curve that is beyond this point (in both tails) is used to determine the likelihood of randomly selecting a sample with the observed correlation (or a stronger correlation), if there is no correlation between the two variables in the population. Only 0.7 per cent of the area under the curve of a t-distribution with 8 degrees of freedom is beyond +/–3.597; in other words, there is only a 0.007 probability of randomly selecting this sample if there is no correlation between years since immigration and annual employment income in the (hypothetical) population. Alternatively, researchers can compare the t-statistic to the critical value of the t-distribution (with the appropriate degrees of freedom) that corresponds to the alpha value they select. Using an alpha value of 0.05, the critical value of a t-distribution with 8 degrees of freedom is +/–2.31. Since the calculated t-statistic of 3.597 is beyond the critical value of 2.31, these results indicate that there is likely a correlation between the number of years since immigration and annual employment income in the (hypothetical) population.

Let's return to the NHS data on recent immigrants to Canada. We want to know whether there is a relationship between the length of time since immigration and employment income in 2010. More formally, the research hypothesis is that there is a correlation between the number of years since a person immigrated to Canada and their employment income in 2010. The null hypothesis is that there is no correlation between the number of years since a person immigrated to Canada and their employment income in 2010. The Pearson's correlation coefficient for the relationship between these two variables is 0.139, substantially lower than in the hypothetical example. There appears to be only a weak positive relationship between the number of years since immigration and employment income in 2010. Using a proportionate reduction in error interpretation, we can assert that the error in predicting recent immigrants' annual employment income can be reduced by 2 per cent ($0.139^2 = 0.019$) if we know how many years ago they immigrated to Canada. The test of statistical significance shows that it would be very unlikely ($p < 0.001$) to randomly select this sample if there is no correlation between the number of years since immigration and annual employment income in the larger population. As a result, we are relatively confident that there is some correlation between the length of time since immigration and annual employment income in the population of recent immigrants to Canada.

These results can also be interpreted in relation to the standard deviation of each variable. The standard deviation of years since immigration is 3.11, and the standard deviation of employment income in 2010 is $23,862. The Pearson's correlation coefficient indicates that a one standard deviation increase in years since immigration is associated with a 0.139 standard deviation increase in employment income in 2010. So, a 3.11 year increase in time since immigration is associated with a $3,317 increase in annual employment income ($23,862 multiplied by 0.139

is $3,317). Or, a one-year increase in time since immigration is associated with a $1,067 increase in annual employment income ($3,317 divided by 3.11 is $1,067). In general, these results show that for every additional year since immigrating to Canada, a recent immigrant earns about another thousand dollars in annual employment income. In practical terms, this isn't a substantial increase: $1,067 per year translates into an additional $89 a month.

Taken together, these findings suggest that acculturation alone is not associated with substantial gains in employment earnings during the first 10 years following immigration to Canada. Although some immigrants may increase their employment income as a result of obtaining Canadian work experience—by moving into higher paying work or receiving raises associated with job seniority—these gains appear to be minimal. Instead, immigrants' employment income remains relatively consistent—with only modest increases—during their first decade after settling in Canada. There are likely other factors that have more influence on recent immigrants' employment earnings than their length of time since immigration, such as whether they have been able to secure full-time work, their level of education, whether their credentials are recognized by Canadian employers, and the occupational prestige of their work.

Pearson's correlation coefficient relies on several assumptions about the variables being assessed. First, both variables must be either continuous variables or, at minimum, count variables with a wide range. And, as noted earlier in this chapter, the variables must have a linear relationship—not a curvilinear or non-linear relationship. Finally, both variables must be approximately normally distributed so that the mean and standard deviation are reasonably good measures of the centre and spread of each variable. When these criteria are not met—such as when a researcher wants to investigate the relationship between variables that are not normally distributed or between ordinal-level variables—a rank-order correlation coefficient should be used instead of Pearson's correlation coefficient.

Step-by-Step: Pearson's Correlation Coefficient

Note: In these calculations, do not include cases that are missing information for either variable.

Step 1: Identify the research hypothesis and the null hypothesis.

Sum of Products

$$SP = \Sigma\left(x_i - \bar{x}\right)\left(y_i - \bar{y}\right)$$

Step 2: Find the mean of the independent variable in the relationship you are assessing ($\bar{x}$).

Step 3: Find the mean of the dependent variable in the relationship you are assessing ($\bar{y}$).

Continued

Step 4: For *each case individually*, subtract the mean of the independent variable (from Step 2) from the value of the case on the independent variable (x_i) to find the deviation from the mean. Complete this calculation once for each case.

Step 5: For *each case individually*, subtract the mean of the dependent variable (from Step 3) from the value of the case on the dependent variable (y_i) to find the deviation from the mean. Complete this calculation once for each case.

Step 6: For *each case individually*, multiply the result of Step 4 by the result of Step 5. Pay attention to the positive/negative sign of each result when you multiply. Complete this calculation once for each case.

Step 7: Add together the results of Step 6 for each case to find the sum of products.

Sum of Squares

$$SS_x = \Sigma\left(x_i - \bar{x}\right)^2$$

$$SS_y = \Sigma\left(y_i - \bar{y}\right)^2$$

Step 8: For *each case individually*, square the result of Step 4. Complete this calculation once for each case.

Step 9: Add together the results of Step 8 for each case to find the sum of squares for the independent variable.

Step 10: For *each case individually*, square the result of Step 5. Complete this calculation once for each case.

Step 11: Add together the results of Step 10 for each case to find the sum of squares for the dependent variable.

Pearson's r

$$r = \frac{SP}{\sqrt{SS_x}\sqrt{SS_y}}$$

Step 12: Find the square root of the sum of squares for the independent variable (from Step 9).

Step 13: Find the square root of the sum of squares for the dependent variable (from Step 11).

Step 14: Multiply the result of Step 12 and Step 13 to find the denominator of the Pearson's r equation.

Step 15: Divide the sum of products (from Step 7) by the result of Step 14 to find Pearson's *r*.

T-statistic for Pearson's r

$$t = r\sqrt{\frac{n-2}{1-r^2}}$$

Step 16: Count the total number of cases (n). Do not include cases that are missing information for either variable.

Step 17: Subtract 2 from the total number of cases (from Step 16) to find the numerator of the fraction in the t-statistic equation.

Step 18: Square Pearson's r (from Step 15).

Step 19: Subtract the result of Step 18 from 1 to find the denominator of the fraction in the t-statistic equation.

Step 20: Divide the result of Step 17 by the result of Step 19.

Step 21: Find the square root of the result of Step 20.

Step 22: Multiply the result of Step 21 by Pearson's r (from Step 15) to find the t-statistic.

Degrees of Freedom

$$df_t = n - 2$$

Step 23: Subtract 2 from the total number of cases (from Step 16) to find the degrees of freedom for the t-statistic for Pearson's r.

Statistical Significance

Step 24: Determine the alpha value you want to use for the t-test for Pearson's correlation coefficient. The most common alpha value used in the social sciences is 0.05.

Step 25: Determine the critical value of the t-statistic, using the alpha value you selected in Step 24, for a t-distribution with the degrees of freedom you found in Step 23. If you are using an alpha value of 0.05 and the degrees of freedom are greater than 500, the critical value is +/-1.96. If you are using a different alpha value and/or have fewer degrees of freedom, use a printed table of the critical values of the Student's t-distribution or an online calculator to find the critical value.

Step 26: Compare the t-statistic (from Step 22) to the critical value (from Step 25). If the t-statistic is beyond the critical value, the result is statistically significant: you can reject the null hypothesis and state that there is likely a correlation between the two variables in the population. If the t-statistic is not beyond the critical value, the result is not statistically significant: you fail to reject the null hypothesis and state that you are not confident that there is a correlation between the two variables in the population.

How Does It Look in SPSS?

Pearson's Correlation Coefficient

The Bivariate Correlate procedure produces results that look like those in Image 10.1. This correlation matrix shows the relationship between two variables. More variables can be added to the matrix; each additional variable generates another set of rows and another column.

Correlations ⒝

			Years since immigration (in 2010)	Employment income in 2010
Years since immigration (in 2010) ⒜	⒞ Pearson Correlation		1	.139**
	Sig. (2-tailed)	⒟		⒡ .000
	N		29471	26621
Employment income in 2010	Pearson Correlation		.139**	1
	Sig. (2-tailed)	⒠	.000	⒢
	N		26621	26621

Ⓗ **. Correlation is significant at the 0.01 level (2-tailed).

Image 10.1 **An SPSS Pearson's Correlation Coefficient Matrix**

A. The variables that the procedure is assessing relationships between are listed in rows. It doesn't matter what order they are in.

B. The variables listed in the rows are also listed in the columns, in the same order as the rows.

C. These labels indicate that each cell in the correlation matrix contains three pieces of information: (1) Pearson's correlation coefficient, (2) the result of a statistical significance test for the correlation coefficient (based on the t-statistic), and (3) the number of cases used in the calculations.

D. The lines in this cell show the relationship between "Years since immigration" (in the row) and "Years since immigration" (in the column). The Pearson correlation coefficient is 1; as expected, the variable matches with itself perfectly. This should not be reported. There were 29,471 cases with valid values used to calculate the correlation coefficient.

E. The lines in this cell show the relationship between "Employment income" (in the row) and "Years since immigration" (in the column). The Pearson correlation coefficient is 0.139, indicating that there is a weak positive relationship between years since immigration and employment income in 2010 among recent immigrants to Canada. The "Sig. (2-tailed)" line shows the p-value associated with the t-statistic for the correlation coefficient (with the appropriate degrees of freedom). That is, it shows the likelihood of randomly selecting a sample with the observed correlation (or a stronger correlation) between the two variables, if no correlation exists in the population. This p-value indicates that there is less than a 0.1 per cent chance of selecting this sample from a population of recent immigrants in which there is no correlation between years since immigration and annual employment income ($p < 0.001$). There were 26,621 cases with valid values for both variables; the number of cases is used to determine the degrees of freedom of the t-statistic.

F. The lines in this cell show the relationship between "Years since immigration" (in the row) and "Employment income" (in the column). The results and interpretation are the same as for E since they show the relationship between the same two variables. Results should only be reported for the cells above the diagonal *or* the cells below the diagonal in a correlation matrix to avoid duplication.

G. The lines in this cell show the relationship between "Employment income" (in the row) and "Employment income" (in the column). The Pearson correlation coefficient is 1; as expected, the variable matches with itself perfectly. This should not be reported. There were 26,621 cases with valid values used to calculate the correlation coefficient.

H. This footnote indicates that two asterisks are used to denote correlations that are statistically significant using an alpha value of 0.01.

Spearman's Rank-Order Correlation Coefficient

The most commonly used rank-order correlation coefficient is **Spearman's rank-order correlation**. It is sometimes called Spearman's rho and denoted using the Greek letter rho (ρ), which looks similar to a lower-case letter *p*. Rho is pronounced the same way as the word *row*. Spearman's rank-order correlation coefficient is similar to Pearson's correlation coefficient except that it is calculated using the rank of each case within a variable instead of the values on each variable. Like Pearson's correlation coefficient, Spearman's correlation coefficient ranges from −1 to +1. But because it uses ranks, and not values, the mean and standard deviation of a variable do not influence Spearman's correlation coefficient. Spearman's correlation coefficient does not rely on any assumptions about the underlying distribution of a variable; thus, it is considered a non-parametric test (like the chi-square test of independence).

To illustrate how Spearman's correlation coefficient is calculated using ranks, let's return to the hypothetical 10-case example we used to calculate Pearson's correlation coefficient. Table 10.2 lists the number of years since immigration and the annual employment income of each person. Let's begin by ranking employment income: each person's employment income is listed in order from smallest to largest, and each is assigned a rank that reflects that person's position in the sequence. So the person with the lowest employment income (Arjun, making $20,000 a year) is assigned a rank of 1. The person with the next highest employment income (Damia,

Spearman's rank-order correlation coefficient (ρ) A non-parametric measure of association that provides information about the strength and direction of the monotonic relationship between two variables.

Table 10.2 Assigning Ranks to Cases (Hypothetical Data)

Person (Case)	Years since Immigration	Rank of Years since Immigration	Employment Income ($)	Rank of Employment Income
Arjun	0	1	20,000	1
Damia	1	2	23,000	2
Dmitri	3	3	25,000	3
Priya	4	4.5	34,000	7
Zhang Li	4	4.5	31,000	6
Yusuf	6	6	27,000	4
Jessica	7	7	36,000	8
Esengul	8	8.5	29,000	5
Kwame	8	8.5	46,000	10
Alejandro	9	10	39,000	9

making $23,000 a year) is assigned a rank of 2, and so on, until all the cases are ranked. The same ranking process is completed separately for the number of years since immigration. The person who immigrated most recently (Arjun) is assigned a rank of 1, and the person who immigrated longest ago (Alejandro) is assigned a rank of 10. When two or more cases have the same value, each case is assigned the average rank of the tied cases. So for Priya and Zhang Li, who each immigrated four years ago, there is no way to decide who should be ranked fourth and who should be ranked fifth. Instead, both cases are assigned the average rank of 4.5 ([4 + 5] ÷ 2). A similar situation results in both Esengul and Kwame being assigned a rank of 8.5.

Once a rank has been assigned to each case for each variable, Spearman's correlation coefficient is calculated in the same way as Pearson's correlation coefficient, but using the ranks instead of the values. So, the numbers in the "Rank of Years since Immigration" column are used in place of the values of each case (x_i), and 5.5 is used as the average (since 5.5 is the average rank). Similarly, the numbers in the "Rank of Employment Income" column are used in place of the values of each case (y_i), and 5.5 is used as the average. When the ranks are used to calculate the sum of products, the result is 67.5. (Try it yourself to make sure you can obtain this result.) And, still using the ranks, the sum of squares for years since immigration is 81.5, and the sum of squares for employment income is 82.5. (When there are no tied ranks, the sum of squares is identical for both variables.) This sum of products and these sums of squares are substituted into the formula for Spearman's correlation coefficient:

$$\rho = \frac{SP}{\sqrt{SS_x}\sqrt{SS_y}} \quad or \quad \rho = \frac{\Sigma(x_i - \bar{x})(y_i - \bar{y})}{\sqrt{\Sigma(x_i - \bar{x})^2}\sqrt{\Sigma(y_i - \bar{y})^2}}$$

$$\rho = \frac{67.5}{\sqrt{81.5}\sqrt{82.5}}$$

$$= \frac{67.5}{(9.03)(9.08)}$$

$$= \frac{67.5}{81.99}$$

$$= 0.823$$

The interpretation of Spearman's correlation coefficient is similar to that of Pearson's correlation coefficient. The sign indicates whether there is a positive or a negative rank-order relationship between the two variables. The (absolute) size of the coefficient indicates whether there is a weak, moderate, or strong rank-order relationship between the two variables. Again, different researchers use varying thresholds for weak, moderate, and strong relationships; the 0.3 (weak), 0.5 (moderate), and 0.7 (strong) thresholds used with Pearson's correlation coefficient can also be used with Spearman's correlation coefficient, although some researchers use lower thresholds. A Spearman's correlation coefficient of 0 indicates that there is no rank-order relationship between the two variables, and a coefficient of 1 indicates that there is a perfect rank-order relationship between the two variables. In the hypothetical example, Spearman's correlation coefficient of 0.823 indicates that there is a strong positive relationship between the rank-order of years since immigration and the rank-order of annual employment income among recent immigrants.

Just like Pearson's correlation coefficient, the square of Spearman's correlation coefficient is also a proportionate reduction in error measure. When Spearman's correlation coefficient is squared, it indicates how much the prediction errors can be reduced by knowing the rank of each case within one of the variables. In the hypothetical example, the square of Spearman's correlation coefficient is 0.677 (0.823^2), which indicates that the error in predicting people's rank on the "Annual employment income" variable can be reduced by 67.7 per cent if you know their rank on the "Years since immigration" variable. Alternatively, the error in predicting people's rank on the "Years since immigration" variable can be reduced by 67.7 per cent if you know their rank on the "Annual employment income" variable.

A key difference between Pearson's and Spearman's correlation coefficients is that Spearman's correlation coefficient cannot be interpreted in relation to the standard deviation of each variable. Since it relies on ranks, Spearman's correlation coefficient has no relationship to the variable standard deviations (although it is related to the standard deviations of the ranks, this doesn't aid in interpretation). There are also two alternate formulas for Spearman's rank-order correlation coefficient: one that is used when there are no tied ranks in the data, and one that is used when there are tied ranks in the data. Although these alternate formulas are more efficient, I do not describe them here since you already know how to implement the Pearson's correlation coefficient formula.

Interpreting T-Statistics for Spearman's Correlation Coefficient

The reliability of a Spearman's correlation coefficient can be assessed using the same approach as Pearson's correlation coefficient. The research hypothesis is that there is a rank-order correlation between the two variables in the population, and the null hypothesis is that no rank-order correlation exists between the two variables in the population ($\rho = 0$). Once again, a t-statistic is produced and is evaluated in relation to a t-distribution with degrees of freedom equal to the number of cases minus 2. Notice that the formula for the t-statistic for Spearman's correlation coefficient is identical to that for the t-statistic for Pearson's correlation coefficient, only r has been replaced with ρ:

t-statistic for Spearman's rho

$$t = \rho\sqrt{\frac{n-2}{1-\rho^2}}$$

Let's use this formula to assess the reliability of the relationship in the hypothetical example, which produced a Spearman's correlation coefficient of 0.823:

$$t = \rho\sqrt{\frac{n-2}{1-\rho^2}}$$

$$t = 0.823\sqrt{\frac{10-2}{1-0.823^2}}$$

$$= 0.823\sqrt{\frac{8}{1-0.677}}$$

$$= 0.823\sqrt{\frac{8}{0.323}}$$

$$= 0.823\sqrt{24.768}$$

$$= 0.823(4.977)$$

$$= 4.096$$

Earlier in this chapter, using an alpha value of 0.05, we established that the critical value of a t-distribution with 8 degrees of freedom is +/−2.31. Since 4.10 is beyond 2.31, there is likely a rank-order correlation between years since immigration and employment income in the (hypothetical) population.

Let's use Spearman's rank-order correlation coefficient to assess the real-world relationship between the number of years since immigration and annual employment income in 2010 for recent immigrants to Canada. Since both of these variables are only approximately normally distributed in the NHS, Spearman's rank-order correlation provides an alternate test of this relationship. The Spearman's correlation coefficient is 0.140, almost the same as the Pearson's correlation coefficient. This result reinforces the idea that there is only a weak positive relationship between the number of years since immigration and annual employment income for recent immigrants. Using Spearman's correlation coefficient, we can assert that the error in predicting the rank of recent immigrants' employment income in 2010 can be reduced by 2 per cent ($0.140^2 = 0.020$) if we know their rank on the "Years since immigration" variable. The result is statistically significant, using an alpha value of 0.05, and thus we are relatively confident that there is a rank-order correlation between the number of years since immigration and annual employment income in the larger population of recent immigrants to Canada.

One main difference between the two correlation coefficients described in this chapter is that Pearson's correlation coefficient assesses the strength of the linear (straight-line) relationship between the two variables, whereas Spearman's correlation coefficient assesses the strength of the monotonic relationship between the two variables. As a result, comparing Pearson's correlation coefficient to Spearman's correlation coefficient for the same pair of variables indicates whether there is a stronger linear relationship or a stronger monotonic relationship. If Pearson's correlation coefficient is larger than Spearman's correlation coefficient, there is a stronger linear relationship between the two variables. In contrast, if Spearman's correlation coefficient is larger than Pearson's correlation coefficient, there is a stronger monotonic relationship between these two variables. In the NHS example, both correlation coefficients are similar, and the two types of relationships have roughly the same strength. When this occurs, researchers typically report Pearson's correlation coefficient since it uses more information about each case and, thus, is considered more precise.

In general, Spearman's rank-order correlation coefficient is useful for assessing relationships between ratio-level variables that are not normally distributed or between ordinal-level variables with many attributes. Because the decision about which correlation coefficient to report depends on the distributions of the two variables, it's important to look closely at the descriptive statistics for each variable as well as a graph showing the relationship between the variables, before making a choice.

Step-by-Step: Spearman's Rank-Order Correlation Coefficient

Step 1: Identify the research hypothesis and the null hypothesis.

Step 2: Manually or electronically arrange (or sort) the cases in order from the lowest attribute to the highest attribute of the independent variable. If there is more than one case with the same attribute, it doesn't matter which one is listed first or last. Do not include cases that are missing information for either variable.

Continued

Step 3: Label the *rank* of each case in the sequence, from 1 to *n*, so that the case with the lowest attribute is assigned a rank of 1, the case with the next lowest attribute is assigned a rank of 2, and so on until every case is assigned a rank.

Step 4: Identify any cases with the *same* attributes for the independent variable. For each group of cases with the same attributes, find the average rank assigned to them. Then, replace the original ranks with the average rank for the group.

Step 5: Repeat Steps 2 to 4 for the dependent variable.

Sum of Products

$$SP = \Sigma\left(x_i - \bar{x}\right)\left(y_i - \bar{y}\right)$$

Step 6: Find the mean rank of the independent variable in the relationship you are assessing ($\bar{x}$).

Step 7: Find the mean rank of the dependent variable in the relationship you are assessing ($\bar{y}$).

Step 8: For *each case individually*, subtract the mean rank of the independent variable (from Step 6) from the rank of the case on the independent variable (x_i) to find the deviation from the mean. Complete this calculation once for each case.

Step 9: For *each case individually*, subtract the mean rank of the dependent variable (from Step 7) from the rank of the case on the dependent variable (y_i) to find the deviation from the mean. Complete this calculation once for each case.

Step 10: For *each case individually*, multiply the result of Step 8 by the result of Step 9. Pay attention to the positive/negative sign of each result when you multiply. Complete this calculation once for each case.

Step 11: Add together the results of Step 10 for each case to find the sum of products.

Sum of Squares

$$SS_x = \Sigma\left(x_i - \bar{x}\right)^2$$

$$SS_y = \Sigma\left(y_i - \bar{y}\right)^2$$

Step 12: For *each case individually*, square the result of Step 8. Complete this calculation once for each case.

Step 13: Add together the results of Step 12 for each case to find the sum of squares for the independent variable.

Step 14: For *each case individually*, square the result of Step 9. Complete this calculation once for each case.

Step 15: Add together the results of Step 14 for each case to find the sum of squares for the dependent variable.

Spearman's rho

$$\rho = \frac{SP}{\sqrt{SS_x}\,\sqrt{SS_y}}$$

Step 16: Find the square root of the sum of squares for the independent variable (from Step 13).

Step 17: Find the square root of the sum of squares for the dependent variable (from Step 15).

Step 18: Multiply the result of Step 16 and Step 17 to find the denominator of the Spearman's rho equation.

Step 19: Divide the sum of products (from Step 11) by the result of Step 18 to find Spearman's rho.

T-statistic for Spearman's rho

$$t = \rho\sqrt{\frac{n-2}{1-\rho^2}}$$

Step 20: Count the total number of cases (n). Do not include cases that are missing information for either variable.

Step 21: Subtract 2 from the total number of cases (from Step 20) to find the numerator of the fraction in the t-statistic equation.

Step 22: Square Spearman's rho (from Step 19).

Step 23: Subtract the result of Step 22 from 1 to find the denominator of the fraction in the t-statistic equation.

Step 24: Divide the result of Step 21 by the result of Step 23.

Step 25: Find the square root of the result of Step 24.

Step 26: Multiply the result of Step 25 by Spearman's rho (from Step 19) to find the t-statistic.

Degrees of Freedom

$$df_t = n - 2$$

Step 27: Subtract 2 from the total number of cases (from Step 20) to find the degrees of freedom of the t-statistic for Spearman's rho.

Continued

Statistical Significance

Step 28:   Determine the alpha value you want to use for the t-test for Spearman's rho. The most common alpha value used in the social sciences is 0.05.

Step 29:   Determine the critical value of the t-statistic, using the alpha value you selected in Step 28, for a t-distribution with the degrees of freedom you found in Step 27. If you are using an alpha value of 0.05 and the degrees of freedom are greater than 500, the critical value is +/−1.96. If you are using a different alpha value and/or have fewer degrees of freedom, find the critical value using a printed table of the critical values of the Student's t-distribution, an online calculator, or the spreadsheet tool available from the Student Resources area of the companion website for this book.

Step 30:   Compare the t-statistic (from Step 26) to the critical value (from Step 29). If the t-statistic is beyond the critical value, the result is statistically significant: you can reject the null hypothesis and state that there is likely a rank-order correlation between the two variables in the population. If the t-statistic is not beyond the critical value, the result is not statistically significant: you fail to reject the null hypothesis and state that you are not confident that there is a rank-order correlation between the two variables in the population.

How Does It Look in SPSS?

Spearman's Rank-Order Correlation Coefficient

The Bivariate Correlate procedure with the "Spearman" correlation coefficients checkbox selected (and the "Pearson" correlation coefficients checkbox deselected) produces results that look like those in Image 10.2. Just as for Pearson's correlations, more variables can be added to the matrix; each additional variable generates another set of rows and another column.

Correlations   ©

			Years since immigration (in 2010)	Employment income in 2010
Spearman's rho (A)	Years since immigration (in 2010) (B)	(D) Correlation Coefficient	1.000	.140**
		Sig. (2-tailed) (E)	.	.000 (G)
		N	33222	30004
	Employment income in 2010	Correlation Coefficient	.140**	1.000
		Sig. (2-tailed) (F)	.000	. (H)
		N	30004	30004

(I) **. Correlation is significant at the 0.01 level (2-tailed).

Image 10.2   **An SPSS Spearman's Correlation Coefficient Matrix**

A. This label indicates that the correlation coefficients shown in the matrix are Spearman's rho (rank-order) correlations.

B. The variables that the procedure is assessing relationships between are listed in rows. It doesn't matter what order they are in.

C. The variables listed in the rows are also listed in the columns, in the same order as the rows.

D. These labels indicate that each cell in the correlation matrix contains three pieces of information: (1) a (Spearman's) correlation coefficient, (2) the result of a statistical significance test for the correlation coefficient (based on the t-statistic), and (3) the number of cases used in the calculations.

E. The lines in this cell show the relationship between "Years since immigration" (in the row) and "Years since immigration" (in the column). The Spearman's correlation coefficient is 1; as expected, the variable matches with itself perfectly. This should not be reported. There were the equivalent of 33,222 cases used to calculate the correlation coefficient; notice that this number is larger than for Pearson's correlation coefficient. For Spearman's correlation coefficients, the number of cases can become artificially inflated when the ranks are weighted. If you are using weights, use another procedure to determine the number of cases.

F. The lines in this cell show the relationship between "Employment income" (in the row) and "Years since immigration" (in the column). The Spearman's correlation coefficient is 0.140, indicating that there is a weak positive relationship between the rank of years since immigration and the rank of employment income in 2010. The "Sig. (2-tailed)" line shows the p-value associated with the t-statistic for the correlation coefficient (with the appropriate degrees of freedom). That is, it shows the likelihood of randomly selecting a sample with the observed rank-order correlation (or a stronger correlation) between the two variables, if no rank-order correlation exists in the population. This p-value indicates that there is less than a 0.1 per cent chance of selecting this sample from a population of recent immigrants in which there is no rank-order correlation between years since immigration and annual employment income ($p < 0.001$). There were the equivalent of 30,004 cases used to calculate the correlation coefficient; again, this number is artificially inflated as a result of weighting the ranks. Because this number is used to calculate the degrees of freedom of the t-statistic, you should be careful when interpreting the results of the significance test for Spearman's rho if you are using a small sample of weighted data.

G. The lines in this cell show the relationship between "Years since immigration" (in the row) and "Employment income" (in the column). The results and interpretation are the same as for F, since they show the relationship between the same two variables. Results should only be reported for the cells above the diagonal *or* the cells below the diagonal in a correlation matrix to avoid duplication.

Continued

H. The lines in this cell show the relationship between "Employment income" (in the row) and "Employment income" (in the column). The Spearman correlation coefficient is 1; as expected, the variable matches with itself perfectly. This should not be reported. There were the equivalent of 30,004 cases used to calculate the correlation coefficient. (See the note about this in E.)

I. This footnote indicates that two asterisks are used to denote correlations that are statistically significant using an alpha value of 0.01.

Analyzing Partial Correlations

In Chapter 9, you learned how the elaboration model is used to develop a more comprehensive understanding of patterns in data by investigating relationships within subgroups. The elaboration model can also be used with correlations: begin by finding the correlation between two variables in the sample overall, and then find the correlation between the same two variables in different subgroups of the sample. The correlation between the two variables in the sample overall is called the zero-order correlation, and the correlations in each subgroup of the sample are called partial correlations. Researchers compare the partial correlations with the zero-order correlation to determine whether replication, specification, interpretation, explanation, suppression, or distortion occurs.

Let's use the elaboration model to assess whether visible minority status influences the relationship between years since immigration and annual employment income for recent immigrants. Recall that the Pearson's correlation coefficient for the relationship between these two variables in the sample overall is 0.139. In the elaboration model, the sample is divided into two subgroups: people who are classified as visible minorities and those who are not. (See Chapter 9 for a discussion of this classification.) Then, the Pearson's correlation coefficient between the number of years since immigration and employment income in 2010 is calculated for each subgroup separately. For people who are visible minorities, the partial correlation for the relationship between years since immigration and employment income in 2010 is 0.157, and for people who are not visible minorities the partial correlation for the same relationship is 0.078. These results suggest that the relationship between length of time since immigration and annual employment income is stronger for people who are visible minorities than for those who are not. Since the relationship between the two variables becomes stronger in one subgroup and weaker in another subgroup, this is another example of specification. For people who are not visible minorities, the number of years since immigration has only a very weak association with employment income in 2010. In contrast, for people who are visible minorities, the number of years since immigration has a slightly stronger association with employment income in 2010; in general, the longer it has been since people who are visible minorities immigrated to Canada, the higher their annual employment income. Even though the relationship between the two variables is weak overall, these findings suggest that the process of settlement and integration into the labour force may be different for immigrants who are racialized compared to those who are not.

Partial correlations can also be interpreted in relation to the standard deviation of each variable, calculated for each subgroup separately. Among people who are visible minorities, the standard deviation of years since immigration is 3.11, and the standard deviation of employment income in 2010 is $23,233. So, a 3.11-year increase in time since immigration is associated with a $3,648 increase in annual employment income (since $23,233 multiplied by 0.157 is $3,648). Or, for people who are visible minorities, each additional year since immigrating to Canada is associated with a $1,173 increase in annual employment income. Among people who are not visible minorities, the standard deviation of years since immigration is 3.09, and the standard deviation of employment income is $25,552. So, among people who are not visible minorities, each additional year since immigrating to Canada is associated with a $645 increase in annual employment income. (See if you can calculate this number yourself.) Presented this way, it becomes clear that the relationship between length of time since immigration and annual employment income is stronger for recent immigrants who are visible minorities than for recent immigrants who are not visible minorities.

Interpreting partial correlation coefficients using standard deviations can become tedious, especially if there are many subgroups or if you are investigating relationships between many variables. Linear regression makes it easier to obtain these results. But before developing a regression model, a careful investigation of partial correlations will show whether or not controlling for a third characteristic helps to better explain a relationship between two variables.

Best Practices in Presenting Results

Writing about Correlations

The word *correlation* is widely misused. In everyday conversation, many people describe the presence of *any* type of relationship between two characteristics as a correlation, making claims such as this: "There is a correlation between gender and religious affiliation." Since gender and religious affiliation are both nominal-level variables, it's simply not possible for there to be a correlation between them. Now that you understand what correlations are, you should only use the term *correlation* in the context of two ratio-level variables that you have assessed using a Pearson's correlation coefficient, or in the context of two ordinal- or ratio-level variables that you have assessed using a Spearman's correlation coefficient.

Writing about correlation coefficients is similar to writing about tests of statistical significance, in that researchers usually present the results in parentheses following a claim, as for a citation. But, typically, only the correlation coefficient and the associated p-value are reported; the t-statistic and the degrees of freedom are not reported. Pearson's and Spearman's correlation coefficients are distinguished from each other by using either an r or a ρ in the citation. For the NHS example in this chapter, a researcher might report that "among recent immigrants, there is a weak positive correlation between the number of years since immigration and employment income in 2010 ($r = 0.139$, $p < 0.001$)." Alternatively, a researcher reporting the rank-order correlation might indicate that "among recent immigrations, there is a weak positive rank-order correlation between the number of years since immigration and employment income in 2010 ($\rho = 0.140$, $p < 0.001$)." Because researchers use different thresholds to define weak, moderate, and strong relationships, it's important to always report the exact correlation coefficient as well as your characterization of the relationship.

A final note of caution about making causal claims based on correlations is warranted. Although this might seem repetitive, numerous research studies conflate correlation and causation. Several websites are dedicated to absurd claims made using correlative evidence. Whenever you write about correlations, guard against using phrases such as "effect of" or "leads to," which might suggest causation. Instead, use more neutral terms that emphasize "associations" and "relationships" between variables.

What You Have Learned

This chapter began by introducing scatterplots and three different ways of characterizing a relationship between two ratio-level variables. You then learned about two measures of association that can be used to assess the magnitude of a relationship between ratio-level variables: Pearson's (product-moment) correlation coefficient and Spearman's rank-order correlation coefficient. You learned how to calculate and interpret each of these measures, as well as their associated test of statistical significance. Finally, you learned how the elaboration method can be extended to correlations, in order to assess whether a bivariate correlation is being influenced by a third variable. In the upcoming chapters, you'll learn to apply these ideas in linear regression models.

The research focus of this chapter was the relationship between the amount of time since immigration and the annual employment income of recent immigrants. Canada relies on immigration to replace its human capital. Finding well-paying employment is an important part of immigrants' integration into Canadian society. In this chapter, we found only a weak correlation between the number of years since immigration and recent immigrants' employment income in 2010. Recent immigrants did not appear to be able to move into better-paying work after a period of acculturation; instead, their annual employment income appears to be relatively consistent during their first decade in Canada. Notably, the relationship between years since immigration and annual employment income is stronger for immigrants who are visible minorities than for those who are not. It's likely that recent immigrants' annual employment earnings are influenced by a wide range of factors, including socio-demographic characteristics, job characteristics, geographic location, and employers' willingness to recognize international educational credentials and work experiences.

Check Your Understanding

Check to see if you understand the key concepts in this chapter by answering the following questions:

1. What are scatterplots and what do they show?
2. What is the difference between a linear relationship and a non-linear relationship?
3. What is the difference between a monotonic relationship and a non-monotonic relationship?
4. What is covariance?
5. How is the covariance related to Pearson's correlation coefficient?
6. How does the calculation of Pearson's correlation coefficient differ from that of Spearman's correlation coefficient?
7. When would you use Spearman's correlation coefficient instead of Pearson's correlation coefficient?
8. What are partial correlations? What can they tell you about the influence of a third variable on a bivariate relationship?

Practice What You Have Learned

Check to see if you can apply the key concepts in this chapter by answering the following questions. Keep two decimal places in any calculations.

1. For each of the four scatterplots in Figure 10.8, determine whether the relationship that is shown is linear or non-linear.
2. For each of the linear relationships that you identified in question 1:

a. Identify which relationships are positive and which are negative. Explain how you know.
b. Identify which relationship is the strongest. Explain how you determined your answer.

3. For each of the four scatterplots in Figure 10.8, determine whether the relationship that is shown is monotonic or non-monotonic.

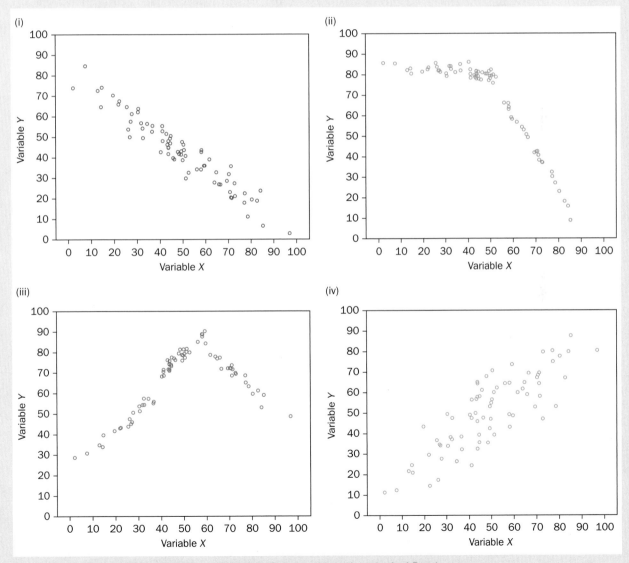

Figure 10.8 **Four Scatterplots Showing Bivariate Relationships (Hypothetical Data)**

Table 10.3   Results from Survey of Immigrants (Hypothetical Data)

Person	Satisfaction with Life in the Town	Number of Close Friends in the Town
Salman	8	4
Farook	18	12
Alisha	11	6
Ayaan	13	1
Dorian	19	11
Kyi Kyi	17	10
Ayesha	15	3
Mélanie	12	10
Michael	16	16
Carlotta	11	2
Felipe	18	13
Mahmoud	10	2

4. A small town wants to attract more immigrants in order to maintain its population size. To bolster a promotional campaign, a local settlement agency surveys a random sample of 12 recent immigrants to get more information. The immigrants are asked to indicate how satisfied they are with life in the town, using a scale ranging from 0 to 20 (where 0 indicates that they are not at all satisfied and 20 indicates they are very satisfied). To measure social integration, they are also asked how many close friends they have in the town, that is, people whom they feel that they can confide in or call on for help. The results are shown in Table 10.3. A settlement agency worker wants to investigate the relationship between the number of close friends that people have and how satisfied they are with life in the town.

    a. Determine which variable should be treated as independent and which should be treated as dependent.

    b. State a non-directional research hypothesis that the agency worker can test.

    c. State the null hypothesis associated with the research hypothesis you identified in (b).

5. Using the information in Table 10.3:

    a. Calculate the sum of products for the relationship between people's number of close friends and their satisfaction with life in the town.

    b. Calculate the sum of squares of the "Satisfaction with life" variable.

    c. Calculate the sum of squares of the "Number of close friends" variable.

6. Using your answers to question 5:

    a. Calculate Pearson's correlation coefficient.

    b. Describe what Pearson's correlation coefficient shows about the magnitude of the relationship between these two variables. Is the relationship weak, moderate, or strong?

    c. Identify whether the relationship is positive or negative. What does this indicate in the context of these two variables?

7. Using the information in Table 10.3 and your answers to question 6(a):

    a. Calculate the t-statistic for Pearson's correlation coefficient.

    b. Calculate the degrees of freedom of the t-statistic for Pearson's correlation coefficient.

    c. Using an alpha value of 0.05, the critical value of the t-statistic for a Pearson's correlation coefficient with these degrees of freedom is +/−2.23. Based on your answer to (a), state whether you reject or fail to reject the null hypothesis. In the population, is there likely to be a relationship between recent immigrants' number of close friends and how satisfied they are with life in the town?

8. Either by hand or using a spreadsheet program, create a scatterplot showing the relationship between recent immigrants' number of close friends and satisfaction with life in the town, using the information in Table 10.3.

9. The settlement agency worker notices that the two variables shown in Table 10.3 are not normally distributed. As a result, the agency worker decides to reassess the relationship between recent immigrants' number of close friends and how satisfied they are

with life in the town, using Spearman's rank-order correlation coefficient.

a. State a non-directional research hypothesis that the agency worker can test using Spearman's rank-order correlation.

b. State the null hypothesis associated with the research hypothesis you identified in (a).

10. Assign each case a rank on each of the two variables in Table 10.3. Be sure to deal appropriately with any tied ranks. Then:

a. Calculate the sum of products for the relationship between people's number of close friends and their satisfaction with life in the town, using the ranks.

b. Calculate the sum of squares of the "Satisfaction with life" variable, using the ranks.

c. Calculate the sum of squares of the "Number of close friends" variable, using the ranks.

11. Using your answers to question 10:

a. Calculate Spearman's correlation coefficient.

b. Describe what Spearman's correlation coefficient shows about the magnitude of the rank-order relationship between these two variables. Is the relationship weak, moderate, or strong?

c. Identify whether the rank-order relationship is positive or negative. What does this indicate in the context of these two variables?

12. Using the information in Table 10.3 and your answers to question 11(a):

a. Calculate the t-statistic for Spearman's correlation coefficient.

b. Calculate the degrees of freedom of the t-statistic for Spearman's correlation coefficient.

c. Using an alpha value of 0.05, the critical value of the t-statistic for a Spearman's correlation coefficient with these degrees of freedom is +/–2.23. Based on your answer to (a), state whether you reject or fail to reject the null hypothesis. In the population, is there likely to be a rank-order relationship between immigrants' number of close friends and their satisfaction with life in the town?

13. Compare the Spearman's correlation coefficient from question 11(a) to the Pearson's correlation coefficient from question 6(a). Determine whether the two variables—"Number of close friends" and "Satisfaction with life"—have a stronger linear relationship or a stronger monotonic relationship.

14. The town's settlement agency decides to partner with organizations in several neighbouring communities in order to collect information from a larger number of recent immigrants to the area. Overall, a random sample of 140 recent immigrants from the region participated in a survey. In the larger sample, Pearson's correlation coefficient for the relationship between recent immigrants' number of close friends and their satisfaction with life is 0.73. The agencies want to know whether this relationship is different for men and women. When the sample is divided by gender, the (partial) Pearson's correlation coefficient for the relationship between number of close friends and satisfaction with life among women is 0.85, and the (partial) Pearson's correlation coefficient for the corresponding relationship among men is 0.67.

a. Compare these two partial correlations to the zero-order relationship. Determine whether the addition of gender as a control variable results in replication, specification, explanation, interpretation, suppression, or distortion, and explain why.

b. Describe how the relationship between immigrants' number of close friends and satisfaction with life changes when gender is taken into account.

15. Canada's Longitudinal Survey of Immigrants to Canada (LSIC) collected information from a random sample of immigrants to Canada. Each person was interviewed three times: once six months after arriving in Canada, then two years after arriving in Canada, and, finally, four years after arriving in Canada. Table 10.4 is excerpted from a Statistics Canada report investigating immigrants' assessment of their life in Canada.

Table 10.4 Descriptive Overview of Subjective Assessments of Life in Canada

	Wave 1, Six Months after Landing	Wave 2, Two Years after Landing	Wave 3, Four Years after Landing
	Per cent		
Satisfaction with life in Canada			
Very/completely satisfied	18.8	—	17.5
Satisfied	54.2	—	55.9
Neither satisfied nor dissatisfied	17.7	—	19.4
Dissatisfied	8.0	—	4.2
Very/completely dissatisfied	1.4	—	2.9
Total	100.0	—	100.0
Met expectations - Life in Canada is . . .			
Much better than expected	13.4	14.4	14.8
Somewhat better than expected	22.9	26.9	28.3
About what expected	39.4	33.8	32.7
Somewhat worse than expected	20.4	20.5	19.2
Much worse than expected	3.9	4.4	5.0
Total	100.0	100.0	100.0
If had to make the decision again would you come to Canada?			
Yes	91.2	88.2	86.5
No	8.8	11.8	13.5
Total	100.0	100.0	100.0

Note: The percentage distribution of assessments for each question may not add to 100% because of rounding.

Source: Houle and Schellenberg 2010, 21.

a. Six months after arriving, what percentage of immigrants would come to Canada if they had to make the decision again? How does this percentage compare to the same percentage four years after arriving?

b. Six months after arriving, what percentage of immigrants report that their life in Canada is somewhat or much worse than they expected?

How does this percentage compare to the same percentage four years after arriving?

c. Are recent immigrants more likely to be satisfied or dissatisfied with their life in Canada? How does the distribution of immigrants' satisfaction with life in Canada change between the first wave and the third wave?

Practice Using Statistical Software (IBM SPSS)

Answer these questions using IBM SPSS and the GSS27.sav or the GSS27_student.sav dataset available from the Student Resources area of the companion website for this book. Weight the data using the "Standardized person weight" [STD_WGHT] variable you created following the instructions in Chapter 5. Report two decimal places in your answers, unless fewer are printed by IBM SPSS. It is imperative that you save the dataset to keep any new variables that you create.

1. The variable "Number of relatives respondent feels close to" [RFE_10C] shows people's answers to this question: "How many relatives do you have who you feel close to (that is, who you feel at ease with, can talk to about what is on your mind, or call on for help)?" Use the Bivariate Correlate procedure to find the Pearson's correlation coefficient for the relationship between this variable and the variable "Number of close friends" [SCF_100C].

   a. Describe the magnitude of the relationship between the two variables. Is the relationship weak, moderate, or strong?

   b. Is the direction of the relationship positive or negative? What does this indicate in the context of these two variables?

2. Consider the test of statistical significance produced in question 1:

   a. State a non-directional research hypothesis for this relationship.

   b. State the null hypothesis associated with the research hypothesis you identified in (a).

   c. Interpret the test of statistical significance in relation to the null hypothesis you identified in (b). In the population, is there likely to be a relationship between the number of relatives people feel close to and the number of close friends that they have?

3. Use the Bivariate Correlate procedure to find the Spearman's rank-order correlation coefficient for the relationship between the "Number of relatives respondent feels close to" [RFE_10C] and the respondent's "Number of close friends" [SCF_100C].

   a. Describe the magnitude of the rank-order relationship between the two variables. Is the relationship weak, moderate, or strong?

   b. Is the direction of the rank-order relationship positive or negative? What does this indicate in the context of these two variables?

4. Consider the test of statistical significance produced in question 3:

   a. State a non-directional research hypothesis for this relationship.

   b. State the null hypothesis associated with the research hypothesis you identified in (a).

   c. Interpret the test of statistical significance in relation to the null hypothesis you identified in (b). In the population, is there likely to be a rank-order relationship between the number of relatives people feel close to and the number of close friends that they have?

5. Use the Chart Builder tool to produce a scatterplot of the relationship between the "Number of relatives respondent feels close to" [RFE_10C] and their "Number of close friends" [SCF_100C]. Decide which variable you will treat as independent and which you will treat as dependent, and place them on the correct axis.

   a. Adjust the scale on both axes so that the maximum is 50 and the major increment is 10.

   b. Describe what your scatterplot shows.

6. Use the Select Cases tool to randomly select 4 per cent of the sample. Then, duplicate the graph you produced in question 5, only with fewer cases. Is the pattern of the relationship easier to see? Explain why or why not.

   After completing this question, use the Select Cases tool to return to using all of the cases.

7. To use the elaboration model, select "Organize output by groups" in the Split File tool and divide the cases by "Sex of respondent" [SEX]. Then, use the Bivariate Correlate procedure to find the Pearson's correlation coefficient for the relationship

between the "Number of relatives respondent feels close to" [RFE_10C] and the "Number of close friends" [SCF_100C] for men and women separately.

a.   Describe how the relationship between the number of relatives people feel close to and the number of close friends that they have changes when sex/gender is taken into account.

b.   Compare the two partial relationships to the zero-order relationship in question 1. Determine whether this an example of replication, specification, explanation, interpretation, suppression, or distortion, and explain why.

After completing this question, use the Split File tool to return to using all the cases in a single group.

8.   The variable "Number of new people met - Past month" [SCP_110] captures people's answers to this question: "In the past month, outside of work or school, how many new people did you meet either face-to-face or online? Include people you had not met before and who you intend to stay in contact with." Use the Bivariate Correlate procedure to find the Pearson's correlation coefficients for the relationships between these three variables and display them in a single correlation matrix: "Number of relatives respondent feels close to" [RFE_10C], "Number of close friends" [SCF_100C], and "Number of new people met - Past month" [SCP_110].

a.   The correlation matrix shows three different relationships. Identify each of them.

b.   Determine which of the three relationships is the weakest. How would you explain this result?

c.   Determine which of the three relationships is the strongest. How would you explain this result?

Key Formulas

Covariance

$$cov(x,y) = \frac{\sum(x_i - \bar{x})(y_i - \bar{y})}{n-1}$$

Pearson correlation coefficient (version 1)

$$r = \frac{SP}{\sqrt{SS_x}\sqrt{SS_y}}$$

Sum of products

$$SP = \sum(x_i - \bar{x})(y_i - \bar{y})$$

Sum of squares (x)

$$SS_x = \sum(x_i - \bar{x})^2$$

Sum of squares (y)

$$SS_y = \sum(y_i - \bar{y})^2$$

Pearson correlation coefficient (version 2)

$$r = \frac{\Sigma (x_i - \bar{x})(y_i - \bar{y})}{\sqrt{\Sigma (x_i - \bar{x})^2} \sqrt{\Sigma (y_i - \bar{y})^2}}$$

T-statistic for Pearson's correlation coefficient

$$t = r \sqrt{\frac{n-2}{1-r^2}}$$

Degrees of freedom of the t-statistic (equal variances; for r or ρ)

$$df_t = n - 2$$

Spearman's rank-order correlation coefficient (version 1)

$$\rho = \frac{SP}{\sqrt{SS_x} \sqrt{SS_y}}$$

(where ranks are used in place of values and means in the calculation of the sum of products and sum of squares)

Spearman's rank-order correlation coefficient (version 2)

$$\rho = \frac{\Sigma (x_i - \bar{x})(y_i - \bar{y})}{\sqrt{\Sigma (x_i - \bar{x})^2} \sqrt{\Sigma (y_i - \bar{y})^2}}$$

(where ranks are used in place of values and means)

T-statistic for Spearman's rank-order correlation coefficient

$$t = \rho \sqrt{\frac{n-2}{1-\rho^2}}$$

References

Citizenship and Immigration Canada. 2015. "Facts and Figures 2014: Immigration Overview: Permanent Residents." Catalogue no. D1–2015. http://www.cic.gc.ca/english/pdf/2014-Facts-Permanent.pdf.

Dean, Jennifer Asanin, and Kathi Wilson. 2009. "'Education? It Is Irrelevant to My Job Now. It Makes Me Very Depressed . . .': Exploring the Health Impacts of Under/unemployment among Highly Skilled Recent Immigrants in Canada." *Ethnicity & Health* 14 (2): 185–204. doi:10.1080/13557850802227049.

Guo, Shibao. 2009. "Difference, Deficiency, and Devaluation: Tracing the Roots of Non-Recognition of Foreign Credentials for Immigrant Professionals in Canada." *The Canadian Journal for the Study of Adult Education* 22 (1): 37–52.

Houle, René, and Grant Schellenberg. 2010. "New Immigrants' Assessments of Their Life in Canada." Catalogue no. 11F0019M — No. 322. Ottawa: Statistics Canada. http://www.statcan.gc.ca/pub/11f0019m/11f0019m2010322-eng.pdf.

Migration Policy Institute. 2015. "Top 25 Destination Countries for Global Migrants over Time." http://www.migrationpolicy.org/programs/data-hub/charts/top-25-destination-countries-global-migrants-over-time.

Picot, Garnett, and Arthur Sweetman. 2012. "Making It in Canada: Immigration Outcomes and Policies." Montreal: Institute for Research on Public Policy. http://irpp.org/wp-content/uploads/assets/research/diversity-immigration-and-integration/making-it-in-canada/IRPP-Study-no29.pdf.

Preston, Valerie, Marshia Akbar, Mai Phan, Stella Park, and Philip Kelly. 2011. "Continuity of Employment for Immigrants during the First Four Years in Canada." Analytical Report 24. Toronto: Toronto Immigrant Employment Data Initiative. http://www.yorku.ca/tiedi/doc/AnalyticalReport24.pdf.

Statistics Canada. 2012. "Population Growth in Canada: From 1851 to 2061." Catalogue no. 98–310-X2011003. http://www12.statcan.gc.ca/census-recensement/2011/as-sa/98–310-x/98–310-x2011003_1-eng.pdf.

———. 2013a. *NHS User Guide—National Household Survey, 2011.* Ottawa: Statistics Canada. http://publications.gc.ca/collections/collection_2013/statcan/CS99-001-2011-1-eng.pdf

———. 2013b. *National Household Survey, 2011.* Ottawa: Statistics Canada. http://www12.statcan.gc.ca/nhs-enm/2011/dp-pd/prof/index.cfm?Lang=E.

United Nations. 2015. "International Migration Flows to and from Selected Countries: The 2015 Revision." http://www.un.org/en/development/desa/population/migration/data/empirical2/migrationflows.shtml#.

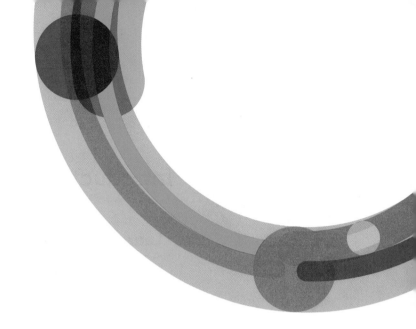

Part III

Modelling Relationships

Introduction to Linear Regression

Introduction

In this chapter, I introduce another way to assess relationships between two ratio-level variables: linear regression. Regression is a powerful tool for understanding relationships because it allows researchers to identify patterns and use them to make predictions. This chapter introduces the most basic type of regression: simple linear regression. In simple linear regression, one ratio-level independent variable is used to predict the value on a ratio-level dependent variable. In chapters 12 and 13, you'll learn how linear regression can be extended to use two or more independent variables.

This chapter has the same research focus as Chapter 10: the relationship between length of time since immigration and annual employment income among recent immigrants to Canada. By using the same research focus, you will be able to see how regression coefficients relate to correlation coefficients. Much like Chapter 10, this chapter uses statistical analysis to discover the following:

- What is the relationship between the length of time since immigration and annual employment income for recent immigrants?
- How much of the variation in recent immigrants' annual employment income can be explained by the length of time since immigration?

Linear Regression Basics

The goal of regression is to identify the general pattern of a relationship. Once a pattern is identified, it can be extrapolated to make predictions about situations that do not occur in the data. Whereas correlation coefficients provide information about the strength and direction of an association, regression allows researchers to predict that people who immigrated "x" number of years ago will have "y" amount of employment income. Researchers usually identify patterns in data collected from a random sample of people and then rely on statistical inference to assess whether or not the same patterns are likely to occur in the population.

This chapter focuses on **linear regression.** As you might guess, linear regression is used to identify a general pattern that takes the shape of a straight line. As a result, it's important to first review a scatterplot of the variables you want to use in a linear regression to confirm that it makes sense to predict a straight-line relationship between them. If the relationship between the variables is non-linear, you will need to use other statistical techniques or more advanced regression techniques to describe the relationship.

Simple linear regression identifies the straight-line pattern that best fits the relationship between two variables that is observed in the data. But which line is the one that fits the best? For many relationships, there are several lines that reasonably illustrate the pattern in the data. In linear regression, the line of best fit is usually defined as the one that minimizes the squared distances of the cases from the line. This way of finding the line of best fit is called the **ordinary least squares** (OLS) method: the goal is to find the line that produces the smallest (least) number when the distance between every case and the line is squared and then summed.

Let's use the same 10 hypothetical cases as in Chapter 10 to illustrate this idea. Once again, years since immigration is treated as the independent variable, and annual employment income is treated as the dependent variable. Figure 11.1 shows two scatterplots, with two different lines that could be used to characterize the relationship between the two variables. The labels printed above each dot show the distance between the case and the line. So, in the scatterplot in the left panel of Figure 11.1, the first dot represents a person who immigrated less than a year ago (recorded as "0 years") and who earned $20,000 in annual employment income. In the line on that scatterplot, "0 years" since immigration corresponds with earning $19,000 in annual employment income. The distance between the case and the line is $1,000 (20,000 − 19,000 = 1,000), so the first dot is labelled $1,000. In the line on the scatterplot in the right panel of Figure 11.1, "0 years" since immigration corresponds with earning $15,000 in annual employment income, so the first dot on that scatterplot is labelled $5,000 (20,000 − 15,000 = 5,000). In both scatterplots, one case is located right on the line, and is thus labelled "0". As well, both scatterplots include some cases that are located below the line, which results in a distance from the line that is a negative number.

To decide which line best fits the pattern in the data, researchers calculate the distance between each case and the line, square each of the distances, and then add together all of the results. The line with the smallest sum is the best. There are

linear regression A type of regression used to predict straight-line relationships between one or more independent variables and a ratio-level dependent variable.

ordinary least squares (OLS) A method of determining the line of best fit by finding the line that produces the smallest (least) number when the distance between every case and the line is squared and then summed.

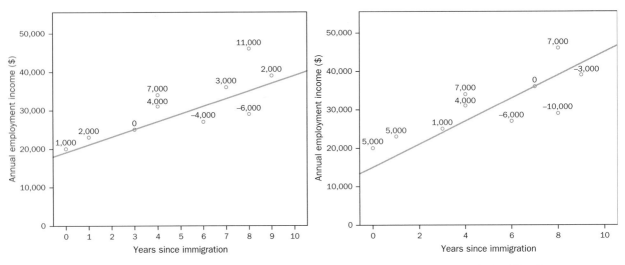

Figure 11.1   Two Lines That Could Be Used to Describe the Pattern of a Relationship (Hypothetical Data)

several advantages to using this approach to find the line of best fit. First, squaring the distances from the line eliminates all of the negative numbers. Second, squaring the distances from the line effectively assigns a larger penalty to cases located farther from the line. Finally, this approach is mathematically consistent with many other statistical techniques; by now, the idea of squaring differences and summing them should be quite familiar. For the line in the scatterplot in the left panel of Figure 11.1, the sum of the squared distances from the line is 256,000,000 (calculated as $1,000^2 + 2,000^2 + 0^2 + 7,000^2 + 4,000^2 + [-4,000]^2 + 3,000^2 + 11,000^2 + [-6,000]^2 + 2,000^2$). For the line in the scatterplot in the right panel, the sum of the squared distances from the line is 310,000,000. Thus, the line on the scatterplot in the left panel fits the data better than the line on the scatterplot in the right panel.

In fact, the straight line that best fits the pattern of these 10 cases is shown in Figure 11.2. Although none of the cases are located exactly on the line of best fit, the farthest distance from the line is reduced to $8,965. (Although two cases—those labelled 47 and −47—are so close to the line that they appear to be on it.) The sum of the squared distances from this line is 215,988,372: less than for the line on the scatterplot in the left panel of Figure 11.1. The line that best fits the pattern of a relationship between variables that is observed in the data—which minimizes the sum of the squared distances between the cases and the line—is called a **regression line**.

regression line The line that best fits the pattern of an observed relationship between variables.

influential cases Cases that substantially affect the location or the direction of a regression line.

Another reason that it is important to review a scatterplot of the variables used in a linear regression is to identify any **influential cases**: those that substantially affect the location or the direction of a regression line. Influential cases typically have unusual values on one or both variables or an unusual combination of values. The scatterplots in Figure 11.3 each show the same 10 cases as in Figure 11.2, plus one additional influential case, and the line of best fit when the influential case is included. The scatterplot in the left panel of Figure 11.3 adds a

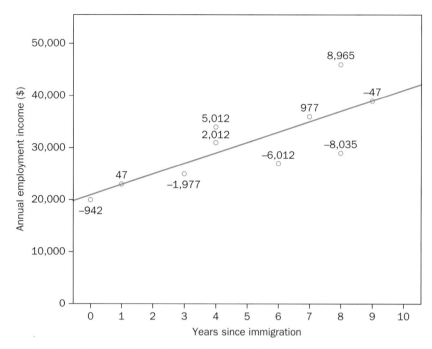

Figure 11.2   **The Line of Best Fit for 10 Hypothetical Cases**

case with an average number of years since immigration but a very high income. When you compare the line of best fit to the one shown in Figure 11.2, you can see that the line has the same angle but is now located above most of the cases. The scatterplot in the right panel of Figure 11.3 adds a case with an unusual combination of values: someone who immigrated less than a year ago but earns $100,000

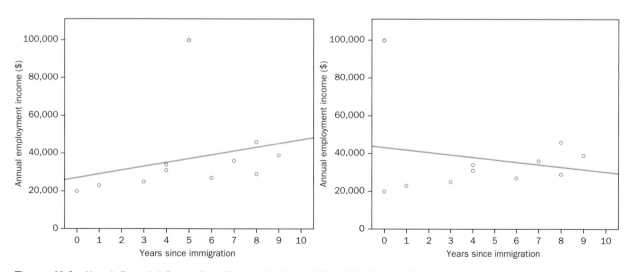

Figure 11.3   **How Influential Cases Can Change the Line of Best Fit (Hypothetical Data)**

in annual employment income. Adding this case changes the direction of the line of best fit so that it now shows a negative relationship: more time since immigration is associated with earning less employment income. Both of these examples illustrate how an influential case can affect the line of best fit. If one or more influential cases appear in the data, you should omit them so that the regression line provides a good representation of the general pattern in the majority of cases. When you present your results, however, you should report that you have omitted any outliers or influential cases.

Describing a Regression Line

The line that best fits the relationship observed in the data—the regression line—can be described mathematically. For linear regressions, a linear function is used to describe how an independent variable (x) is related to a dependent variable (y). At the secondary-school level, linear functions are often presented in the form $y = mx + b$. The linear function used to describe a regression line relies on the same concepts but uses slightly different notation: $y = a + bx$.

Any straight line can be described using only two pieces of information: a slope coefficient and a constant coefficient. A **slope coefficient** provides information about the angle of the line. In other words, it indicates whether an increase in the independent variable is associated with an increase or a decrease in the dependent variable and what the size of that increase or decrease is. The size of a slope coefficient provides information about the magnitude of a relationship between two variables. You may have learned about the slope as the "rise" over the "run," where the "rise" refers to the change in the dependent variable, located on the vertical axis of a scatterplot, and the "run" refers to the change in the independent variable, located on the horizontal axis of a scatterplot. In regression, the slope coefficient is denoted by the letter b, and it shows the size of the increase (or decrease) in the dependent variable that is associated with a one-unit increase in the independent variable. (See Figure 11.4.)

The slope coefficient of the line in Figure 11.4 is 1 since a one-unit increase in the independent variable (x) is associated with exactly a one-unit increase in the dependent variable (y). In other words, the value on the dependent variable is equal to the value on the independent variable, multiplied by 1; thus, the equation that describes the line in Figure 11.4 is $y = (1)x$, where 1 is the slope coefficient. Lines with positive slope coefficients will be angled upwards from left to right, like those shown in the left panel of Figure 11.5; a positive slope coefficient indicates that there is a positive relationship between the variables. The larger a slope coefficient, the closer the line will be to a vertical line (although it will never become perfectly vertical). A line with a slope coefficient of 0 will be perfectly horizontal, indicating that there is no relationship between the two variables. A line with a negative slope coefficient will be angled downwards from left to right, like those shown in the right panel of Figure 11.5; a negative slope coefficient indicates that there is a negative relationship between the variables.

slope coefficient Provides information about the angle of a regression line; it is reported in the same units as the dependent variable.

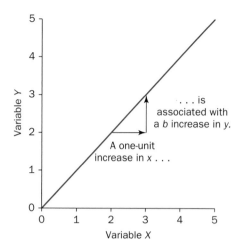

Figure 11.4   Conceptualizing the Slope Coefficient (b)

The second piece of information used to describe a line is a **constant coefficient**, which shows the value on the dependent variable (y) when the independent variable (x) equals 0. In other words, it's the place where the line crosses the vertical axis (y-axis). The constant coefficient is sometimes called the intercept because it is the point where the line intercepts the vertical axis. In regression, the constant coefficient is denoted by the letter a. The constant coefficient of the line in Figure 11.4 is 0 because when the independent variable is 0, the dependent variable is also 0. The left panel of Figure 11.6 shows a line with the same slope coefficient as the line in

constant coefficient Provides information about where a regression line crosses the vertical axis; it is reported in the same units as the dependent variable.

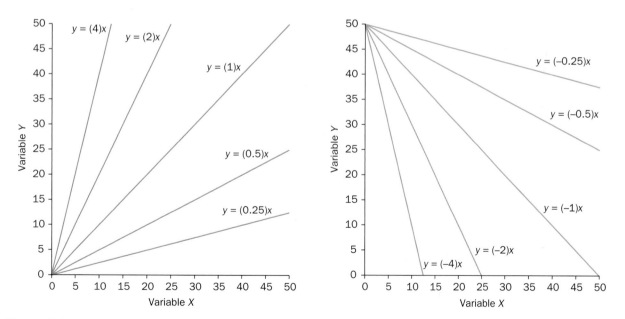

Figure 11.5   An Illustration of Various Positive and Negative Slope Coefficients

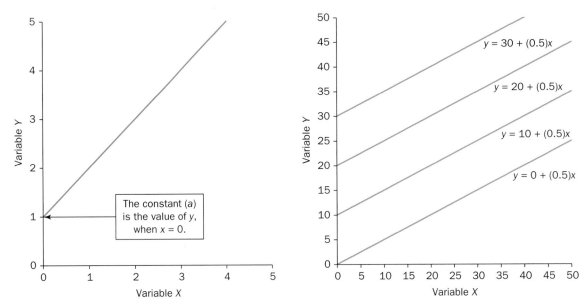

Figure 11.6 An Illustration of Various Constant Coefficients

Figure 11.4 but with a constant coefficient of 1. The equation that describes the line in the left panel of Figure 11.6 is $y = 1 + (1)x$. The right panel of Figure 11.6 shows four lines, all with the same slope coefficient (0.5) but with different constant coefficients, which change the vertical location of the line.

Sometimes constant coefficients are negative. Although this might seem counterintuitive, Figure 11.6 shows only a portion of the larger graph space. Figure 11.7 illustrates the situation when a constant coefficient is negative: the line crosses the vertical axis below the horizontal axis. One line in Figure 11.7 crosses the y-axis at −5, and the other line crosses the y-axis at −15. In fact, all of the lines graphed in Figures 11.4, 11.5, and 11.6 theoretically continue into the negative areas on the x-axis and y-axis, they just aren't depicted as doing so.

The constant coefficient (a) and the slope coefficient (b) of a line can be combined into an equation that describes the line of best fit. The equation is a mathematical representation of the relationship between an independent variable (x) and a dependent variable (y). Because the equation represents the general pattern of the relationship, it can be used to make predictions about a dependent variable, using information from an independent variable. A "hat" symbol, or caret, is placed over the y in the equation (pronounced y-hat) to denote that the value on the dependent variable is a prediction, or an estimate, instead of the actual value of a specific case:

simple linear regression predictions

$$\hat{y} = a + bx$$

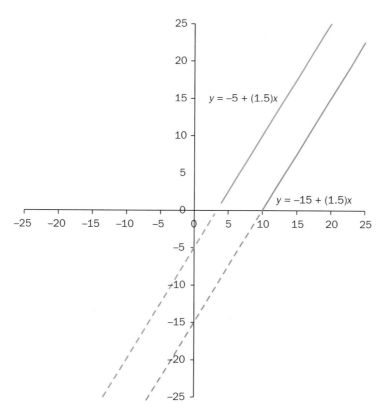

Figure 11.7 Understanding a Negative Constant

Let's return to the hypothetical 10-case example. Using the ordinary least squares method, I identified the straight line that best fits the pattern of the relationship between the two variables in the data. That line—the regression line—can be described using a slope coefficient and a constant coefficient: the slope coefficient is $2,012 (more precisely, $2,011.63), and the constant coefficient is $20,942 (more precisely, $20,941.85). Figure 11.8 shows how these coefficients reflect the line of best fit. Because the dependent variable is measured in dollars, both coefficients are reported in dollars; slope and constant coefficients are always reported in the same units as the dependent variable. Technically, slope coefficients are reported in the units of the dependent variable divided by the units of the independent variable (in other words, the rise over the run)—but since the interpretation of slope coefficients focuses on the change in the dependent variable, it's easiest to think about them as being in the same units as the dependent variable.

The constant coefficient is the value that people (or cases) with a "0" value on the independent variable are predicted to have on the dependent variable. So, given the pattern in the hypothetical example, we predict that people who immigrated less than a year ago (who have been in Canada for 0 years) will earn an annual employment income of $20,942. The slope coefficient shows the amount of change

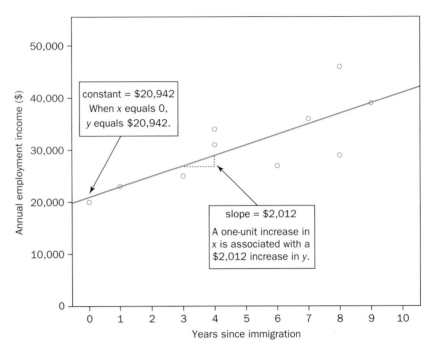

Figure 11.8   **Regression Line Example Showing the Slope and Constant Coefficients (Hypothetical Data)**

in the dependent variable that is predicted to go along with a one-unit increase in the independent variable; it's the amount of "rise" associated with a one-unit increase in the "run." The meaning of a "one-unit" increase in the independent variable depends on how the variable is measured. In this example, the independent variable is measured in years, so the slope represents the predicted change in employment income (+$2,012) that is associated with having one more year since immigrating. Notice that because the pattern of the relationship is defined by a straight line, the slope (or the rise) is the same for every one-unit difference in the independent variable. So, the predicted difference in employment income between people who immigrated two years ago and people who immigrated three years ago is the same as the predicted difference in employment income between people who immigrated eight years ago and people who immigrated nine years ago (or any other one-year difference).

Once the general pattern of a relationship is identified using a regression line, the constant and slope coefficients of that line can be substituted into a linear regression prediction equation. For instance, using the slope and constant coefficients of the regression line in Figure 11.8, I can predict recent immigrants' annual employment income using the following equation:

Predicted annual employment income ($) = 20941.85 + 2011.63(*years since immigration*)

To predict the annual employment income of people who immigrated eight years ago, "years since immigration" is replaced with an 8:

$$Predicted\ annual\ employment\ income\ (\$) = 20941.85 + 2011.63(8)$$
$$= 20941.85 + 16093.04$$
$$= 37034.89$$

So, this regression predicts that people who immigrated eight years ago earn $37,035 in annual employment income.

Because the predictions rely on the general pattern in the data, we can also make them for people with values on the independent variable that do not appear in the data. So, even though no one in the example dataset actually immigrated two years ago, we can use the pattern to predict that people who immigrated two years ago earn $24,965 in annual employment income:

$$Predicted\ annual\ employment\ income\ (\$) = 20941.85 + 2011.63(2)$$
$$= 20941.85 + 4023.26$$
$$= 24965.11$$

The ability to make predictions using the pattern of a relationship that is observed in sample data is one of the most powerful features of regression and one of the main reasons that social scientists rely heavily on this statistical technique. However, it is only reasonable to make predictions about situations within the range of the independent variable in the data. In the hypothetical example, the independent variable ranges from 0 to 9. As a result, it's not reasonable to use the pattern in these data to make predictions for people who immigrated 20 years ago, 15 years ago, or any time longer than 9 years ago.

Step-by-Step: Predicted Value on the Dependent Variable (Simple Linear Regression)

$$\hat{y} = a + bx$$

Step 1: Identify the constant coefficient from the linear regression (a).

Step 2: Identify the slope coefficient from the linear regression (b).

Step 3: Identify the value on the independent variable (x) for the specific case you want to make a prediction for.

Step 4: Multiply the slope coefficient (from Step 2) by the value on the independent variable for the specific case (from Step 3).

Step 5: Add the result of Step 4 to the constant coefficient (from Step 1) to find the predicted value on the dependent variable for the specific case.

Calculating Slope and Constant Coefficients

So far, you have learned that the regression line is the line that best fits the general pattern of the observed relationship between variables. The best-fitting line is defined as the one that minimizes the squared distances between each case and the line. You have also learned that a straight line of best fit can be described mathematically using two pieces of information: a slope coefficient and a constant coefficient.

One way to find a regression line is to test different lines until you eventually find the one with the least squared differences from the line, but this is a tedious process, especially if there are many cases. Fortunately, it's relatively simple to calculate the slope and constant coefficient of the best-fitting straight line for a bivariate relationship.

Let's start with the slope coefficient since we need this information to find the constant coefficient. The formula for a slope coefficient incorporates two of the elements used to calculate Pearson's correlation coefficient: the sum of products (SP), and the sum of squares for the independent variable (SS_x). Recall that the sum of products captures the amount of covariation between two variables by finding how far a case is above or below the mean of the first variable, then finding how far the same case is above or below the mean of the second variable, and then multiplying the two deviations from the means. This process is repeated for every case, and the results are summed. The sum of squares for the independent variable captures the amount of variation in the independent variable alone (as opposed to the amount of covariation with the dependent variable). In simple linear regression, the formula for a slope coefficient is:

slope coefficient
(simple linear regression)

$$b = \frac{SP}{SS_x} \quad or \quad b = \frac{\Sigma(x_i - \bar{x})(y_i - \bar{y})}{\Sigma(x_i - \bar{x})^2}$$

As usual, it's easiest to implement this formula using a table, such as the one in Figure 11.9, which incorporates several columns from the tables in Figures 10.6 and 10.7. (Go back to Chapter 10 to see how they compare.) To save space, the mean of each variable is no longer in a separate column; instead it is listed at the bottom of the columns for the independent and dependent variables. If you find it easier to complete the calculation with the means in columns, just add them back in. The sum of products (173,000) and sum of squares for the independent variable (86) are the same as in Chapter 10.

The sum of products and the sum of squares for the independent variable are substituted into the formula for the slope coefficient:

$$b = \frac{SP}{SS_x} \quad or \quad b = \frac{\Sigma(x_i - \bar{x})(y_i - \bar{y})}{\Sigma(x_i - \bar{x})^2}$$

$$b = \frac{173000}{86}$$

$$= 2011.63$$

$$b = \frac{\Sigma(x_i - \bar{x})(y_i - \bar{y})}{\Sigma(x_i - \bar{x})^2}$$

Person (Case)	Years since Immigration x_i	Employment Income ($) y_i	Deviation from Mean of Years since Immigration $(x_i - \bar{x})$	Squared Deviation from Mean of Years since Immigration $(x_i - \bar{x})^2$	Deviation from Mean of Employment Income $(y_i - \bar{y})$	Product of Deviations from Means $(x_i - \bar{x})(y_i - \bar{y})$
Arjun	0	20,000	−5	25	−11,000	55,000
Damia	1	23,000	−4	16	−8,000	32,000
Dmitri	3	25,000	−2	4	−6,000	12,000
Priya	4	34,000	−1	1	3,000	−3,000
Zhang Li	4	31,000	−1	1	0	0
Yusuf	6	27,000	1	1	−4,000	−4,000
Jessica	7	36,000	2	4	5,000	10,000
Esengul	8	29,000	3	9	−2,000	−6,000
Kwame	8	46,000	3	9	15,000	45,000
Alejandro	9	39,000	4	16	8,000	32,000
	$\bar{x} = 5$	$\bar{y} = 31,000$		column sum: 86		column sum: 173,000

The sigma symbols tell you to add the result of the equations for all of the cases.

Figure 11.9   Translating the Slope Coefficient Formula into a Table (Hypothetical Data)

Let's take a moment to consider how the slope coefficient is related to Pearson's correlation coefficient. In Chapter 10, you learned how to interpret Pearson's correlation coefficient in relation to the standard deviation of each variable. Recall that for this 10-case example, we used Pearson's correlation coefficient (0.786) and the standard deviation of the two variables (3.09 for years since immigration, and $7,916 for employment income) to say that a 3.09 year (one standard deviation) increase in the length of time since immigration is associated with a $6,222.16 increase in employment income (0.786 of the standard deviation of $7,916.23). Then, to make the result easier to understand, we divided $6,222.16 by 3.09 in order to be able to say that a one-year increase in the length of time since immigration is associated with a $2,013.64 increase in annual employment income. At the time, I noted that if we had retained all of the decimals in the calculation, we would have come up with a result of $2,011.63—which is the same as the slope coefficient of the line of best fit. So, the slope coefficient of a regression line just provides a simpler way to capture this information.

To find the constant coefficient, we rely on an important feature of a regression line: it always passes through the point where the means of the two variables intersect. People (or cases) with an average value on the independent variable will always be predicted to have an average value on the dependent variable. This is illustrated in Figure 11.10, which uses dashed lines to indicate the mean of each variable and the

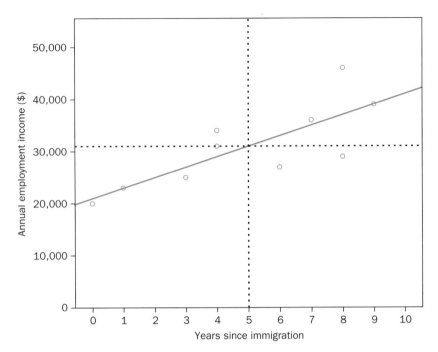

Figure 11.10   The Regression Line Always Passes through the Point Where the Variable Means Intersect (Hypothetical Data)

point where they intersect. So, in the hypothetical example, people who immigrated an average number of years ago (five years ago) are predicted to earn an average annual employment income ($31,000). Notice that this *does not* hold for values other than the overall averages of each variable. We do not predict that each set of cases defined by the independent variable will have the average value for those cases. For instance, the example includes two people who immigrated four years ago, who collectively have an average annual employment income of $32,500, but the regression line does not predict that people who immigrated four years ago will earn this amount.

Because the regression line always passes through the point where the variable means intersect, a case located at the mean of the independent variable will always be predicted to have the mean value of the dependent variable. The prediction equation for the point where the two variable means intersect in a simple linear regression is:

$$\bar{y} = a + b(\bar{x})$$

Since the mean of each variable and the slope coefficient can be calculated from the data, the only unknown in this equation is the constant coefficient. To make the calculation easier, the equation can be rearranged to explicitly solve for the constant coefficient: subtracting $b(\bar{x})$ from both sides of the equation, gives us $\bar{y} - b(\bar{x}) = a$. For readability, the two sides of the equation are reversed to obtain the formula for the constant coefficient in a simple linear regression:

$$a = \bar{y} - b(\bar{x})$$

constant coefficient
(simple linear regression)

When the variable means and the slope coefficient from the hypothetical 10-case example are substituted into the equation, the constant coefficient is calculated as:

$$a = \bar{y} - b(\bar{x})$$
$$a = 31000 - 2011.63(5)$$
$$= 31000 - 10058.15$$
$$= 20941.85$$

The slope (b) and constant (a) coefficients obtained using these calculations are then substituted into the simple regression prediction equation in order to describe the line that best fits the general pattern of the observed relationship between the two variables:

$$\hat{y} = a + bx$$

$$\hat{y} = 20941.85 + 2011.63(x)$$

The regression line defined by these coefficients matches the line of best fit shown in Figure 11.8. Of course, with large datasets, researchers usually rely on statistical software to identify the line of best fit and to calculate regression coefficients.

Step-by-Step: Slope and Constant Coefficients (Simple Linear Regression)

Note: In these calculations, do not include cases that are missing information for either variable used in the regression.

Slope Coefficient

$$b = \frac{SP}{SS_x} \qquad or \qquad b = \frac{\Sigma(x_i - \bar{x})(y_i - \bar{y})}{\Sigma(x_i - \bar{x})^2}$$

Step 1: Find the mean of the independent variable in the relationship you are assessing ($\bar{x}$).

Step 2: Find the mean of the dependent variable in the relationship you are assessing ($\bar{y}$).

Continued

Step 3: For *each case individually*, subtract the mean of the independent variable (from Step 1) from the value on the independent variable for the case (x_i) to find the deviation from the mean. Complete this calculation once for each case.

Step 4: For *each case individually*, subtract the mean of the dependent variable (from Step 2) from the value on the dependent variable for the case (y_i) to find the deviation from the mean. Complete this calculation once for each case.

Step 5: For *each case individually*, multiply the result of Step 3 by the result of Step 4. Pay attention to the positive/negative sign of each result when you multiply. Complete this calculation once for each case.

Step 6: Add together the results of Step 5 for each case to find the sum of products, which is the numerator of the slope coefficient equation.

Step 7: For *each case individually*, square the result of Step 3. Complete this calculation once for each case.

Step 8: Add together the results of Step 7 for each case to find the sum of squares for the independent variable, which Is the denominator of the slope coefficient equation.

Step 9: Divide the result of Step 6 by the result of Step 8 to find the slope coefficient.

Constant Coefficient

$$a = \bar{y} - b(\bar{x})$$

Step 10: Multiply the slope coefficient (from Step 9) by the mean of the independent variable (from Step 1).

Step 11: Subtract the result of Step 10 from the mean of the dependent variable (from Step 2) to find the constant coefficient.

How Well Does the Line Fit?

A regression line shows the general pattern of an observed relationship between an independent variable and a dependent variable. Statisticians like to think of this general pattern as the explained variation. That is, the regression line shows the variation in the dependent variable that can be explained by the independent variable. But not all of the variation in the dependent variable is explained by the general pattern. If it were, all of the cases would be located directly on the regression line. The regression line for the hypothetical 10-case example (in Figure 11.2) shows that while the general pattern is a good predictor of annual employment income, it isn't perfect. In fact, none of the cases are located exactly on the regression line. The distance between each case and the regression line represents the amount of unexplained variation. In other words, it's the variation in annual employment income that cannot be explained by the amount of time since immigration. Some of the unexplained variation can likely be attributed to factors such as people's level of education, the skill level of the job, the number of hours people work, and more. Some of the unexplained variation is likely random—that is, the variation represents random variation in people's annual employment income: even if two recent immigrants have exactly the same job, with the same hours of work and educational credentials, they may not have exactly the same annual employment income.

In regression, the unexplained variation—the difference between the predicted value on the dependent variable for each case and its actual value—is called a **residual**. Residuals represent the amount of error in the regression predictions. Once a regression line is identified, a residual can be established for each case. If a case has a residual of 0, it indicates that the value on the dependent variable is perfectly predicted by the value on the independent variable for that case; it is located right on the regression line. If a case has a large residual, it indicates that the value on the independent variable does a poor job of predicting the value on the dependent variable for that case; it is located far from the regression line. Residuals are denoted using the letter e, for error. Sometimes, the linear regression equation is shown with an e_i at the end to capture the idea that each case has a residual and that the exact value on the dependent variable is equal to the predicted value, plus any unexplained variation or error (e_i). While the slope (b) and the constant (a) in the regression equation are the same for every case, the value on the independent variable (x_i) and the error term (e_i) are potentially different for every case.

> **residual** The difference between the predicted value on the dependent variable and the actual value, for each case; the unexplained variation in a regression model.

$$y_i = a + bx_i + e_i$$

> **simple linear regression**

One way to assess how well a regression line fits the data is to investigate the residuals. (Most statistical software programs allow you to save the residuals in a new variable.) If the variables used in the regression are normally distributed, the residuals will be normally distributed and will centre on 0. Most cases will have a residual close to 0, although some may have larger residuals. Cases with huge residuals may be outliers or high-influence cases. The distribution of the residuals can be analyzed in relation to each of the variables used in the regression, in order to determine whether its predictions are better for some types of people (or things) than for others. You'll learn more about analyzing residuals in Chapter 13.

The Coefficient of Determination: R-Squared

The statistic that is most commonly used to assess how well a regression line fits the data is formally called *the coefficient of determination*—although researchers almost always refer to it as **R-squared (R^2)**. The coefficient of determination (R^2) shows how much of the variation in the dependent variable (DV) can be explained

> **R-squared (R^2)** Shows the proportion of the variation in the dependent variable that can be explained by the independent variable(s) in a linear regression.

by the independent variable (IV). The explained variation is shown as a proportion of the total variation in the dependent variable:

$$R^2 = \frac{variation\ in\ the\ DV\ explained\ by\ the\ IV}{total\ variation\ in\ the\ DV}$$

null model A hypothetical regression model with only a dependent variable (no independent variables); the mean of the dependent variable is predicted for every case.

The total variation in the dependent variable is captured in what's called a **null model**, which is a hypothetical regression model with only a dependent variable, that is, there are no independent variables. Since there are no independent variables, in the null model the mean of the dependent variable is predicted for every case. In other words, if researchers have no additional information, the best prediction they can make is the average value. The null model illustrates a situation in which the null hypothesis is true: by predicting the same (average) value on the dependent variable for every case, it implies that there is no relationship between the independent variable and the dependent variable.

Figure 11.11 shows a dashed horizontal line at the mean of the dependent variable ($31,000) in the hypothetical example. The distances between each case (y_i) and the dashed line ($\bar{y}$) are squared and the results are summed to determine the total variation in the dependent variable. This is the same process used to find the sum of squares for the dependent (y) variable in the calculation of Pearson's correlation coefficient.

The total variation in the dependent variable can be conceptually divided into two parts: (1) the variation that is explained by the independent variable and (2) the variation that remains unexplained (the residual). The explained variation is reflected in the distance between the prediction made by the regression line and the prediction made by the null model (the mean). In other words, the explained variation captures how much better the predictions of the regression line are than the predictions of the null model. In a routine that should be familiar by now, the

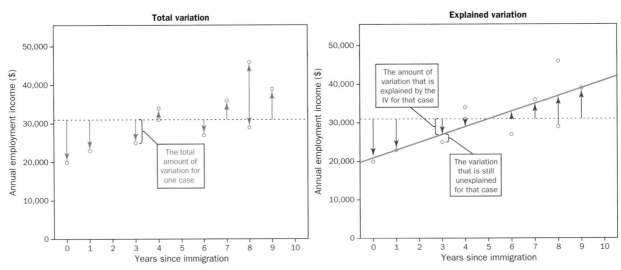

Figure 11.11   **Conceptualizing the Total Variation, Explained Variation, and Unexplained Variation in a Dependent Variable (Hypothetical Data)**

distances between the predicted values produced by the regression ($\hat{y}_i$) and the predicted values in the null model ($\bar{y}$) for each case are squared, and the results are summed to obtain the amount of explained variation. Mathematically, the explained variation and the total variation are calculated as:

$$\text{variation in the DV explained by the IV} = \Sigma(\hat{y}_i - \bar{y})^2$$

$$\text{total variation in the DV} = \Sigma(y_i - \bar{y})^2$$

To calculate R-squared, simply divide the explained variation by the total variation:

$$R^2 = \frac{\Sigma(\hat{y}_i - \bar{y})^2}{\Sigma(y_i - \bar{y})^2}$$

R² (coefficient of determination)

The result shows the variation in the dependent variable that is explained by the independent variable as a proportion (or percentage) of the total variation in the dependent variable. Figure 11.12 shows how to implement the R^2 formula in a table. The fourth column, showing the predicted employment income, was

$$R^2 = \frac{\Sigma(\hat{y}_i - \bar{y})^2}{\Sigma(y_i - \bar{y})^2}$$

Person (Case)	Years since Immigration x_i	Actual Employment Income ($) y_i	Predicted Employment Income ($) $\hat{y}_i$	Deviation of Predicted Employment Income from Mean $(\hat{y}_i - \bar{y})$	Squared Deviation of Predicted Employment Income from Mean $(\hat{y}_i - \bar{y})^2$	Deviation of Actual Employment Income from Mean $(y_i - \bar{y})$	Squared Deviation of Actual Employment Income from Mean $(y_i - \bar{y})^2$
Arjun	0	20,000	20,942	−10,058	101,163,364	−11,000	121,000,000
Damia	1	23,000	22,953	−8,047	64,754,209	−8,000	64,000,000
Dmitri	3	25,000	26,977	−4,023	16,184,529	−6,000	36,000,000
Priya	4	34,000	28,988	−2,012	4,048,144	3,000	9,000,000
Zhang Li	4	31,000	28,988	−2,012	4,048,144	0	0
Yusuf	6	27,000	33,012	2,012	4,048,144	−4,000	16,000,000
Jessica	7	36,000	35,023	4,023	16,184,529	5,000	25,000,000
Esengul	8	29,000	37,035	6,035	36,421,225	−2,000	4,000,000
Kwame	8	46,000	37,035	6,035	36,421,225	15,000	225,000,000
Alejandro	9	39,000	39,047	8,047	6,4754,209	8,000	64,000,000
	$\bar{x} = 5$	$\bar{y} = 31,000$			column sum: 348,027,722		column sum: 564,000,000

The sigma symbols tell you to add the result of the equations for all of the cases.

Figure 11.12 Translating the R² Formula to a Table (Hypothetical Data)

obtained by substituting the "Years since immigration" for each case into the regression prediction equation with the constant and slope coefficient that we calculated for this example.

The explained variation and the total variation are substituted into the R-squared equation to obtain:

$$R^2 = \frac{\Sigma\left(\hat{y}_i - \overline{y}\right)^2}{\Sigma\left(y_i - \overline{y}\right)^2}$$

$$R^2 = \frac{348027722}{564000000}$$

$$= 0.617$$

Since R-squared represents a proportion of a whole, it is typically reported as a percentage: for the hypothetical example, the R-squared indicates that 61.7 per cent of the variation in annual employment income can be explained by the length of time since immigration. R-squared ranges from 0 to 1 (or 0 per cent to 100 per cent), where 0 indicates that none of the variation in the dependent variable is explained by the independent variable (there is no relationship) and where 1 indicates that 100 per cent of the variation in the dependent variable is explained by the independent variable (there is a perfect relationship). In practice, R-squared shows how much better the regression line predictions are compared to simply predicting the average of the dependent variable for every case. As a result, the coefficient of determination is also a proportionate reduction in error (PRE) measure—that is, it shows how much the error in predicting the values on the dependent variable can be reduced if we know the values on the independent variable. For the hypothetical example, we can assert that the errors in predicting recent immigrants' annual employment income would be reduced by 61.7 per cent if we knew how long ago they immigrated.

As it turns out, for a simple linear regression, the R-squared calculated using the explained and total variation in the dependent variable matches the squared value of Pearson's correlation coefficient (r). Recall that for this hypothetical 10-case example, the Pearson's correlation coefficient is 0.786, which is 0.618 when squared. (There's a slight discrepancy because of rounding.) It's useful to know how to calculate R-squared using a more generic approach, though, because when regressions use more than one independent variable, R-squared can't be obtained by squaring Pearson's correlation coefficient.

In general, a regression with a higher R-squared is considered to fit the data better. In the social sciences, researchers typically aim to obtain an R-squared higher than 0.1 (or 10 per cent) before reporting on a regression. This varies, however, depending on a researcher's topic and the accuracy of the measurement. For those things that are difficult to predict or measure, a lower R-squared can sometimes be acceptable. But for things that are more straightforward to predict or that can be precisely measured, an R-squared of 0.1 is considered too low.

Step-by-Step: R^2

Note: In these calculations, do not include cases that are missing information for any variable used in the regression.

$$R^2 = \frac{\Sigma(\hat{y}_i - \bar{y})^2}{\Sigma(y_i - \bar{y})^2}$$

Step 1: Find the mean of the dependent variable in the relationship you are assessing ($\bar{y}$).

Step 2: For *each case*, find the predicted value on the dependent variable ($\hat{y}_i$).

Step 3: For *each case individually*, subtract the mean of the dependent variable (from Step 1) from the predicted value on the dependent variable for the case (from Step 2). Complete this calculation once for each case.

Step 4: For *each case individually*, square the result of Step 3. Complete this calculation once for each case.

Step 5: Add together the results of Step 4 for each case to find the explained variation, which is the numerator of the R^2 equation.

Step 6: For *each case individually*, subtract the mean of the dependent variable (from Step 1) from the actual value on the dependent variable for the case (y_i) to find the deviation from the mean. Complete this calculation once for each case.

Step 7: For *each case individually*, square the deviations from the mean (from Step 6). Complete this calculation once for each case.

Step 8: Add together the results of Step 7 for each case to find the total variation, which is the denominator of the R^2 equation.

Step 9: Divide the result of Step 5 by the result of Step 8 to find the R^2.

Statistical Inference in Linear Regression

So far, I have only described how regression is used to find the line that best fits the pattern of a relationship observed in a sample of cases. But, as you know, researchers are usually interested in using sample data to make claims about a larger population. To illustrate how statistical inference works in the context of linear regression, let's turn to one final simulation example. I created a dataset of 15,000 cases that represent an entire population, such as all of the employees of a large company. The dataset includes a variable that captures the number of months that people have worked at the company and a variable that captures their weekly earnings. People's weekly earnings range from $98 to $200, with a mean of $150 (s.d. = $14), and people's length of employment ranges from 11 months (just less than a year) to

88 months (more than seven years), with a mean of 50 months (slightly more than four years) (s.d. = 10). Across the entire simulated population of 15,000 employees, the general pattern of the relationship between months of employment and weekly earnings has a slope coefficient of 1. That is, every additional month worked is associated with a $1 increase in weekly earnings. The general pattern of the relationship in the simulated population also shows that people who just started working at the company (who have worked for 0 months) are predicted to earn $100 a week. In the entire population of 15,000 employees, the R-squared is 0.5; that is, half of the variation in weekly earnings is predicted by the length of employment.

By now, you can likely guess what is coming next. I randomly select a single sample of 500 employees from the population and use linear regression to determine the general pattern of the observed relationship between months of employment and weekly earnings. The constant coefficient is $97 and the slope coefficient is $1.04. Using these sample results, I predict that people just starting at the company will earn $97 per week and that each additional month worked is associated with a $1.04 increase in weekly earnings. But, as you know, this is only one sample that can be selected from the population of 15,000 employees. So, I randomly select a second sample of 500 employees from the population and conduct another linear regression. This time, I predict that people just starting at the company will earn $102 per week and that each additional month worked is associated with a $0.95 increase in weekly earnings. This result is slightly different than the result from the first sample because the pattern was established using a different group of 500 cases. But, just as in previous simulations, all cases have an equal chance of being selected into a sample, regardless of whether they were part of a previous sample. I continue on to select a third sample of 500 employees from the population and conduct yet another linear regression. Using the third sample, I predict that people just starting at the company will earn $99 per week and that each additional month worked is associated with a $1.01 increase in weekly earnings. I continue on to select 500 samples from the population of 15,000 people and use linear regression to generate 500 slope coefficients and 500 constant coefficients predicting the relationship between length of employment and weekly earnings. And, as you might anticipate, when I plot the 500 slope coefficients and the 500 constant coefficients in histograms, a familiar shape appears. (See Figure 11.13.)

In previous chapters, you learned about sampling distributions of means and proportions. The histograms in Figure 11.13 are approximations of the sampling distributions of regression coefficients. In theory, the sampling distribution of a regression coefficient shows the distribution of the coefficients from every possible unique sample of the same size that can be selected from a population. The sampling distribution of a regression coefficient takes the shape of a t-distribution. Recall that for large samples (more than 500 cases), the t-distribution is approximately the same shape as the normal distribution. And, because of the central limit theorem and the law of large numbers, the sampling distribution of a regression coefficient centres on the population parameter. Notice that the histograms in Figure 11.3 centre on the slope coefficient ($1) and the constant coefficient ($100) in the simulated population.

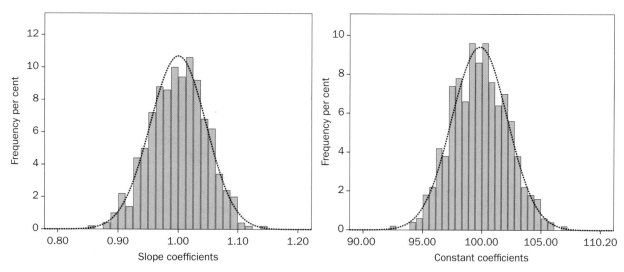

Figure 11.13 Histograms of Slope Coefficients and Constant Coefficients in 500 Samples from the Same Population (Simulated Data)

The standard error of a regression coefficient is the standard deviation of the sampling distribution of that coefficient. Like the standard error of a mean or a proportion, the standard error of a regression coefficient can be estimated using only information from the sample. Unlike the standard error of a mean or a proportion, though, calculating the standard error of a regression coefficient is a two-step process:

1. The first step in the process is to calculate the standard error of the regression overall, which is sometimes called the **standard error of the estimate**. The standard error of the estimate captures the overall accuracy of the predictions made by a regression.

2. The second step in the process is to calculate the standard error of the individual slope or constant coefficient. There are slightly different formulas for calculating the standard error of slope coefficients and the standard error of constant coefficients, but both incorporate the standard error of the estimate.

standard error of the estimate A statistic that captures the overall accuracy of the predictions made by a linear regression.

The standard error of the estimate captures how well a regression fits the data overall by determining how far the predicted value on the dependent variable ($\hat{y}_i$) is from the actual value for each case in the sample. This distance is the same as the residual for each case. In a familiar process, these residuals are squared to eliminate the negative values and then divided by the number of cases minus 2. Dividing by the number of cases minus 2, as opposed to the number of cases minus 1, accounts for the fact that the regression is predicting two parameters: (1) the constant and (2) the slope. Finally, the square root of the result is taken, to compensate for the earlier squaring. More formally, the formula for the standard error of the estimate is:

standard error of the estimate

$$se_{est} = \sqrt{\frac{\sum \left(y_i - \hat{y}_i\right)^2}{n-2}}$$

Once again, it is easiest to implement this formula using a table. For the hypothetical 10-case example of the relationship between years since immigration and annual employment income, we have already found the predicted employment income for each case to calculate the R^2. (See Figure 11.12; I've retained two decimals, though, in this calculation.) Figure 11.14 illustrates the process of calculating the sum of the squared distances from the predicted values (or the sum of the squared residuals).

The sum of the squared residuals can then be substituted into the numerator of the equation for the standard error of the estimate. For our hypothetical 10-case example, the standard error of the estimate is:

$$se_{est} = \sqrt{\frac{\sum \left(y_i - \hat{y}_i\right)^2}{n-2}}$$

$$se_{est} = \sqrt{\frac{215988372.09}{10-2}}$$

$$= \sqrt{26998546.51}$$

$$= 5196.01$$

$$\Sigma(y_i - \hat{y}_i)^2$$

Person (Case)	Actual Employment Income ($) y_i	Predicted Employment Income ($) $\hat{y}_i$	Deviation of Predicted Employment Income from Actual Employment Income $(y_i - \hat{y}_i)$	Squared Deviation of Predicted Employment Income from Actual Employment Income $(y_i - \hat{y}_i)^2$
Arjun	20,000	20,941.85	−941.85	887,081.42
Damia	23,000	22,953.48	46.52	2,164.11
Dmitri	25,000	26,976.74	−1,976.74	3,907,501.03
Priya	34,000	28,988.37	5,011.63	25,116,435.26
Zhang Li	31,000	28,988.37	2,011.63	4,046,655.26
Yusuf	27,000	33,011.63	−6,011.63	36,139,695.26
Jessica	36,000	35,023.26	976.74	954,021.03
Esengul	29,000	37,034.89	−8,034.89	64,559,457.31
Kwame	46,000	37,034.89	8,965.11	80,373,197.31
Alejandro	39,000	39,046.52	−46.52	2,164.11
			column sum:	215,988,372.09

The sigma symbol tells you to add the result of the equation for all of the cases.

Figure 11.14 Using a Table to Find the Sum of the Squared Distances from the Predicted Values (Hypothetical Data)

Once you have found the standard error of the estimate for the regression over-all, you can find the standard error of a slope coefficient. The standard error of a slope coefficient is calculated by dividing the standard error of the estimate by the amount of variation in the independent variable alone. This is captured by finding the square root of the sum of the squared deviations from the mean (or the square root of the sum of squares) of the independent variable. The formula for the standard error of a regression coefficient is:

$$se_b = \frac{se_{est}}{\sqrt{SS_x}} \quad or \quad se_b = \frac{se_{est}}{\sqrt{\Sigma(x_i - \bar{x})^2}}$$

standard error of a slope coefficient

We already know that the sum of the squared deviations from the mean of the independent variable in the hypothetical 10-case example is 86. (See the calculation of the slope coefficient in Figure 11.9.) So the standard error of the slope coefficient in the example is:

$$se_b = \frac{se_{est}}{\sqrt{SS_x}} \quad or \quad se_b = \frac{se_{est}}{\sqrt{\Sigma(x_i - \bar{x})^2}}$$

$$se_b = \frac{5196.01}{\sqrt{86}}$$

$$= \frac{5196.01}{9.27}$$

$$= 560.52$$

The standard error of a constant coefficient in a simple linear regression is calculated using the following formula, which also incorporates the sum of squares for the independent variable:

$$se_a = se_{est}\sqrt{\frac{1}{n} + \frac{(\bar{x})^2}{SS_x}} \quad or \quad se_a = se_{est}\sqrt{\frac{1}{n} + \frac{(\bar{x})^2}{\Sigma(x_i - \bar{x})^2}}$$

standard error of the constant coefficient (simple linear regression)

In the hypothetical 10-case example, the mean of the independent variable is five years and the sum of squares for the independent variable is 86. Thus, the standard error of the constant coefficient is calculated as:

$$se_a = se_{est}\sqrt{\frac{1}{n} + \frac{(\bar{x})^2}{SS_x}} \quad or \quad se_a = se_{est}\sqrt{\frac{1}{n} + \frac{(\bar{x})^2}{\Sigma(x_i - \bar{x})^2}}$$

$$se_a = 5196.01\sqrt{\frac{1}{10} + \frac{5^2}{86}}$$

$$= 5196.01 \sqrt{\frac{1}{10} + \frac{25}{86}}$$

$$= 5196.01 \sqrt{0.10 + 0.29}$$

$$= 5196.01 \sqrt{0.39}$$

$$= 5196.01 (0.62)$$

$$= 3221.53$$

So now we know that the slope coefficient in the hypothetical example is $2,011.63 and that its standard error is estimated to be 560.52. Similarly, we know that the constant coefficient in the hypothetical example is $20,941.85 and that its standard error is estimated to be 3,221.53. Knowing these standard errors makes it possible to calculate a t-statistic for each regression coefficient.

Since sampling distributions of regression coefficients have the shape of a t-distribution, t-statistics are used to estimate the likelihood of randomly selecting a sample with the observed relationship (or one of greater magnitude), if no relationship exists in the population. For slope coefficients, the research hypothesis is that there is a relationship between the two variables in the population; in other words, the research hypothesis is that the slope coefficient is not equal to 0. The null hypothesis is that no relationship exists between the two variables in the population; in other words, the null hypothesis is that the slope coefficient is equal to 0 and the line of best fit is perfectly horizontal. Recall that the calculation of the t-statistic relies on using standard errors as a unit of measure in order to estimate how far a statistic is from the centre of a sampling distribution that is centred on 0, that is, a sampling distribution for a population where the null hypothesis is true. To find the t-statistic for a regression coefficient, simply divide the coefficient by its standard error:

t-statistic for regression coefficients

$$t = \frac{b}{se_b} \quad or \quad t = \frac{a}{se_a}$$

The degrees of freedom of the t-distribution for a regression coefficient is the number of cases minus the number of independent variables minus 1. The amount of the area under the curve that is beyond the value of the t-statistic on the horizontal axis of the t-distribution shows the probability of randomly selecting a sample with a regression coefficient of the observed size or larger, if no relationship exists in the population.

For the slope coefficient in the hypothetical example, the t-statistic is:

$$t = \frac{b}{se_b}$$

$$t = \frac{2011.63}{560.52}$$

$$= 3.59$$

For the hypothetical example, the t-distribution has 8 degrees of freedom: 10 cases minus 1 independent variable minus 1. In Chapter 9, we established that, using an alpha value of 0.05, the critical value of the t-statistic for a t-distribution with 8 degrees of freedom is +/–2.31. Since 3.59 is beyond 2.31, I am relatively confident that there is a relationship between the number of years since immigration and annual employment income in the (hypothetical) population.

Let's briefly return to the simulation example and the sampling distribution of slopes in Figure 11.13 to illustrate how the t-statistic is interpreted in the context of regression. If no relationship exists between months of employment and weekly earnings in the hypothetical company (therefore, the null hypothesis is true), then the slope coefficient is 0 in the population. In the approximation of the sampling distribution of slope coefficients in the left panel of Figure 11.13, *all* of the samples had a slope coefficient between 0.85 and 1.15. There is an extremely low probability of randomly selecting a sample with a slope coefficient of 0 from the simulated population; the coefficient from such a sample would be located far into the left tail of the sampling distribution. Conversely, if the sampling distribution of slope coefficients is centred on 0, as would be the case if the null hypothesis is true, there would be an extremely low probability of randomly selecting a sample with a large slope coefficient (either positive or negative); the coefficient from such a sample would be located in the tails of the sampling distribution. The t-statistic shows the likelihood of randomly selecting a sample with the observed slope coefficient (or a larger slope coefficient) from a population where the sampling distribution of slope coefficients is centred on 0, using standard errors as units of measure.

T-statistics can also be calculated for constant coefficients, but the results are much less useful. For a constant coefficient, the research hypothesis is that a "0" value on the independent variable is related to a non-"0" value on the dependent variable; in other words, the research hypothesis is that the constant coefficient is not equal to 0 in the population. The null hypothesis is that a "0" value on the independent variable is related to a "0" value on the dependent variable; in other words, the null hypothesis is that the constant coefficient is equal to 0 in the population. Except in very specific research situations, this information isn't particularly useful to know, and so researchers usually do not report the results of t-tests for constant coefficients.

Step-by-Step: Standard Errors of Regression Coefficients

Note: In these calculations, do not include cases that are missing information for any variable used in the regression.

Standard Error of the Estimate

$$se_{est} = \sqrt{\frac{\sum \left(y_i - \hat{y}_i\right)^2}{n-2}}$$

Continued

Step 1: Count the total number of cases (n). Do not include cases that are missing information for any variable used in the regression.

Step 2: For *each case*, find the predicted value on the dependent variable ($\hat{y}_i$).

Step 3: For *each case individually*, subtract the predicted value on the dependent variable (from Step 2) from the actual value on the dependent variable for the case (y_i) to find the deviation from the predicted value. Complete this calculation once for each case.

Step 4: For *each case individually*, square the deviations from the predicted values (from Step 3). Complete this calculation once for each case.

Step 5: Add together the results of Step 4 for each case to find the numerator of the fraction in the standard error of the estimate equation.

Step 6: Subtract 2 from the total number of cases (from Step 1) to find the denominator of the fraction in the standard error of the estimate equation.

Step 7: Divide the result of Step 5 by the result of Step 6.

Step 8: Find the square root of the result of Step 7 to find the standard error of the estimate.

Standard Error of the Slope

$$se_b = \frac{se_{est}}{\sqrt{\Sigma(x_i - \bar{x})^2}}$$

Step 9: Find the mean of the independent variable in the relationship you are assessing ($\bar{x}$).

Step 10: For *each case individually*, subtract the mean of the independent variable (from Step 9) from the value on the independent variable for the case (x_i) to find the deviation from the mean. Complete this calculation once for each case.

Step 11: For *each case individually*, square the deviations from the mean (from Step 10). Complete this calculation once for each case.

Step 12: Add together the results of Step 11 for each case.

Step 13: Find the square root of the result of Step 12 to find the denominator of the standard error of the slope equation.

Step 14: Divide the standard error of the estimate (from Step 8) by the result of Step 13 to find the standard error of the slope coefficient.

Standard Error of the Constant

$$se_a = se_{est}\sqrt{\frac{1}{n} + \frac{(\bar{x})^2}{\Sigma(x_i - \bar{x})^2}}$$

Step 15: Square the mean of the independent variable (from Step 9)

Step 16: Divide the result of Step 15 by the result of Step 12.

Step 17: Divide 1 by the total number of cases (from Step 1).

Step 18: Add together the result of Step 16 and Step 17.

Step 19: Find the square root of the result of Step 18.

Step 20: Multiply the result of Step 19 by the standard error of the estimate (from Step 8) to find the standard error of the constant coefficient.

Step-by-Step: T-Statistics for Regression Coefficients

Step 1: Identify the research hypothesis and the null hypothesis for the coefficient that you are testing.

T-Statistic

$$t = \frac{b}{se_b} \quad \text{or} \quad t = \frac{a}{se_a}$$

Step 2: Identify the regression coefficient that you want to assess (a or b).

Step 3: Find the standard error of the coefficient that you identified in Step 2.

Step 4: Divide the coefficient by its standard error to find the t-statistic.

Degrees of Freedom

$$df_t = n - k - 1$$

Step 5: Count the total number of cases (n). Do not include cases that are missing information for any variable in the regression.

Step 6: Count the number of independent variables in the regression (k).

Step 7: Subtract 1 from the number of independent variables (from Step 6).

Step 8: Subtract the result of Step 7 from the total number of cases (from Step 5) to find the degrees of freedom of the t-statistic for a regression coefficient.

Statistical Significance

Step 9: Determine the alpha value you want to use for the t-test for the regression coefficient. The most common alpha value used in the social sciences is 0.05.

Step 10: Determine the critical value of the t-statistic, using the alpha value you selected in Step 9, for a t-distribution with the degrees of freedom you found in Step 8. If you are using an alpha value of 0.05 and the degrees of freedom are greater than 500, the critical value is +/−1.96. If you are using a different alpha value and/or have fewer

Continued

degrees of freedom, find the critical value using a printed table of the critical values of the Student's t-distribution, an online calculator, or the spreadsheet tool available from the Student Resources area of the companion website for this book.

Step 11: Compare the t-statistic (from Step 4) to the critical value (from Step 10). If the t-statistic is beyond the critical value, the result is statistically significant: you can reject the null hypothesis and state that there is likely a relationship between the independent variable and the dependent variable in the population. (If you are testing a constant coefficient, state that a 0 value on the independent variable is not likely to be associated with a 0 on the dependent variable in the population.) If the t-statistic is not beyond the critical value, the result is not statistically significant: you fail to reject the null hypothesis and state that you are not confident that there is a relationship between the independent variable and the dependent variable in the population. (If you are testing a constant coefficient, state that you are not confident that a 0 value on the independent variable is associated with a non-0 value on the dependent variable in the population.)

Confidence Intervals for Regression Coefficients

Another way that researchers use the standard errors of regression coefficients is to establish 95 per cent confidence intervals, which show the range that the population regression coefficient is likely to be within, with 95 per cent confidence. For large samples (greater than 500), the 95 per cent confidence interval for a regression coefficient is established by multiplying the standard error of a coefficient by 1.96 (which corresponds to 95 per cent of the area under the normal distribution/ t-distribution), then adding the result to the coefficient to determine the upper bound of the confidence interval, and subtracting the result from the coefficient to determine the lower bound of the confidence interval. These formulas should be familiar:

95% confidence intervals for regression coefficients

$$95\%CI_b = b \pm 1.96\left(se_b\right)$$

$$95\%CI_a = a \pm 1.96\left(se_a\right)$$

It's worth taking a moment to consider what the 95 per cent confidence intervals for slope and constant coefficients actually show. The 95 per cent confidence interval for the slope coefficient shows the bounds of the predicted angle of the line

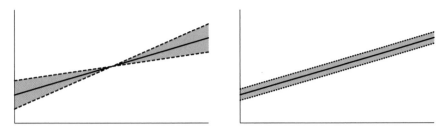

Figure 11.15 Understanding the 95 Per Cent Confidence Interval for Slope and Constant Coefficients

in the population. In the left panel of Figure 11.15, the solid line is the regression line produced by sample data, and the two dashed lines show the lines with slope coefficients at the upper and lower bounds of the 95 per cent confidence interval. The width of the 95 per cent confidence interval for the slope becomes wider at the ends of the line as a result of the different angles established by the upper and lower bounds. (Looking at Figure 11.5 again may help you to understand how this works.) So, the 95 per cent confidence interval for the slope coefficient indicates that the angle of the regression line in the larger population is likely to be somewhere within the purple shaded area in the left panel of Figure 11.15.

The 95 per cent confidence interval for the constant coefficient shows the upper and lower bounds of where the regression line is predicted to cross the vertical axis in the population. In the right panel of Figure 11.15, the solid line shows the height of the regression line produced by sample data, and the two dotted lines show the height of the lines at the upper and lower bounds of the 95 per cent confidence interval. (Looking at Figure 11.6 again may help you to understand how this works.) So, the 95 per cent confidence interval for the constant coefficient indicates that the height of the regression line in the larger population is likely to be somewhere within the purple shaded area in the right panel of Figure 11.15.

Figure 11.16 combines the two panels of Figure 11.15 to show the 95 per cent confidence interval for the regression line overall. Near the centre of the line, the width of the confidence interval captures the uncertainty in predicting the constant coefficient in the population. Near the ends of the line, the width of the confidence

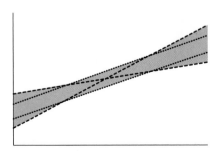

Figure 11.16 The 95 Per Cent Confidence Interval for a Regression Line (Slope and Constant Coefficients Combined)

interval captures the uncertainty in predicting the slope coefficient in the population. The combined effect of the confidence intervals for the slope and constant coefficients makes the 95 per cent confidence interval for a regression line overall appear curved, and wider at the ends of the line than near the centre.

Linear Regression in Action

So far, I've used a hypothetical 10-case example and a simulation example to illustrate the principles of simple linear regression. In this section, I use linear regression to investigate whether the length of time since immigration is related to recent immigrants' annual employment income, using the data from the 2011 National Household Survey introduced in Chapter 10. The research hypothesis is that there is a relationship between the number of years since immigration and employment income in 2010 among recent immigrants to Canada. The null hypothesis is that no relationship exists between the number of years since immigration and employment income in 2010 among recent immigrants to Canada. Table 11.1 shows the results of a simple linear regression that identifies the general pattern of the relationship between these two variables. Often, researchers do not report the standard error or t-statistic with regression results, but I have included them here.

The constant coefficient shows that people who immigrated to Canada less than a year ago (recorded as "0" years) are predicted to earn $25,558 in annual employment income. In the second row of Table 11.1, the slope coefficient of the "Years since immigration" variable indicates that every additional year since immigrating is associated with earning $1,066 more in annual employment income. The t-statistics for both coefficients are statistically significant, using an alpha value of 0.05. Thus, it is very unlikely that a sample with this observed relationship could have been randomly selected from a population in which no relationship exists between years since immigration and employment income in 2010. Of course, the slope and constant coefficients produced by this sample likely do not match exactly with the pattern of the straight-line relationship in the population. The upper and

Table 11.1   Results of a Simple Linear Regression Using NHS Data

Dependent variable: Employment income in 2010 ($; n = 26,621)

	Coefficient	Standard Error	T-Statistic	95% Confidence Interval	
				Lower Bound	Upper Bound
Constant	25,558.46*	281.45	90.81	25,006.80	26,110.11
Years since immigration (in 2010)	1,066.47*	46.60	22.89	975.13	1,157.81
R^2	0.019				

*indicates that results are statistically significant at the p < 0.05 level.
Source: Author generated; Calculated using data from Statistics Canada, 2013.

lower bounds of the 95 per cent confidence interval show the range that each of the regression coefficients is likely to be within for the regression line that represents the pattern of the relationship in the larger population. The width of the confidence intervals give researchers an idea of how precise their estimates are. In this example, the confidence intervals are narrow; thus, we are relatively confident that an additional year since immigration is associated with earning about $1,000 more in annual employment income in the population of recent immigrants to Canada.

The R-squared for this regression is 0.019, which indicates that only 1.9 per cent of the variation in recent immigrants' employment income in 2010 can be

How Does It Look in SPSS?

Linear Regression

The Linear Regression procedure, with 95 per cent "Confidence intervals" selected in the Statistics option, produces results that look like those in Image 11.1. Above these results, SPSS prints a "Variables Entered/Removed" box (not shown), which lists the independent variables that were used or entered in the regression and the independent variables that were removed; review this information to make sure that all of the requested variables were used.

Model Summary

Model	(B) R	(C) R Square	(D) Adjusted R Square	(E) Std. Error of the Estimate
1(A)	.139[a]	.019	.019	23630.67245

(F) a. Predictors: (Constant), Years since immigration (in 2010)

ANOVA[a]

Model		Sum of Squares	df	Mean Square	F	Sig.
1	Regression	292446135678.928	1	292446135678.928	523.713	.000[b]
(G)	Residual	14864363698020.623	26619	558408680.464		
	Total	15156809833699.550	26620			

(F) a. Dependent Variable: Employment income in 2010

b. Predictors: (Constant), Years since immigration (in 2010)

Coefficients[a]

Model		Unstandardized Coefficients (I) B	Unstandardized Coefficients (J) Std. Error	Standardized Coefficients (K) Beta	(L) t	(M) Sig.	(N) 95.0% Confidence Interval for B Lower Bound	(N) 95.0% Confidence Interval for B Upper Bound
1(H)	(Constant)	25558.458	281.450		90.810	.000	25006.802	26110.114
	Years since immigration (in 2010)	1066.472	46.602	.139	22.885	.000	975.130	1157.814

(F) a. Dependent Variable: Employment income in 2010

Image 11.1   An SPSS Linear Regression with One Independent Variable

Continued

A. Each regression model is assigned a number. These results are for the first (and only) model in this regression.

B. The "R" column for a simple linear regression shows the Pearson's correlation coefficient (r) for the two variables. This result ($r = 0.139$) indicates that there is a weak positive correlation between years since immigration and annual employment income for recent immigrants.

C. The "R Square" column shows the coefficient of determination for the regression. An R^2 of 0.019 indicates that the number of years since immigration explains 2 per cent of the variation in recent immigrants' annual employment earnings.

D. This column shows a variation on R^2; you'll learn more about the adjusted R^2 in Chapter 13.

E. This column shows the standard error of the estimate, which is used to calculate the standard errors of regression coefficients.

F. The footnotes below each table list the independent variables (predictors) and the dependent variable used in the regression. This regression includes only one independent variable (predictor).

G. The "Regression" sum of squares shows the amount of variation in the dependent variable that is explained by the independent variable; it is the numerator of the R^2 equation. The "Total" sum of squares shows the total variation in the dependent variable; it is the denominator in the R^2 equation. The "Residual" sum of squares shows the amount of unexplained variation in the regression; it is obtained by subtracting the explained variation from the total variation. The results of the ANOVA test show whether or not, as a group, the independent variables are likely to be related to the dependent variable in the population. These results are not usually reported.

H. The "Coefficient" table always begins with a row for the "Constant." The independent variables are listed in rows below the constant row.

I. The first row of the "B" column shows the constant coefficient (a). This regression predicts that people who immigrated less than a year ago (0 years ago) earn $25,558 in annual employment income. The subsequent rows in the "B" column show the slope coefficient of each independent variable. Each additional year since immigration is associated with earning $1,066 more in annual employment income.

J. This column shows the standard errors of the regression coefficients. The first row shows the standard error of the constant coefficient, and the subsequent rows show the standard error of each slope coefficient. The standard errors are used to calculate the t-statistic and the confidence intervals.

K. The "Beta" column shows the standardized slope coefficient of each independent variable. You'll learn more about standardized slope coefficients in Chapter 12.

L. This column shows the t-statistic for each coefficient. The t-statistic is calculated by dividing the coefficient by its standard error. Each t-statistic is evaluated in relation to a t-distribution, with degrees of freedom equal to the number of cases minus the number of independent variables minus 1.

M. The "Sig." column shows the p-value associated with each t-statistic. That is, it shows the likelihood of randomly selecting a sample with the observed relationship (or one of a greater magnitude), if no relationship exists between the independent variable and the dependent variable in the population. The p-value in the "Years since immigration" row indicates that there is less than a 0.1 per cent chance of selecting this sample from a population of recent immigrants in which no relationship exists between years since immigration and annual employment income ($p < 0.001$). The p-value in the "(Constant)" row shows the likelihood of selecting this sample if a "0" value on the independent variable is associated with a "0" value on the dependent variable in the population. The significance test is not usually discussed for the constant coefficient.

N. These columns show the 95 per cent confidence intervals for each regression coefficient. The width of the confidence interval indicates the range that the regression coefficient is likely to be within in the population.

explained by the length of time since immigration. This low R-squared suggests that there are likely many other things that influence recent immigrants' annual employment earnings beyond the length of time since immigration. Researchers have shown that the earnings of recent immigrants to Canada are influenced by gender, language skills, educational credentials (and whether they are recognized), country of origin, and place of residence (Frank et al. 2013, Phythian, Walters, and Anisef 2011).

Centring Independent Variables

At some point in reading this chapter, you may have noticed that it is awkward to predict the annual employment income of people who immigrated zero years ago. Because the constant coefficient shows the predicted value on the dependent variable when the value on the independent variable is "0", it can sometimes be difficult to discuss meaningfully. For instance, in a regression that uses the number of people living in a household to predict household income, the constant predicts the income of a household with zero people living in it. Similarly, in a regression that uses age as an independent variable, the constant predicts the value on the dependent variable for people who are age zero.

A strategy that some researchers use to avoid this interpretation problem is called *centring*. Centring refers to shifting the values on an independent variable so that "0" represents a common value or characteristic (the "Hands-on Data Analysis" box in this chapter illustrates how to do this). Then, the constant becomes the predicted value on the dependent variable for people with that common characteristic. Sometimes researchers centre variables on the mean or median of the variable, and other times they centre variables on another common and theoretically useful value (typically a whole number).

Hands-on Data Analysis

Centring a Variable

Researchers centre variables in order to make a regression constant coefficient more meaningful and easier to interpret. The first step in centring a variable is deciding which value will be used as the centre. This central value is assigned the value "0", and all of the other values are adjusted relative to it. Researchers usually select a central value by generating descriptive statistics for the variable and choosing a whole number near the mean or the median. (Alternatively, you can select a common or theoretically meaningful value to centre on.)

To create a centred variable, compute a new variable that is equal to the value on the original variable for each case, minus the value that you are centring on. In the example in this chapter, I centred the variable "Years since immigration" on 5, using this formula:

$$\text{YEARS\_IN\_CANADA\_CENTRED} = \text{YEARS\_IN\_CANADA} - 5$$

Before using your new centred variable in a regression, check that it was created correctly. To do this, find the mean, standard deviation, and number of valid cases for both the original variable and the new variable. The difference between the two variable means should be exactly equal to the value you centred on. The standard deviations for the two variables should be exactly the same, as should the number of valid cases. Once you are confident that the new variable is correctly centred, use it instead of the original uncentred variable in regression.

Let's illustrate how centring works by making the NHS example easier to interpret. The independent variable is the number of years since immigrating to Canada, which has a mean of 5.2. Because the exact mean of 5.2 years is awkward to discuss, it makes more sense to centre this variable on five years. To do this, the researcher assigns the value "0" to people who immigrated five years ago. Then, the values of all the other cases are adjusted in relation to the central value: people who immigrated six years ago are assigned the value "1" because they are one unit above the central value; people who immigrated seven years ago are assigned the value "2" because they are two units above the central value; and so on. Similarly, people who immigrated four years ago are assigned the value "–1" because they are one unit below the central value; people who immigrated three years ago are assigned the value "–2" because they are two units below the central value; and so on.

When a regression that uses a centred independent variable is compared to the same regression that uses an uncentred version of the same independent variable, the slope coefficients are identical. (See Table 11.2.) Only the constant coefficients are different. The constant coefficients show the predicted value on the dependent variable when the value on the independent variable is "0", and centring the independent variable changes what the value "0" represents in the data. The constant coefficient of the centred variable, shown on the right of Table 11.2, now

Table 11.2   Changes to Regression Coefficients after Centring the Independent Variable

Dependent variable: Employment income in 2010 ($; n = 26,621)

	Original Independent Variable			Centred Independent Variable		
		95% Confidence Interval			95% Confidence Interval	
	Coefficient	Lower Bound	Upper Bound	Coefficient	Lower Bound	Upper Bound
Constant	5,558.46*	25,006.80	26,110.11	30,890.82*	30,606.47	31,175.16
Years since immigration (in 2010)	1,066.47*	975.13	1,157.81	1,066.47*	975.13	1,157.81
R^2	0.019			0.019		

*indicates that results are statistically significant at the $p < 0.05$ level.
Source: Author generated; Calculated using data from Statistics Canada, 2013.

shows the predicted employment income for people who immigrated five years ago. To predict the employment income of people who immigrated six years ago, multiply the slope coefficient by 1 (because six years is one unit above the central value), and add it to the constant coefficient. To predict the employment income of people who immigrated four years ago, multiply the slope coefficient by –1 (because four years is one unit below the central value), and add it to the constant coefficient—in effect, subtracting the slope coefficient from the constant because multiplying a positive number and a negative number produces a negative number.

Figure 11.17 may help to illustrate the effect of centring the independent variable in this regression. The left panel of Figure 11.17 shows the regression line for the relationship between years since immigration and employment income in 2010 using the original (uncentred) version of the independent variable. The right panel of Figure 11.17 shows the regression line for the relationship between years since immigration and employment income in 2010 using the centred version of the independent variable. Notice that the relative position of the cases and the slope of the

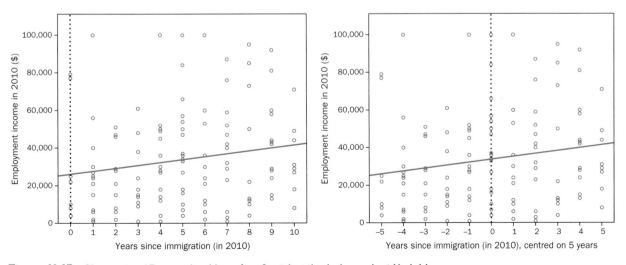

Figure 11.17   Change to a Regression Line after Centring the Independent Variable
Source: Author generated; Calculated using data from Statistics Canada, 2013.

regression line is the same in both panels, but the place where the regression line crosses 0 is different. In the right panel of Figure 11.17, the constant coefficient now shows the predicted employment income in 2010 for people who immigrated five years ago (which is now recorded as "0" in the data).

As you learn to build more complex regression models, you can use centring techniques to ensure that the constant coefficient provides a meaningful prediction, that is, it makes a prediction about people or cases with a meaningful set of characteristics. Adjusting independent variables in order to make the results easier to interpret is part of the skill of regression modelling.

Some Regression Assumptions

Linear regression relies on a series of assumptions. These assumptions can sometimes be worked around, or even ignored, but you should know what they are and how they might influence your results. First, as I indicated at the beginning of this chapter, linear regression is premised on the notion that a straight line is the best way to represent the pattern of a relationship. It does not detect non-linear patterns that might actually be a better fit.

Linear regression also relies on several assumptions about the errors, or residuals, that are produced. First, the errors are assumed to be normally distributed. Regression methods are relatively forgiving when this assumption is violated; however, if the distribution of the residuals is very far from normal, then you should pay attention to what is producing this result. When residuals are not normally distributed, it usually indicates that one or more of the variables in the regression are very far from being normally distributed. In this situation, researchers can mathematically transform the variable so that it becomes closer in shape to a normal distribution, or they can use a different type of regression technique.

The errors, or residuals, produced by a regression are also assumed to be independent of one another. In other words, the error associated with predicting the value on the dependent variable for one case is assumed to be completely unrelated to the error that is associated with predicting the value on the dependent variable for another case. Correlated errors occur most often when cases are clustered together in some way in the data but this is not accounted for in the analysis. For instance, if there are four people who all live in the same household, and a researcher attempts to predict each of their household incomes using a household characteristic (such as the number of people living in the household), the errors associated with the four cases will be correlated (in fact, they will be identical). Correlated errors also occur when data are collected over time, such as in panel studies. Specialized regression techniques must be used in situations when the errors are correlated.

Another assumption about the errors produced by linear regression is that the variation or dispersion of the errors is consistent across all of the values on the dependent variable (i.e., the relationship is homoscedastic). When the spread of the errors is related to the values on the dependent variable, this suggests that a regression makes systematically better predictions for some types of people (or things) than others. For instance, in the NHS example used in this chapter, it would be a concern if the prediction errors were very small for recent immigrants with low employment incomes but very large for recent immigrants with high employment incomes. A related assumption is that the variation or dispersion of the errors is consistent across all of the values on the independent variable. Whenever these assumptions are violated, a common solution is to divide the sample into subgroups, and run two or more separate regressions in order to ensure that the spread of the errors is consistent within each one.

Finally, the errors should not be correlated with an independent variable. Such an occurrence typically indicates that important predictors (that are themselves correlated with that independent variable) are omitted from the regression. You'll learn more about how to use more than one independent variable in a regression in Chapter 12.

These assumptions—a linear relationship, normally distributed errors, independent errors, homoscedastic errors, and errors uncorrelated with the independent variable—are key to making inferences using linear regression. When these assumptions are violated, the predictions made by linear regressions can become biased or inaccurate (although not always).

Best Practices in Presenting Results

Writing about Regression Results

One strength of regression is that it allows researchers to make predictions about the general pattern of a relationship in a population. When writing about regression results, however, you must acknowledge that the statistical results produced by sample data are only estimates or predictions, and use language that reflects this. Consider the following reports of a constant coefficient:

> *Incorrect:* People who immigrated to Canada less than a year ago earn $25,558 in annual employment income.
> *Correct:* People who immigrated to Canada less than a year ago are predicted to earn $25,558 in annual employment income.
> *Correct:* This regression estimates that people who immigrated to Canada less than a year ago earn $25,558 in annual employment income.

The first statement is incorrect because it is too definitive; is does not reflect the idea that coefficient reflects a prediction based on a general pattern in the data. In addition, when writing about regression coefficients, be careful to avoid language that implies a causal relationship; use the language of association instead. Consider the following reports of a slope coefficient:

> *Incorrect:* Among recent immigrants to Canada, each additional year since immigration leads to earning $1,066 more in annual employment income.
> *Correct:* Among recent immigrants to Canada, each additional year since immigration is associated with earning $1,066 more in annual employment income.

Again, the first statement is incorrect because it is too definitive and suggests that a change in the independent variable leads directly to a change in the dependent variable.

Remember that when an independent variable is centred, it only affects the interpretation of the constant coefficient, not the slope coefficient. As a result, indicating that a variable is centred is not necessary when reporting slope coefficients. But you *must* include information about centring when reporting constant coefficients for regressions with centred independent variables. Here are some examples of how to incorrectly and correctly report the regression results on the right of Table 11.2:

> *Incorrect:* Among recent immigrants to Canada, each additional year since immigration, centred on five, is associated with earning $1,066 more in annual employment income.
>
> *Incorrect:* After five years since immigrating to Canada, each additional year since immigration is associated with earning $1,066 more in annual employment income.
>
> *Correct:* Among recent immigrants to Canada, each additional year since immigration is associated with earning $1,066 more in annual employment income.
>
> *Correct:* People who immigrated to Canada five years ago are predicted to earn $30,891 in annual employment income.

The first two statements are incorrect because they suggest that the centring on five years is key to interpreting the slope coefficient. The second statement is particularly problematic because it suggests that the length of time since immigration is only related to annual employment income for people who immigrated at least five years ago. The second two statements are correct because they report the effect of centring in the appropriate context: there is no change to the interpretation of the slope coefficient, but the interpretation of the constant coefficient does change.

As you learned in Chapter 6, some caution is also warranted when writing about 95 per cent confidence intervals. When reporting 95 per cent confidence intervals for regression coefficients, be sure to indicate that they represent the potential range of the coefficient in 95 per cent of *samples*, not 95 per cent of *people* in the population. Here are some examples of incorrect and correct ways to report 95 per cent confidence intervals:

> *Incorrect:* The confidence interval for the slope coefficient shows that 95 per cent of recent immigrants to Canada earn between $975 and $1,158 more in annual employment income for each additional year since immigration.
>
> *Incorrect:* Among recent immigrants to Canada, each additional year since immigration is associated with earning $975 to $1,158 more in annual employment income, 95 per cent of the time.

Correct: Among recent immigrants to Canada, each additional year since immigration is associated with earning more annual employment income; we are 95 per cent confident that the amount of additional employment income associated with each additional year since immigration is between $975 and $1,158.

Correct: Among recent immigrants to Canada, each additional year since immigration is associated with earning $1,066 more in annual employment income (95% CI: $975–$1,158).

The first two statements are incorrect because they suggest that the confidence interval ($975 to $1,158) shows variation between people or variation over time. The final two statements are better because they correctly indicate that the confidence interval shows the amount of uncertainty associated with the prediction.

Finally, it is easy to misinterpret the R^2 of a regression. Recall that the R^2 illustrates how well a regression line fits the data by showing the percentage of the variation in the dependent variable that can be explained by the independent variable. When reporting the R^2 statistic, you must refer to the variation within the dependent variable, not the actual values on the dependent variable. Once again, here are some examples of correct and incorrect ways to present this information:

Incorrect: The R^2 shows that we can predict 1.9 per cent of the annual employment income of recent immigrants if we know how long ago they immigrated to Canada.

Incorrect: The R^2 shows that we can predict the annual employment income of 1.9 per cent of recent immigrants if we know how long ago they immigrated to Canada.

Correct: The R^2 shows that we can predict 1.9 per cent of the variation in the annual employment income of recent immigrants if we know how long ago they immigrated to Canada.

The first statement is incorrect because it suggests that we can predict a proportion of recent immigrants' income, not the variation—or differences—in income. The second statement is incorrect because it suggests that we can predict the income of a specific proportion of recent immigrants. Only the final statement correctly indicates that knowing the value on the independent variable helps to explain a proportion (or a percentage) of the variation in the dependent variable.

At first, writing about regression results using these strategies might seem awkward. It's easiest to begin by just modifying these sentences to incorporate information from the regression you are reporting on. With practice, though, you will be able to re-word and re-structure these claims. Being aware of the common errors that people make when reporting regression results will help you to avoid repeating these mistakes in your own work, as you develop your statistical writing skills.

What You Have Learned

This chapter introduced simple linear regression and explained how it is used to understand the general pattern of a relationship between two ratio-level variables. You learned how a line of best fit is established and how to calculate and interpret the slope and constant coefficients that define this line. Slope coefficients are used to assess the magnitude of a relationship between two variables. You also learned to assess the reliability of regression coefficients using the t-statistic, and you were introduced to R^2 as a measure of the overall fit of a regression. Finally, you learned some correct and incorrect ways to report regression results. Together, these ideas and skills provide the conceptual background needed to understand multiple linear regression, which is introduced in Chapter 12.

This chapter used the same research example as Chapter 10 in order to illustrate the conceptual relationship between correlation and regression. The regression results support the conclusion made in Chapter 10: that the length of time since immigration is not associated strongly with recent immigrants' annual employment income. Indeed, the length of time since immigration only explains about 2 per cent of the variation in recent immigrants' annual employment income. And, although it appears that more time since immigration is associated with higher annual employment incomes, the magnitude of the relationship is relatively modest: about $1,000 per additional year since immigration. As noted in Chapter 10, it's very likely that other socio-demographic and job characteristics influence the employment earnings of recent immigrants to Canada. In chapters 12 and 13, we'll use multiple linear regression to investigate how several other characteristics are related to people's employment income.

Check Your Understanding

Check to see if you understand the key concepts in this chapter by answering the following questions:

1. What does it mean to use the "ordinary least squares" method to determine a line of best fit?
2. What does a constant coefficient show? What does a slope coefficient show?
3. What does the explained variation in a regression refer to?
4. What is a residual? How do you find the residual for a specific case?
5. What is the logic of R^2? Why is R^2 a proportionate reduction in error measure?
6. What do the 95 per cent confidence intervals for regression coefficients show?
7. Why do researchers centre independent variables before using them in regression?

Practice What You Have Learned

Check to see if you can apply the key concepts in this chapter by answering the following questions. Keep two decimal places in any calculations.

1. As described in question 4 of the "Practice What You Have Learned" section in Chapter 10, a local settlement agency in a small town surveyed a random sample of 12 recent immigrants to get more information to bolster a campaign to attract more immigrants to the town. In addition to collecting information about satisfaction and social integration, the agency also asked recent immigrants how many months they have lived in the town, how many years of work experience they have (both in and outside of Canada), and their hourly wages. The results are shown in Table 11.3. A settlement agency worker wants to investigate whether or not people who have lived in the town longer earn higher hourly wages.

a. Determine which variable should be treated as independent and which should be treated as dependent.

b. State a non-directional research hypothesis that the agency worker can test.

c. State the null hypothesis associated with the research hypothesis you identified in (b).

Table 11.3 Additional Results from Survey of Immigrants (Hypothetical Data)

Person	Months Living in the Town	Years of Work Experience	Hourly Wage
Salman	7	16	$10.00
Farook	35	2	$20.50
Alisha	24	13	$11.00
Ayaan	9	11	$21.00
Dorian	34	1	$29.50
Kyi Kyi	25	2	$25.00
Ayesha	22	3	$19.00
Mélanie	13	10	$16.50
Michael	9	4	$25.00
Carlotta	14	10	$16.00
Felipe	26	6	$40.50
Mahmoud	16	12	$30.00

2. Using the information in Table 11.3:

a. Calculate the slope coefficient of the regression line that captures the general pattern of the relationship between the number of months living in the town and hourly wages.

b. Explain what the slope coefficient shows. How would you describe the magnitude of the relationship between these two variables?

3. Using the information in Table 11.3 and your answer to question 2(a):

a. Calculate the constant coefficient of the regression line that captures the general pattern of the relationship between the number of months living in the town and hourly wages.

b. Explain what the constant coefficient shows.

4. The agency worker wants to use the regression results to make some predictions about people's hourly wages,

based on the length of time they have lived in the town. Using your answers to questions 2(a) and 3(a):

a. Calculate the predicted hourly wages for a recent immigrants who have lived in the town for exactly 12 months.

b. Calculate the predicted hourly wages for recent immigrants who have lived in the town for exactly 24 months.

5. As it turns out, Alisha has lived Canada for exactly 24 months. Using your answer to question 4(b), find the residual of Alisha's case.

6. Add a column to the end of Table 11.3 that lists the predicted hourly wages of all 12 people, based on the number of months they have lived in the town, calculated using your answers to questions 2(a) and 3(a).

7. The agency worker wants to know whether the relationship between immigrants' number of months living in the town and their hourly wages is likely to exist in the population.

a. Using your answer to question 6, find the standard error of the estimate.

b. Using your answer to (a), find the standard error of the slope coefficient.

8. Using your answer to question 7(b):

a. Calculate the t-statistic associated with the slope coefficient.

b. Determine the degrees of freedom of the t-distribution for this slope coefficient.

c. For a t-statistic with this many degrees of freedom, the critical value is +/–2.23. Based on your answer to (a), state whether you reject or fail to reject the null hypothesis. Is the relationship between immigrants' number of months living in the town and their hourly wages likely to exist in the population?

9. The settlement agency worker is also interested in knowing whether recent immigrants with more years of work experience (both in and outside of Canada) earn higher hourly wages.

a. Determine which variable should be treated as independent and which should be treated as dependent.

b. State a non-directional research hypothesis that the agency worker can test.

c. State the null hypothesis associated with the research hypothesis you identified in (b).

10. Using the information in Table 11.3:

a. Calculate the slope coefficient of the regression line that captures the general pattern of the relationship between years of work experience and hourly wages.

b. Explain what the slope coefficient shows. How would you describe the magnitude of the relationship between these two variables?

11. Using the information in Table 11.3 and your answer to question 10(a):

a. Calculate the constant coefficient of the regression line that captures the general pattern of the relationship between years of work experience and hourly wages.

b. Explain what the constant coefficient shows.

12. Add a final column to the end of Table 11.3 that lists the predicted hourly wages of all 12 people, based on their years of work experience, calculated using your answers to questions 10(a) and 11(a).

13. The agency worker wants to know whether the relationship between immigrants' years of work experience and their hourly wages is likely to exist in the population.

a. Using your answer to question 12, find the standard error of the estimate.

b. Using your answer to (a), find the standard error of the slope coefficient.

14. Using your answer to question 13(b):

a. Calculate the t-statistic associated with the slope coefficient.

b. Determine the degrees of freedom of the t-distribution for this slope coefficient.

c. For a t-statistic with this many degrees of freedom, the critical value is +/–2.23. Based on your answer to (a), state whether you reject or fail to reject the null hypothesis. Is the relationship between immigrants' years of work experience and their hourly wages likely to exist in the population?

15. The agency worker wants to know how much of the variation in hourly wages can be explained by people's years of work experience. Calculate the R^2 of this regression. Explain what the result shows.

16. The graph in Figure 11.18, excerpted from a Statistics Canada publication, shows how the low-income rate of immigrants compares to those of the Canadian-born. Recent immigrants are divided into three groups: people who have been in Canada

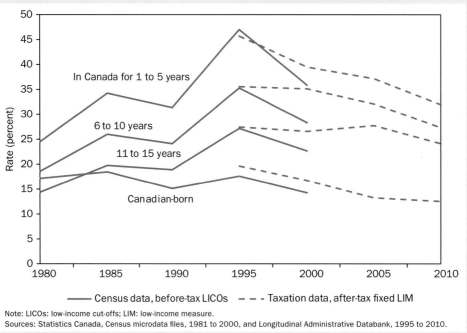

Note: LICOs: low-income cut-offs; LIM: low-income measure.
Sources: Statistics Canada, Census microdata files, 1981 to 2000, and Longitudinal Administrative Databank, 1995 to 2010.

Figure 11.18   **Low-Income Rates of Immigrants, Canada, 1980 to 2010**

for 1 to 5 years, 6 to 10 years, and 11 to 15 years. The graph displays the per cent of people below the low-income cut-off (LICO) for the years 1980–2000, and the per cent of people below the low-income measure (LIM) for the years 1995–2010. (See Chapter 1 for a description of the differences between these two measures.)

a. Overall, how does the low-income rate of recent immigrants compare to that of the Canadian-born?

b. How does the number of years since immigration affect recent immigrants' low-income rate? How has this relationship changed over time?

Practice Using Statistical Software (IBM SPSS)

Answer these questions using IBM SPSS and the GSS27. sav or the GSS27_student.sav dataset available from the Student Resources area of the companion website for this book. Weight the data using the "Standardized person weight" [STD_WGHT] variable you created following the instructions in Chapter 5. Report two decimal places in your answers, unless fewer are printed by IBM SPSS. It is imperative that you save the dataset to keep any new variables that you create.

1. Researchers often need to modify variables before using them in regression. In the GSS 27, people are asked whether they were a member of or participant in many different types of groups during the past year: unions, professional associations, political parties, sports teams, hobby groups, religious groups, school/neighbourhood groups, civic/community associations, service clubs, seniors' groups, youth groups, ethnic associations or any other type of organization or group. People who indicated that they participated in at least one type of group were asked how many different groups they participated in, and the responses are recorded in the variable "Number of groups - 12 months" [GRP_10C]. People who did not participate in any type of group were assigned a "Valid skip" (96) attribute on this variable.

    a. Use the Recode into Different Variables tool to recode the original variable "Number of groups - 12 months" [GRP_10C] into a new variable called "Number of groups - 12 months (recoded)" [GRP_10C_RECODED]. Assign the old value 96 ("Valid skip") the new value 0, and copy the old value for all of the other values.

    b. In the *Variable View*, alter the missing values of the "Number of groups - 12 months (recoded)" [GRP_10C_RECODED] variable so that the values "97" through "99" are treated as missing.

    c. Produce frequency distributions of the original variable "Number of groups - 12 months" [GRP_10C] and the new variable "Number of groups - 12 months (recoded)" [GRP_10C_RECODED], and compare them to be sure that the recoding is correct.

2. Use the Linear Regression procedure to produce a regression of the independent variable "Number of groups - 12 months (recoded)" [GRP_10C_RECODED] on the dependent variable "Number of close friends" [SCF_100C].

    a. Interpret the constant and slope coefficient of the regression. What do they show about the relationship between the number of groups people participate in and their number of close friends?

    b. Interpret the R^2 of this regression.

    c. Interpret the t-statistic and its associated p-value for the slope coefficient. Is there likely to be a relationship between the number of groups people participate in and their number of close friends in the population?

3. Use the Statistics option in the Linear Regression procedure to generate 95 per cent confidence intervals for the coefficients of the regression of the independent variable "Number of groups – 12 months (recoded)" [GRP_10C_RECODED] on the dependent variable "Number of close friends" [SCF_100C]. Explain what the confidence intervals show.

4. Use the Chart Builder tool to produce a scatterplot of the relationship between people's "Number of groups – 12 months (recoded)" [GRP_10C_RECODED] and their "Number of close friends" [SCF_100C]. Place the independent and dependent variable on the correct axes.

   a. Adjust the scale on the axis showing "Number of close friends" so that the maximum is 50, and the major increment is 10.
   b. Use the Chart Editor to add a "Fit line at total" to the graph. This displays the regression line.
   c. The label on the fit line shows the equation that defines the line. Describe how the linear equation is related to the regression coefficients that you produced in question 2.

5. Use the Linear Regression procedure to produce a regression of the independent variable "Number of new people met - Past month" [SCP_110] on the dependent variable "Number of close friends" [SCF_100C].

   a. Interpret the constant and slope coefficient of the regression. What do they show about the relationship between the number of new people that a person meets and their number of close friends?
   b. Interpret the t-statistic and its associated p-value for the slope coefficient. Is there likely to be a relationship between the number of new people that a person meets and their number of close friends in the population?

6. For the regression you produced in question 5:

   a. Interpret the R^2.
   b. Explain how the R^2 is related to the corresponding Pearson's correlation coefficient that you produced in question 8 of "Practice Using Statistical Software" in Chapter 10.

7. Use the Statistics option in the Linear Regression procedure to generate 95 per cent confidence intervals for the coefficients of the regression of the independent variable "Number of new people met - Past month" [SCP_110] on the dependent variable "Number of close friends" [SCF_100C]. Explain what the confidence intervals show.

8. Create a new variable that centres the original variable "Number of new people met - Past month" [SCP_110] on its mean.

   a. Find the mean of "Number of new people met - Past month" [SCP_110]. Round the mean to the nearest whole number; this is the value that you will centre the variable on.
   b. Use the Compute Variable tool to create a new variable, called "Number of new people met - Past month (centred)" [SCP_110_CENTRED]. The value on the new variable for each case should be the value on the original variable, minus the mean (rounded to the nearest whole number).
   c. Use the Means procedure to find the mean of the original variable "Number of new people met - Past month" [SCP_110] and the new variable "Number of new people met - Past month (centred)" [SCP_110_CENTRED]. Compare the results to be sure that the centring is correct: the difference between the means should be the same as the whole number that you centred the variable on, and the number of cases and the standard deviations should be the same for both variables.

9. Use the Linear Regression procedure to produce a regression of the independent variable "Number of new people met - Past month (centred)" [SCP_110_CENTRED] on the dependent variable "Number of close friends" [SCF_100C].

   a. Compare the results to those you produced in question 5. Determine which parts of the regression output are identical and which are different.
   b. Interpret the constant and slope coefficient of the regression. What do they show about the relationship between the number of new people that a person meets and their number of close friends?

Key Formulas

Simple linear regression predictions

$$\hat{y} = a + bx$$

Simple linear regression with residuals

$$y_i = a + bx_i + e_i$$

Slope coefficient, simple linear regression (version 1)

$$b = \frac{SP}{SS_x}$$

Slope coefficient, simple linear regression (version 2)

$$b = \frac{\Sigma(x_i - \bar{x})(y_i - \bar{y})}{\Sigma(x_i - \bar{x})^2}$$

Constant coefficient, simple linear regression

$$a = \bar{y} - b(\bar{x})$$

R^2 (coefficient of determination)

$$R^2 = \frac{\Sigma(\hat{y}_i - \bar{y})^2}{\Sigma(y_i - \bar{y})^2}$$

Standard error of the estimate

$$se_{est} = \sqrt{\frac{\Sigma(y_i - \hat{y}_i)^2}{n - 2}}$$

Standard error of a slope coefficient (version 1)

$$se_b = \frac{se_{est}}{\sqrt{SS_x}}$$

Standard error of a slope coefficient (version 2)

$$se_b = \frac{se_{est}}{\sqrt{\Sigma(x_i - \bar{x})^2}}$$

Standard error of the constant coefficient, simple linear regression (version 1)

$$se_a = se_{est}\sqrt{\frac{1}{n} + \frac{(\bar{x})^2}{SS_x}}$$

Standard error of the constant coefficient, simple linear regression (version 2)	$se_a = se_{est}\sqrt{\dfrac{1}{n}+\dfrac{(\bar{x})^2}{\Sigma(x_i-\bar{x})^2}}$
95% confidence intervals for regression coefficients (large samples)	$95\%CI_b = b \pm 1.96(se_b)$ $95\%CI_a = a \pm 1.96(se_a)$
T-statistic for regression coefficients	$t=\dfrac{b}{se_b}$ or $t=\dfrac{a}{se_a}$
Degrees of freedom of the t-statistic for regression coefficients	$df_t = n-k-1$

References

Frank, Kristyn, Kelli Phythian, David Walters, and Paul Anisef. 2013. "Understanding the Economic Integration of Immigrants: A Wage Decomposition of the Earnings Disparities between Native-Born Canadians and Recent Immigrant Cohorts." *Social Sciences* 2 (2): 40–61. doi:10.3390/socsci2020040.

Houle, René, and Grant Schellenberg. 2010. *New Immigrants' Assessments of Their Life in Canada.* Research Paper Series/Analytical Studies no. 322. Ottawa: Statistics Canada.

Phythian, Kelli, David Walters, and Paul Anisef. 2011. "Predicting Earnings among Immigrants to Canada: The Role of Source Country: Immigrant Origins and Employment Earnings." *International Migration* 49 (6): 129–54. doi:10.1111/j.1468-2435.2010.00626.x.

Picot, Garnett, and Feng Hou. 2014. "Immigration, Low Income and Income Inequality in Canada: What's New in the 2000s?" Catalogue no. 11F0019M—No. 364. Ottawa: Statistics Canada. http://www.statcan.gc.ca/pub/11f0019m/11f0019m2014364-eng.pdf.

Statistics Canada. 2013. *National Household Survey, 2011.* Ottawa: Statistics Canada. http://www12.statcan.gc.ca/nhs-enm/2011/dp-pd/prof/index.cfm?Lang=E

Linear Regression with Multiple Independent Variables

Learning Objectives

In this chapter, you will learn:

- What it means to "control for" variables in a regression

- How to interpret regression coefficients when there is more than one independent variable

- How to calculate and interpret standardized slope coefficients

- What dummy variables are and how they are used in regression

- How to interpret the slope coefficients of dummy variables

- How to write about regression results

Introduction

In this chapter, you will learn about one of the most powerful features of regression, which is the ability to simultaneously use two or more independent variables to make predictions about a dependent variable. In Chapter 11, you were introduced to simple linear regression and learned how slope and constant coefficients are used to describe the general pattern of a relationship between one ratio-level independent variable and a ratio-level dependent variable. This approach can be extended to describe the general pattern of a relationship between several independent variables and a ratio-level dependent variable. Because it uses more than one independent variable, this type of regression is called multiple linear regression. Researchers use multiple linear regression to create complex models in order to better understand the social world.

Once again, the research focus of this chapter is employment income—specifically wages—but the sample includes all employees in Canada, not just recent immigrants. In this chapter and the next, we will consider how several different characteristics are related to people's annual wages. First, we will investigate how the number of hours that people work at their job(s) and their seniority with an employer are related to their annual wages, in order to illustrate the logic of controlling for variables. Then, we will focus on how having a disability is related to annual wages, even after controlling for the number of hours that people work, their seniority, and their level of education.

As the World Health Organization notes, "disability is part of the human condition—almost everyone will be temporarily or permanently impaired at some point in life" (WHO 2011). In 2014, in Canada, two in ten people aged 25 to 64 (21 per cent)—or 4.9 million people—were substantially limited in their daily activities. The most common types of disabilities are those related to pain, flexibility, and mobility (Arim 2015). Although some people are born with disabilities, most are acquired later in life as a result of ageing, accidents, or the onset of illness; among people with disabilities aged 15 to 64, only 13 per cent reported that their disability had existed at birth (Turcotte 2014). The Canadian Human Rights Act (1985) explicitly prohibits discrimination on the grounds of disability, and the federal Employment Equity Act (1995) identifies people with disabilities as a group that is disadvantaged in the workplace. In 2014, only 60 per cent of working-age people with disabilities were employed, compared to 86 per cent of working-age people without disabilities. Advocacy groups, such as the Council of Canadians with Disabilities, make the link between employment discrimination and higher levels of poverty among people with disabilities (Crawford 2010).

Defining what it means to "have a disability" is difficult. Some disability advocates reject the label *disability* entirely because it implies reduced abilities or an absence of ability; they prefer the term *differently abled*. Others use the word *disabled* to foreground the fact that people are actively made less able (or are disabled) by social and physical structures that do not account for the full range of individual human variation. This approach reflects a social model of disability, which asserts that disabilities result from the way that able-bodied society organizes its physical, political, economic, and social relationships (MacKenzie, Hurst, and Crompton 2009, 55–56). The United Nations Convention on the Rights of Persons with Disabilities specifies that persons with disabilities include those who have "long-term physical, mental, intellectual, or sensory impairments" that interact with various attitudinal and environmental barriers to "hinder their full and effective participation in society on an equal basis with others" (United Nations 2008). In Canadian surveys, the measurement of disability relies on assessing how much difficulty a person has doing everyday tasks and how often these difficulties limit their daily activities; this measurement does not consider whether a respondent self-identifies as a person with a disability.

The analyses in this chapter use data from the Canadian Income Survey (CIS), which identifies people with disabilities in five main areas:

1. Sensory disabilities (related to hearing or seeing)
2. Physical disabilities (related to mobility, flexibility, dexterity, or pain)
3. Cognitive disabilities (related to learning, memory, or development)
4. Mental-health disabilities (related to emotional or psychological conditions)
5. Other long-term health problems or conditions

Only conditions that have lasted or are expected to last for six months or more are recorded. People whose daily activities are limited "Sometimes," "Often" or "Always" by these conditions, as well as people whose daily activities are limited

"Rarely" but who experience "A lot" of difficulty or "Cannot do" something are considered to have a disability. The most commonly reported disabilities are physical; relatively few people have cognitive disabilities, and less than 1 per cent of adults in Canada have developmental disabilities (Arim 2015).

Crucially, the statistical analyses in this chapter include only people who are employees and exclude people who are unemployed, self-employed, or not in the labour market. The experiences of people with disabilities who have not been able to obtain employment or who have left the labour market are not captured in these results. Since people with more severe disabilities are less likely to be formally employed (Turcotte 2014), these results likely reflect the influence of having a mild or moderate disability on employment earnings. Unfortunately, the survey data used in these analyses do not include a measure of the severity of a disability.

In this chapter, statistical analysis is used to discover the following:

- Are people's annual wages related to whether or not they have a disability?
- How does the relationship between people's disability status and their annual wages change once the number of hours they work, their seniority, and their level of education are taken into account?

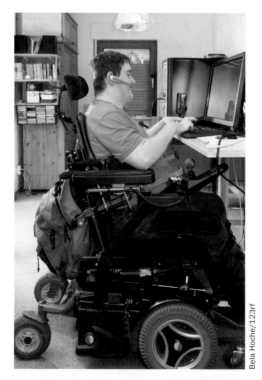

Photo 12.1   **Approximately 2.4 million employees in Canada have long-term disibilities or health problems that limit their daily activities.**

Multiple Linear Regression

In Chapter 11, you learned that simple linear regression is used to identify the general pattern of a relationship between two ratio-level variables. But social science researchers are usually interested in things that are influenced by more than one characteristic or social process. One way to assess how multiple independent variables are related to a dependent variable is to perform a series of simple linear regressions that each predict the relationship between a different independent variable and the same dependent variable. To illustrate this approach and its limitations, let's look at a series of regressions that predict people's annual wages, using data from the Canadian Income Survey. (See the "Spotlight on Data" box for more information.)

Spotlight on Data

The Canadian Income Survey

The Canadian Income Survey (CIS) is designed to provide information about the income sources of people and households in Canada. In addition to income data, the CIS collects information about topics such as housing and disabilities. Households are selected to participate in the CIS from the Labour Force Survey sample (described in Chapter 4). The data include some variables taken from answers to Labour Force Survey questions (such as those relating to education and job characteristics) and some variables derived from people's income-tax records (such as those relating to the amount of income from various sources and the use of government benefits).

CIS data are collected using computer-assisted telephone interviewing immediately after a person completes the Labour Force Survey. The CIS does not collect information from people living in the territories, residents of institutions, people living on reserves or other Aboriginal settlements, or members of the Canadian Forces living in military camps (Statistics Canada 2016). CIS data about the 2014 year was collected between January and March 2015. In 2014, 70 per cent of households that were selected to participate in the CIS actually did so. The questions relating to disabilities, which are used in the analyses in this chapter, are asked about one randomly selected person aged 16 or older in each household. The data are weighted to match the distribution of the larger Canadian population in relation to age, sex, household size, and economic family size for each province, as well as the wage distribution in the larger population (generated from Canada Revenue Agency's taxation files).

The statistical analyses in this chapter include only people who were formally employees during the 2014 year, and exclude the self-employed and unpaid family workers. Annual wages refer to people's wages and salaries in all jobs in 2014, before deductions. The 1 per cent of respondents who report earning more than $200,000 in wages and salaries are excluded from this analysis, as are people who reported earning $0 in wages and salaries. The "Total usual hours worked at all jobs" variable captures the total number of hours that people usually worked for pay at all of their jobs during 2014, but excludes both paid and unpaid overtime hours. About 5 per cent of employees in Canada hold more than one job. For readability, throughout this chapter I generally refer to people as working at a single job, although people with multiple jobs are included in the analyses. Seniority is measured as the number of consecutive months that people have worked for their current employer (or their most recent employer). For people who changed employers during 2014, the job that seniority is recorded for may not be the job that most wages were earned from in 2014. The CIS data used in this chapter were accessed at a Statistics Canada Research Data Centre. (See the "Spotlight on Data" box in Chapter 1 for more information about Research Data Centres.)

On average, the employees included in this analysis earn $45,477 (s.d. = $34,970) per year, working in one or more jobs. Let's begin by investigating the relationship between the number of hours that people work at their job each year and their annual wages. It's reasonable to think that, in general, people who work more hours will have higher annual wages—although earnings are also affected by people's hourly wage or salary—so a high-wage employee who works part-time might still earn more than a low-wage employee who works full-time. Table 12.1 shows the results of two regressions that use the number of hours that people work at their job to predict their annual wages. In the table on the left, the slope coefficient shows that working one additional hour is associated with earning $30 more in annual wages. The constant coefficient is difficult to interpret because it shows the predicted annual wages for people who work 0 hours and is a negative number. (And people can't earn negative wages.) To make the constant coefficient easier to interpret, I centred the "Total usual hours worked at all jobs" variable on 2,080 (see Chapter 11 for a discussion of centring variables), which is equivalent to working 40 hours per week, 52 weeks per year. In this sample, 2,080 is the median number of hours worked by people who are employed full-time, for the full year. The table on the right shows the same regression as the table on the left, but using the centred independent variable. As expected, the slope coefficient does not change, but the constant coefficient does. The constant coefficient shows that people who work 2,080 hours per year (40 hours per week for 52 weeks) are predicted to earn $58,636. The R^2 indicates that 36 per cent of the variation in annual wages can be explained by the number of hours that people work at their job.

Now, let's consider the relationship between people's seniority with an employer and their annual wages. Typically, the longer people have worked

Table 12.1   **Results of Two Simple Linear Regressions Predicting Annual Wages, Using People's Total Usual Hours Worked at All Jobs**

Dependent variable: Annual wages and salaries ($), before deductions (n = 12,896)

Without Centring		With Centring	
	Coefficient		Coefficient
Constant	−3,919.13*	Constant	58,635.84*
Total usual hours worked at all jobs	30.08*	Total usual hours worked at all jobs (centred on 2,080)	30.08*
R^2	0.36	R^2	0.36

*Indicates that results are statistically significant at the p < 0.05 level.
Source: Author generated; Calculated using data from Statistics Canada, 2016.

for an employer, the higher their wages. Some companies provide annual (or semi-annual) wage increases as a way of rewarding employees' loyalty and recognizing their growing on-the-job experience. Some people move through job ranks over time and earn higher wages as their career develops. In the CIS dataset, seniority was originally measured in months, but I rescaled it so that it is measured in years in order to make the results easier to interpret. (See the "Hands-on Data Analysis" box later in this chapter to learn about rescaling.) Table 12.2 shows the results of two regressions that use people's years of seniority with an employer to predict their annual wages. The slope coefficient shows that each additional year working for an employer is associated with earning $1,460 more in annual wages. In the table on the left, the constant coefficient shows the predicted annual wages ($34,519) for people with zero years of seniority. In the table on the right, the "Years of employment in current or main job" variable is centred on 1, and so the constant coefficient shows the predicted annual wages ($35,979) for people who have spent one year working for the same employer. The R^2 indicates that 13 per cent of the variation in annual wages can be explained by people's years of seniority with an employer.

Table 12.2   **Results of Two Simple Linear Regressions Predicting Annual Wages, Using People's Years of Employment**

Dependent variable: Annual wages and salaries ($), before deductions (n = 12,896)

Without Centring		With Centring	
	Coefficient		Coefficient
Constant	34,518.98*	Constant	35,979.09*
Years of employment in current or main job	1,460.11*	Years of employment in current or main job (centred on 1)	1,460.11*
R^2	0.13	R^2	0.13

*Indicates that results are statistically significant at the p < 0.05 level.
Source: Author generated; Calculated using data from Statistics Canada, 2016.

These results can be incorporated into regression prediction equations to estimate people's annual wages using the number of hours they work at their job, or to estimate people's annual wages using their years of seniority with an employer. But if we want to incorporate information about *both* the number of hours that people work and their years of seniority into the prediction, we can't combine the slope coefficients from different regressions into a single prediction equation. The reason why the slope coefficients can't be combined is because doing so doesn't account for the relationship that exists *between* the number of hours that people work and their years of seniority. In general, people with more seniority tend to work more hours, and people with less seniority tend to work fewer hours. Employees with more seniority might get to choose their shifts first or be more in-demand in their workplaces. The Pearson's correlation between the variable capturing the number of hours that people work and the variable capturing their years of seniority is 0.28, which shows that there is a moderate positive relationship between the two variables.

In the regression that uses the number of hours that people work at their job to predict their annual wages, the slope coefficient captures not only the direct relationship between the number of hours that people work and their wages, it also inadvertently captures the influence of any other variables that are related to both people's hours of work and their wages. In other words, some of the variation in wages is because of the number of hours that people work, but some of the variation in wages is because of other characteristics that are related to the number of hours that people work but that aren't accounted for in the regression. Similarly, in the regression that uses people's years of seniority with an employer to predict their annual wages, the slope coefficient captures not only the direct relationship between seniority and wages, it also inadvertently captures the influence of any other variables that are related to both people's seniority and their wages, but that aren't accounted for in the regression.

Controlling for Independent Variables in Regression

Multiple regression takes into account the relationships between the independent variables and, thus, allows researchers to sort out how much of the variation in the dependent variable is attributable to each independent variable alone. In this example, a multiple linear regression will allow us to determine how much of the variation in annual wages can be attributed to the number of hours that people work at their job—separate from their years of seniority with an employer—and how much of the variation in annual wages can be attributed to people's years of seniority—separate from the number of hours they work. Statisticians refer to this process as "controlling for" the other independent variables used in a regression. In other words, a multiple regression predicts the relationship between each independent variable and the dependent variable, while controlling for—or holding constant—the other independent variables in the regression.

It's easiest to understand the idea of "controlling for" or "holding constant" a second variable using an illustration. Recall that in a simple linear regression, the general pattern that is predicted takes the form of a one-dimensional straight line, in two-dimensional space. In a multiple linear regression with two independent variables, the general pattern that is predicted is the equivalent of a two-dimensional flat plane, in three-dimensional space. (See Figure 12.1.) The predicted value on the dependent variable—annual wages—is shown on the vertical axis. Since there is only a single vertical axis, there is still only one constant coefficient in the regression—and it is located at the point where *both* independent variables are equal to zero. One edge of the plane shows the slope associated with the number of hours that people work at their job while holding their years of seniority with an employer constant. In other words, it shows the change in wages associated with a one-hour increase in time worked while accounting for seniority. The parallel gridlines on the plane show that the slope of the relationship between the number of hours that people work at their job and their annual wages is the same for people with each different amount of seniority (or at each different point on the years of seniority axis). As a result, the slope coefficient shows the predicted change in wages associated with working one additional hour, *controlling for* people's seniority. That is, the prediction takes into account employees' years of seniority with an employer when it considers the relationship between the number of hours they work at their job and their annual wages.

The other edge of the plane shows the slope associated with people's years of seniority with an employer while holding the number of hours they work at their job constant. In other words, it shows the change in wages associated with a one-year increase in seniority while accounting for the number of hours that people work at their job. Again, the parallel gridlines on the plane show that the

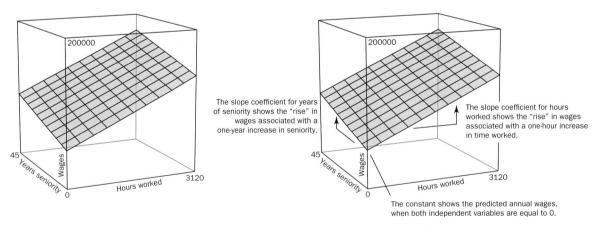

Figure 12.1   A Conceptual Illustration of Multiple Linear Regression with Two Independent Variables, without and with Annotation

Source: Author generated; Calculated using data from Statistics Canada, 2016.

slope of the relationship between years of seniority and annual wages is the same for people at each different point on the hours-worked axis. As a result, the slope coefficient shows the predicted change in wages associated with having an additional year of seniority, *controlling for* the number of hours that people work. That is, the prediction takes into account the number of hours that people work at their job when it considers the relationship between their years of seniority with an employer and their annual wages.

Just like the line of best fit in a simple linear regression, the plane of best fit for a multiple linear regression (such as that in Figure 12.1) is the one that minimizes the sum of the squared distances of each case from the plane. Although the cases aren't displayed in Figure 12.1, you can imagine a cloud of dots above and below the plane, with each dot representing the position of a single case on all three variables. As in simple linear regression, residuals or errors can be calculated for each case that capture the distance between people's actual annual wages and the annual wages that are predicted using the number of hours they work at their job and their years of seniority with an employer. And, even though this multiple regression predicts a plane of best fit, it is still considered a linear regression because it predicts straight-line relationships between variables.

In Figure 12.1, the dependent variable—annual wages—is located on the vertical axis, and people's annual wages are predicted by finding the height of the plane at the point where the number of hours they work at their job and their years of seniority with an employer intersect. Recall that in simple linear regression, the equation used to predict the value on a dependent variable using information from an independent variable is:

$$\hat{y} = a + bx$$

In multiple linear regression, predicting the value on a dependent variable uses a similar approach. As in simple linear regression, there is only one constant coefficient. But since there is more than one independent variable (x), there is more than one slope coefficient (b). Subscript numbers are used to keep track of the different independent variables and their respective slope coefficients. So, in a multiple linear regression with two independent variables, the first independent variable is denoted x_1 and its slope coefficient is denoted b_1; the second independent variable is denoted x_2 and its slope coefficient is denoted b_2:

$$\hat{y} = a + b_1 x_1 + b_2 x_2$$

Table 12.3 shows the results of a multiple linear regression that uses two independent variables to predict people's annual wages: one capturing the number of hours that they work at their job and one capturing their years of seniority with an employer. The constant coefficient of $51,698 shows the predicted annual wages of people with a "0" value on *both* independent variables. Both variables are centred, so a "0" value on the "Total usual hours worked at all jobs" variable indicates working full-time, full-year (2,080 hours) and a "0" value on the "Years of employment in current or main job" variable indicates one year of seniority. Thus, this regression predicts that people with one year of seniority who work full-time, full-year, earn $51,698 in wages each year.

The first slope coefficient in Table 12.3 shows that each additional hour that people work at their job is associated with earning $27 more in annual wages, after controlling for their years of seniority with an employer. This is slightly smaller than in the simple linear regression, which shows that each additional hour people work at their job is associated with earning $30 more in annual wages, without accounting for their years of seniority. (See Table 12.1.) The slightly smaller slope coefficient in the multiple regression indicates that some of the variation in annual wages that was attributed to the number of hours that people work at their job in the simple linear regression is actually attributable to seniority. Once seniority is accounted for, the change in wages associated with working an additional hour is slightly smaller.

Similarly, the second slope coefficient in Table 12.3 shows that each additional year of seniority with an employer is associated with earning $868 more in annual wages after controlling for the number of hours that people work at their job. This is substantially smaller than in the simple linear regression, which shows that each additional year of seniority with an employer is associated with earning $1,460 more in annual wages. (See Table 12.2.) These results suggest that a fair bit of the variation in annual wages that was being attributed to years of seniority with an employer was actually just capturing the fact that people with more seniority also tend to work more hours at their job. Once people's number of work hours are accounted for in the regression, the difference in wages associated with having an additional year of seniority with an employer is much smaller.

The R^2 indicates that 40 per cent of the variation in annual wages can be explained by the number of hours that people work at their job and their years of seniority with an employer. As in simple linear regression, the R^2 shows how much better the regression predictions are than the predictions of a null model, which is a model where the average annual wage is predicted for every case. In other words, by

Table 12.3   **Results of a Multiple Linear Regression with Two Independent Variables**

Dependent variable: Annual wages and salaries ($), before deductions (n = 12,896)

	Coefficient
Constant	51,697.72*
Total usual hours worked at all jobs (centred on 2,080)	27.13*
Years of employment in current or main job (centred on 1)	868.37*
R^2	0.40

*Indicates that results are statistically significant at the p < 0.05 level.
Source: Author generated; Calculated using data from Statistics Canada, 2016.

taking into account information about both the number of hours that people work at their job and their years of seniority with an employer, the errors in predicting annual wages are reduced by 40 per cent. Notice that the amount of variation in the dependent variable that is explained by both independent variables in the multiple linear regression (0.40) is less than the sum of the two R^2 for the two corresponding simple linear regressions (0.13 + 0.36 = 0.49). This is because the multiple regression determines how much of the variation in the dependent variable is uniquely attributable to each independent variable alone, and accounts for the correlation between them. For multiple linear regressions, R^2 is calculated using the same formula as shown in Chapter 11, except that all of the independent variables are used to find the predicted value on the dependent variable for each case. In other words, a multiple linear regression prediction equation is used to determine the predicted value on the dependent variable for each case, which is then substituted into the R^2 calculation.

By now, I hope that you are beginning to see the advantage of using multiple regression to assess how several independent variables are related to a single dependent variable. The main advantage of multiple regression is that it lets researchers sort out the unique relationship between each independent variable and the dependent variable, while accounting for the relationships between all of the independent variables. In this multiple linear regression example, I used only two independent variables—but multiple regressions can incorporate many independent variables. The regression prediction equation is simply extended to add as many independent variables as needed:

$$\hat{y} = a + b_1 x_1 + b_2 x_2 + b_3 x_3 + b_4 x_4 + b_5 x_5 + \cdots$$

multiple linear regression predictions

Even when there are many independent variables, the interpretation of the regression coefficients remains the same: there will be a single constant coefficient, which shows the predicted value on the dependent variable when *all* of the independent variables are zero. And each independent variable will have a slope coefficient, which shows the change in the dependent variable associated with a one-unit increase in the independent variable, after controlling for all of the other independent variables in the regression (or when all of the other independent variables are held constant).

Creating Regression Models

The process of **model specification** refers to the decisions that researchers make about which independent variables to use in a regression. Regressions are commonly referred to as models because they implicitly establish a model of how the social world operates or how social processes work in relation to the dependent variable. Researchers who use social statistics strive to build regression models that effectively capture all of the predictors of a dependent variable but exclude non-meaningful independent variables.

model specification The decisions that researchers make about which independent variables to use (or not use) in a regression model.

Step-by-Step: Predicted Value on the Dependent Variable (Multiple Linear Regression)

$$\hat{y} = a + b_1x_1 + b_2x_2 + b_3x_3 + b_4x_4 + b_5x_5 + \cdots$$

Step 1: Identify the constant coefficient from the multiple linear regression (a).

Step 2: Identify the slope coefficient of the first independent variable (b_1) in the multiple linear regression.

Step 3: Identify the value on the first independent variable (x_1) for the specific case you want to make a prediction for.

Step 4: Multiply the slope coefficient of the first independent variable (from Step 2) by the value on the first independent variable for the specific case (from Step 3).

Step 5: Repeat Steps 2 to 4 for each additional independent variable in the regression.

Step 6: Add together the results of Step 4 for *each independent variable* in the regression.

Step 7: Add the result of Step 6 to the constant coefficient (from Step 1) to find the predicted value on the dependent variable for the specific case.

Researchers typically make their model specification choices based on a theoretical understanding of what might influence something in the population. The starting point for selecting independent variables should always be the most current knowledge about what the dependent variable is related to. Often, however, researchers are limited by the variables that are available in a dataset. But a theoretical framework can help researchers to identify the difference between a legitimate relationship between an independent variable and a dependent variable, and a **spurious relationship**, which is a relationship between two variables that disappears once additional variables are controlled for.

As part of their model specification, researchers also strive to avoid **omitted variable bias**. This bias occurs when an important independent variable is omitted from a regression, and it affects the slope coefficients of one or more of the independent variables in the regression. Typically, omitting an important independent variable from a regression makes the slope coefficients of other independent variables higher or lower than they would be if the variable were not omitted. This occurs because the omitted variable is correlated with one or more of the independent variables (such as how the number of hours that people work at their job is correlated with their years of seniority with an employer); thus, the slope coefficient of an independent variable in the regression captures both the influence of the variable itself and the influence of the omitted variable with which it is correlated. Concerns about model specification and omitted variable bias motivate researchers to carefully consider which independent variables they will use (or not use) in a regression.

spurious relationship A relationship between two variables that disappears once additional variables are controlled for.

omitted variable bias When an important independent variable is omitted from a regression and it affects the slope coefficients of one or more of the independent variables in the regression.

Calculating Multiple Linear Regression Coefficients

The mathematical calculations associated with multiple linear regression are typically more complex than those for simple linear regression; thus, most researchers rely on statistical software to complete them. As a result, in this chapter and the next I primarily emphasize understanding and interpreting regression coefficients, as opposed to calculating them. But learning to calculate the slope coefficients of a multiple linear regression with two independent variables can help you to understand what it means to "control" for other variables.

In Chapter 10, you learned how the elaboration model is used to calculate partial correlation coefficients, which show the correlation between two variables after controlling for group membership, defined by a third variable. In multiple linear regression, the slope coefficients also show a partial relationship; that is, they show the relationship between an independent variable and a dependent variable, after controlling for the other independent variables. Because of this, they are sometimes referred to as partial slope coefficients.

In Chapter 11, you learned how Pearson's correlation coefficients are related to regression slope coefficients. A Pearson's correlation coefficient provides the same information as a simple linear regression slope coefficient, but the amount of change is measured in standard deviations. Specifically, Pearson's correlation coefficient shows the amount of change in the dependent variable (measured in standard deviations) associated with a one standard deviation increase in the independent variable.

Calculating a partial slope coefficient relies on the correlation between each independent variable and the dependent variable *and* accounts for (or controls for) the correlation between the independent variables. In a linear regression with two independent variables, calculating the partial slope coefficient of the first independent variable begins with the Pearson's correlation between that independent variable and the dependent variable. But in order to account for the correlation between the two independent variables, two additional relationships are taken into consideration: (1) the Pearson's correlation between the two independent variables, and (2) the Pearson's correlation between the second independent variable and the dependent variable. By accounting for the correlation between the two independent variables and the correlation between the second independent variable and the dependent variable, researchers can isolate the unique influence of the first independent variable on the dependent variable.

It's easiest to understand the logic of calculating partial slope coefficients using an example. Table 12.4 shows the Pearson's correlations between the three variables in the regression shown in Table 12.3. Since these bivariate correlations don't account for any other variables, they are called zero-order correlations. As you might expect, both of the independent variables (capturing the total number of hours that people work at their job and their years of seniority with an employer) are positively correlated with the dependent variable, annual wages. But, as I noted earlier in this chapter, there is also a positive correlation between the two

Table 12.4   Pearson's Correlations between the Variables in the Regression in Table 12.3 (n = 12,896)

	Annual Wages and Salaries ($), before Deductions	Total Usual Hours Worked at All Jobs (Centred on 2,080)	Years of Employment in Current or Main Job (Centred on 1)
Annual wages and salaries ($), before deductions	1.0000	0.5967	0.3614
Total usual hours worked at all jobs (centred on 2,080)		1.0000	0.2751
Years of employment in current or main job (centred on 1)			1.0000

Source: Author generated; Calculated using data from Statistics Canada, 2016.

independent variables: people with more years of seniority with an employer tend to work more hours at their job. The calculation of the partial slope coefficient takes this relationship into account.

In this example, I'll refer to the "Total usual hours worked at all jobs" variable as x_1 and the "Years of employment in current or main job" variable as x_2. As usual, the dependent variable, "Annual wages and salaries, before deductions," will be referred to as y. For each bivariate correlation, subscript is used to identify the two variables being correlated. So, the correlation between the "Total usual hours worked at all jobs" variable (x_1) and the "Annual wages and salaries, before deductions" variable (y) is referred to as $r_{x_1,y}$ and is equal to 0.5967. The correlation between "Years of employment in current or main job" variable (x_2) and the "Annual wages and salaries, before deductions" variable (y) is referred to as $r_{x_2,y}$ and is equal to 0.3614. Finally, the correlation between the "Total usual hours worked at all jobs" variable (x_1) and "Years of employment in current or main job" variable (x_2) is referred to as r_{x_1,x_2}, and is equal to 0.2751.

Since the Pearson's correlations used in the calculation of partial slope coefficients are measured in standard deviations, the standard deviations of each variable are also incorporated into the formula in order to translate the results back into the same units as the dependent variable. In this example, the standard deviation of the "Total usual hours worked at all jobs" variable (s_{x_1}) is 693.84, the standard deviation of the "Years of employment in current or main job" variable (s_{x_2}) is 8.65, and the standard deviation of the "Annual wages or salaries, before deductions" variable (s_y) is 34,970.38.

The formula used to calculate the partial slope coefficient of the first independent variable is:

partial slope coefficient
(multiple linear regression)

$$b_{x_1} = \left(\frac{r_{x_1,y} - \left[r_{x_1,x_2} \right]\left[r_{x_2,y} \right]}{1 - \left[r_{x_1,x_2} \right]^2} \right)\left(\frac{s_y}{s_{x_1}} \right)$$

Although this formula might seem daunting, that's primarily because of the subscripts used to identify each of the Pearson correlation coefficients. Notice how the indirect influence of the second independent variable is accounted for in the numerator of the first fraction by multiplying the correlation between the two independent variables and the correlation between the second independent variable and the dependent variable, and then subtracting the result from the correlation between the first independent variable and the dependent variable. The result is simple to calculate once we substitute in each of the values:

$$b_{x_1} = \left(\frac{r_{x_1,y} - \left[r_{x_1,x_2} \right]\left[r_{x_2,y} \right]}{1 - \left[r_{x_1,x_2} \right]^2} \right)\left(\frac{s_y}{s_{x_1}} \right)$$

$$b_{x_1} = \left(\frac{0.5967 - [0.2751][0.3614]}{1 - [0.2751]^2} \right)\left(\frac{34970.38}{693.84} \right)$$

$$= \left(\frac{0.5967 - 0.0994}{1 - 0.0757} \right)\left(\frac{34970.38}{693.84} \right)$$

$$= \left(\frac{0.4973}{0.9243} \right)\left(\frac{34970.38}{693.84} \right)$$

$$= (0.5380)(50.4012)$$

$$= 27.12$$

The result of 27.12 corresponds to the partial slope coefficient shown in Table 12.3 (although there's a slight difference in the decimals because of rounding). The calculation of the partial slope coefficient of the second variable follows the same pattern, although the position of the two independent variables, and their associated correlations, is reversed:

$$b_{x_2} = \left(\frac{r_{x_2,y} - \left[r_{x_1,x_2} \right]\left[r_{x_1,y} \right]}{1 - \left[r_{x_1,x_2} \right]^2} \right)\left(\frac{s_y}{s_{x_2}} \right)$$

$$b_{x_2} = \left(\frac{0.3614 - [0.2751][0.5967]}{1 - [0.2751]^2} \right)\left(\frac{34970.38}{8.65} \right)$$

$$= \left(\frac{0.3614 - 0.1642}{1 - 0.0757} \right)\left(\frac{34970.38}{8.65} \right)$$

$$= \left(\frac{0.1972}{0.9243} \right)\left(\frac{34970.38}{8.65} \right)$$

$$= (0.2134)(4042.8185)$$

$$= 862.74$$

Again, this roughly corresponds to the slope coefficient shown in Table 12.3, with a slight difference in the decimals because of rounding. Of course, once you have found the partial slope coefficients of a multiple linear regression, finding the constant coefficient becomes easy. As described in Chapter 11, in a linear regression the line (or plane) of best fit will always pass through the point where the means of the independent variable(s) and the dependent variable intersect, regardless of how many independent variables are used in the regression. Thus, if you know the variable means and the partial slope coefficients, you can calculate the constant. In this example, the variable means are less intuitive because the variables are centred: the mean of the "Annual wages and salaries, before deductions" variable is $45,476.81, the mean of the "Total usual hours worked at all jobs" variable is −437.55 (it's negative because of the centring), and the mean of the "Years of employment in current or main job" variable is 6.50. Since we know that:

constant coefficient
(multiple linear regression)

$$\bar{y} = a + b_1\bar{x}_1 + b_2\bar{x}_2 \qquad or \qquad a = \bar{y} - b_1\bar{x}_1 - b_2\bar{x}_2$$

The constant is found by solving for a in this formula:

$$\bar{y} = a + b_1\bar{x}_1 + b_2\bar{x}_2$$
$$45476.81 = a + (27.12)(-437.55) + (862.74)(6.50)$$
$$45476.81 = a + (-11866.36) + (5607.81)$$
$$45476.81 = a + (-6258.55)$$
$$45476.81 + 6258.55 = a + (-6258.55) + 6258.55$$
$$51735.36 = a$$
$$a = 51735.36$$

Once again, this result is close to the constant coefficient shown in Table 12.3, although it does not match exactly because the slope coefficients we calculated do not match exactly. But, most importantly, understanding how partial slope coefficients are calculated helps to illustrate how the influence of other independent variables is controlled for in a multiple regression. It becomes possible to estimate the portion of the relationship between an independent variable and a dependent variable that is unique to that independent variable alone, by accounting for the correlations between each of the independent variables and the correlations between the *other* independent variables and the dependent variable.

Step-by-Step: Slope and Constant Coefficients (Multiple Linear Regression with Two Independent Variables)

Note: In these calculations, do not include cases that are missing information for either variable used in the regression.

Partial Slope Coefficient (First Independent Variable)

$$b_{x_1} = \left(\frac{r_{x_1,y} - \left[r_{x_1,x_2} \right]\left[r_{x_2,y} \right]}{1 - \left[r_{x_1,x_2} \right]^2} \right)\left(\frac{s_y}{s_{x_1}} \right)$$

Step 1: Find the correlation between the first independent variable and the dependent variable (r_{x_1},y).

Step 2: Find the correlation between the second independent variable and the dependent variable (r_{x_2},y).

Step 3: Find the correlation between the first independent variable and the second independent variable (r_{x_1},x_2).

Step 4: Find the standard deviation of the dependent variable (s_y).

Step 5: Find the standard deviation of the first independent variable (s_{x_1}).

Step 6: Multiply the correlation between the two independent variables (from Step 3) by the correlation between the second independent variable and the dependent variable (from Step 2).

Step 7: Subtract the result of Step 6 from the correlation between the first independent variable and the dependent variable (from Step 1).

Step 8: Square the correlation between the two independent variables (from Step 3).

Step 9: Subtract the result of Step 8 from 1.

Step 10: Divide the result of Step 7 by the result of Step 9.

Step 11: Divide the standard deviation of the dependent variable (from Step 4) by the standard deviation of the first independent variable (from Step 5).

Step 12: Multiply the result of Step 10 and Step 11 to find the slope coefficient of the first independent variable.

Partial Slope Coefficient (Second Independent Variable)

$$b_{x_2} = \left(\frac{r_{x_2,y} - \left[r_{x_1,x_2} \right]\left[r_{x_1,y} \right]}{1 - \left[r_{x_1,x_2} \right]^2} \right)\left(\frac{s_y}{s_{x_2}} \right)$$

Step 13: Find the standard deviation of the second independent variable (s_{x_2}).

Continued

Step 14: Multiply the correlation between the two independent variables (from Step 3) by the correlation between the first independent variable and the dependent variable (from Step 1).

Step 15: Subtract the result of Step 14 from the correlation between the second independent variable and the dependent variable (from Step 2).

Step 16: Divide the result of Step 15 by the result of Step 9.

Step 17: Divide the standard deviation of the dependent variable (from Step 4) by the standard deviation of the second independent variable (from Step 13).

Step 18: Multiple the result of Step 16 and Step 17 to find the slope coefficient of the second independent variable.

Constant Coefficient

$$\bar{y} = a + b_1\bar{x}_1 + b_2\bar{x}_2 \qquad or \qquad a = \bar{y} - b_1\bar{x}_1 - b_2\bar{x}_2$$

Step 19: Find the mean of the first independent variable in the regression ($\bar{x}_1$).

Step 20: Find the mean of the second independent variable in the regression ($\bar{x}_2$).

Step 21: Find the mean of the dependent variable in the regression (y).

Step 22: Multiply the slope coefficient of the first independent variable (from Step 12) by the mean of the first independent variable (from Step 19).

Step 23: Multiply the slope coefficient of the second independent variable (from Step 18) by the mean of the second independent variable (from Step 20).

Step 24: Subtract the result of Step 23 from the result of Step 22.

Step 25: Subtract the result of Step 24 from the mean of the dependent variable (from Step 21) to find the constant coefficient.

Standardized Slope Coefficients

When a regression uses more than one independent variable, researchers often want to know which independent variable has the strongest relationship with the dependent variable. If the independent variables are all measured in the same units—that is, they are all in hours, or they are all in dollars, or they are all in some other unit—the slope coefficients can be compared directly to determine which has the strongest relationship with the dependent variable.

Most of the time, though, the independent variables used in a regression are each measured in different units. For instance, in the regression shown in Table 12.3, time spent working is measured in hours, and seniority is measured in years. In this situation, researchers use **standardized slope coefficients** to assess which independent variable has the strongest relationship with the dependent variable. Standardized slope coefficients are commonly called **beta coefficients** and are denoted using the lower-case Greek letter beta (β), which looks like a stylized capital letter B.

standardized slope coefficient A regression slope coefficient that is reported in standard deviations instead of the original units of each variable.

beta coefficient (ß) A common name for a standardized slope coefficient.

Unstandardized slope coefficients show the change in the dependent variable associated with a one-unit increase in the independent variable. They are expressed in the units of the dependent variable, divided by the units of the independent variable—in other words, the "rise" over the "run." In regression, the "run" is always held constant at 1, which corresponds to a one-unit increase in the independent variable. Thus, the slope coefficient of the "Total usual hours worked at all jobs" variable from the regression shown in Table 12.3 can be shown as:

$$b = \frac{rise \ (in \ the \ DV)}{run \ (of \ the \ IV)} \quad or \quad \frac{\$27.13}{1 \ hour}$$

In Chapter 4, you learned how to create standardized scores, called z-scores, which represent the distance of a case from the mean using standard deviations as the unit of measurement. A similar approach—using standard deviations as the unit of measurement for each independent variable and the dependent variable—is used to calculate standardized slope coefficients. A standardized slope coefficient shows the "rise" in the dependent variable, measured in standard deviations, associated with a one standard deviation increase ("run") in the independent variable. So, to calculate a standardized slope coefficient, we show this same fraction in standard deviations instead of in dollars and hours.

The numerator shows the rise in people's annual wages in dollars ($27.13). The standard deviation of the variable capturing people's annual wages is $34,970.38. If one standard deviation is equal to $34,970.38, how many standard deviations is $27.13 equivalent to? Dividing $27.13 by $34,970.38 shows that it is equivalent to 0.000776 of a standard deviation. (Because the numbers are small, I'll show six decimals for the calculations in this section.) The same approach to standardization is used with the denominator, which is one hour. The standard deviation of the variable capturing the number of hours that people work at their job is 693.84. If one standard deviation is equal to 693.84, how many standard deviations is 1 equivalent to? By dividing 1 by 693.84, we find that 1 hour is equivalent to 0.001441 of a standard deviation. So, the slope coefficient of the "Total usual hours worked at all jobs" variable from the regression shown in Table 12.3, reported in standard deviations—or the standardized slope coefficient—is:

$$\beta = \frac{rise \ (in \ the \ DV)}{run \ (of \ the \ IV)} \quad or \quad \frac{0.000776 \ s.d.}{0.001441 \ s.d.}$$

In order to compare the size of the standardized slope coefficients of different variables, it's easiest if they all have the same denominator. Let's use multiplication to make the denominator of the standardized slope coefficient equal to one standard deviation (just as the denominator of the unstandardized slope coefficient is one unit). Because of how the denominator was calculated, we already know what number it needs to be multiplied by to make it equivalent to 1—it's the standard deviation of the independent variable (693.84 in this example). But, following the rules for multiplying fractions, *both* the top and the bottom of the fraction must be multiplied by the same value (effectively multiplying by 1). So, in this instance, we can multiply by 693.84/693.84, which is equivalent to 1. Then, the two numerators are multiplied together to get 0.538420, and the two denominators are multiplied together to get 1:

$$\beta = \frac{0.000776}{0.001441} \times \frac{693.84}{693.84}$$
$$= \frac{0.538420}{1}$$

The numerator of the fraction, 0.538420, is typically reported as the beta coefficient. (Much like the unstandardized slope coefficient, the denominator is not shown.) This standardized slope coefficient indicates that a one standard deviation increase in the number of hours that people work at their job is associated with a 0.538 standard deviation increase in their annual wages. Although this information isn't meaningful on its own, it makes it possible to compare the relative strength (or magnitude) of the relationship between different independent variables and the dependent variable.

Understanding the logic behind the calculation of standardized slope coefficients is useful. But once you understand this logic, there's a more efficient way to calculate a standardized slope coefficient. The formula used to calculate a standardized slope coefficient incorporates the same three pieces of information: (1) the slope coefficient of an independent variable (b_x), (2) the standard deviation of that independent variable (s_x) and (3) the standard deviation of the dependent variable (s_y).

standardized slope coefficient

$$\beta_x = b_x \left(\frac{s_x}{s_y} \right)$$

Let's use this formula to calculate the beta coefficients of both of the independent variables in the regression shown in Table 12.3. The unstandardized slope coefficients of the variables capturing the total number of hours that people work at their job and their years of seniority with an employer are $27.13 and $868.37, respectively. Recall that the standard deviation of the variable capturing the total number of hours that people work at their jobs is 693.84, the standard deviation of the variable capturing people's years of seniority with an employer is 8.65, and the standard deviation of the variable capturing people's annual wages is $34,970.38. So, for the variable capturing the total number of hours that people work at their job, the standardized slope coefficient is calculated:

$$\beta_x = b_x\left(\frac{s_x}{s_y}\right)$$

$$\beta_x = 27.13\left(\frac{693.84}{34970.38}\right)$$

$$= 27.13\,(0.019841)$$

$$= 0.538$$

For the variable capturing people's years of seniority with their employer, the standardized slope coefficient is calculated:

$$\beta_x = b_x\left(\frac{s_x}{s_y}\right)$$

$$\beta_x = 868.37\left(\frac{8.65}{34970.38}\right)$$

$$= 868.37\,(0.000247)$$

$$= 0.214$$

Most statistical software programs report both unstandardized and standardized slope coefficients by default. Table 12.5 shows both types of coefficients for the regression shown in Table 12.3. The standardized slope coefficients provide information about the increase or decrease in the dependent variable, in standard deviations, associated with a one standard deviation increase in each independent variable. As a result, the two beta coefficients can be compared to each other, even though the two independent variables are measured in different units.

In this example, the beta coefficients show that the number of hours that people work at their job has a stronger relationship with their annual wages than their years of seniority with an employer, because the former variable has a larger standardized slope coefficient. This might seem surprising because the "Total usual hours worked at all jobs" variable has a smaller unstandardized slope coefficient than the "Years of employment in current or main job" variable. But because the two independent variables have different spreads—or different standard deviations—they have different potential influences on wages. For instance, the predicted difference in annual wages between people with a very high number of years of seniority with an employer—say 50 years—and people with no seniority is \$43,419 (the slope coefficient of \$868.37 multiplied by 50). In contrast, the predicted difference in annual wages between people who work a very high number of hours at their job—say 80 hours a week for 52 weeks, or 4,160 hours a year—and people who work 1 hour a year is \$112,834 (the slope coefficient of 27.13 multiplied by 4,159 hours). Thus, despite its larger unstandardized slope coefficient, the potential influence of people's years of seniority on their annual wages is actually smaller than the potential influence of the number of hours that they work. As a result, the standardized regression

Table 12.5   **Results of a Multiple Linear Regression with Two Independent Variables, Showing Unstandardized and Standardized Slope Coefficients**

Dependent variable: Annual wages and salaries ($), before deductions (n = 12,896)

	Unstandardized Coefficient	Standardized Coefficient
Constant	51,697.72*	–
Total usual hours worked at all jobs (centred on 2,080)	27.13*	0.538
Years of employment in current or main job (centred on 1)	868.37*	0.214
R^2	0.40	

*Indicates that results are statistically significant at the p < 0.05 level.
Source: Author generated; Calculated using data from Statistics Canada, 2016.

coefficient of the variable capturing people's years of seniority with an employer is smaller than the standardized regression coefficient of the variable capturing the number of hours that people work at their job.

Whether an unstandardized slope coefficient is positive or negative doesn't matter; if an unstandardized slope coefficient is negative, then the corresponding standardized slope coefficient will be negative, and if an unstandardized slope coefficient is positive then the corresponding standardized slope coefficient will be positive. The independent variable with the largest *absolute* standardized slope coefficient (regardless of the sign) has the strongest relationship with the dependent variable. Similarly, the independent variable with the smallest absolute standardized slope coefficient has the weakest relationship with the dependent variable.

Step-by-Step: Standardized Slope Coefficients

$$\beta_x = b_x \left(\frac{s_x}{s_y} \right)$$

Step 1: Find the unstandardized slope coefficient of the independent variable you want to find the standardized slope coefficient of (b_x).

Step 2: Find the standard deviation of the independent variable you want to find the standardized slope coefficient of (s_x).

Step 3: Find the standard deviation of the dependent variable (s_y).

Step 4: Divide the standard deviation of the independent variable (from Step 2) by the standard deviation of the dependent variable (from Step 3).

Step 5: Multiply the result of Step 4 by the unstandardized slope coefficient of the independent variable (from Step 1) to find the standardized slope coefficient.

Categorical Variables as Independent Variables in Regression

So far, I have only discussed using ratio-level variables in regression. But researchers often want to use categorical variables as independent variables in a regression. To do so, the categorical variable must be transformed into one or more dichotomous variables. In Chapter 1, you learned that dichotomous variables have only two attributes and two values ("0" and "1"), which indicate the presence or absence of some characteristic. You might recall that although dichotomous variables are nominal-level variables, I noted that some of the time they can be treated as ratio-level variables. In the context of regression, a dichotomous variable can be used in the same way as a ratio-level independent variable.

When dichotomous variables are used in regression, they are usually called **dummy variables**. This is just another name for a dichotomous variable with only two values: "0" and "1". You already know that a regression slope coefficient shows the change in the dependent variable associated with a one-unit or one-step increase in an independent variable. When that independent variable is a dummy variable, the regression slope coefficient shows the change in the dependent variable associated with *having* that particular characteristic (the one-step difference between the values "0" and "1").

To illustrate how dummy variables are used in regression, let's assess the relationship between having a disability (or not) and people's annual wages. In the original disability status variable the value "1" indicates that, "Yes", the respondent has a disability; and the value "2" indicates that, "No", the respondent does not have a disability. Even though there are only two non-missing values, they are not "0" and "1", and so this variable cannot be used in a regression without modification. To make this variable into a dummy variable, I recoded it so that "0" indicates that, "No", the respondent does not have a disability, and "1" indicates that, "Yes", the respondent does have a disability. (See the "Hands-on Data Analysis" box in Chapter 3 for a description of recoding variables.) Whenever a variable is recoded to create a dummy variable, make sure that cases with missing values aren't inadvertently assigned a "0" value; cases that are designated as missing in the original categorical variable should also be designated as missing in the corresponding dummy variable.

Overall, 15 per cent of employees have a disability, as defined by their limited activities. On average, employees with a disability earn \$40,090 (s.d. = \$31,825) in wages each year, and employees without a disability earn \$46,460 (s.d. = \$35,428). The difference between the two group means is statistically significant (F = 59.09; $df_1 = 1$, $df_2 = 12,894$; p < 0.001), and thus we are relatively confident that there is a relationship between people's disability status and their annual wages in the population of employees in Canada.

Table 12.6 shows the results of a simple linear regression where the dummy variable capturing whether or not people have a disability is used as

dummy variables Dichotomous variables that are used to incorporate a categorical variable as an independent variable in a regression.

Table 12.6   Results of a Simple Linear Regression with a Dummy Variable as an Independent Variable

Dependent variable: Annual wages and salaries ($), before deductions (n = 12,896)

	Unstandardized Coefficient	Standardized Coefficient
Constant	46,460.02*	–
Has a disability (dummy variable)	–6,369.95*	–0.066
R^2	0.004	

*Indicates that results are statistically significant at the $p < 0.05$ level.
Source: Author generated; Calculated using data from Statistics Canada, 2016.

the independent variable and people's annual wages are used as the dependent variable. The constant coefficient shows the predicted annual wages when the independent variable equals 0. In this example, a "0" value on the "Has a disability" dummy variable represents people without a disability, so the constant coefficient shows that people without a disability are predicted to earn $46,460 in wages each year. Since there is only one independent variable in this regression, the predicted annual wages for people without a disability is the same as the average annual wages for people without a disability. The slope coefficient shows that every one-unit increase in the "Has a disability" dummy variable is associated with earning $6,370 less (–6,370) in annual wages. In this example, a one-unit increase in the "Has a disability" dummy variable shows the difference between having a "1" compared to having a "0"; that is, it shows how people who have a disability compare to people who do not have a disability. So, the slope coefficient of –6,370 shows that people with disabilities are predicted to earn $6,370 less in annual wages compared to people without disabilities. A regression prediction equation can be used to calculate that the predicted annual wages for people with disabilities are $40,090 (the constant coefficient of $46,460 minus the slope coefficient of $6,370). Once again, the predicted annual wages for people with disabilities is the same as the average annual wages for people with disabilities since this regression does not account for any other characteristics besides people's disability status.

Let's add another independent variable to this regression. There are undoubtedly many other characteristics related to having a disability that also affect wages. For instance, in general, people with disabilities tend to work fewer hours than people without disabilities. In fact, the most common workplace accommodation required by people with disabilities is modified or reduced hours of work (Till et al. 2015). In this sample, employees with disabilities work an average of 1,557 hours at their job each year (s.d. = 732), while employees without disabilities work an average of 1,658 hours at their job each year (s.d. = 686). The difference between the average number of hours that people in each group work is statistically significant and, thus, likely also exists in the population (F = 35.92; df_1 = 1; df_2 = 12,894; $p < 0.001$). It's possible that people with disabilities have lower annual wages only because they work fewer hours at their job.

Table 12.7 shows the results of a regression that uses two independent variables: one capturing the number of hours that people work at their job and a dummy variable capturing whether or not they have a disability (annual wages is used as the dependent variable). The constant coefficient shows the predicted annual wages for people with a "0" value on both independent variables. Since the variable capturing the number of hours that people work at their job is centred, the constant coefficient shows that people who work full-time, full-year (2,080 hours) and who do not have a disability are predicted to earn \$59,111 in annual wages. The first slope coefficient shows that each additional hour that people work at their job is associated with earning \$30 more each year, after controlling for whether or not they have a disability. The second slope coefficient shows that people with disabilities are predicted to earn \$3,335 less in wages each year, after controlling for the number of hours that they work at their job. Taking into account the number of hours that people work at their job makes the slope coefficient of the disability status dummy variable much smaller than in the simple linear regression, which used *only* disability status as a predictor (in Table 12.6). Some of the difference in annual wages between people with disabilities and those without disabilities is accounted for by the different number of hours that people in the two groups work. The standardized slope coefficients show that the number of hours that people work at their job has more influence on their annual wages than their disability status because the standardized slope coefficient of the "Total usual hours worked at all jobs" variable (0.595) is larger than that of the "Has a disability" dummy variable (−0.034).

The regression results in Table 12.7 show that for two employees who each work the same number of hours at their job, an employee with a disability is predicted to earn \$3,335 less in wages each year than an employee without a disability. Let's use the regression prediction equation to illustrate this by calculating the predicted annual wages for two employees who each work 2,080 hours a year and, thus, have

Table 12.7 Results of a Multiple Linear Regression with Two Independent Variables (Including One Dummy Variable)

Dependent variable: Annual wages and salaries (\$), before deductions (n = 12,896)

	Unstandardized Coefficient	Standardized Coefficient
Constant	59,110.55*	–
Total usual hours worked at all jobs (centred on 2,080)	29.98*	0.595
Has a disability (dummy variable)	−3,335.00*	−0.034
R^2	0.36	

*Indicates that results are statistically significant at the $p < 0.05$ level.
Source: Author generated; Calculated using data from Statistics Canada, 2016.

a "0" value on the centred "Total usual hours worked at all jobs" variable. The prediction equation for the regression shown in Table 12.7 is:

$$\hat{y} = a + b_1 x_1 + b_2 x_2$$

$$Predicted\ annual\ wages\ (\$) = 59110.55 + 29.98(usual\ hours\ worked\ centred)$$

$$- 3335.00\ (has\ a\ disability)$$

The predicted annual wages for people without disabilities who work 2,080 hours a year are:

$$Predicted\ annual\ wages\ (\$) = 59110.55 + 29.98(0) - 3335.00\ (0)$$

$$= 59110.55$$

And the predicted annual wages for people with disabilities who work 2,080 hours a year are:

$$Predicted\ annual\ wages\ (\$) = 59110.55 + 29.98(0) - 3335.00\ (1)$$

$$= 59110.55 - 3335.00$$

$$= 55775.55$$

Notice the key difference between these two calculations: people without disabilities have a "0" value on the "Has a disability" dummy variable and so the slope coefficient of that variable doesn't affect their predicted wages; people with disabilities have a "1" value on the "Has a disability" dummy variable and so the slope coefficient of that variable results in lower predicted wages. The predicted difference in wages between people with disabilities and those without disabilities is consistent across all hours of work. You can visualize this as two parallel regression lines that each show the relationship between the number of hours that people work at their job and their annual wages: one for people with disabilities and one for people without disabilities. (See Figure 12.2.) The distance between the two lines corresponds to the slope coefficient of the disability status dummy variable (−$3,335).

Using Categorical Variables with More Than Two Attributes

Categorical variables with more than two attributes can also be used as independent variables in a regression, using more dummy variables. When a categorical variable has three or more attributes, a dummy variable must be created for *each* of the attributes. In other words, there will be as many new dummy variables as there are attributes in the original categorical variable (excluding those attributes that indicate missing information). For each of the dummy variables, each case is assigned the value "1" if it has that attribute and the value "0" if it does not. I'll illustrate this process with a categorical variable that measures people's highest educational credential, in four groups: "Less than high school graduation," "High school graduation only," "Non-university certificate/diploma," or "University degree/certificate."

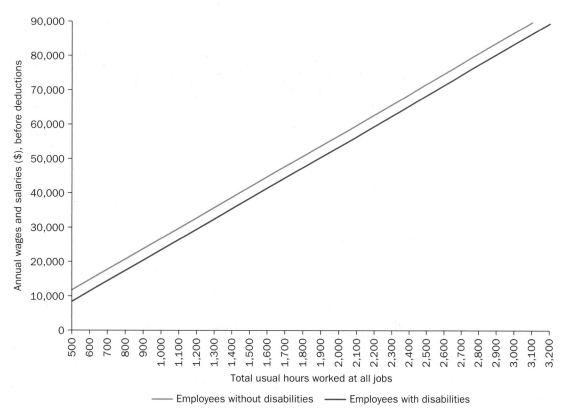

Figure 12.2   Annual Wages Predicted by the Regression in Table 12.7

Source: Author generated; Calculated using data from Statistics Canada, 2016.

Table 12.8 shows the percentage of employees in Canada with each level of education. About one in ten employees (9 per cent) did not graduate high school. An additional 28 per cent have only graduated from high school. Finally, 34 per cent have a non-university post-secondary certificate or diploma (such as a college or trade certificate), and 29 per cent have a university degree or certificate. It's important to look at the frequency distribution of categorical variables and, if necessary, collapse any small categories before creating dummy variables for regression. (Groups with less than about 10 per cent of cases are a cause for concern, especially in smaller samples.) Not surprisingly, having a higher level of education is associated with earning higher annual wages. Notably, lower levels of education are also associated with less variation in wages, suggesting that the range of wages available to people with lower levels of education is more constrained. The differences between the average annual wages for people with each level of education are statistically significant (F = 494.40; df_1 = 3; df_2 = 12,892; $p < 0.001$) and, thus, are likely to occur in the population of employees in Canada.

To use this variable capturing people's highest educational credential in a regression, I need to create four dummy variables: one for each level of education. Each person is assigned the value "1" on the dummy variable that corresponds with their highest level of education and the value "0" on the other

Table 12.8 **Frequency Distribution of Highest Educational Credential and Average Annual Wages and Salaries (n = 12,896)**

Highest Educational Credential	Percentage	Annual Wages and Salaries ($), Before Deductions	
		Mean	Standard Deviation
Less than high school graduation	9.3	24,906.46	26,288.21
High school graduation only	28.1	34,726.85	30,098.39
Non-university certificate/diploma	33.7	48,777.11	32,095.41
University degree/certificate	28.9	58,729.57	38,629.14
Total/Overall	*100.0*	*45,476.81*	*34,970.38*

Source: Author generated; Calculated using data from Statistics Canada, 2016.

education dummy variables. An example of how this is put into practice is shown in Table 12.9. The cases are listed in rows, and the original education variable and the four newly created dummy variables are displayed in the columns. There is one dummy variable for each level of education. The dummy variable has a "1" value if the person's highest educational credential corresponds with the variable, and the dummy variable has a "0" value if the person's highest educational credential does not correspond with the variable. Researchers typically create dummy variables by recoding the original variable into a new variable multiple times and altering which groups are assigned the values of "0" and "1"; some statistical software automates this process.

To assess the relationship between people's highest educational credential and their annual wages, all but one of these dummy variables are used as independent variables in a regression. One of the dummy variables must be left out because the value on the final dummy variable can always be perfectly predicted by the other dummy variables. For instance, even if you cover up the final column ("University degree/certificate") in Table 12.9, you can still figure out which people have a

Table 12.9 **How Dummy Variables Are Constructed for a Categorical Variable with Four Attributes**

Case	Highest Educational Credential	Dummy Variable: Less than high school graduation	Dummy Variable: High school graduation only	Dummy Variable: Non-university certificate/ diploma	Dummy Variable: University degree/ certificate
1	Less than high school graduation	1	0	0	0
2	High school graduation only	0	1	0	0
3	Non-university certificate/diploma	0	0	1	0
4	University degree/certificate	0	0	0	1
5	High school graduation only	0	1	0	0
6	University degree/certificate	0	0	0	1
7	Less than high school graduation	1	0	0	0
8	Non-university certificate/diploma	0	0	1	0

university degree/certificate: they're the cases with a "0" value on all of the other dummy variables, and so by default they must be in the remaining category. The attribute or category that corresponds to the dummy variable that is omitted from a regression is called the **reference group**. The prediction for people who are in the reference group is captured in the constant coefficient as it predicts the value on the dependent variable when all of the independent variables equal 0. And, as you just learned, when all of the dummy variables are equal to 0, the case must have the last, omitted attribute.

The omission of the dummy variable for the reference group is why researchers only need to create one dummy variable for categorical variables with only two attributes. For categorical variables with only two attributes, a researcher could create two dummy variables: one to indicate that people have a characteristic and one to indicate they do not; the researcher would then omit one of the two dummy variables from the regression. But since researchers almost always use the dummy variable indicating that people have a characteristic (and omit the dummy variable indicating they do not), they usually don't go through this process. But, as you saw in the simple linear regression using the disability status dummy variable to predict people's annual wages (in Table 12.6), the predicted value on the dependent variable for cases in the omitted category (people who do not have a characteristic) is captured in the constant coefficient.

Researchers consider several things when they decide which attribute or category to use as the reference group. A relatively high proportion of cases should have the attribute. Choosing an attribute that it is theoretically useful to make comparisons with is also important. And you should avoid using catchall categories, such as "Other" or "Something else" as the reference group because it's hard to know who is in these groups. In this example, "High school graduation only" is used as the reference group because high school graduation marks the end of universal public education, and thus most people have the opportunity to graduate from high school.

Table 12.10 shows the results of a regression that uses people's highest educational credential, captured in a series of dummy variables, to predict their annual wages. The constant coefficient shows the predicted annual wages of people with a "0" value on all of the independent variables in the regression. People with a "0" value on all three of the education dummy variables have only graduated from high school as this is the dummy variable that was omitted and, thus, is used as the reference group. So, this regression predicts that people who only graduated from high school will earn $34,727 in annual wages. The first slope coefficient shows that people who did not graduate from high school are predicted to earn $9,820 less in annual wages than people who only graduated from high school. The second slope coefficient shows that people with a non-university post-secondary certificate/diploma are predicted to earn $14,050 more in annual wages than people who only graduated from high school. The third slope coefficient shows that people with a university degree/certificate are predicted to earn $24,003 more in annual wages than people who only graduated from high school. Notice that the slope coefficient of each of these dummy variables is always interpreted in relation to the reference

reference group The attribute or category corresponding to the dummy variable that must be omitted when a group of dummy variables (capturing the attributes of a categorical variable) are used as independent variables in a regression.

group—that is, it shows how much higher or lower the predicted annual wages are for people with each educational credential, compared to people who only graduated from high school (the reference group).

Visualizing the predicted differences in annual wages between people with different levels of education on a number line, such as the one shown in Figure 12.3, can be useful. The reference group—in this example, people who only graduated from high school—is situated at a point on the line. Then, the slope coefficients of the remaining dummy variables are used to situate the predictions for each group relative to the reference group. The predicted differences in wages between people with different attributes is calculated by taking into account how far each one is from the reference group. For instance, on the number line in Figure 12.3, the distance between the dot representing a people with a non-university certificate/diploma and the dot representing a people with a university degree/certificate is 9,952.46, which is calculated by taking the slope coefficient of the "University degree/certificate" dummy variable (24,002.72) and subtracting the slope coefficient of the "Non-university certificate/diploma" dummy variable (14,050.26). This distance shows that people with a university degree/certificate are predicted to earn $9,952 more in annual wages than people with only a non-university certificate/diploma. Segments of the number line can also be added together to find the predicted differences in annual wages between people with different levels of education; for instance, to compare people who did not graduate from high school with people who have a non-university certificate/diploma, add together the length of the first two segments of the number line: 9,820.39 + 14,050.26. The sum of $23,870.65 is the predicted difference in annual wages between people who did not graduate from high school and people with a non-university certificate/diploma. The length of all three segments of the number line are added together to find the predicted difference in annual wages between people who did not graduate from high school and people with a university degree/certificate: $33,823.11 (9,820.39 + 14,050.26 + 9,952.46).

A multiple regression prediction equation can also be used to calculate the predicted differences in annual wages between people with different levels of

Table 12.10 **Results of a Multiple Linear Regression with Three Independent Variables (All Dummy Variables)**

Dependent variable: Annual wages and salaries ($), before deductions (n = 12,896)

	Unstandardized Coefficient	Standardized Coefficient
Constant	34,726.85*	–
Education (ref: high school only)		
Less than high school graduation (dummy variable)	−9,820.39*	−0.082
Non-university certificate/diploma (dummy variable)	14,050.26*	0.190
University degree/certificate (dummy variable)	24,002.72*	0.311
R^2	0.10	

*Indicates that results are statistically significant at the $p < 0.05$ level.
Source: Author generated; Calculated using data from Statistics Canada, 2016.

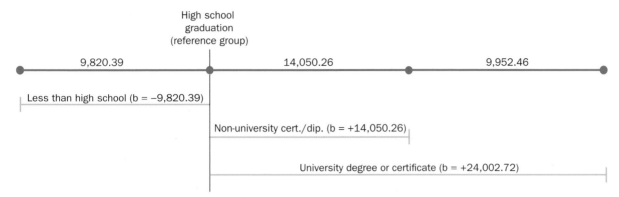

Figure 12.3   Visualizing the Slope Coefficients of Multiple Dummy Variables on a Number Line

Source: Author generated; Calculated using data from Statistics Canada, 2016.

education. The prediction equation is similar to the simple dummy variable example used earlier in this chapter, only there are more slope coefficients. Each slope coefficient is multiplied by 0 or 1, depending on which educational group the prediction is for. The prediction equation for the regression shown in Table 12.10 is:

$$\hat{y} = a + b_1 x_1 + b_2 x_2 + b_3 x_3 + b_4 x_4$$

$$Predicted\ annual\ wages\ (\$) = 34726.85 - 9820.39 \left(less\ than\ high\ school\right)$$

$$+ 14050.26 \left(non-university\ certificate\right) + 24002.72 \left(university\ degree\right)$$

The predicted annual wages for people who have not graduated from high school are:

$$Predicted\ annual\ wages\ (\$) = 34726.85 - 9820.39(1) + 14050.26(0) + 24002.72(0)$$

$$= 34726.85 - 9820.39$$

$$= 24906.46$$

The predicted annual wages for people who have only graduated from high school are:

$$Predicted\ annual\ wages\ (\$) = 34726.85 - 9820.39(0) + 14050.26(0) + 24002.72(0)$$

$$= 34726.85$$

The predicted annual wages for people with a non-university certificate/diploma are:

$$Predicted\ annual\ wages\ (\$) = 34726.85 - 9820.39(0) + 14050.26(1) + 24002.72(0)$$

$$= 34726.85 + 14050.26$$

$$= 48777.11$$

And the predicted annual wages for people with a university degree/certificate are:

$$Predicted\ annual\ wages\ (\$) = 34726.85 - 9820.39(0) + 14050.26(0) + 24002.72(1)$$
$$= 34726.85 + 24002.72$$
$$= 58729.57$$

The differences between the predicted annual wages for people with each different level of education correspond with the results calculated using the number line; this is just another way to find the same information. And because this regression only accounts for people's highest educational credential, and nothing else, the predicted annual wages for people with each level of education match the average annual wages shown in Table 12.8. This will change when more independent variables are added to the regression.

Now that you know how to interpret dummy variables for categorical variables with more than two attributes, let's add an independent variable capturing the number of hours that people work at their job to the regression, as well as the dummy variables capturing people's highest educational credential. (See Table 12.11.) The constant coefficient shows the predicted annual wages for people with a "0" value on all of the independent variables: in this instance, people who work full-time, full-year (2,080 hours a year) and who have only graduated from high school are predicted to earn $50,681 in wages each year. Each additional hour that people work at their job is associated with earning $28 more in wages each year, after controlling for their highest educational credential. People who have not graduated from high school are predicted to earn $2,066 less in annual wages than people who have only graduated from high school, after controlling for the number of hours that they work at their job. People with a non-university certificate or diploma are predicted to earn $6,653 more in annual wages than people who have only graduated from high school, and people with a university degree or certificate are predicted to earn $17,734 more in annual wages than people who have only graduated from high school, after controlling for the number of hours that they work at their job. The standardized slope coefficients show that the number of hours that people work at their job has more influence on their annual wages than their level of education.

These regression results show that the predicted influence of education is substantially reduced once the number of hours that people work at their job is controlled for, illustrating the importance of accounting for many different characteristics when predicting wages. Some of the variation in annual wages that was being attributed to people's level of education was actually capturing the fact that people with higher educational credentials tend to work more hours at their job, on average.

The predicted relationship between the number of hours that people work at their job and their annual wages, for people with each of the four different levels

Table 12.11 **Results of a Multiple Linear Regression with Four Independent Variables (Three Dummy Variables)**

Dependent variable: Annual wages and salaries ($), before deductions (n = 12,896)

	Unstandardized Coefficient	Standardized Coefficient
Constant	50,681.47*	–
Total usual hours worked at all jobs (centred on 2,080)	28.28*	0.561
Education (ref: high school only)		
Less than high school graduation (dummy variable)	–2,066.03*	–0.017
Non-university certificate/diploma (dummy variable)	6,653.10*	0.090
University degree/certificate (dummy variable)	17,734.01*	0.230
R^2	0.40	

*Indicates that results are statistically significant at the $p < 0.05$ level.

Source: Author generated; Calculated using data from Statistics Canada, 2016.

of education, is depicted in Figure 12.4. The results clearly show that—after controlling for the number of hours that people work at their job—having a university degree/certificate is associated with earning higher annual wages. This suggests that the cost of obtaining a post-secondary education may be a good investment.

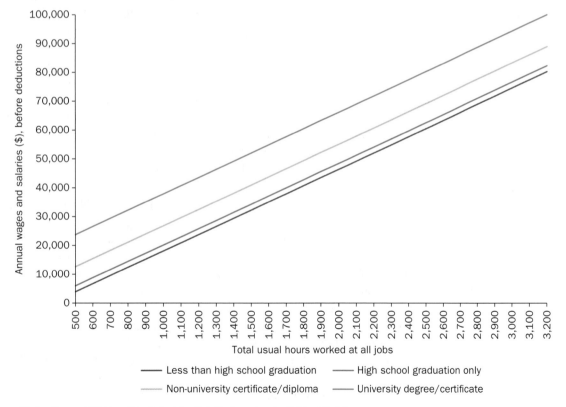

Figure 12.4 **Annual Wages Predicted by the Regression in Table 12.11**

Source: Author generated; Calculated using data from Statistics Canada, 2016.

Multiple Linear Regression in Action

Researchers spend a substantial amount of time working to develop and refine multiple regression models that effectively capture the social characteristics and processes related to the dependent variable. Let's combine the various techniques described in this chapter to begin developing a more nuanced regression to predict annual wages. As well, let's return to the main research question in this chapter, which is the relationship between having a disability (or not) and people's wages. So far, we've discovered that the number of hours that people work at their job, their years of seniority with an employer, and their level of education are all related to their annual wages. The question remains: Does having a disability still influence people's wages, even after all of these other things are accounted for? Although people with disabilities work fewer hours, on average, than people without disabilities, they tend to have more seniority with their employer. But people with disabilities also tend to have lower levels of education than those without disabilities. The results of a multiple linear regression will help us begin to assess how much of the wage differential between people with disabilities and those without disabilities is attributable to the different circumstances of people with disabilities, and how much of the wage differential may be the result of discrimination.

Table 12.12 shows the results of a regression predicting people's annual wages, using the number of hours they work at their job, their years of seniority with an employer, their highest educational credential, and their disability status as independent variables. The R^2 shows that, overall, these four characteristics account for 45 per cent of the variation in annual wages. The constant coefficient shows that people who work full-time, full-year (2,080 hours), have one year of seniority, have only graduated from high school, and do not have a disability, are predicted to earn \$43,901 in annual wages. Each additional hour that people work at their job is associated with earning \$25 more each year, and each additional year of seniority with an employer is associated with earning \$888 more each year, after controlling for the other variables in the model. As expected, people with post-secondary educational credentials are predicted to earn higher wages than people who have only graduated from high school. People with a non-university certificate or diploma are predicted to earn \$6,689 more in annual wages each year than people who have only graduated from high school, and people with a university degree or certificate are predicted to earn \$17,993 more in annual wages each year, after controlling for the number of hours they work at their job, their seniority with an employer, and whether or not they have a disability. The slope coefficient of the "Less than high school graduation" dummy variable is not statistically significant, and thus we are not confident that there is a difference in annual wages between people who have graduated from high school and those who have not in the larger population, after controlling for the number of hours they work at their job, their seniority with an employer, and whether or not they have a disability.

The slope coefficient of the "Has a disability" dummy variable is the most useful for answering our research question. It shows that after controlling for the number of hours that people work at their job, their years of seniority with an employer, and their highest educational credential, people with disabilities are predicted to earn \$3,042

Table 12.12 Results of a Multiple Linear Regression Predicting Annual Wages

Dependent variable: Annual wages and salaries ($), before deductions (n = 12,896)

	Unstandardized Coefficient	Standardized Coefficient
Constant	43,901.28*	–
Total usual hours worked at all jobs (centred on 2,080)	25.20*	0.500
Years of employment in current or main job (centred on 1)	887.53*	0.220
Education (ref: high school only)		
Less than high school graduation (dummy variable)	−1,605.12	−0.013
Non-university certificate/diploma (dummy variable)	6,688.64*	0.090
University degree/certificate (dummy variable)	17,993.31*	0.233
Has a disability (dummy variable)	−3,041.55*	−0.031
R^2	0.45	

*Indicates that results are statistically significant at the $p < 0.05$ level.
Source: Author generated; Calculated using data from Statistics Canada, 2016.

less in wages each year. That is, for two employees who have the same level of education, the same years of seniority with an employer, and who work the same number of hours at their job, where one person has a disability and the other does not, the person with the disability is predicted to earn roughly $3,000 less in wages each year. These results suggest that systematic discrimination against people with disabilities persists in the labour force and that it is likely to affect their wages. This is particularly a concern since people with disabilities often require additional resources in order to access aids and assistive devices that enhance their daily lives. More promisingly, though, the standardized slope coefficients indicate that the "Has a disability" dummy variable has the second-weakest relationship with annual wages, among all of the independent variables in the regression. Compared to the previous regressions shown in this chapter, these regression results suggest that the lower wages of people with disabilities are partly attributable to the fact that they work fewer hours and have lower levels of education. One way to potentially increase the annual wages of people with disabilities is to ensure that they have access to a full range of educational opportunities.

How Does It Look in SPSS?

Multiple Linear Regression

When more than one independent variable is used, the Linear Regression procedure produces results that look like those in Image 12.1. Above these results, SPSS prints a "Variables Entered/Removed" box (not shown), which lists the independent variables that were used or entered in the regression and the independent variables that were removed; review this information to make sure that all of the requested variables were used.

Continued

Model Summary

Model	R	R Square	Adjusted R Square	Std. Error of the Estimate
1 Ⓐ	.667[a]	.445	.444	26067.469

Ⓑ over R, Ⓒ over R Square, Ⓓ over Adjusted R Square, Ⓔ over Std. Error of the Estimate

Ⓕ a. Predictors: (Constant), Has a disability, Total usual hours worked at all jobs (centred on 2,080), University degree/certificate, Years of employment in current or main job (centred on 1), Less than high school graduation, Non-university certificate/diploma

ANOVA[a]

Model		Sum of Squares	df	Mean Square	F	Sig.
1	Regression	7011493775317.112	6	1168582295886.185	1719.735	.000[b]
Ⓖ	Residual	8758353879655.507	12889	679512935.714		
	Total	15769847654972.620	12895			

Ⓕ a. Dependent Variable: Annual wages and salaries before deductions

b. Predictors: (Constant), Has a disability, Total usual hours worked at all jobs (centred on 2,080), University degree/certificate, Years of employment in current or main job (centred on 1), Less than high school graduation, Non-university certificate/diploma

Coefficients[a]

Model		Unstandardized Coefficients		Standardized Coefficients		
		B	Std. Error	Beta	t	Sig.
1 Ⓗ	(Constant)	43901.275	531.062		82.667	.000
	Total usual hours worked at all jobs (centred on 2,080)	25.198	.354	.500	71.175	.000
	Years of employment in current or main job (centred on 1)	887.530	27.592	.220	32.166	.000
	Less than high school graduation	-1605.119	874.073	-.013	-1.836	.066
	Non-university certificate/diploma	6688.642	592.854	.090	11.282	.000
	University degree/certificate	17993.309	613.363	.233	29.335	.000
	Has a disability	-3041.548	638.009	-.031	-4.767	.000

Column markers: Ⓘ over Unstandardized Coefficients, Ⓙ over Std. Error, Ⓚ over Standardized Coefficients, Ⓛ over t, Ⓜ over Sig.

Ⓕ a. Dependent Variable: Annual wages and salaries before deductions

Image 12.1 An SPSS Linear Regression with Multiple Independent Variables

A. Each regression model is assigned a number. These results are for the first (and only) model in this regression.

B. The "R" column shows the square root of the R^2 value. It is not easily interpretable for a multiple regression and, thus, is typically not reported.

C. The "R Square" column shows the coefficient of determination for the regression. An R^2 of 0.445 indicates that, collectively, the independent variables explain 45 per cent of the variation in annual wages.

D. This column shows a variation on R^2; you'll learn more about the adjusted R^2 in Chapter 13.

E. This column shows the standard error of the estimate, which is used to calculate the standard errors of regression coefficients.

F. The footnotes below each table list the independent variables (predictors) and the dependent variable used in the regression.

G. The "Regression" sum of squares shows the amount of variation in the dependent variable explained by the independent variables; it is the numerator of the R^2 equation. The "Total" sum of squares shows the total variation in the dependent variable; it is the denominator in the R^2 equation. The "Residual" sum of squares shows the amount of unexplained variation in the regression; it is obtained by subtracting the explained variation from the total variation. The results of the ANOVA test show whether or not, as a group, the independent variables are likely to be related to the dependent variable in the population. These results are not usually reported.

H.   The "Coefficient" table always begins with a row for the "Constant." The independent variables are listed in rows below the constant row.

I.   The first row of the "B" column shows the constant coefficient (*a*). This regression predicts that people who work 2,080 hours at their job each year, have one year of seniority with their employer, have only graduated from high school, and have no disability earn $43,901 in annual wages. The subsequent rows in the "B" column show the slope coefficient of each independent variable. Each additional hour that people work at their job is associated with earning $25 more in annual wages, and each additional year of seniority with an employer is associated with earning $888 more in annual wages, after controlling for their highest educational credential and whether or not they have a disability. People with a disability are predicted to earn $3,042 less in annual wages than people without a disability, after controlling for the number of hours they work at their job, their seniority with an employer, and their highest educational credential. The slope coefficients of the education dummy variables show that, compared to people who only graduated from high school, people who did not graduate from high school are predicted to earn $1,605 less in annual wages, people with a non-university certificate or diploma are predicted to earn $6,689 more, and people with a university degree or certificate are predicted to earn $17,993 more, after controlling for the other variables in the model.

J.   This column shows the standard errors of the regression coefficients. The first row shows the standard error of the constant coefficient, and the subsequent rows show the standard error of each slope coefficient. The standard errors are used to calculate the t-statistic.

K.   The "Beta" column shows the standardized slope coefficient of each independent variable. These results show that the number of hours that people work at their job has the strongest relationship with their annual wages, as it is the independent variable with the largest absolute standardized coefficient (0.500). The "Less than high school graduation" dummy variable has the weakest relationship with people's annual wages as it is the independent variable with the smallest absolute standardized coefficient (−0.013).

L.   This column shows the t-statistic for each coefficient. The t-statistic is calculated by dividing the coefficient by its standard error. Each t-statistic is evaluated in relation to a t-distribution, with degrees of freedom equal to the number of cases, minus the number of independent variables, minus 1.

Continued

M. The "Sig." column shows the p-value associated with each t-statistic. That is, it shows the likelihood of randomly selecting a sample with the observed relationship (or one of a greater magnitude), if no relationship exists between each independent variable and the dependent variable in the population. Since the p-value of every independent variable except the "Less than high school graduation" dummy variable is less than 0.001, there is less than a 0.1 per cent chance of selecting this sample from a population in which no relationship exists between each of these variables and annual wages. The p-value of 0.07 for the "Less than high school graduation" dummy variable shows that there is a 7 per cent chance of selecting this sample from a population in which people who did not graduate from high school and people who only graduated from high school have the same annual wages, after controlling for the number of hours that they work at their job, their years of seniority with an employer, and whether or not they have a disability. Since the p-value of 0.07 is higher than the alpha value of 0.05, we are not confident that people with these two attributes have different annual wages in the population. The p-value in the "(Constant)" row shows the likelihood of selecting this sample if employees who work 2,080 hours each year, have one year of seniority, have only graduated from high school, and do not have a disability earn zero dollars in annual wages in the population. Since this situation is not expected, the significance test is not usually discussed for the constant coefficient.

Hands-on Data Analysis

Rescaling a Variable

Researchers sometimes adjust the units that an independent variable is measured in before using it in a regression. Altering the unit of a variable, usually by multiplying or dividing each of the values, is called rescaling. When a variable is rescaled, it changes the interpretation of the slope coefficient of the variable because it changes what a one-unit increase in the independent variable means. Researchers typically rescale variables to make slope coefficients easier to interpret.

For instance, the original CIS variable measuring seniority with an employer was reported in months. To make the regression results easier to interpret, I rescaled the variable so that it shows the number of years of seniority with an employer, by dividing the number of months by 12 (because there are 12 months in a year). I did this by using this formula to compute a new variable:

YEARS_OF_EMPLOYMENT = MONTHS_OF_EMPLOYMENT/12

Table 12.13 shows how the regression coefficients change when the variable capturing seniority with an employer is rescaled to years instead of months. Just as when mean-centring is used, when an independent variable is rescaled, the constant coefficient doesn't change and neither does the R^2; only the slope coefficient is different. When the original variable is used as the independent variable, the slope coefficient shows that each additional month of seniority with an employer is associated with earning $122 more in annual wages. When the rescaled variable is used as the independent variable, the slope coefficient shows that each additional year of seniority with an employer is associated with earning $1,460 more in annual wages. Notice that the slope coefficient of the rescaled seniority variable (measured in years) divided by 12 is equal to the slope coefficient of original seniority variable (measured in months). Rescaling doesn't change the size or direction of a relationship between two variables; rather, it is used to make slope coefficients easier to understand and interpret.

Table 12.13   How Rescaling Affects Regression Coefficients

Dependent variable: Annual wages and salaries ($), before deductions (n = 12,896)

Without Rescaling		With Rescaling	
	Unstandardized Coefficient		Unstandardized Coefficient
Constant	34,518.98*	Constant	34,518.98*
Months of employment in current or main job	121.68*	Years of employment in current or main job	1,460.11*
R^2	0.13	R^2	0.13

*Indicates that results are statistically significant at the p < 0.05 level.
Source: Author generated; Calculated using data from Statistics Canada, 2016.

Rescaling techniques are often useful when an independent variable is measured in dollars. The slope coefficient associated with a one-dollar increase (in say, annual income) sometimes appears to be 0.000. This slope coefficient does not indicate that the independent variable has no relationship with the dependent variable (especially if it has a p-value less than 0.05). Instead, this slope coefficient may indicate that the variable needs to be rescaled to make the relationship visible; typically there are non-zero digits in the later decimals. By rescaling the variable so that the slope coefficient shows the change in the dependent variable associated with a 1,000-dollar increase or a 10,000-dollar increase in the independent variable, the regression results become easier to interpret.

Best Practices in Presenting Results

More on Writing about Regression Results

In Chapter 11, you learned the basics of writing about regression results. In this chapter, you learned how to interpret multiple linear regressions. Reporting the results of these regressions requires some additional writing strategies. First, when reporting multiple regression results, you should always note that each slope coefficient predicts the relationship between the respective independent variable and the dependent variable, *controlling* for the other independent variables in the regression. However, listing all of the independent variables that are being controlled for each time that you report a slope coefficient can become unwieldy. One typical tactic is to list all of the independent variables the first time that you mention *controlling* in a paragraph, and then to simply say "controlling for the other variables in the model" in the remainder of the paragraph. In addition, if you are using dummy variables to incorporate a categorical independent variable into a regression, you only need to list the original categorical variable and not each of the dummy variables individually:

> *Incorrect:* Each additional hour that people work is associated with earning $28 more in annual wages, controlling for having less than a high school graduation, having a non-university certificate or diploma, and having a university degree or certificate.
> *Correct:* Each additional hour that people work is associated with earning $28 more in annual wages, controlling for their 'highest educational credential.

Writing about the slope coefficients of dummy variables is similar to writing about any other regression slope coefficient, with one key difference: you must identify the reference group to whom comparisons are being made. Without this key piece of information, a reader is unable to fully understand the results.

> *Incorrect:* People with disabilities are predicted to earn $3,042 less in annual wages, after controlling for the number of hours that they work at their job, their seniority with an employer, and their highest educational credential.
> *Correct:* People with disabilities are predicted to earn $3,042 less in annual wages than people without disabilities, after controlling for the number of hours that they work at their job, their seniority with an employer, and their highest educational credential.
> *Incorrect:* People with a university degree or certificate are predicted to earn $17,993 more in annual wages, after controlling for the number of hours that they work at their job, their seniority with an employer, and whether or not they have a disability.

> *Correct:* People with a university degree or certificate are predicted to earn $17,993 more in annual wages than people who only graduated from high school, after controlling for the number of hours that they work at their job, their seniority with an employer, and whether or not they have a disability.

Finally, standardized regression coefficients (beta coefficients) show the amount of change in the dependent variable (measured in standard deviations) associated with a one standard deviation increase in the independent variable. But because these details are hard to intuitively understand, especially without corresponding information about the standard deviations of each of the variables in the regression, researchers typically do not report the exact size of beta coefficients. Instead, they only report the relative magnitude of the relationship between each independent variable and the dependent variable; that is, they report which independent variables have the strongest and weakest relationships with the dependent variable.

What You Have Learned

In this chapter, you learned how linear regression can be extended to incorporate multiple independent variables. The main advantage of this approach is that it allows researchers to predict the relationship between each independent variable and the dependent variable, while controlling for, or holding constant, all of the other independent variables. In addition, you learned how to calculate and interpret standardized slope coefficients in order to assess which independent variable has the strongest relationship with the dependent variable. Finally, you learned how to create dummy variables in order to use categorical variables as independent variables in a regression. Together, these skills allow you to create more complex regression models that make more accurate predictions.

This chapter focused on how the annual wages of employees with disabilities compare to those of employees without disabilities. People with disabilities have a different employment profile than people without disabilities: they tend to be older and, thus, have more seniority, but they have lower levels of education and are less likely to work full-time. When only disability status is considered, people with disabilities are predicted to earn $6,370 less in annual wages than people without disabilities. But, once the number of hours that people work at their job, their years of seniority with an employer, and their highest educational credential are accounted for, people with disabilities are predicted to earn only $3,042 less in annual wages than people without disabilities. These results suggest that people with disabilities still experience some wage discrimination but that some of the difference in annual wages between people with disabilities and those without disabilities is attributable to differences in the number of hours that people work at their jobs and their level of education. Thus, one strategy for reducing the wage differential between people with disabilities and those without is to provide the workplace accommodations necessary to ensure that people with disabilities can work full-time hours, if they choose to do so. Further, reducing institutional barriers to post-secondary education for people with disabilities is another way to potentially reduce the wage differential over time.

Check Your Understanding

Check to see if you understand the key concepts in this chapter by answering the following questions:

1. How is multiple linear regression different than simple linear regression?
2. What does it mean to "control for" an independent variable?
3. What is omitted variable bias, and why do researchers strive to avoid it?
4. What do standardized slope coefficients show?

5. What modifications are required to incorporate categorical variables as independent variables in a linear regression?
6. How do you interpret the slope coefficient of a dummy variable?
7. How do you interpret the constant coefficient when a linear regression uses dummy variables?
8. What is rescaling? Why do researchers rescale a variable before using it in a linear regression?

Practice What You Have Learned

Check to see if you can apply the key concepts in this chapter by answering the following questions. Keep *four* decimal places in all of the calculations in this chapter.

1. A community agency that helps people to find employment is developing its annual fundraising campaign. Last year, the agency collected information from 500 people who had made donations, including their age, weekly income, and how much they donated. Now the agency wants to use a multiple linear regression to predict how the independent variables "Age" and "Weekly income" are related to donation amounts. A preliminary statistical analysis shows that:

   - The average age of donors was 45 (s.d. = 20).
   - The average weekly income of donors was $900 (s.d. = $400).
   - The average donation was $1,200 (s.d. = $500).
   - The Pearson's correlation (r) between "Age" and "Weekly income" is 0.48.
   - The Pearson's correlation (r) between "Age" and "Amount donated" is 0.12.
   - The Pearson's correlation (r) between "Weekly income" and "Amount donated" is 0.31.

   a. Calculate the partial slope coefficient that captures the general pattern of the relationship between people's age and how much they donated, controlling for their weekly income.

   b. Explain what the partial slope coefficient shows. Be sure to pay attention to the idea of "controlling."

2. Using the information in question 1:

   a. Calculate the partial slope coefficient that captures the general pattern of the relationship between people's weekly income and how much they donated, controlling for their age.

   b. Explain what the partial slope coefficient shows.

3. Using the information in question 1, and your answers to questions 1(a) and 2(a):

   a. Calculate the constant coefficient of the regression.

   b. Explain what the constant coefficient shows.

4. Using the information in question 1, and your answers to questions 1(a) and 2(a):

   a. Calculate the standardized partial slope coefficient capturing the general pattern of the relationship between people's age and how much they donated, controlling for their weekly income.

   b. Calculate the standardized partial slope coefficient capturing the general pattern of the relationship between people's weekly income and how much they donated, controlling for their age.

   c. Using your answers to (a) and (b), determine whether age or weekly income has a stronger relationship with donation amounts.

5. The community agency is interested in investigating who donates more money to charity among Canadians overall (and not just the agency's previous donors). The agency's research team uses information from the 2013 General Social Survey conducted by Statistics Canada to produce the multiple linear regression shown in Table 12.14, which predicts the total amount of money that people donated to charitable organizations in the past 12 months, using two independent variables: "Age (in years, centred on 45)" and "Annual personal income" (rescaled to be in thousands of dollars). (Note: the 1.2 per cent of people who report donating more than $5,000 in the past 12 months are excluded from this analysis.)

   a. Explain what the unstandardized slope coefficient of the "Age" variable shows. Be sure to pay attention to the idea of "controlling."
   b. Explain what the unstandardized slope coefficient of the "Annual personal income" variable shows.
   c. Explain what the constant coefficient shows.

Table 12.14 Results of a Multiple Linear Regression Predicting the Total Financial Donations to Charitable Organizations ($) in the Past 12 Months (GSS 2013)

Dependent variable: Total financial donations to charitable organizations ($) in the past 12 months (n = 14,538)

	Unstandardized Coefficient	Standardized Coefficient
Constant	187.07*	–
Age (in years, centred on 45)	5.04*	0.141
Annual personal income (in thousands of dollars)	3.33*	0.161
R^2	0.05	

*Indicates that results are statistically significant at the p < 0.05 level.
Source: Author generated; Calculated using data from Statistics Canada, 2015.

6. Using the information in Table 12.14:
   a. Determine which independent variable has the strongest relationship with the dependent variable, using the standardized slope coefficients.
   b. Explain what the R^2 shows.

7. The agency wants to make some predictions so that it can target potential donors. Using the information in Table 12.14, calculate how much money each of the following types of people is predicted to donate to charitable organizations in a 12-month period. (Be sure to account for the centring of the "Age" variable and the scaling of the "Annual personal income" variable in your calculations.)

   a. A 45-year-old who earns $40,000 per year
   b. A 25-year-old who earns $0 per year
   c. A 65-year-old who earns $80,000 per year

8. The agency is also interested in understanding how people's level of education is related to their charitable giving. The agency's research team begins by producing the simple linear regression shown in Table 12.15, which predicts the total amount of money that people donated to charitable organizations in the past 12 months, using a dummy variable indicating whether or not they have a post-secondary education.

   a. Explain what the unstandardized slope coefficient of the "Has a post-secondary education" dummy variable shows.
   b. Explain what the constant coefficient shows.

Table 12.15 Results of a Simple Linear Regression Predicting the Total Financial Donations to Charitable Organizations ($) in the Past 12 Months (GSS 2013)

Dependent variable: Total financial donations to charitable organizations ($) in the past 12 months (n = 13,658)

	Unstandardized Coefficient	Standardized Coefficient
Constant	223.50*	–
Has a post-secondary education (dummy variable)	168.80*	0.128
R^2	0.02	

*Indicates that results are statistically significant at the p < 0.05 level.
Source: Author generated; Calculated using data from Statistics Canada, 2015.

9. In order to investigate the independent influence of people's annual personal income and having post-secondary education on charitable giving, the agency's research team uses both independent variables in a multiple linear regression predicting the total amount of money that people donated to charitable organizations in the past 12 months. The results are shown in Table 12.16.

    a. Explain what the unstandardized slope coefficient of the "Annual personal income" variable shows.

    b. Explain what the unstandardized slope coefficient of the "Has a post-secondary education" dummy variable shows, now that annual personal income is being controlled for.

    c. Explain why there is a difference between the unstandardized slope coefficient of the "Has a post-secondary education" dummy variable in the two regressions shown in Tables 12.15 and 12.16.

10. Using the information in Table 12.16:

    a. Explain what the constant coefficient shows.

    b. Explain why there is a difference between the constant coefficient in the two regressions shown in Tables 12.15 and 12.16.

11. Either by hand or using a spreadsheet program, use the information in Table 12.16 to create a graph showing the predicted donations of people with different annual personal incomes, ranging from $0 to $120,000 (i.e. $10,000, $20,000, $30,000 and so on), with and without a post-secondary education.

12. The agency's research team decides that it wants to develop a more detailed understanding of how people's level of education relates to their charitable giving. Instead of using a single dummy variable indicating whether or not people have a post-secondary education, the team creates four dummy variables, each associated with a different level of education: "Less than

Table 12.16   Results of a Multiple Linear Regression Predicting the Total Financial Donations to Charitable Organizations ($) in the Past 12 Months (GSS 2013)

Dependent variable: Total financial donations to charitable organizations ($) in the past 12 months (n = 13,658)

	Unstandardized Coefficient	Standardized Coefficient
Constant	129.29*	–
Annual personal income (in thousands of dollars)	3.36*	0.163
Has a post-secondary education (dummy variable)	105.36*	0.080
R^2	0.04	

*Indicates that results are statistically significant at the $p < 0.05$ level.
Source: Author generated; Calculated using data from Statistics Canada, 2015.

Table 12.17   Results of a Multiple Linear Regression Predicting the Total Financial Donations to Charitable Organizations ($) in the Past 12 Months (GSS 2013)

Dependent variable: Total financial donations to charitable organizations ($) in the past 12 months (n = 13,658)

	Unstandardized Coefficient	Standardized Coefficient
Constant	243.32*	–
Education (ref: high school only)		
Less than high school (dummy variable)	−67.01*	−0.034
Post-secondary diploma (dummy variable)	75.17*	0.054
University degree (dummy variable)	242.00*	0.161
R^2	0.03	

*Indicates that results are statistically significant at the $p < 0.05$ level.
Source: Author generated; Calculated using data from Statistics Canada, 2015.

high school," "High school only," "Post-secondary diploma," and "University degree." Three of the dummy variables are used as independent variables in a multiple linear regression predicting the total amount of money that people donated to charitable organizations in the past 12 months. ("High school only" is used as the reference group.) The results are shown in Table 12.17.

a. Explain what the constant coefficient shows.
b. Explain what the unstandardized slope coefficients of the three education dummy variables show.

13. Use the information in Table 12.17 to calculate how much money people with each of the four levels of education are predicted to donate to charitable organizations in a 12-month period. Plot the predicted values and the associated slope coefficients on a number line, like the one in Figure 12.3.

14. Table 12.18 shows a more complex multiple linear regression, which uses age, sex/gender, level of education, and annual personal income to predict the total amount of money that people donated to charitable organizations in the past 12 months.

a. Explain how people's age and sex/gender are related to the amount of money they donate to charitable organizations.
b. The unstandardized slope coefficient of the "Women" dummy variable is not statistically significant at the $p < 0.05$ level. Explain what this result indicates.
c. Explain how people's level of education and annual personal income are related to the amount of money they donate to charitable organizations.
d. Explain what the R^2 shows.

15. Using the information from Table 12.16 and your answer to question 14, describe the advice that you would give to the local agency about developing its fundraising campaign. What characteristics are associated most strongly with the amount of money that people donate?

Table 12.18 Results of a Multiple Linear Regression Predicting the Total Financial Donations to Charitable Organizations ($) in the Past 12 Months (GSS 2013)

Dependent variable: Total financial donations to charitable organizations ($) in the past 12 months (n = 13,658)

	Unstandardized Coefficient	Standardized Coefficient
Constant	153.26*	–
Age (in years, centred on 45)	5.19*	0.145
Women (dummy variable)	20.29	0.016
Annual personal income (in thousands of dollars)	2.67*	0.130
Education (ref: high school only)		
Less than a high school education (dummy variable)	−57.75*	−0.030
Post-secondary diploma (dummy variable)	37.94*	0.027
University degree (dummy variable)	180.51*	0.120
R^2	0.07	

*Indicates that results are statistically significant at the $p < 0.05$ level.
Source: Author generated; Calculated using data from Statistics Canada, 2015.

16. The graph in Figure 12.5 is excerpted from a Statistics Canada report on the relationship between employment and disabilities (Turcotte 2014). It shows the employment rate—that is, the percentage of people in each group who are employed—in relation to people's highest level of education and severity of disability. Use the information in the graph to write a paragraph describing the general pattern of the relationships between employment rates and level of education, and employment rates and severity of disability.

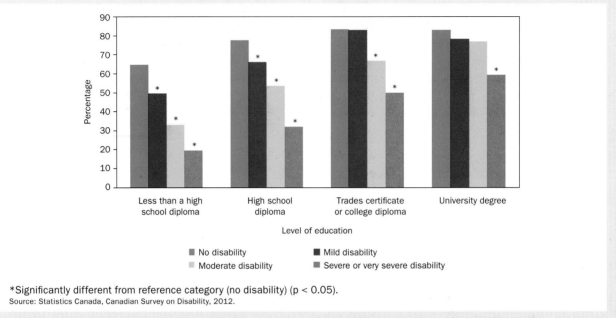

*Significantly different from reference category (no disability) (p < 0.05).
Source: Statistics Canada, Canadian Survey on Disability, 2012.

Figure 12.5   **Employment Rate Adjusted for Age, by Education Level and by Severity of Disability, 2011**

Source: Turcotte 2014, 4.

Practice Using Statistical Software (IBM SPSS)

Answer these questions using IBM SPSS and the GSS27. sav or the GSS27_student.sav dataset available from the Student Resources area of the companion website for this book. Weight the data using the "Standardized person weight" [STD_WGHT] variable you created following the instructions in Chapter 5. Report two decimal places in your answers, unless fewer are printed by IBM SPSS. It is imperative that you save the dataset to keep any new variables that you create.

1. Use the Linear Regression procedure to produce a regression of the independent variables "Number of groups – 12 months (recoded)" [GRP_10C_ RECODED] (which you created in Chapter 11) and "Number of new people met - Past month" [SCP_110] on the dependent variable "Number of close friends" [SCF_100C].

    a. Explain what the constant coefficient shows.
    b. Explain what the two unstandardized slope coefficients show. Be sure to pay attention to the idea of "controlling."

    c. Determine which independent variable has the strongest relationship with the dependent variable, using the standardized slope coefficients.

2. For the regression you produced in question 1:

    a. Explain what the R^2 shows.
    b. Describe how the R^2 of this multiple linear regression compares to the R^2 of the two simple linear regressions you produced using these independent variables in questions 2 and 5 of "Practice Using Statistical Software" in Chapter 11.

3. Use the Recode into Different Variables tool to recode the "Sex of respondent" [SEX] variable into a "Women" [WOMEN] dummy variable. Assign the value "1" to women (females) and assign the value "0" to men (males) in the new variable. Produce frequency distributions of the original variable "Sex of respondent" [SEX] and the new variable "Women" [WOMEN], and compare them to be sure that the recoding is correct.

4. Use the Linear Regression procedure to produce a regression of the independent variable "Women" [WOMEN] on the dependent variable "Number of close friends" [SCF_100C].

   a. Explain what the constant coefficient shows.
   b. Explain what the unstandardized slope coefficient of the dummy variable shows.

5. Use the Linear Regression procedure to produce a regression of the independent variables "Number of groups – 12 months (recoded)" [GRP_10C_RECODED], "Number of new people met - Past month" [SCP_110], and "Women" [WOMEN] on the dependent variable "Number of close friends" [SCF_100C].

   a. Explain what the constant coefficient shows.
   b. Explain what each of the three unstandardized slope coefficients show. Be sure to pay attention to the idea of "controlling."
   c. Determine which independent variable has the strongest relationship with the dependent variable, using the standardized slope coefficients.

6. Create three dummy variables to capture marital status. Divide marital status into three groups: people who are in relationships (married [1] or common-law [2]), people who were previously in relationships (widowed [3], separated [4], or divorced [5]), and people who are single (6). Use the Recode into Different Variables tool to recode the original variable "Marital status of respondent" [MARSTAT] into dummy variables as follows:

   a. Create the new dummy variable "In a relationship" [IN_RELATIONSHIP] by assigning the old values "1" and "2", the new value "1"; and assigning the old values "3" through "6", the new value "0". (The remaining values can be designated as "System-missing" in the new variable.)
   b. Create the new dummy variable "Previous relationship" [PREVIOUS_RELATIONSHIP] by assigning the old values "3" through "5", the new value "1"; and assigning the old values "1", "2" and "6", the new value "0". (The remaining values can be designated as "System-missing" in the new variable.)

   c. Create the new dummy variable "Single" [SINGLE] by assigning the old value "6", the new value "1"; and assigning the old values "1" through "5", the new value "0". (The remaining values can be designated "System-missing" in the new variable.)
   d. Produce frequency distributions of the original variable "Marital status of respondent" [MARSTAT] and each of the three new dummy variables ("In a relationship" [IN_RELATIONSHIP], "Previous relationship" [PREVIOUS_RELATIONSHIP], and "Single" [SINGLE]), and compare them to be sure the recoding is correct.

7. Use the Linear Regression procedure to produce a regression of the independent variables "Previous relationship" [PREVIOUS_RELATIONSHIP] and "Single" [SINGLE] on the dependent variable "Number of close friends" [SCF_100C]. "In a relationship" [IN_RELATIONSHIP] is used as the reference group.

   a. Explain what the constant coefficient shows.
   b. Explain what each of the two unstandardized slope coefficients show. Be sure to make comparisons to the reference group.

8. Use the Linear Regression procedure to produce a regression of the independent variables "Number of groups – 12 months (recoded)" [GRP_10C_RECODED], "Number of new people met - Past month" [SCP_110], "Women" [WOMEN], "Previous relationship" [PREVIOUS_RELATIONSHIP], and "Single" [SINGLE] on the dependent variable "Number of close friends" [SCF_100C].

   a. Explain what the constant coefficient shows.
   b. Explain what each of the unstandardized slope coefficients show. Be sure to pay attention to the idea of "controlling."
   c. Determine which independent variable has the strongest relationship with the dependent variable, using the standardized slope coefficients.
   d. Explain what the R^2 shows.

9. Rescale a variable to make the regression coefficient easier to interpret. Use the Compute Variable tool to create a new variable called "Number of new people met - Past month (scaled to 10)" [SCP_110_

RESCALED]. The new variable is equal to the value of "Number of new people met - Past month" [SCP_110] divided by 10. As a result, a one-unit increase in the rescaled variable indicates meeting an additional 10 new people each month. Use the Means procedure to find the mean of the original variable "Number of new people met - Past month" [SCP_110] and the new variable "Number of new people met - Past month (scaled to 10)" [SCP_110_RESCALED]. Compare the two results to be sure that the rescaling is correct: the number of cases should be exactly the same, and the mean and standard deviation of the rescaled variable should be exactly 10 times the size of the mean and standard deviation of the original variable.

10. Use the Linear Regression procedure to produce a regression of the independent variables "Number of new people met - Past month (scaled to 10)" [SCP_110_RESCALED] on the dependent variable "Number of close friends" [SCF_100C].

a. Explain what the constant and slope coefficients show. What do they indicate about the relationship between the number of new people met and people's number of close friends?

b. Compare the regression results to those you produced in question 5 of "Practice Using Statistical Software" in Chapter 11. Determine which parts of the regression output are identical and which are different.

Key Formulas

Multiple linear regression predictions	$\hat{y} = a + b_1 x_1 + b_2 x_2 + b_3 x_3 + b_4 x_4 + b_5 x_5 + \cdots$
Multiple linear regression with residuals	$y_i = a + b_1 x_{1i} + b_2 x_{2i} + b_3 x_{3i} + b_4 x_{4i} + b_5 x_{5i} + \cdots + e_i$
Partial slope coefficients (multiple linear regression with two independent variables)	$b_{x_1} = \left(\dfrac{r_{x_1,y} - \left[r_{x_1,x_2} \right] \left[r_{x_2,y} \right]}{1 - \left[r_{x_1,x_2} \right]^2} \right) \left(\dfrac{s_y}{s_{x_1}} \right)$ $b_{x_2} = \left(\dfrac{r_{x_2,y} - \left[r_{x_1,x_2} \right] \left[r_{x_1,y} \right]}{1 - \left[r_{x_1,x_2} \right]^2} \right) \left(\dfrac{s_y}{s_{x_2}} \right)$
Constant coefficient (multiple linear regression with two independent variables)	$a = \bar{y} - b_1 \bar{x}_1 - b_2 \bar{x}_2$
Standardized slope coefficient (Beta)	$\beta_x = b_x \left(\dfrac{s_x}{s_y} \right)$

References

Arim, Rubab. 2015. "A Profile of Persons with Disabilities among Canadians Aged 15 Years or Older, 2012." Catalogue no. 89-654-X. Ottawa: Statistics Canada. http://epe.lac-bac.gc.ca/100/201/301/weekly_checklist/2015/internet/w15-11-F-E.html/collections/collection_2015/statcan/89-654-x2015001-eng.pdf.

Canadian Human Rights Act. 1985. *Revised Statutes of Canada*. Vol. H-6. http://laws-lois.justice.gc.ca.

Crawford, Cameron. 2010. "Disabling Poverty and Enabling Citizenship: Understanding the Poverty and Exclusion of Canadians with Disabilities." Council of Canadians with Disabilities. http://www.ccdonline.ca/en/socialpolicy/poverty-citizenship/demographic-profile/understanding-poverty-exclusion.

Employment Equity Act. 1995. *Statutes of Canada*. Vol. 44. http://laws-lois.justice.gc.ca.

MacKenzie, Andrew, Matt Hurst, and Susan Crompton. 2009. "Defining Disability in the Participation and Activity Limitation Survey." Catalogue no. 11-008-X. Ottawa: Statistics Canada. http://www.statcan.gc.ca/pub/11-008-x/2009002/article/11024-eng.pdf.

Statistics Canada. 2015. General Social Survey, 2013: Cycle 27, Giving, Volunteering and Participating. *Public Use Microdata File*. Ottawa, ON: Statistics Canada.

———. 2016. "Canadian Income Survey (CIS) 2014." http://www23.statcan.gc.ca/imdb/p2SV.pl?Function=getSurvey&SDDS=5200.

Till, Matthew, Tim Leonard, Sebastian Yeung, and Gradon Nicholls. 2015. "A Profile of the Labour Market Experiences of Adults with Disabilities among Canadians Aged 15 Years or Older, 2012." Catalogue no. 89-654-X2015005. Ottawa: Statistics Canada. http://www.statcan.gc.ca/pub/89-654-x/89-654-x2015005-eng.pdf.

Turcotte, Martin. 2014. "Persons with Disabilities and Employment." Catalogue no. 75-006-X. Ottawa: Statistics Canada. http://www.statcan.gc.ca/pub/75-006-x/2014001/article/14115-eng.pdf.

United Nations. 2008. *Convention on the Rights of Persons with Disabilities*. https://www.un.org/development/desa/disabilities/convention-on-the-rights-of-persons-with-disabilities.html.

World Health Organization (WHO). 2011. "World Report on Disability: Summary." Geneva, Switzerland. http://apps.who.int/iris/bitstream/10665/70670/1/WHO_NMH_VIP_11.01_eng.pdf.

Building Linear Regression Models

13

Learning Objectives

In this chapter, you will learn:

- How to use blocks to group independent variables in a regression
- What nested regressions are and how to interpret them
- Strategies for determining which independent variables to use in a regression and which to omit
- How to analyze regression residuals
- What a Q-Q plot is and what it tells you
- What to report about regression modelling

Introduction

In Chapter 12, you learned how to interpret linear regressions that use more than one independent variable. In this chapter, you'll add to those skills by learning how to interpret nested regressions. I'll also describe some strategies for choosing which independent variables to use in a regression and which to omit. In addition, you'll learn how to analyze regression residuals in order to detect bias and refine linear regression models.

As in Chapter 12, the research focus of this chapter is people's wages. Using data from the 2016 Labour Force Survey, which was introduced in Chapter 4, multiple linear regression is used to consider how people's ascribed and achieved characteristics are related to their wages. Anthropologist Ralph Linton was the first to distinguish between ascribed status and achieved status: he defines ascribed status as being assigned to people at the moment of birth, whereas achieved status is attained as a result of individual effort (1936). Since then, the concept of ascribed status has been expanded to include both status that is assigned at birth and status that is assumed involuntarily later in life. Although the meaning of an ascribed characteristic can change depending on people's social and cultural context, within a specific context it is difficult (although not always impossible) for individuals to change their ascribed characteristics. Some ascribed characteristics that social scientists commonly study are age, sex/gender, race/racialization, ethnicity, disability status, and country of birth. In contrast, people can acquire achieved characteristics through their individual efforts and choices. Some achieved characteristics that social scientists commonly study are level of education, marital status, number of children, occupation, and income.

In theory, the Canadian labour market is structured as a meritocracy. That is, people who get ahead and who are successful are presumed to do so based on their individual merit or their individual skills, efforts, and abilities. Merit-based systems are designed to allow for maximum social mobility; that is, people's social status is not necessarily tied to their circumstances at birth or to their family of origin. Social scientists have, however, presented substantial evidence that the idea of meritocracy is a myth and that ascribed characteristics, as well as social and cultural capital, affect people's ability to get ahead (MacNamee and Miller 2014). Discrimination related to ascribed characteristics, such as sex/gender, age, racialization, disability status, and immigrant status, results in employment outcomes that perpetuate inequality and limit social mobility. This discrimination occurs in various ways: some people are streamed into particular types of jobs, some people are paid less to do the same job, and some jobs are valued or paid less (or more) because they are done by people from specific social groups (Peterson and Saporta 2004). For instance, Castilla (2008) found that within one large organization, among people with similar performance evaluations, women, racialized people, and immigrants received lower merit-based pay increases; that is, even among people with similar "merit," people's ascribed characteristics affected their rewards.

The statistical analyses in this chapter investigate how ascribed and achieved characteristics are related to people's usual weekly wages in their main job, that is, the job that they spend the most time working at (Statistics Canada 2016a). Only people who are currently employed and who are formally employees (not self-employed) are included in the analyses. As well, the 0.3 per cent of people who earn more than $3,000 per week (equivalent to more than $156,000 per year) in their main job are excluded. The average usual weekly wages are $914 (s.d. = $567), or the equivalent of $47,528 for people who work 52 weeks per year. The ascribed characteristics used in these analyses are age and sex/gender. Even though some people change their sex/gender, people's initial sex/gender is typically ascribed at birth. Although it would be ideal to incorporate the ascribed characteristics of racialization, immigrant status, and disability status into these analyses, no variables measuring these characteristics are available in the Labour Force Survey.

The achieved characteristics used in these analyses are people's highest level of education and their marital status. People typically achieve educational credentials through their own effort, often for the express purpose of increasing their earning potential. The ladder system of education in Canada is structured around the idea that people with higher educational credentials have access to higher paying jobs. In addition, people's marital status is an achieved characteristic that reflects their efforts to find and retain a partner. Substantial evidence demonstrates that marriage is associated with a wage premium, especially for men (Killewald and Gough

2013, Budig and Lim 2016). The higher wages associated with marriage are often attributed to the fact that people can specialize within a couple so that the higher earning partner can devote more time to working for pay, while the lower earning partner can devote more time to the domestic work needed to maintain a household. Another explanation for the marriage wage premium is that it reflects selection bias, whereby higher earning people (or those with higher earning potential) are more likely to attract a mate and enter into marriage. Finally, employers may favour married workers because they are perceived as more dependable, and employers may pay married workers more in an effort to retain them. In the analyses in this chapter, people who are either married or in common-law relationships are compared to people who are single (never married) and to people who are divorced, widowed, or separated.

Finally, it's important to control for the characteristics of people's jobs that affect their earnings. In these analyses, the job characteristics that are controlled for are the number of hours that people usually work at their job each week, their years of seniority with an employer, whether a job is permanent (as opposed to temporary, casual, or seasonal), and whether a job is unionized. After controlling for job characteristics, in a meritocratic system, achieved characteristics should have more influence on people's wages than ascribed characteristics.

In this chapter, statistical analysis is used to investigate the following:

- How are ascribed characteristics related to wages?
- How are achieved characteristics related to wages?
- How are job characteristics related to wages?
- Do ascribed characteristics or achieved characteristics have more influence on wages after job characteristics are taken into account?

Nested Regressions

As you know, linear regression is used to identify the general pattern of a relationship between one or more independent variables and a ratio-level dependent variable. One main strength of regression is that it allows researchers to identify the unique influence of each independent variable on a dependent variable, after controlling for the other independent variables in the regression.

nested regressions Two or more regressions that use the same dependent variable, where each subsequent regression adds new independent variables without removing any of the previous independent variables.

As part of the model-building process, researchers often generate a series of linear regressions that use the same dependent variable, adding more and more independent variables each time. These are called **nested regressions**. In order for two or more regressions to be considered nested, they must have the same dependent variable, and each subsequent regression must add new independent variables without removing any of the previous independent variables. Nested regressions are particularly useful for understanding how controlling for additional characteristics changes the relationship between each independent variable and the dependent variable. In other words, researchers used nested regressions to assess how the partial slope coefficient of each independent variable changes as additional variables or groups of variables are accounted for.

Most statistical software allows researchers to add independent variables to a regression in "blocks" or groups. This allows researchers to determine how a group of variables is associated with the dependent variable. It also makes it possible for researchers to more explicitly link their regression models to theoretical models. In a nested regression with two blocks, one set of regression coefficients is calculated as though the variables in the first block are the only independent variables in the regression. Then a second set of regression coefficients is calculated using the independent variables in both the first and the second blocks. This makes it easy to see how the slope coefficients of the independent variables in the first block change once the variables in the second block are controlled for. One useful feature of nested regression is that cases that are missing information on any independent variable, in any block, are omitted at the start; thus, exactly the same cases are used in each regression.

Let's illustrate how nested regressions work by investigating how ascribed and achieved characteristics are related to wages. The ascribed characteristics of age and sex/gender are entered as the first block of independent variables in a multiple linear regression predicting people's usual weekly wages. The regression results for the first block of independent variables, or the first model, are shown in Table 13.1. The constant coefficient indicates that 40-year-old men are predicted to earn $1,023 in wages each week. For each additional year older that people are, they are predicted to earn $10 more in weekly wages, after controlling for sex/gender. This result reflects young people's relative disadvantage in the labour market, compared to older workers. Gender inequalities are also apparent in these results: women are predicted to earn $231 less in weekly wages, after controlling for age. The standardized regression coefficients (not shown), indicate that age has a stronger relationship with weekly earnings than sex/gender. Taken together, these two ascribed characteristics account for 11 per cent of the variation in weekly wages.

As illustrated in Chapter 12, these initial slope coefficients are likely to change once additional variables are controlled for. In a meritocracy, achieved characteristics, such as level of education, are expected to have more influence

Table 13.1 Results of a Multiple Linear Regression with One Block of Independent Variables (Model 1 of a Nested Regression)

Dependent variable: Usual weekly wages ($), main job (n = 532,353)

	Model 1 Unstandardized Coefficient
Constant	1,023.38*
Ascribed Characteristics	
Age (in years, centred on 40)	10.42*
Women (ref: men)	−230.84*
R^2	0.11

*Indicates that results are statistically significant at the $p < 0.05$ level.
Source: Author generated; Calculated using data from Statistics Canada, 2016b.

Table 13.2   Results of a Multiple Linear Regression with Two Blocks of Independent Variables (Models 1 and 2 of a Nested Regression)

Dependent variable: Usual weekly wages ($), main job (n = 532,353)

	Model 1 Unstandardized Coefficient	Model 2 Unstandardized Coefficient
Constant	1,023.38*	947.66*
Ascribed Characteristics		
Age (in years, centred on 40)	10.42*	4.94*
Women (ref: men)	−230.84*	−266.22*
Achieved Characteristics		
Highest educational credential (ref: high school graduation only)		
Less than high school graduation		−134.57*
Post-secondary certificate/diploma		181.85*
University degree		424.23*
Marital status (ref: married or common-law)		
Single, never married		−211.53*
Widowed, separated, or divorced		−35.90*
R^2	0.11	0.25

*Indicates that results are statistically significant at the $p < 0.05$ level.
Source: Author generated; Calculated using data from Statistics Canada, 2016b.

on people's earnings than ascribed characteristics. The dummy variables corresponding to people's achieved characteristics of highest educational credential and marital status are entered as the second block of independent variables in the multiple linear regression predicting people's usual weekly wages. Table 13.2 shows the regression results for the first and second block of independent variables, or the first and second models. The regression coefficients in the "Model 1" column are identical to those shown in Table 13.1, and the "Model 2" column shows how the regression coefficients change when people's level of education and their marital status are accounted for.

Comparing the regression coefficients in the "Model 1" and "Model 2" columns in Table 13.2 shows that the relationship between age and wages is weaker once people's level of education and marital status are accounted for. In the first model, each additional year of age is associated with a $10 increase in weekly wages, whereas in the second model each additional year of age is associated with only a $5 increase in weekly wages. This is because young people are more likely to be single (never married) than older people, and some of the variation in wages that was attributed to age is now being attributed to marital status. More notably, however, the relationship between sex/gender and wages is stronger—in other words, being a woman is associated with an even greater wage penalty in the second model. This is because women are more likely to have a university degree than men. Once people's highest level of education is taken into account, the effect of gender discrimination on wages becomes even more apparent. Compared to men who are the same age, have the same marital status, and have the same level of education, women are predicted to earn $266 less in weekly wages (the equivalent of $13,843 less per year).

In the second model shown in Table 13.2, the constant coefficient of $948 indicates the predicted weekly wages of people with a "0" value on all of the independent variables: 40-year-old men who are married or in common-law relationships and have only graduated from high school. As expected, having a higher level of education is associated with earning higher wages. The reference group for the education dummy variables is people who have only graduated from high school. People with a post-secondary certificate or diploma are predicted to earn $182 more in weekly wages than people who have only graduated from high school, and people with a university degree are predicted to earn $424 more in weekly wages than people who have only graduated from high school, after controlling for age, sex/gender, and marital status.

These regression results also support the idea of a marriage wage premium. The reference group for marital status is people who are married or in common-law relationships. People who are single and who have never been married are predicted to earn $212 less in wages each week than people who are married or in common-law relationships, after controlling for age, sex/gender, and highest educational credential. People who are widowed, separated, or divorced are also predicted to earn slightly less than people who are married or in common-law relationships: $36 less each week, after controlling for the other variables in the regression.

In the second model shown in Table 13.2, the variable with the largest standardized slope coefficient is "University degree" (an achieved characteristic), and the variable with the second-largest standardized slope coefficient is "Women" (an ascribed characteristic). In the first model, the R^2 of 0.11 indicates that 11 per cent of the variation in weekly wages can be explained by the two ascribed characteristics. In the second model, the R^2 of 0.25 indicates that 25 per cent of the variation in weekly wages can be explained by the ascribed characteristics and the achieved characteristics together. In other words, adding achieved characteristics to the regression explains 14 per cent more of the variation in weekly wages, compared to ascribed characteristics alone. These results suggest that, collectively, the achieved characteristics included in this model account for slightly more of the variation in wages than the ascribed characteristics.

Whenever you compare the R^2 of nested regressions, be aware that the results will change depending on which order the blocks of variables are entered in. For instance, when the achieved characteristics are entered as the first block of independent variables and the ascribed characteristics are entered as the second block, the R^2 of the first model is 0.19 and the R^2 of the second model is 0.25. In this context, a researcher might conclude that achieved characteristics account for 19 per cent of the variation in weekly wages (not 14 per cent) and that ascribed characteristics only account for an additional 6 per cent of the variation in weekly wages (not 11 per cent). Ultimately, your theoretical model, or the argument that you are striving to develop, should determine the order in which variables or blocks of variables are added to the regression.

The results of the analyses in Chapter 12 show that job characteristics have a substantial influence on wages—particularly the number of hours that people work at their job. Having more years of seniority with an employer is also related to higher wages. In addition, people in permanent jobs tend to earn higher wages than those who are in temporary, contract, or casual jobs. Finally, people in jobs that are unionized or covered by a collective agreement typically earn higher wages than people in non-unionized jobs, as a result of collective bargaining outcomes (people whose job is covered by a collective agreement are treated as if they are unionized in this analysis). Each of these characteristics is a feature of the job itself, not a characteristic of the person who holds the job. Nonetheless, some researchers treat job characteristics as achieved characteristics because they reflect people's choices or efforts to obtain a job with desirable features. Here, however, I am more interested in job characteristics as control variables. The four variables corresponding to job characteristics are used as the third block of independent variables in the multiple linear regression predicting people's usual weekly wages. My decision to add the three blocks of variables into the regression in this order reflects people's typical progression over time: first, their age and sex/gender are established (ascribed characteristics); then, they gain an education and marital status (achieved characteristics); and, finally, they enter into a job with specific characteristics (job characteristics).

Table 13.3 shows the results of all three nested regressions in a single table. Reading across each row shows how the unstandardized slope coefficient associated with each independent variable changes as additional variables are controlled for. The constant coefficient of the third model, shown in the final column of Table 13.3, indicates that 40-year-old men, who are married or in common-law relationships, with only a high school education, work 40 hours a week, and have one year of seniority in a temporary, non-unionized job, are predicted to earn $829 in weekly wages. Each additional hour that people work at their job is associated with earning $23.50 more in weekly wages, after controlling for people's age, sex/gender, highest educational credential, marital status, years of seniority with an employer, job permanency and unionization. Each additional year of seniority with an employer is associated with earning $15 more in weekly wages, after controlling for the other variables in the regression. The slope coefficients of the third model also show that people in unionized jobs are predicted to earn $97 more each week than people in non-unionized jobs, and people in permanent jobs are predicted to earn $74 more per week than people in non-permanent jobs, after controlling for people's age, sex/gender, highest educational credential, marital status, the number of hours they work at their job each week, and their years of seniority with an employer. As expected, job characteristics explain a substantial amount of the variation in weekly wages. The change in the R^2 from 0.25 in the second model to 0.47 in the third model indicates that an additional 22 per cent of the variation in weekly wages can be explained by job characteristics.

It is interesting to examine how the slope coefficients of the ascribed characteristics in the third model compare to those in the first and second models. The

Table 13.3 Results of a Multiple Linear Regression with Three Blocks of Independent Variables (Models 1, 2, and 3 of a Nested Regression)

Dependent variable: Usual weekly wages ($), main job (n = 532,353)

	Model 1 Unstandardized Coefficient	Model 2 Unstandardized Coefficient	Model 3 Unstandardized Coefficient
Constant	1,023.38*	947.66*	829.43*
Ascribed Characteristics			
Age (in years, centred on 40)	10.42*	4.94*	−1.15*
Women (ref: men)	−230.84*	−266.22*	−152.88*
Achieved Characteristics			
Highest educational credential (ref: high school graduation only)			
Less than high school graduation		−134.57*	−45.09*
Post-secondary certificate/diploma		181.85*	123.56*
University degree		424.23*	364.79*
Marital status (ref: married or common-law)			
Single, never married		−211.53*	−131.35*
Widowed, separated, or divorced		−35.90*	−17.31*
Job Characteristics			
Usual hours worked per week at main job (centred on 40)			23.50*
Years of employment in current or main job (centred on 1)			15.23*
Unionized job (ref: not unionized)			96.50*
Permanent job (ref: temporary)			74.47*
R^2	0.11	0.25	0.47

*Indicates that results are statistically significant at the $p < 0.05$ level.
Source: Author generated; Calculated using data from Statistics Canada, 2016b.

slope coefficient of age becomes very small, and negative, once achieved characteristics and job characteristics are taken into account. Each additional year of age is now associated with earning $1 less per week. The relationship between gender and weekly wages also becomes weaker but is still substantial in real-world terms: even after achieved characteristics and job characteristics are accounted for, women are predicted to earn $153 less each week than men (the equivalent of about $7,950 less each year).

The slope coefficients of the dummy variables capturing people's level of education and marital status also become smaller once job characteristics are accounted for, suggesting that some of the variation in wages that was attributed to level of education and marital status is actually attributable to job characteristics. For instance, people with a post-secondary education are more likely to be employed in unionized, permanent jobs; they also work more hours at their job each week and have more years of seniority with their employer than people who do not have a post-secondary education. Similarly, people who are single are more likely to be employed in non-unionized, temporary jobs; they also work fewer hours at their job each week and have fewer years of seniority with their employer than people who are married or in common-law relationships.

The standardized slope coefficients of the third model show that the variable with the strongest relationship with weekly wages is the "Usual hours worked per week at main job," followed by "University degree," "Years of employment in current or main job," and "Women." Together, these three nested regressions help us to assess the overall influence of ascribed characteristics, achieved characteristics, and job characteristics in predicting wages. Social scientists strive to build regression models that accurately reflect how individual characteristics and social processes influence people's lives. A more comprehensive investigation might also incorporate information about the occupations and industries that people work in to assess whether the wage penalty associated with gender persists even after taking into account the types of work that people do.

Strategies for Selecting Independent Variables

Researchers aim to construct the best regression models possible with the available data. In other words, they try to create models that use independent variables that are theoretically meaningful and useful for making predictions about the dependent variable and that omit independent variables that are not theoretically meaningful and do not help to make predictions about the dependent variable. Overall, researchers strive to generate models that are **parsimonious**. A parsimonious regression model is one that efficiently predicts the values on a dependent variable, without any extraneous independent variables. In other words, it is the simplest plausible regression model. But the more simplistic a model becomes, the more assumptions that researchers need to make about how the social world operates; thus, researchers must balance the goal of parsimony with how confident they are in their assumptions.

parsimonious The simplest and most efficient plausible explanation or model.

You might compare the process of building a parsimonious regression model to the adventures of Goldilocks in the fairy tale *Goldilocks and the Three Bears*. In the popular children's story, Goldilocks ventures into the forest and comes upon an abandoned house, belonging to the titular three bears. As Goldilocks investigates the house, she tries three versions of various household items, finding that the first has too much of something, the second has too little of something, and the third is "just right." (For instance, Goldilocks tries three bowls of porridge, finding that the first is too hot, the second is too cold, and the third is just the right temperature; she similarly finds that the first bed she tries is too hard, the second is too soft, and the third is just right.) A parsimonious regression model is one that is "just right"—it's not too complex but also not too simple; it uses just the right combination of independent variables—not too many and not too few—to predict the dependent variable. There are no extra independent variables that capture spurious relationships in the regression, but there are also no omitted variables, leading to omitted variable bias.

The starting point for selecting the independent variables in a regression should always be the most current knowledge and theories about predictors of the dependent variable. But researchers must often make additional choices about which independent variables to use and which to omit. Sometimes, the theoretically appropriate variables simply aren't available in the data; thus, researchers must substitute one, or more, less than ideal variables as proxies. Sometimes researchers need to limit the number of independent variables that they use in a regression because they don't have enough cases to support a complex model. (At absolute minimum, the number of cases should be 10 times the number of independent variables in a regression, dummy variables included.) There are several strategies that researchers use to help them decide which independent variables to include in a regression model.

Adjusted R^2

How does a researcher decide whether adding another independent variable to a linear regression makes enough of a difference to justify its inclusion? So far, you've learned how to use R^2 to assess how well a linear regression fits the observed data, by determining how much of the variation in the dependent variable is explained by the independent variables. Researchers also sometimes use the change in the R^2 to assess whether adding an independent variable to a regression explains more variation in the dependent variable. But this approach raises two concerns. First, it's important to use nested regressions to make this assessment so that the same cases are used in each regression. This ensures that any differences in the R^2 are because of improvements in the predictive power of the model, not just because some cases were excluded from a subsequent regression due to missing values on the newly added independent variables. More crucially, though, the R^2 increases a little bit whenever *any* independent variable is added to a linear regression—even if there is no plausible relationship between that independent variable and the dependent variable. Unless there is exactly zero correlation between the newly added independent variable and the dependent variable (a very unlikely situation), the additional independent variable will always explain just a bit more of the unexplained variation in the dependent variable. As a result, the R^2 will always increase whenever a new independent variable is added to a regression. This feature of R^2 makes it less useful for deciding whether it is worthwhile adding an additional independent variable to a regression model.

As you might guess, the **adjusted R^2** is a variation on R^2. It compensates for the fact that R^2 doesn't distinguish between whether an independent variable is a good predictor or a poor predictor of the dependent variable. The adjusted R^2 is calculated using R^2, the number of cases (n), and the number of independent variables (k) used in a linear regression, with this formula:

adjusted R^2 A variation on R^2 that penalizes linear regressions that use independent variables that are poor predictors of the dependent variable.

$$Adjusted\ R^2 = 1 - \left[\frac{\left(1 - R^2\right)\left(n - 1\right)}{n - k - 1} \right]$$

Adjusted R^2

By accounting for the number of independent variables in the denominator of the fraction ($n - k - 1$), the adjusted R^2 balances out the inevitable increase in R^2 associated with the increase in the number of independent variables. In practice, the adjusted R^2 invokes a penalty when independent variables that are poor predictors—that is, variables that do not explain much more of the variation in the dependent variable—are added to a regression. So, while the original R^2 will always increase when a new independent variable is added to a linear regression, the adjusted R^2 may increase *or* decrease. If the independent variable that is added is a good predictor of the dependent variable, the adjusted R^2 will increase, but if it is a poor predictor, the adjusted R^2 will stay the same or decrease. Most statistical software displays both the R^2 and the adjusted R^2 for linear regressions. In most published regression results, researchers only report the adjusted R^2, instead of the original R^2, to avoid overstating the predictive power of the model.

Let's use an example to illustrate. Table 13.4 shows a nested regression with two blocks of independent variables for a random 0.1 per cent sample of cases from the LFS data. Because the number of cases is used to calculate the adjusted R^2, the differences between the R^2 and the adjusted R^2 are more apparent when there are fewer cases. The first model uses only the number of hours that people work at their job each week to predict their weekly wages. The second model adds a variable that contains a random number ranging from 0 to 1,000. Since the variable contains random numbers, it is guaranteed to be a poor predictor of people's weekly wages. The Pearson's correlation between the random variable and weekly wages is -0.01 ($p = 0.876$). Table 13.4 shows the original R^2 and the adjusted R^2 of each regression. In the second model, which adds the random variable, the original R^2 increases (a tiny bit), but the adjusted R^2 decreases.

The lower adjusted R^2 in the second model indicates that the random variable is not a good predictor of weekly wages. Whenever the adjusted R^2 decreases in a subsequent nested linear regression, it indicates that the independent variable or variables that were added are not particularly good predictors of the

Table 13.4 Change in the R^2 and Adjusted R^2 between Two Nested Linear Regressions, One Including a Random Variable as an Independent Variable

Dependent variable: Usual weekly wages ($), main job (n = 532)

	Model 1 Unstandardized Coefficient	Model 2 Unstandardized Coefficient
Constant	1,113.77	1,129.14*
Usual hours per week at main job (centred on 40)	31.22*	31.24*
Random variable		−0.03
R^2	0.279	0.280
Adjusted R^2	0.279	0.277

*Indicates that results are statistically significant at the p < 0.05 level.
Source: Author generated; Calculated using data from Statistics Canada, 2016b.

dependent variable. As a result, researchers typically use the change in the adjusted R^2 to assess whether the independent variables that they have added to a nested regression improve the overall fit between the model and the observed data. If adding an independent variable does not improve the overall model fit, researchers sometimes will choose to omit that variable from their final regression. But even if an independent variable does not improve the overall model fit, if it is theoretically important, it should still be included in the regression model. In these situations, researchers should show the slope coefficients of the theoretically important variable and discuss the lack of a relationship in the context of their theoretical model.

Collinearity

As you know, one of the main features of multiple linear regression is that it estimates the unique relationship between each independent variable and the dependent variable while controlling for, or holding constant, the influence of other independent variables. Because of this, it becomes problematic when the independent variables in a regression are highly correlated with each other. In fact, this is one of the assumptions that multiple linear regression depends on: the independent variables are not

Step-by-Step: Adjusted R^2

$$\text{Adjusted } R^2 = 1 - \left[\frac{\left(1 - R^2\right)\left(n - 1\right)}{n - k - 1} \right]$$

Step 1: Find the total number of cases (n) used in the multiple linear regression. Do not include cases that are missing information for any variable in the regression.

Step 2: Count the number of independent variables (k) used in the multiple linear regression.

Step 3: Find the R^2 of the multiple linear regression.

Step 4: Subtract 1 from the total number of cases (from Step 1).

Step 5: Subtract the R^2 (from Step 3) from 1.

Step 6: Multiply the results of Step 4 and Step 5 to find the numerator of the fraction in the adjusted R^2 equation.

Step 7: Subtract the number of independent variables (from Step 2) from the result of Step 4 to find the denominator of the fraction in the R^2 equation.

Step 8: Divide the result of Step 6 by the result of Step 7.

Step 9: Subtract the result of Step 8 from 1 to find the adjusted R^2.

collinearity Occurs when the linear relationship between one independent variable and the dependent variable in a regression is very similar to the linear relationship between another independent variable and the dependent variable.

correlated with one another; or, more specifically, the value on any one independent variable cannot be predicted from the value on another independent variable (or a group of other independent variables in combination). When there is a strong correlation between two independent variables used in a regression, it is difficult to determine how much of the variation in the dependent variable is attributable to each variable alone. This is commonly referred to as a problem of **collinearity** or multicollinearity. For instance, imagine a regression where both people's age and their year of birth are used to make predictions about a dependent variable. It would be difficult to assess how much of the variation in the dependent variable is attributable to age, separate from people's year of birth, and how much of the variation in the dependent variable is attributable to year of birth, separate from people's age. This problem is called collinearity—or co-linearity—because the regression line predicting the relationship between age and the dependent variable is almost exactly the same as the regression line predicting the relationship between year of birth and the dependent variable.

There are several ways to assess the collinearity among the independent variables in a regression. The simplest way is to find the Pearson's correlation (r) between each of the independent variables. Although Pearson's correlation coefficient is intended to assess correlations between ratio-level variables, it can also provide information about correlations between dummy variables in the context of examining collinearity (though Pearson's correlation coefficients should not be used with dichotomous or categorical variables in bivariate analyses).

Table 13.5 shows a matrix of the Pearson's correlations between all of the independent variables in the third model in Table 13.3. Each independent variable is listed on the left and is also assigned a number. The top row lists the numbers that correspond to each of the variables, to avoid having to repeat the variable labels (and to avoid making the table too wide). The diagonal cells contain ones, indicating that each variable is perfectly correlated with itself. The remainder of the correlations are shown above the diagonal line of ones; the cells below the diagonal line are blank because they are a mirror image of those above the diagonal and because the table is easier to read when they are omitted.

There are two correlation coefficients above 0.5, which indicates that there is a moderate relationship between the pair of independent variables. "Age" and "Single" have a moderate negative relationship ($r = -0.59$). This coefficient is located at the intersection of the row for "Age" and column 6 (which represents the "Single" variable). The older people are, the less likely they are to be single (never married). "Age" and "Years of employment" (column 9) also have a moderate positive relationship ($r = 0.53$); as expected, people who are older tend to have more years of seniority with an employer.

tolerance Shows how much of the variation in an independent variable cannot be explained by the other independent variables in a regression.

variance inflation factor (VIF) Shows how much the variance of a slope coefficient is likely to be inflated as a result of an independent variable being correlated with the other independent variables in a regression.

Strong correlations between independent variables (typically, more than 0.7) suggest the need to further investigate whether or not a regression is affected by collinearity. To do this, researchers typically use two related pieces of information: the **tolerance** and the **variance inflation factor** (VIF). The tolerance shows how much of the variation within each independent variable is *not* predicted by the other independent variables in the regression. In other words, it provides information

Table 13.5   Correlations between the Independent Variables in the Third Model in Table 13.3

	1.	2.	3.	4.	5.	6.	7.	8.	9.	10.	11.
1. Age	1.00	0.01	−0.08	0.09	0.04	**−0.59**	0.24	0.20	**0.53**	0.12	0.18
2. Woman		1.00	−0.05	0.00	0.06	−0.04	0.09	−0.23	0.02	0.04	−0.02
3. Less than HS graduation			1.00	−0.23	−0.20	0.12	0.00	−0.14	−0.06	−0.07	−0.07
4. Post-secondary certificate				1.00	−0.48	−0.08	0.04	0.08	0.07	0.07	0.06
5. University degree					1.00	−0.13	−0.04	0.08	0.02	0.06	0.03
6. Single						1.00	−0.25	−0.20	−0.35	−0.10	−0.18
7. Widowed, sep., div.							1.00	0.03	0.10	0.04	0.03
8. Usual hours per week								1.00	0.18	0.04	0.23
9. Years of employment									1.00	0.24	0.24
10. Unionized job										1.00	0.02
11. Permanent job											1.00

Source: Author generated; Calculated using data from Statistics Canada, 2016b.

about how much of the variation in each independent variable is unique to that variable and cannot be explained by the other independent variables in the regression. Tolerance ranges from 0 to 1, where 0 indicates that an independent variable does not have any additional variation that cannot be explained by the other independent variables, and 1 indicates that an independent variable is completely unrelated to the other independent variables. As a general guideline, a tolerance less than 0.1 suggests that there may be a collinearity problem, and one or more independent variables should be omitted from the regression.

Tolerance is easiest to understand by illustrating how it is calculated. To find the tolerance of any single independent variable, an alternative linear regression is produced where the variable that the tolerance is being obtained for is used as the *dependent* variable, and the remainder of the independent variables in the original regression are used as independent variables. The dependent variable in the original linear regression is omitted from the alternative regression. The (unadjusted) R^2 of the alternative linear regression shows how much of the variation in the variable of interest can be explained by the other independent variables in the original regression. Since the R^2 is a proportion, the remaining variation (out of 1, or 100 per cent) is the amount of variation in the variable of interest that cannot be explained by the other independent variables in the original regression: this is the tolerance. In other words, the tolerance is calculated as $1 − R^2$ for a linear regression that uses the independent variable of interest as the dependent variable, and the remaining independent variables in the original regression as predictors.

For example, to find the tolerance of the "Age" variable in the third model in Table 13.3, I generate an alternative linear regression that uses "Age" as the dependent variable, and the remaining variables as independent variables: the dummy variables for sex/gender, the three variables for educational credentials, and the two variables for marital status, as well as the variables capturing the number of hours that people work at their job each week, their years of seniority with an employer, and the dummy variables for having a permanent job, and working in a unionized job. (The dependent variable "Usual weekly wages" is not used to calculate the tolerance.) The constant and slope coefficients of the alternative regression aren't particularly useful and can be ignored. But the R^2 of 0.484 indicates that 48.4 per cent of the variation in age can be predicted by the other independent variables in the third model in Table 13.3. Conversely, 51.6 per cent of the variation in age is *not* predicted by the other independent variables. Thus, the tolerance of the "Age" variable is 0.516 (1 − 0.484). This same process is repeated for each independent variable in a regression. Table 13.6 lists the tolerance of each of the independent variables in the third model in Table 13.3.

The variance inflation factor (VIF) is simply the inverse of the tolerance: that is, it is calculated as 1 divided by the tolerance. In Table 13.6, the tolerance of the "Age" variable is 0.516 (actually 0.516164). To find the variance inflation factor of age, divide 1 by 0.516164 to get 1.937. Similarly, the tolerance of the "Women" dummy variable is 0.923 (actually 0.922948). To find the variance inflation factor of the "Women" dummy variable, divide 1 by 0.922948 to get 1.083.

Not surprisingly, given its name, the variance inflation factor provides information about how much the variance of a slope coefficient is likely to be inflated because an independent variable is correlated with the other independent variables in a regression. When the variance of a slope coefficient is inflated, its standard error is also inflated. As a result, the 95 per cent confident intervals associated with that slope coefficient are wider because the standard error is used to calculate confidence

Table 13.6 Collinearity Diagnostics for the Independent Variables in the Third Model in Table 13.3

	Tolerance	Variance Inflation Factor
Age (in years, centred on 40)	0.516	1.937
Women	0.923	1.083
Highest educational credential		
Less than high school graduation	0.813	1.230
Post-secondary certificate/ diploma	0.641	1.560
University degree	0.638	1.568
Marital status		
Single, never married	0.604	1.655
Widowed, separated, or divorced	0.912	1.097
Usual hours per week at main job (centred on 40)	0.841	1.190
Years of employment in current or main job (centred on 1)	0.659	1.518
Unionized job	0.925	1.081
Permanent job	0.897	1.115

Source: Author generated; Calculated using data from Statistics Canada, 2016b.

intervals. The test of statistical significance associated with the slope coefficient is also affected because the standard error is used to calculate the t-statistic. The variance inflation factor of a slope coefficient can range from 1 to infinity, where 1 indicates that the variance and standard errors are not inflated, and higher numbers indicate that they are very inflated. As a general guideline, variance inflation factors greater than 10 suggest that there may be a collinearity problem, and one or more variables should be omitted from the regression.

The collinearity diagnostics shown in Table 13.6 make it possible to determine whether or not the correlation between "Age" and "Single" and the correlation between "Age" and "Years of employment" are a concern for the third regression model shown in Table 13.3. The slightly higher variance inflation factor of the "Age" variable is a good reminder that its standard error, confidence interval, and t-statistic are being influenced by its correlation with the other independent variables. But the tolerance suggests that despite these correlations, the "Age" variable still has enough unique variation to be a useful addition to the regression. About half of the variation (52 per cent) in age cannot be explained by the other independent variables in the regression.

In general, researchers use the tolerances and variance inflation factors of the independent variables, along with Pearson's correlations, to determine whether a regression has collinearity problems. When collinearity problems do occur, the most common solution is to omit one of the collinear variables from the regression. Researchers' decisions about which independent variables to keep and which to omit are usually based on their judgement about which variable provides the best measure of the social phenomenon they are trying to model in the regression.

How to Analyze Regression Residuals

Another way that researchers assess how well a linear regression model fits the observed data is by analyzing the errors, or the residuals. As you might recall from Chapter 10, residuals show the amount of error in the predictions made by a regression and thus are often denoted using the letter e. The residual captures the difference between the *predicted* value on the dependent variable for a single case, calculated using the regression coefficients and the characteristics of that case, and the *actual* value on the dependent variable for that case in the data.

For example, the constant coefficient in the third model in Table 13.3 shows that 40-year-old men, who are married or in common-law relationships, have only a high school education, work 40 hours a week, and have one year of seniority in a temporary, non-unionized job, are predicted to earn $829 each week. It's entirely possible that the dataset includes information collected from someone with exactly this set of personal and job characteristics—but that person reported usual weekly wages of $900. The difference between the weekly wages that the person reported and the weekly wages predicted by the regression ($71 in this example) is the residual for that case. If a residual is a positive number, it indicates that the regression prediction is too low; the residual is added to the prediction to make it match the reported value on the dependent variable. If a residual is a negative number, it indicates that the regression prediction is too high; the residual is subtracted from the prediction to make it match the reported value on the dependent variable.

A predicted value on the dependent variable and a residual can be calculated for each case that is used in a regression. Statistical software usually automates this process and saves the predicted values and the residuals as new variables in the dataset. The ideal value of a residual is 0, since a residual of 0 indicates that regression prediction is exactly perfect. Since the social world is complex, and there are limits to what can be measured and how accurately, residuals of 0 rarely occur. But researchers strive to build regression models that produce the smallest possible residuals.

A first step in assessing the overall fit of a linear regression is to identify cases with particularly large residuals. These cases should be examined to determine whether or not they are outliers in the data and, if so, how they affect the regression coefficients. One common strategy is to filter out the cases with very large residuals, and then re-run the regression to see how the coefficients change. If the addition (or omission) of a few cases results in substantial changes to the regression coefficients, this suggests that they are influential cases, and researchers should consider omitting them from their final regression model.

You might recall from Chapter 11 that linear regression rests on the assumption that the residuals, or errors, are normally distributed. Normally distributed residuals are an indicator that the unexplained variation in the data is the result of real-world random variation, as opposed to the result of another key characteristic (or variable) that the regression does not account for. Residuals that are not normally distributed indicate that there may be one or more important independent variables that have been omitted from the regression.

Once the residual associated with each case has been saved in a new variable, the distribution of the residuals can be examined. The easiest way to assess whether regression residuals are normally distributed is by graphing them in a histogram with a normal curve superimposed on it. Another visual tool that researchers use to gauge whether or not a variable is normally distributed is called a **quantile–quantile plot**, or a Q-Q plot for short. A Q-Q plot is a special type of scatterplot that is used to assess whether the values on a variable could plausibly be randomly selected from a normal distribution. Q-Q plots do this by comparing two sets of quantiles to each other: the first set of quantiles is calculated using the observed data, and the second set of quantiles is calculated using a perfect normal distribution.

As you learned in Chapter 3, a quantile is a cut-off point that is used to divide a set of data into percentiles. The first step in producing a quantile–quantile plot is to find the quantile that corresponds to the position of each case within a variable. Then, plot the value of the case in comparison to the value of a case that is located at the *same quantile* in a perfect normal distribution. In other words, the value of a case located at the twenty-fifth percentile of a variable in the observed data is plotted against the value of a case located at the twenty-fifth percentile of a perfectly normally distributed variable. Similarly, a case located at the fortieth percentile of a variable in the observed data is plotted against the value of a case located at the fortieth percentile of a perfectly normally distributed variable, and so on. If the variable being evaluated is normally distributed, all of the cases in the scatterplot will be located on the diagonal line, where the observed value equals the expected value from a normal distribution. Cases located far from the diagonal line are cases where the values on the variable in the observed data do not match a normal distribution.

quantile–quantile plot (Q-Q plot) A special type of scatterplot that compares the value of each case on an observed variable to the value of a case located at the corresponding quantile in a perfect normal distribution.

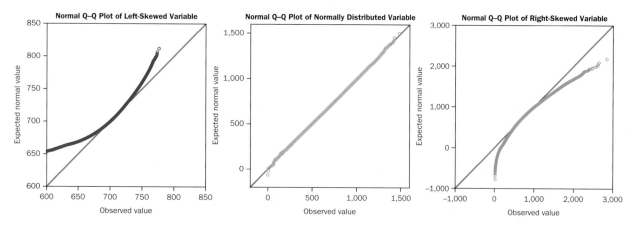

Figure 13.1   Quantile-Quantile Plots for a Left-Skewed Variable, a Normally Distributed Variable, and a Right-Skewed Variable (Hypothetical Data)

Figure 13.1 shows three Q-Q plots. In each plot, the observed values are plotted on the horizontal axis (x-axis), and values from the corresponding quantile of a normal distribution are plotted on the vertical axis (y-axis). For a normally distributed variable, shown in the plot in the centre panel of Figure 13.1, most of the cases are located right on the diagonal line. (By chance, there is a bit of deviation from the line at the ends of the distribution.) In contrast, the Q-Q plot for a left-skewed variable (in the left panel of Figure 13.1) shows an upward curve at the ends of the distribution, and the Q-Q plot for a right-skewed variable (in the right panel of Figure 13.1) shows a downward curve at the ends of the distribution, indicating that these variables are not normally distributed.

Figure 13.2 shows a histogram and a corresponding Q-Q plot for the residuals from the third regression model in Table 13.3. The histogram of residuals is centred on 0, as expected, and shows that many cases have relatively small residuals, which is an encouraging result. But the right tail of the histogram shows that for some cases, weekly wages are under-predicted by $1,000 or more. The Q-Q plot of the same residuals also shows that they are approximately normally distributed, but there is some deviation from the normal distribution at the upper end of the variable; the line of dots representing the cases drops well below the diagonal for residuals higher than $1,000. This corresponds to the higher than expected proportion of residuals (compared to a normal distribution) that are above $1,000 in the histogram.

In addition to visually assessing whether or not regression residuals are normally distributed, researchers can also use a statistical test to make this determination. For large samples, the Kolmogorov-Smirnov (K-S) goodness-of-fit test is commonly used to determine whether a sample of variable values could plausibly be randomly selected from a specific distribution. In the context of linear regression, the Kolmogorov-Smirnov test is used to determine whether a sample of residuals could plausibly be randomly selected from a normal distribution. A discussion of how to calculate the Kolmogorov-Smirnov test is beyond the scope of this text, but the test is produced by most statistical software. (Be sure to compare the residual variable to a normal

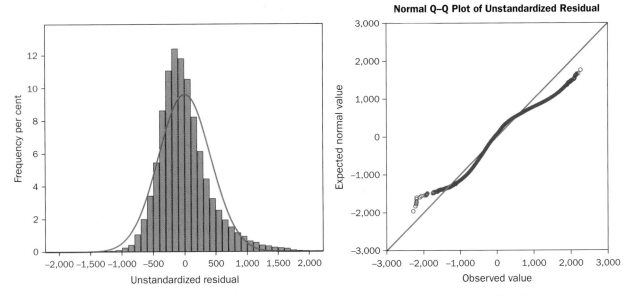

Figure 13.2   A Histogram and a Q-Q Plot Are Used to Assess Whether the Residuals from the Third Model in Table 13.3 Are Normally Distributed

Source: Author generated; Calculated using data from Statistics Canada, 2016b

distribution, not another distribution.) The key result of a Kolmogorov-Smirnov test is the p-value, or the result of the statistical significance test. P-values that are less than 0.05 indicate that the distribution of the variable being tested *is significantly different* than a normal distribution, whereas p-values of 0.05 or greater indicate that the distribution of the variable being tested is not significantly different than a normal distribution.

For the residuals from the third model in Table 13.3, the Kolmogorov-Smirnov test produces a p-value that is less than 0.001 (D[532,353] = 0.077, p < 0.001). This indicates that the distribution of these residuals is significantly different than a normal distribution. As a result, we know that the regression model needs further refinement in order to account for additional characteristics that influence people's weekly wages.

Using Residuals to Assess Bias

Researchers can also assess whether or not a regression makes better (or worse) predictions for different groups of people by analyzing the residuals. In a well-fitting model, the residuals should *not* be related to the dependent variable, or to any of the independent variables. In statistical analyses, most of the time researchers are looking for patterns and relationships. But when analyzing regression residuals, the ideal situation is the *absence* of a pattern. If the residuals are unrelated to any of the variables used in a regression, it suggests that they are randomly distributed and, thus, represent random variation in the sample.

Although it's also possible to use any of the bivariate analysis techniques described earlier in this book to assess the relationships between regression residuals and other variables, residual analysis is typically done by visually assessing graphs.

For example, the scatterplot in Figure 13.3 shows the relationship between the residuals from the third model in Table 13.3 (on the vertical axis) and the dependent variable, "Usual weekly wages" (on the horizontal axis). To make the graph more readable, only a randomly selected one per cent of cases are displayed. (See the "Hands-on Data" Analysis" box in Chapter 10 for more information about how to select or filter cases.) Since the ideal residual is 0, I've displayed a horizontal line at 0 on the vertical axis. Residuals located on this 0 line represent cases where the value on the dependent variable predicted by the regression is equal to the reported value on the dependent variable.

The scatterplot in Figure 13.3 shows that there *is* a relationship between the value on the dependent variable and the residuals. This indicates that this model still needs some improvement. The group of cases located below the 0 line on the left of the graph show that the regression tends to consistently over-predict the wages of people making less than $1,000 per week (because these residuals tend to be negative). The group of cases located above the 0 line on the right of the graph show that the regression tends to consistently under-predict wages of people making more than $1,500 per week (because these residuals tend to be positive). As a result, we know that this regression model needs to include additional independent variables that will help it to better predict wages, especially for people earning less than $1,000 per week or more than $1,500 per week. For instance, it may be useful to add some independent variables that capture occupational prestige to the regression, which would help to distinguish between high-status and low-status jobs, which tend to be associated with different pay scales.

Similarly, graphs can be used to assess whether or not regression residuals are related to any of the independent variables. As you learned in Chapter 11, linear

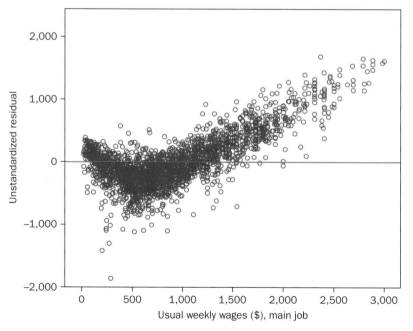

Figure 13.3 A Scatterplot Showing the Relationship between the Residuals from the Third Model in Table 13.3 and the Dependent Variable

Source: Author generated; Calculated using data from Statistics Canada, 2016b.

regression rests on the assumptions that the errors are not correlated with an independent variable and that the variation or dispersion of the errors is consistent across all of the values on an independent variable. The scatterplot in the left panel of Figure 13.4 shows the relationship between the regression residuals (on the vertical axis) and age (on the horizontal axis). Here, we see an ideal result: the scatterplot shows no clear pattern. This shows that the regression predicts wages equally well (or equally badly) for people of different ages; it is not biased in relation to age. In addition, the relationship is homoscedastic; that is, the spread of the errors is consistent across all of the ages. The boxplots in the right panel of Figure 13.4 shows the relationship between the regression residuals (on the vertical axis) and people's highest educational credential (on the horizontal axis). You might recall that the original educational credential variable was transformed into a series of dummy variables in order to use it in the regression. This graph uses the original categorical variable capturing people's highest educational credential (and not the dummy variables) so that all four levels of education are displayed on the same graph. The box plots show that the median residual is near 0 for people with each level of education. But the range and inter-quartile range of the regression residuals are wider for people who have a university degree than for people with other levels of education. This indicates that the relationship is heteroscedastic, that is, the spread of the errors is different across the different levels of education. The greater variation in the regression residuals for people with a university degree indicates that, in general, the regression predictions are worse for these people than for people with other levels of education. One reason for this result might be the range of credentials that people with a university degree might have: this group includes people with a bachelor's degree only but also people with master's degrees, people with doctoral degrees, and people with professional degrees (such as law or medical degrees). The fit of the regression might be improved by further

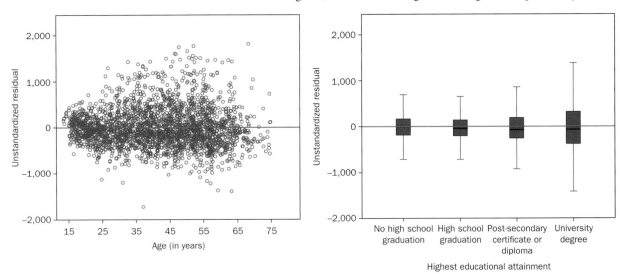

Figure 13.4 Graphs Showing the Relationship between the Residuals from the Third Model in Table 13.3 and a Ratio-Level Independent Variable, and between the Residuals and a Categorical Independent Variable

Source: Author generated; Calculated using data from Statistics Canada, 2016b.

dividing this level of education to distinguish between people with a bachelor's degree only and those with additional or higher degrees.

A full regression residual analysis would include an assessment of the relationship between the residuals and the dependent variable, all of the independent variables, and any other meaningful characteristics in the population. The process of building a regression model typically involves ongoing refinement, where researchers use a theoretical model to select an initial group of independent variables and then proceed to add and omit variables in order to develop the best possible empirical model. The best possible linear regression model is not necessarily the one with the highest adjusted R^2 but one with a relatively high adjusted R^2 that does not exhibit collinearity or bias, and that matches well with a theoretical framework. Sometimes, a residual analysis can indicate that producing entirely separate regression models for different groups of people—such as men and women, or people with low incomes and people with high incomes—makes the most sense. To return to the ideas introduced at the beginning of this section, the goal is to create parsimonious regression models—those that predict as much of the variation in the dependent variable as possible, using the simplest combination of independent variables.

How Does It Look in SPSS?

Nested Regression with Collinearity and Residual Statistics

When the independent variables are entered in three blocks, "Collinearity diagnostics" are selected in the Statistics option, and "Unstandardized" residuals are selected in the Save option, the Linear Regression procedure produces results that look like those in Image 13.1. Above these results, SPSS prints a "Variables Entered/Removed" box (not shown), which lists the independent variables that were used or entered in each block and the independent variables that were removed. Review this information to make sure that all of the requested variables were used. SPSS also prints an ANOVA table (not shown); this table is described in the "How Does It Look in SPSS?" box in chapters 11 and 12. Below the regression coefficients, SPSS prints an "Excluded Variables" table (not shown), which provides information about the variables excluded from earlier nested regressions, and a "Collinearity Diagnostics" table (not shown), which provides more information about the relationships between the independent variables.

A. Each regression model is assigned a number. There are three models, and model fit information is provided for each separately. A description of how to interpret the information in the "R," "R Square" and "Std. Error of the Estimate" columns is provided in Chapter 12.

Continued

Model Summary[d]

Model	R	R Square	Adjusted R Square	Std. Error of the Estimate
1 (A)	.327[a]	.107	.107	536.23158
2	.502[b]	.252	(C) .252	490.86832
3	.687[c]	.472	.472	412.47986

(B) a. Predictors: (Constant), Women, Age (in years, centred on 40)

b. Predictors: (Constant), Women, Age (in years, centred on 40), University degree, Less than high school graduation, Widowed, separated or divorced, Post secondary certificate/diploma, Single, never married

c. Predictors: (Constant), Women, Age (in years, centred on 40), University degree, Less than high school graduation,

Usual hours per week at main job (centred on 40), Years of employment in current or main job (centred on 1)

d. Dependent Variable: Usual weekly wages, main job

Coefficients[a] (G)

	Model	Unstandardized Coefficients B	Std. Error	Standardized Coefficients Beta	t	Sig.	Collinearity Statistics (H) Tolerance	(I) VIF
1	(Constant)	1023.380	1.038		985.921	.000		
(D)	Age (in years, centred on 40)	10.423	.052	.258	199.031	.000	1.000	1.000
	Women	-230.839	1.470	-.203	-157.039	.000	1.000	1.000
2	(Constant)	947.658	1.696		558.898	.000		
(E)	Age (in years, centred on 40)	4.944	.060	.122	82.200	.000	.635	1.575
	Women	-266.222	1.356	-.235	-196.361	.000	.985	1.015
	Less than high school graduation	-134.566	2.605	-.068	-51.650	.000	.820	1.220
	Post secondary certificate/diploma	181.855	1.753	.153	103.741	.000	.650	1.538
	University degree	424.233	1.828	.342	232.050	.000	.647	1.545
	Single, never married	-211.527	1.780	-.180	-118.854	.000	.614	1.629
	Widowed, separated or divorced	-35.898	2.410	-.018	-14.897	.000	.914	1.094
3	(Constant)	829.433	2.157		384.454	.000		
(F)	Age (in years, centred on 40)	-1.147	.056	-.028	-20.455	.000	.516	1.937
	Women	-152.878	1.177	-.135	-129.897	.000	.923	1.083
	Less than high school graduation	-45.086	2.198	-.023	-20.511	.000	.813	1.230
	Post secondary certificate/diploma	123.557	1.483	.104	83.307	.000	.641	1.560
	University degree	364.793	1.548	.294	235.709	.000	.638	1.568
	Single, never married	-131.352	1.507	-.112	-87.145	.000	.604	1.655
	Widowed, separated or divorced	-17.314	2.027	-.009	-8.542	.000	.912	1.097
	Usual hours per week at main job (centred on 40)	23.502	.059	.432	397.254	.000	.841	1.190
	Years of employment in current or main job (centred on 1)	15.233	.103	.181	147.338	.000	.659	1.518
	Unionized job	96.500	1.290	.077	74.799	.000	.925	1.081
	Permanent job	74.467	1.733	.045	42.974	.000	.897	1.115

a. Dependent Variable: Usual weekly wages, main job

(J) **Residuals Statistics[a]**

		Minimum	Maximum	Mean	Std. Deviation	N
Predicted Value	(K)	-403.9330	2892.3159	913.5868	389.64527	532353
Residual	(L)	-2433.10498	2266.76636	.00000	412.47559	532353
Std. Predicted Value	(M)	-3.381	5.078	.000	1.000	532353
Std. Residual		-5.899	5.495	.000	1.000	532353

a. Dependent Variable: Usual weekly wages, main job

Image 13.1   **An SPSS Linear Regression with Three Blocks of Independent Variables (Nested Regressions)**

B. The footnotes below this table lists the independent variables (predictors) used in each regression. Footnote (a) lists the independent variables in Model 1, footnote (b) lists the independent variables in Model 2, and footnote (c) lists the independent variables in Model 3. The final footnote identifies the dependent variable.

C. This column shows the adjusted R^2 of the three models. The adjusted R^2 of the first model (0.107) indicates that, collectively, the independent variables in the first regression (ascribed characteristics) explain 11 per cent of the variation in weekly wages, after adjusting for the predictive power of the variables in the model. The adjusted R^2 of the second model (0.252) indicates that, collectively, the independent variables in the second regression (both ascribed and achieved characteristics) explain 25 per cent of the variation in weekly wages, after adjusting for the predictive power of the variables in the model. The adjusted R^2 of the third model (0.472) indicates that, collectively, the independent variables in the third regression (ascribed, achieved, and job characteristics) explain 47 per cent of the variation in weekly wages, after adjusting for the predictive power of the variables in the model. When the adjusted R^2 and the unadjusted R^2 of a model are the same, as they are for these three models, it suggests that all of the independent variables in the regression are good predictors of the dependent variable.

D. The rows in the Model 1 block show the results of a linear regression when only these two variables are used as independent variables.

E. The rows in the Model 2 block show the results of a linear regression when these seven variables are used as independent variables: the same two variables as the first model and the five new variables that are added. The constant coefficient and the slope coefficients of the first two variables can be compared to those in the first model to assess how the relationship between each independent variable and the dependent variable changes as additional variables are controlled for.

F. The rows in the Model 3 block show the results of a linear regression when these eleven variables are used as independent variables: the same seven variables as the second model and the four new variables that are added. The constant coefficient and the slope coefficients of the first seven variables can be compared to those in the second model to assess how the relationship between each independent variable and the dependent variable changes as additional variables are controlled for.

G. These columns show the unstandardized and standardized regression coefficients, as well as the results of statistical significance tests. A description of how to interpret the information in these columns is provided in Chapter 12.

H. This column shows the tolerance of each independent variable. The tolerance of some variables becomes smaller as more independent variables are added to the regression. A tolerance less than 0.1 indicates potential collinearity problems.

I. This column shows the variance inflation factor (VIF) of each independent variable. The VIF of some variables becomes larger as more independent

Continued

variables are added to the regression. A VIF greater than 10 indicates potential collinearity problems.

J. This table provides information about the residuals or errors that are produced by the final (third) regression model. This information helps researchers assess how well the regression fits the data. The final column of the table shows that 532,353 cases are used in the regression.

K. This row provides information about the values on the dependent variable that are predicted by the regression. The minimum predicted weekly wages are −$404, and the maximum predicted weekly wages are $2,892. The average predicted weekly wages are $914, which corresponds to the actual average weekly wages in the data.

L. This row provides information about the regression residuals or errors. The lowest residual is −$2,433, which indicates that the regression predicts a weekly wage that is $2,433 higher than the person actually reported. The highest residual is $2,267, which indicates that the regression predicts a weekly wage that is $2,267 lower than the person actually reported. The average residual will always be 0.

M. These two rows provide information about the standardized (or normalized) versions of the predicted values and the residuals. In other words, the distribution of the predicted values and the residuals are mapped onto a standard normal curve. Since the standard normal curve has a mean of 0 and a standard deviation of 1, standardized variables will always have a mean of 0 and a standard deviation of 1. Since 99.7 per cent of the area under the normal curve is between −3.0 and +3.0, standardized predicted values or residuals that are far beyond these points may indicate unusual or influential cases. Although this regression produces some larger standardized predicted values and residuals, there are very few cases at these extremes.

Hands-on Data Analysis

Creating Pseudo-Continuous Variables by Recoding to Midpoints

The publicly available datasets from Statistics Canada and other large statistical agencies often contain many ordinal-level variables and relatively few ratio-level or continuous variables. Sometimes, ordinal-level variables result from efforts to make reporting easier: identifying the range that their income is in is easier for people to do than reporting their exact income. Other times, exact information is collected from respondents, but the responses are collapsed into class intervals to ensure confidentiality, resulting in ordinal-level variables. Ordinal-level variables can sometimes be challenging to use in regressions because the researcher must create many dummy variables and then interpret each of their slope coefficients.

If an ordinal-level variable has attributes that are class intervals that mask an underlying continuous distribution, researchers sometimes opt to transform the ordinal-level variable into a pseudo-continuous variable before using it in a regression. To do this, the researcher recodes the variable so that each value on the original ordinal-level variable is assigned the midpoint of the class interval that the value represents. The result is a variable that can be used in a regression in the same way as if it were a ratio-level or continuous variable. The "Age" variable used in the regressions in this chapter is a pseudo-continuous variable that was constructed in this way.

An example will help to illustrate this process. The original ordinal-level "Age" variable shown on the left of Table 13.7 has nine attributes and, thus, would require eight dummy variables to use in a regression. But these nine attributes are class intervals that mask an underlying continuous distribution—that is, people's ages are truly continuous, but they have been artificially collapsed into categories in this variable. Because we are confident that there is an underlying continuous distribution, this variable can be transformed into a pseudo-continuous variable by recoding it. Each original value is assigned a new value that is equal to the midpoint of the class interval that it represents. The new values that represent the midpoint of each age interval are shown in the final column of Table 13.7. In practice, a pseudo-continuous variable is created by assigning everyone in the 18- to 19-year-old group an age of 18.5, and assigning everyone in the 20- to 24-year-old group an age of 22 and so on. When the new, recoded variable is used in a regression, additional error is introduced because some of the people who are treated as if they are 22 are actually 20, 21, 23, or 24. However, the additional predictive power that is gained by using a pseudo-continuous variable in a regression typically makes up for this extra error (unless the class intervals in the original variable are unusually large).

Table 13.7 An Example of How to Create a Pseudo-Continuous Variable by Recoding an Ordinal-Level Variable to the Midpoints of Class Intervals

Original Variable: Age (ordinal-level)			New Variable: Age (pseudo-continuous)
Original Value	Original Attribute		New Value
1	18 to 19 years	→	18.5
2	20 to 24 years	→	22
3	25 to 29 years	→	27
4	30 to 34 years	→	32
5	35 to 39 years	→	37
6	40 to 49 years	→	44.5
7	50 to 59 years	→	54.5
8	60 to 69 years	→	64.5
9	70 years and up	→	74.5

Continued

Transforming an ordinal-level variable with attributes that are class intervals into a pseudo-continuous variable is easy to do. First, find the mid-point of each interval by adding together the lower bound of the interval and the upper bound of the interval, and then dividing by 2 (so 18 + 19 = 37 and 37 ÷ 2 = 18.5). Then, use statistical software to recode the original variable into a new variable, and replace the original value associated with each class interval with the value at the midpoint of that interval. For the example shown in Table 13.7, the original value "1" (representing 18 to 19 years) is replaced with the midpoint value "18.5", and the original value "2" (representing 20 to 24 years) is replaced with the midpoint value "22", and so on for all of the class intervals.

Sometimes variables have open-ended class intervals, such as "70 years and up," as their highest or lowest attributes. When this occurs, the typical practice is to use the same distance between the lower (or upper) bound of interval and the midpoint as the next-closest class interval. For instance, the class interval just below "70 years and up" is "60 to 69 years," and it is recoded into 64.5 years. The midpoint of "64.5" years is 4.5 years above the lower bound of the interval. The same distance (4.5 years) is added to the lower bound of the "70 years and up" class interval, and thus it becomes "74.5" when it is recoded. Sometimes, the lower bound of a class interval is implicitly 0. For instance, an age category of "17 years or less" implicitly ranges from 0 to 17 (or from 15 to 17, depending on the age of the youngest person surveyed). In these situations, be sure to think critically and review the survey documentation to determine what the lowest (or highest) possible bound of a class interval is, and then use that bound to calculate the midpoint.

As with all recoding, once you have created a new, pseudo-continuous variable, compare a frequency distribution of the old and new variables to ensure that all values were correctly recoded and that missing values were treated appropriately. Because pseudo-continuous variables are treated as ratio-level variables, no value labels are needed because the values represent the quantities themselves (that is, "18.5" stands for 18.5 years old). Once you are confident that the new variable has been correctly recoded, it can be used and interpreted in the same was as any other ratio-level variable in a regression. In your description of the methodology or the analysis, however, you should note whether any variables were created in this way.

Best Practices in Presenting Results

Writing about Regression Modelling

Although researchers often spend a substantial amount of time fitting and refining regression models, they typically do not describe this process in detail when writing about regressions. In part, this is because researchers are most interested in presenting and interpreting regression results, and most non-technical audiences aren't interested in reading about the process of modelling.

Sometimes, researchers will briefly describe alternative independent variables or models that they tried in the process of developing their final regression model. More often, researchers just report that they assessed the collinearity between the independent variables and that it was within acceptable limits. They may also briefly describe the results of any residual analyses that they conducted, especially if those analyses prompted changes to the regression model. For a non-technical audience, the goal is to provide enough information about the model-building process so that a reader has confidence that the regression you are presenting is the best one possible; at the same time, the discussion of model-fitting is usually brief. Focus on describing any substantial changes to the original regression that were made as a result of your efforts to improve the model.

What You Have Learned

This chapter introduced nested regressions, which add independent variables in blocks in order to investigate how the results change when additional characteristics are controlled for. You also learned several strategies for assessing how well a linear regression model fits the observed data and how to make decisions about which independent variables to use in a regression and which to omit. In particular, you learned how to check for the presence of collinearity among independent variables and how to use residual analysis to check for bias. Used together, these techniques will help you to build parsimonious regression models.

The statistical analyses in this chapter focused on how ascribed and achieved personal characteristics are related to people's wages, after controlling for job characteristics. The results show that achieved characteristics predict slightly more of the variation in weekly wages than ascribed characteristics. In the final regression model presented in this chapter, the three characteristics with the strongest relationship with weekly wages are the number of hours that people work at their job each week, their level of education (particularly having a university degree), and their years of seniority with an employer. Notably, however, gender is still related to weekly wages, even after achieved characteristics and job characteristics are accounted for. These results suggest that the Canadian labour market has some elements of a meritocracy but that wages are still influenced by individuals' personal characteristics. Including additional ascribed characteristics—such as race/racialization, immigrant status, and disability status—in the regression might allow researchers to develop a more nuanced understanding of wage discrimination. A residual analysis shows that this regression model still needs refinement in order to better predict the earnings of both high-wage and low-wage employees.

The overall goal of this book has been to introduce you to the descriptive and inferential statistical techniques that are commonly used in the social sciences. In each chapter, I have used these techniques to describe and illustrate social inequalities in Canada. I hope that these examples have demonstrated how statistical analyses are useful to social science researchers, community and non-profit organizations, and social activists. Immense amounts of data on a variety of topics are now publicly available. This book has outlined both the theoretical knowledge and the hands-on skills needed to analyze and draw insights from these data. Many schools, organizations, and companies offer more advanced statistics courses that you may want to take advantage of. But one of the easiest ways to further develop your skills is by regularly practicing statistical analyses with different data and continuing to use the techniques introduced in this text to describe and think critically about the world.

Check Your Understanding

Check to see if you understand the key concepts in this chapter by answering the following questions:

1. Why is it useful to group the independent variables in a regression into blocks?
2. What is the difference between the R^2 and the adjusted R^2?
3. What is collinearity, and why is it a concern in regression?
4. What do the tolerance and the variance inflation factor of each independent variable show?
5. How can a researcher assess whether or not regression residuals are normally distributed?
6. What is the ideal residual value? What is the ideal relationship between the residuals and each of the variables in a regression?

Practice What You Have Learned

Check to see if you can apply the key concepts in this chapter by answering the following questions:

1. The community agency described in the "Practice What You Have Learned" section in Chapter 12 wants to develop its analysis of people's charitable giving even further. In particular, the agency wants to know whether people who are more engaged in their local communities are likely to donate more. The agency uses two indicators of people's community engagement: whether or not they volunteered their time in the past 12 months, and whether they participate in religious activities/ services at least once a month. The agency's research team produces a multiple linear regression predicting the total amount of money that people donated to charitable organizations in the past 12 months, which has three blocks of independent variables: one for personal characteristics, one for socio-economic characteristics, and one for community engagement characteristics; the results are shown in Table 13.8. (The variables in the first two blocks are the same as those described in "Practice What You Have Learned" in Chapter 12, although income is centred.)

    a. Explain what the two unstandardized slope coefficients in the first model show.
    b. Explain what the constant coefficient in the first model shows.
    c. Explain what the R^2 and the adjusted R^2 of the first model show.

2. Using the information in Table 13.8:

    a. Explain what the first two unstandardized slope coefficients (of the "Age" and "Women" variables) in the second model show.
    b. In your own words, explain why the slope coefficient of the "Age" variable in the second model is smaller than in the first model.

3. Using the information in Table 13.8:

    a. Explain what the unstandardized slope coefficient of the "Annual personal income" variable in the second model shows.
    b. Explain what the unstandardized slope coefficients of the three education dummy variables in the second model show. Describe the general pattern of the relationship between people's level of education and the amount of money they donate to charitable organizations.

4. Using the information in Table 13.8:

    a. Explain what the unstandardized slope coefficient of the "Volunteered in the past 12 months" dummy variable in the third model shows.

Table 13.8 **Results of a Multiple Linear Regression Predicting the Total Financial Donations to Charitable Organizations ($) in the Past 12 months (GSS 2013)**

Dependent variable: Total financial donations to charitable organizations ($) in the past 12 months (n = 13,459)

	Model 1	Model 2	Model 3
Constant	327.08*	262.29*	106.40*
Personal Characteristics			
Age (in years, centred on 45)	5.87*	5.20*	4.72*
Women	–17.10	18.85	–7.61
Socio-economic Characteristics			
Annual personal income (in thousands of dollars, centred on 40)		2.66*	2.77*
Highest educational credential (ref: high school graduation only)			
Less than high school graduation		–58.89*	–95.11*
Post-secondary diploma		38.47*	–30.60
University degree		183.27*	136.22*
Community Engagement Characteristics			
Volunteered in the past 12 months			171.61*
Participates in religious activities/services once a month or more often			443.57*
R^2	0.027	0.068	0.182
Adjusted R^2	0.027	0.068	0.182

*Indicates that results are statistically significant at the $p < 0.05$ level.
Source: Author generated; Calculated using data from Statistics Canada, 2015.

b. Explain what the unstandardized slope coefficient of the "Participates in religious activities/services once a month or more often" dummy variable in third model shows.
c. Describe how the slope coefficients of the other independent variables change between the second and third model.

5. Using the information in Table 13.8, describe how the constant coefficient changes between the three models. Explain why this change occurs.
6. Using the information in Table 13.8:
   a. Describe how the R^2 changes between the three models. Explain what these changes show.
   b. Compare the R^2 to the adjusted R^2 of each model. Explain what these comparisons indicate about the independent variables in each model.

7. Table 13.9 shows the tolerance and variance inflation statistics for each of the independent variables used in the third model shown in Table 13.8.
   a. Which three variables show the most signs of collinearity? Explain how you know.
   b. Do any of the variables in this model show a level of collinearity that suggests a problem in this regression? Explain how you know.

8. The histogram in the left panel of Figure 13.5 shows the distribution of the unstandardized residuals from the third model shown in Table 13.8.
   a. Describe the shape of the distribution of the unstandardized residuals, in comparison to a normal distribution.
   b. In practical terms, what does this distribution indicate about the predictions that the regression makes?

Table 13.9   Collinearity Diagnostics for the Independent Variables in the Third Model in Table 13.8

	Tolerance	Variance Inflation Factor
Age	0.937	1.067
Women	0.934	1.070
Annual personal income	0.808	1.237
Highest educational credential		
Less than high school graduation	0.794	1.260
Post-secondary diploma	0.700	1.429
University degree	0.670	1.493
Volunteered in the past 12 months	0.939	1.065
Religious activities/ services once a month or more often	0.956	1.046

Source: Author generated; Calculated using data from Statistics Canada, 2015.

9. The Q-Q plot in the right panel of Figure 13.5 shows the distribution of the unstandardized residuals from the third model shown in Table 13.8.

a. What does the Q-Q plot illustrate about the distribution of the residuals? How does it compare to a normal distribution? How do you know?

b. Explain how the Q-Q plot is related to the histogram in the left panel of Figure 13.5. Describe how the two plots correspond with each other.

10. The scatterplot in the left panel of Figure 13.6 shows the relationship between the unstandardized residuals from the third model shown in Table 13.8 and the dependent variable: "Total financial donations to charitable organizations in the past 12 months."

a. Describe the ideal relationship between regression residuals and the dependent variable.

b. Compare the scatterplot results to the ideal relationship. Describe what the scatterplot shows, and explain what it indicates about the regression model.

11. The box plots in the right panel of Figure 13.6 show the relationship between the unstandardized residuals from the third model shown in Table 13.8 and the original categorical variable capturing people's highest educational credential (which was used to

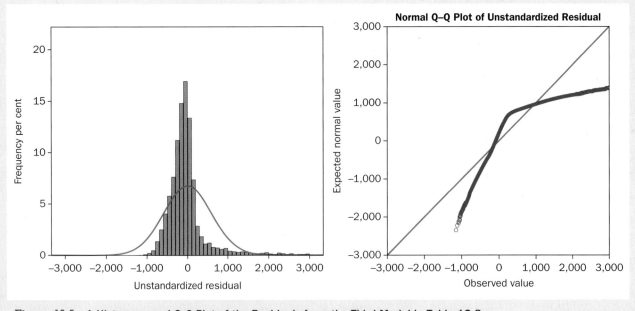

Figure 13.5   A Histogram and Q-Q Plot of the Residuals from the Third Model in Table 13.8

Source: Author generated; Calculated using data from Statistics Canada, 2015.

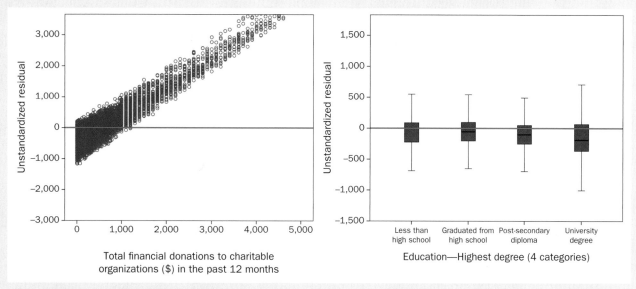

Figure 13.6   **Graphs Showing the Relationship between the Residuals from the Third Model in Table 13.8 and the Dependent Variable, and the Relationship between the Residuals and an Independent Variable**

Source: Author generated; Calculated using data from Statistics Canada, 2015.

create the dummy variables used as independent variables in the regression).

a.   Describe the ideal relationship between regression residuals and an independent variable.

b.   Compare the box plot results to the ideal relationship. Describe what the box plots show, and explain what they indicate about the regression model.

12. If you wanted to further refine this regression model in order to better explain people's charitable giving, what other independent variables would you include? Explain why you think your proposed variables will be related to the amount of money that people donate to charitable organizations.

13. The multiple linear regression in Table 13.10, excerpted from a Statistics Canada report, predicts life satisfaction among people who live in urban areas (Lu et al., 2015). The dependent variable is people's self-reported level of life satisfaction, measured with this question: "Using a scale of 0 to 10, where 0 means "Very dissatisfied" and 10 means "Very satisfied," how do you feel about your life as a whole right now?" Thus, higher values indicate more life satisfaction and lower values indicate less life satisfaction. All of the independent variables are dummy variables, with

the exception of "Age," which is a ratio-level variable and centred on the average age. Geography—or place of residence—is also controlled for in this regression but not displayed in this table. Using the Model 1 results, describe how sex/gender, age, immigrant status, and Aboriginal status are each predicted to relate to life satisfaction.

14. Using the Model 1 results in Table 13.10:

a.   Describe the pattern of the predicted relationship between people's level of education and their life satisfaction.

b.   Describe the pattern of the predicted relationship between people's health status and their life satisfaction.

c.   Describe the pattern of the predicted relationship between people's household income and their life satisfaction.

15. In Model 2 in Table 13.10, one more independent variable is added: a dummy variable that captures whether people have a feeling of belonging to their local community.

a.   Describe how having a feeling of belonging to your local community is predicted to be related to life satisfaction.

Table 13.10   Linear Regression Model Results on Life Satisfaction, Canada, 2009 to 2013

	Model 1	Model 2
	coefficient	
Women	0.079***	0.083***
Age	−0.048***	−0.041***
Immigrants	−0.037***	−0.028***
Marital status (reference: married)		
Living common-law	−0.181***	−0.141***
Widowed	−0.458***	−0.442***
Divorced or separated	−0.608***	−0.507***
Single	−0.491***	−0.468***
Education (reference: university degree)		
Some post-secondary	0.037***	0.069***
High school graduate	0.046***	0.033***
Less than high school	0.131***	0.134***
Health status (reference: good health)		
Excellent	1.008***	0.992***
Very good	0.522***	0.504***
Fair	−0.727***	−0.756***
Poor	−1.793***	−1.791***
Employment status (reference: employed)		
Unemployed	−0.541***	−0.421***
Not in labour force	−0.022**	−0.052***
Household income (reference: $100,000 to $150,000)		
Lowest: less than $30,000	−0.372***	−0.321***
Lower middle: $30,000 to $59,999	−0.186***	−0.154***
Middle: $60,000 to $99,999	−0.066***	−0.039***
High: More than $150,000	0.110***	0.100***
Aboriginal persons	0.042**	0.151***
Community belonging	—	0.438***
Intercept	8.616***	8.210***
Number of observations	337,420	278,980
Adjusted R-squared	0.180	0.196

** Significantly different from reference category (p < 0.01)

*** Significantly different from reference category (p < 0.001)

Note: All models include census metropolitan area and economic region fixed effects.

Sources: Statistics Canada, General Social Survey, 2009 to 2013, and Canadian Community Health Survey, 2009 to 2012.

Source: Excerpt from Lu et al. 2015, 6.

b. Describe how the slope coefficients of the other independent variables in the second model compare to those in the first model. What do these results indicate?

16. Based on the Model 2 results in Table 13.10, what types of people overall are predicted to have the highest levels of life satisfaction? Do these results surprise you? Why or why not?

Practice Using Statistical Software (IBM SPSS)

Answer these questions using IBM SPSS and the GSS27.sav or the GSS27_student.sav dataset available from the Student Resources area of the companion website for this book. Weight the data using the "Standardized person weight" [STD_WGHT] variable you created following the instructions in Chapter 5. Report two decimal places in your answers, unless fewer are printed by IBM SPSS. It's imperative that you save the dataset to keep any new variables that you create.

1. Create a pseudo-continuous variable for age by recoding to midpoints. Then, centre the variable on age 45.

    a. Use the Recode into Different Variables tool to recode the original variable "Age group of respondent (groups of 10)" [AGEGR10] into the new variable "Age" [AGE]. Assign each old value a new value that is equivalent to the midpoint of its class interval. Produce frequency distributions of the original variable "Age group of respondent (groups of 10)" [AGEGR10] and the new variable "Age" [AGE], and compare them to be sure the recoding is correct.

    b. Use the Compute Variable tool to create a new variable, called "Age (centred)" [AGE_CENTRED]. The value on the new variable for each case should be the value on the variable "Age" [AGE] minus 45. Use the Means procedure to find the mean of the variables "Age" [AGE] and "Age (centred)" [AGE_CENTRED], and compare the two results to be sure that the centring is correct.

2. Use the Recode into Different Variables tool to recode the "Visible minority status of the respondent" [VISMIN] variable into a "Visible minority" [IS_VISMIN] dummy variable. Assign the value "1" to people who are visible minorities and assign the value "0" to people who are not visible minorities in the new variable. The remaining values can be designated as system-missing in the new variable. Produce frequency distributions of the original variable "Visible minority status of the respondent" [VISMIN] and the new variable "Visible minority" [IS_VISMIN], and compare them to be sure that the recoding is correct.

3. Create four dummy variables to capture people's highest level of education. Use the Recode into Different Variables tool to recode the variable "Education - Highest degree (4 categories)" [DH1GED] into dummy variables as follows:

    a. Create the new dummy variable "Less than high school" [LTHS] by assigning the old value "1" the new value "1", and assigning the old values "2" through "4" the new value "0". (The remaining values can be designated as system-missing in the new variable.)

    b. Create the new dummy variable "High school only" [HS] by assigning the old value "2" the new value "1", and assigning the old values "1", "3", and "4" the new value "0". (The remaining values can be designated as system-missing in the new variable.)

    c. Create the new dummy variable "Post-secondary diploma" [DIPLOMA] by assigning the old value "3" the new value "1", and assigning the old values "1", "2", and "4" the new value "0". (The remaining values can be designated as system-missing in the new variable.)

    d. Create the new dummy variable "University degree" [UNI] by assigning the old value "4" the new value "1", and assigning the old values "1" through "3" the new value "0". (The remaining values can be designated as system-missing in the new variable.)

e. Produce frequency distributions of the original variable "Education - Highest degree (4 categories)" [DH1GED] and each of the four new dummy variables "Less than high school" [LTHS], "High school only" [HS], "Post-secondary diploma" [DIPLOMA], and "University degree" [UNI], and compare them to be sure the recoding is correct.

4. Use the Linear Regression procedure to produce a regression of the independent variables "Age (centred)" [AGE_CENTRED], "Visible minority" [IS_VISMIN], and "Women" [WOMEN] on the dependent variable "Number of paid hours worked per week - All jobs" [WKWEHRC]. (You created the "Women" variable in question 3 of "Practice Using Statistical Software" in Chapter 12.)

a. Explain what the constant coefficient shows.

b. Explain what each of the unstandardized slope coefficients show.

c. Determine which independent variable has the strongest relationship with the dependent variable, using the standardized slope coefficients.

d. Explain what the R^2 and the adjusted R^2 show.

5. Add a second block to the linear regression you produced in question 4. In the second block, add three dummy variables to capture people's level of education: "Less than high school" [LTHS], "Post-secondary diploma" [DIPLOMA], and "University degree" [UNI]. ("High school only" is the reference group.) Also add two dummy variables to capture marital status: "Previous relationship" [PREVIOUS_RELATIONSHIP] and "Single" [SINGLE]. (You created these variables in question 6 of "Practice Using Statistical Software" in Chapter 12; "In a relationship" [IN_RELATIONSHIP] is the reference group.) Use the Statistics option to request collinearity diagnostics and use the Save option to save the unstandardized residuals in a new variable.

a. Explain what the constant coefficient of the second model shows.

b. Explain what the unstandardized slope coefficients of the three education dummy variables show.

c. Explain what the unstandardized slope coefficients of the two marital status dummy variables show. Be sure to comment on the statistical significance of these variables.

d. Determine which independent variable in the second model has the strongest relationship with the dependent variable, using the standardized slope coefficients.

e. Explain what the R^2 and the adjusted R^2 of the second model show.

6. Compare the results of the two regression models you produced in question 5.

a. Describe how constant coefficient changes in the second model, compared to the first model, and explain why.

b. Describe how the unstandardized slope coefficients of the "Age," "Women," and "Visible minority" variables change in the second model, compared to the first model, and explain why.

c. Describe how the R^2 and the adjusted R^2 of the second model compare to those of the first model, and explain what this result shows.

7. For the regression you produced in question 5, determine whether any of the independent variables have a tolerance or variance inflation factor that indicates a collinearity problem. Explain how you know. Which three variables have the lowest tolerance (or the highest variance inflation factor)?

8. Analyze the unstandardized residuals from the regression you produced in question 5, which are saved in the variable "Unstandardized residual" [RES_1].

a. Use the Chart Builder tool to produce a histogram of the "Unstandardized residual" [RES_1] variable, and add a normal curve to use as a reference. Use the histogram to determine whether the distribution of the residuals is substantially different than normal, and, if so, explain how it differs.

b. Use the Q-Q Plots procedure to produce a Q-Q plot of the "Unstandardized residual"

[RES_1] variable. Describe how the Q-Q plot relates to the histogram you produced in (a). Use the Q-Q plot to determine whether the distribution of the residuals is substantially different than normal, and, if so, explain how it differs.

9. Investigate the relationship between the unstandardized residuals and the dependent and independent variables used in the regression.

   a. Use the Chart Builder tool to produce a scatterplot of the relationship between the unstandardized

residuals (on the y-axis) and the dependent variable "Number of paid hours worked per week - All jobs" [WKWEHRC] (on the x-axis). Explain what the results show.

   b. Use the Chart Builder tool to produce a box plot of the relationship between the unstandardized residuals (on the y-axis) and the independent variable "Education - Highest degree (4 categories)" [DH1GED] (on the x-axis). Explain what the results show.

Key Formulas

Adjusted R^2

$$Adjusted\ R^2 = 1 - \left[\frac{(1-R^2)(n-1)}{n-k-1} \right]$$

References

Budig, Michelle J., and Misun Lim. 2016. "Cohort Differences and the Marriage Premium: Emergence of Gender-Neutral Household Specialization Effects: Marriage Premiums, Specialization, Cohort Change." *Journal of Marriage and Family* 78 (5): 1352–70. doi:10.1111/jomf.12326.

Castilla, Emilio J. 2008. "Gender, Race, and Meritocracy in Organizational Careers." *American Journal of Sociology* 113 (6): 1479–1526. doi:10.1086/588738.

Killewald, Alexandra, and Margaret Gough. 2013. "Does Specialization Explain Marriage Penalties and Premiums?" *American Sociological Review* 78 (3): 477–502. doi:10.1177/0003122413484151.

Linton, Ralph. 1936. *The Study of Man: An Introduction.* Student's Edition. New York: Appleton-Century Crofts.

Lu, Chaohui, Grant Schellenberg, Feng Hou, and John Helliwell. 2015. "How's Life in the City? Life Satisfaction across Census Metropolitan Areas and Economic Regions in Canada." Catalogue no. 11–626-X—No. 046. Ottawa: Statistics Canada. http://www.statcan.gc.ca/pub/11–626-x/11–626-x2015046-eng.pdf.

McNamee, Stephen J., and Robert K. Miller. 2014. *The Meritocracy Myth.* 3rd ed. Lanham, MD: Rowman and Littlefield.

Petersen, Trond, and Ishak Saporta. 2004. "The Opportunity Structure for Discrimination." *American Journal of Sociology* 109 (4): 852–901. doi:10.1086/378536.

Statistics Canada. 2015. General Social Survey, 2013: Cycle 27, Giving, Volunteering and Participating. *Public Use Microdata File.* Ottawa, ON: Statistics Canada.

———. 2016. "Guide to the Labour Force Survey." Catalogue no. 71–543-G. http://www.statcan.gc.ca/pub/71-543-g/71-543-g2016001-eng.pdf.

———. 2016b. "Labour Force Survey 2016." http://www23.statcan.gc.ca/imdb/p2SV.pl?Function=getSurvey&Id=331692.

APPENDIX

A Brief Math Refresher

Many people are concerned about whether or not their math skills are strong enough to be able to do statistical data analysis. Be assured that all the math you need in order to understand this book is covered in grades 4 to 8 in most jurisdictions. This appendix provides a brief refresher of some basic mathematical concepts and skills; I encourage you seek out additional print and online sources for more explanation if needed. Other mathematical techniques are introduced throughout the text.

Types of Numbers

To complete the statistical calculations in this book, you need to be familiar with several different types of numbers:

1. *Whole numbers*: The first numbers that you learned were probably the counting numbers: 1, 2, 3, 4, 5, and so on. These counting numbers, with the addition of 0, are referred to as whole numbers. Conceptually, they represent a quantity or an amount.
2. *Negative numbers*: Negative numbers are numbers that are less than 0. They are denoted by a negative sign ("–") in front of the number, and conceptually, they represent the opposite or absence of a quantity. Figure A.1 shows a number line: 0 is positioned at the centre, positive numbers are depicted to the right of 0, and negative numbers are depicted to the left of 0. (Numbers are assumed to be positive unless they are explicitly marked as negative by a "–" sign.)
3. *Fractions*: Fractions are used to represent a part of a whole. The number in the numerator (top) of the fraction shows the number of parts, and the number in the denominator (bottom) of the fraction shows how many equal-size parts are in the whole. (See Figure A.2.)

Figure A.1   **A Number Line**

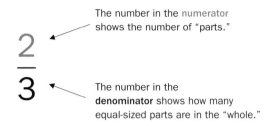

The number in the numerator shows the number of "parts."

The number in the **denominator** shows how many equal-sized parts are in the "whole."

Figure A.2   A Fraction

Most statistical reporting uses decimal notation to denote parts of a whole. To convert a fraction into a decimal, divide the numerator by the denominator:

$$\frac{2}{3} \quad \textit{is equal to} \quad 2 \div 3$$

Place Values

There are only 10 digits used in numbers: 0, 1, 2, 3, 4, 5, 6, 7, 8, and 9. The placement of a digit within a number determines its value. We use a base-10 number system. That means that each place in a number has a value that is 10 times bigger than the place immediately to the right. Figure A.3 shows some place values and their names.

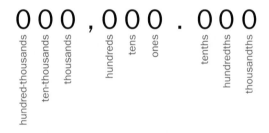

Figure A.3   Place Values in the Base-10 Number System

Whole numbers are shown to the left of the decimal point, and parts of a whole are shown to the right of the decimal point. So:

5,793.42 = 5 *thousands* + 7 *hundreds* + 9 *tens* + 3 *ones* + 4 *tenths* + 2 *hundredths*
= 5000 + 700 + 90 + 3 + 0.4 + 0.02

Rounding Decimal Places

Sometimes a number includes more decimal places than a researcher wants to report (and the decimals of some numbers continue infinitely). When this occurs, researchers "round" the result.

Figure A.4   Choosing Which Digit to Assess When Rounding Decimals

To round a number, assess the digit in the place immediately to the right of the last decimal place that you want to retain:

- If the digit is 4 or lower, round down: drop the digit you assessed and any digits further to the right, but do not make any other changes. So, 0.2499 becomes 0.2 when it is rounded to one decimal place.
- If the digit is 5 or higher, round up: add one to the digit in the place immediately to the left of the digit you assessed, and then drop the digit you assessed and any digits further to the right. So, 0.2499 becomes 0.25 when it is rounded to two decimal places.
- *A special situation:* if the digit is 5 or higher *and* the digit in the place immediately to the left is a 9, add one to the digit in the place two to the left, and make the digit in the place immediately to the left a 0 (and drop the digit you assessed and any digits further to the right). So, 0.2499 becomes 0.250 when it is rounded to three decimal places.

Mathematical Operations

There are four basic mathematical operations: addition, subtraction, multiplication, and division. Statistical calculations often require two more operations: exponents and square roots.

Addition: Addition is indicated using the plus sign (+). Addition corresponds to moving to the right on a number line. So:

1 + 3 tells you to begin at 1, and move three units to the right (to get 4).
−2 + 3 tells you to begin at −2, and move three units to the right (to get 1).

Subtraction: Subtraction is indicated using the minus sign (−). Subtraction corresponds to moving to the left on a number line. So:

4 − 3 tells you to begin at 4, and move three units to the left (to get 1).
−2 − 3 tells you to begin at −2, and move three units to the left (to get −5).

Multiplication: Multiplication is indicated using an "x," a dot (·), an asterisk (∗) or two parentheses adjacent to each other. The following expressions all indicate that 10 should be multiplied by 3:

$$10 \times 3 \qquad 10 \cdot 3 \qquad 10 * 3 \qquad (10)(3)$$

Multiplication can be thought of as a special type of addition. The expression 10×3 indicates that 10 should be added to itself, three times. Since:

$$10 \times 3 = 10 + 10 + 10 \qquad \textit{thus} \qquad 10 \times 3 = 30$$

Similarly, since:

$$10 \times 4 = 10 + 10 + 10 + 10 \qquad \textit{thus} \qquad 10 \times 4 = 40$$
$$10 \times 5 = 10 + 10 + 10 + 10 + 10 \qquad \textit{thus} \qquad 10 \times 5 = 50$$

Division: Division is indicated with a ÷ sign, a horizontal bar, or a slash (/). The following expressions all indicate that 30 should be divided by 10:

$$30 \div 10 \qquad \frac{30}{10} \qquad 30/10$$

Division can be thought of as a special type of subtraction. The expression $30 \div 10$ asks how many times 10 can be subtracted from 30 until there is nothing left over. Since:

$$30 - 10 - 10 - 10 \qquad = 0 \qquad \textit{thus} \qquad 30 \div 10 = 3$$

Similarly, since:

$$40 - 10 - 10 - 10 - 10 \qquad = 0 \qquad \textit{thus} \qquad 40 \div 10 = 4$$
$$50 - 10 - 10 - 10 - 10 - 10 = 0 \qquad \textit{thus} \qquad 50 \div 10 = 5$$

Exponents: Exponents are used to show repeated multiplication. They are denoted using superscript numbers and indicate that a number should be multiplied by itself several times. The number that is multiplied is called the base number, and the exponent (or the power) shows how many times the base number is multiplied. For example:

$$5^2 = 5 \times 5 \qquad \textit{thus} \qquad 5^2 = 25$$
$$5^3 = 5 \times 5 \times 5 \qquad \textit{thus} \qquad 5^3 = 125$$
$$5^4 = 5 \times 5 \times 5 \times 5 \qquad \textit{thus} \qquad 5^4 = 625$$

Many statistical calculations rely on multiplying a number by itself, indicated using the exponent "2." This is called "squaring" a number; the result is called the "square" of the original number.

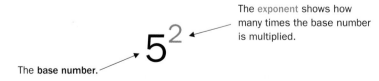

The exponent shows how many times the base number is multiplied.

The base number.

Figure A.5   An Exponent

Square roots: Roots (or radicals) are conceptually the opposite of exponents; finding a square root is the opposite to squaring a number. Square roots are denoted using the radical symbol ($\sqrt{\ }$), with a number below the symbol. To find the square root of a number, find the number that can be multiplied by itself in order to produce the number below the radical symbol. Since:

$$5 \times 5 = 25 \quad thus \quad \sqrt{25} = 5$$
$$6 \times 6 = 36 \quad thus \quad \sqrt{36} = 6$$
$$7 \times 7 = 49 \quad thus \quad \sqrt{49} = 7$$

Often, the square root of a number is not a whole number and, thus, includes decimals.

The Order of Operations

The result of a mathematical equation can change depending on the order you complete the operations in. To ensure that people solving the same equation always get the same result, mathematicians have adopted a universally agreed-upon order of operations: first complete anything in parentheses (or brackets), followed by any exponents, followed by division, multiplication, addition, and subtraction, in that order. (You may have learned this using the acronym PEDMAS or BEDMAS; see Figure A.6.)

Here's an example of the difference that parentheses make:

$$10 \times 3 + 4 = 34 \qquad but$$
$$10 \times (3 + 4) = 70$$

In the first equation, 10 is multiplied by 3, and then 4 is added to the result. In the second equation, the parentheses indicate that 3 and 4 should be added together *before* multiplying the result by 10.

The order of operations stipulates that division and multiplication must always be completed before addition and subtraction, even if the addition and subtraction appear first in the sequence when reading from left to right. For example:

$$7 + 5 \times 2 \neq 12 \times 2 \qquad and$$
$$7 + 5 \times 2 \neq 24$$

The multiplication must be completed first, so instead:

$$7 + 5 \times 2 = 7 + 10 \qquad thus$$
$$7 + 5 \times 2 = 17$$

It's important to follow the order of operations when you use the statistical formulas in this book.

Parentheses	Brackets
Exponents	Exponents
Division	Division
Multiplication	Multiplication
Addition	Addition
Subtraction	Subtraction

or (between the two columns)

Figure A.6   The PEDMAS and BEDMAS Acronyms

Multiplying and Dividing by 10 and 100

In the base-10 number system, you can multiply or divide by 10 and by 100 (as well as other powers of 10) by moving the location of the decimal point. Multiplying and dividing by 100 is regularly used to convert between proportions and percentages in statistical calculations.

To multiply a number by 10, move the decimal point one place to the right. (See the first line of Figure A.7.) To multiply a number by 100, move the decimal point two places to the right (see the second line of Figure A.7). Insert a zero into any places that do not have a number in them.

A similar approach is used to divide by 10 and 100 (and other powers of 10). To divide a number by 10, move the decimal point one place to the left; to divide a number by 100, move the decimal point two places to the left. (See lines 3 and 4 of Figure A.7.)

Multiplying Positive and Negative Numbers

Statistical calculations often require multiplying negative numbers—either by themselves (in other words, squaring them) or by other numbers.

If two numbers are on the *same side* of the number line, when they are multiplied together the result is always a *positive* number. So, if both numbers are to the right of 0 on the number line (positive numbers), the result is a positive number. If both numbers are to the left of 0 on the number line (negative numbers), the result is also a positive number. So:

+	*multiplied by* +	*equals* +	*thus*	$3 \times 3 = 9$
−	*multiplied by* −	*equals* +	*thus*	$-3 \times -3 = 9$

$$0.67 \times 10 \ = 0.67 \ = 6.7$$

$$0.67 \times 100 = 0.670 = 67.0$$

$$0.67 \div 10 \ = 0.67 \ = 0.067$$

$$0.67 \div 100 = 00.67 = 0.0067$$

Figure A.7   Multiplying and Dividing by 10 and by 100 by Moving the Decimal Point

If two numbers are on *opposite sides* of the number line, when they are multiplied together the result is always a *negative* number. So, if one number is to the right of 0 (a positive number) and one number is to the left of 0 (a negative number), the result is a negative number. So:

$+$	multiplied by	$-$	equals	$-$		thus		$3 \times -3 = -9$
$-$	multiplied by	$+$	equals	$-$		thus		$-3 \times 3 = -9$

Of course, when any number is multiplied by 0, the result is 0 (0 is neither positive nor negative).

Solving Equations

The equals sign (=) indicates that two expressions are equivalent to one another. This makes it possible to solve for unknown values, which are often designated using a letter (such as x).

So the following equation:

$$10 + 2 = ?$$

. . . is written more formally as:

$$10 + 2 = x$$

Since the left-hand side of the equation is equal to the right-hand side, x must equal 12.

When the unknown value is not alone on one side of the equation, use mathematical operations to isolate it. To maintain the equivalency, however, you must *do the same thing* to both sides of the equation. So the following equation:

$$x + 6 = 10$$

. . . can be solved by subtracting 6 from both sides of the equation:

$$x + 6 - 6 = 10 - 6$$
$$x = 4$$

Multiplication and division can also be used to isolate x on one side of the equation. So the following equation:

$$5x = 15$$

. . . can be solved by dividing both sides of the equation by 5:

$$\frac{5x}{5} = \frac{15}{5}$$
$$x = 3$$

This approach is used in some statistical calculations, such as those for rates.

"All together, these are the basic mathematical skills that you need to understand and use statistical formulas. The key to using any statistical formula is to work methodically through what the formula is asking you to do, and to complete each mathematical operation in sequence. The instructions provided in the "Step-by-Step" boxes located throughout this book help you to do this."

APPENDIX

SPSS Basics

Part I: SPSS Windows and Views

IBM SPSS Statistics is specialized software that is used to do statistical analysis. Statistical procedures can be requested using drop-down menus and windows or a written command syntax. In this appendix, I describe how to produce common statistics using the drop-down menus and windows because this approach is easiest to learn. This appendix provides an overview of basic SPSS procedures only. For further instruction, more comprehensive guides to SPSS are widely available, both online and in print.

SPSS has two main windows: a Data Editor and an Output Viewer. Each window displays a different type of file, designated using different file extensions: SPSS data files end with ".sav" and SPSS output files end with ".spv." The screenshots in this appendix are from SPSS Statistics version 25 for Windows. Other versions may have a slightly different appearance, structure, and/or features.

The Menu Bar

Like many other computer programs, there is a menu bar at the top of each SPSS window. The menus listed on the bar, and the items in each menu vary depending on the window or the item that is selected, and also depending on which edition of SPSS you are using (base, standard, professional, or premium).

The most commonly used menus are: **File**, **Data**, **Transform**, **Analyze**, and **Graphs**.

- The **File** menu provides access to a series of typical commands for opening and saving files. Data files (.sav files) and output files (.spv files) must each be saved separately in their respective windows: saving a data file does not save the corresponding output, and saving output does not save the corresponding data.
- The **Data** and **Transform** menus provide access to data management tools. Part IV of this appendix illustrates how to use several of these tools.
- The **Analyze** menu provides access to most of the statistical procedures described in this book. They are described in Part II of this appendix.
- The **Graphs** menu provides access to SPSS's graphing capabilities. Part III of this appendix describes how to create and edit basic graphs.

Figure B.1 **The Menu Bar**

	RECID	AGEGR10	SEX	MARSTAT	HSDSIZEC	AGEPRGR0
1	1	4	1	1	5	4
2	2	4	1	1	3	3
3	3	3	2	1	4	4
4	4	6	2	3	2	95
5	5	4	1	6	1	95
6	6	5	2	3	1	95
7	7	1	2	6	4	95
8	8	3	1	1	3	2
9	9	2	1	1	3	2
10	10	4	2	6	1	95

Figure B.2   The Data View

The Data Editor

The main window in SPSS is the Data Editor. It is divided into two sub-screens—a *Data View* and a *Variable View*—that are accessed using tabs located in the bottom left of the window. In the bottom right corner of the Data Editor is a status area, which provides messages about the SPSS processor and any special conditions that are affecting the data, such as weighting, filtering, or splitting. The width of columns in both the data view and the variable view can be adjusted by dragging the edge of the column where it crosses the blue label bar at the top of the screen.

The Data View The *Data View* lists the cases in rows and the variables in columns. The cases are numbered sequentially in the blue column down the left side, and the variable names are listed in the blue row at the top of each column. Hover over the variable name at the top of each column to see the full variable label. Right-clicking on a variable name provides access to a pop-up menu that allows you to sort cases by the values on that variable, get information about the variable, and produce some basic descriptive statistics.

The value on each variable, for each case, is shown in each of the white cells. Clicking the "Value Labels" icon on the toolbar (in the red square in Figure B.3) will display the labels associated with each value in the *Data View*. Typically, only categorical variables have value labels.

The Variable View The *Variable View* lists each variable in rows. The first column lists the name of each variable. Variable names cannot include spaces or special characters and are usually short. More detailed information about what each variable captures is in the "Label" column, which does not have these restrictions. Right-clicking on the row number in the blue column to the left of the variable names provides access to a pop-up menu that allows you to get information about the variable and produce some basic descriptive statistics.

The second column of the *Variable View* lists the variable type. The two most common types are numeric variables and string variables. Numeric variables contain only numbers. As a result, the values of numeric variables can be mathematically transformed (for instance, divided by 10). String variables can contain

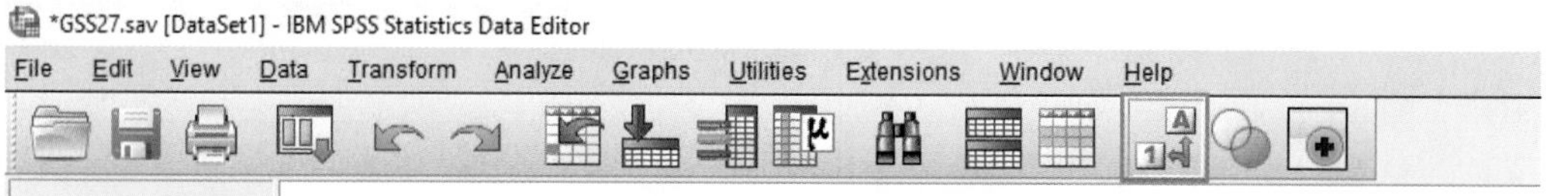

	RECID	AGEGR10	SEX	MARSTAT	HSDSIZEC	AGEPRGR0
1	1	45 to 54 years	Male	Married	5	45 to 54 years
2	2	45 to 54 years	Male	Married	3	35 to 44 years
3	3	35 to 44 years	Female	Married	4	45 to 54 years
4	4	65 to 74 years	Female	Widowed	2	No spouse/partner in the household
5	5	45 to 54 years	Male	Single, never married	1	No spouse/partner in the household
6	6	55 to 64 years	Female	Widowed	1	No spouse/partner in the household
7	7	15 to 24 years	Female	Single, never married	4	No spouse/partner in the household
8	8	35 to 44 years	Male	Married	3	25 to 34 years
9	9	25 to 34 years	Male	Married	3	25 to 34 years
10	10	45 to 54 years	Female	Single, never married	1	No spouse/partner in the household

Figure B.3 The Data View with Value Labels Displayed

letters, numbers, and other text characters. Because they can contain letters and other text characters, string variables cannot be mathematically transformed. Other types of variables include date variables (for capturing calendar dates) and dollar/currency variables. Most datasets contain primarily numeric variables.

The "Width" column provides information about the total number of characters allowed in each cell. For numeric and currency variables, the "Decimals" column determines how many decimals are displayed.

The "Values" column provides information about the attributes that are associated with each value. To view or edit the value labels for a specific variable, click on the ". . ." button at the end of the cell in the "Values" column to launch the Value Labels window. (See the left of Figure B.5.) The Value Labels window can be used to add or remove value labels from a variable, and to change existing value labels.

The "Missing" column shows which values on a variable are designated as missing information and, thus, excluded from any statistical calculations that use that variable. To view or change the missing values for a specific variable, click on the ". . ." button at the end of the cell in the "Missing" column to launch the Missing Values window. (See the right of Figure B.5.) Select "Discrete missing values" to specify up to three non-contiguous missing values (such as "0," "97," and "99"). Select "Range plus one" to specify a range of contiguous missing values (such as

	Name	Type	Width	Decimals	Label	Values	Missing	Columns	Align	Measure
1	RECID	Numeric	5	0	Record identification	{99996, Valid skip}...	99996 - 99999	9	Right	Scale
2	AGEGR10	Numeric	2	0	Age group of respondent (groups of 10)	{1, 15 to 24 years}...	96 - 99	13	Right	Ordinal
3	SEX	Numeric	1	0	Sex of respondent	{1, Male}...	6 - 9	8	Right	Nominal
4	MARSTAT	Numeric	2	0	Marital status of respondent	{1, Married}...	96 - 99	15	Right	Nominal
5	HSDSIZEC	Numeric	2	0	Household size of respondent	{6, 6 or more}...	96 - 99	9	Right	Scale
6	AGEPRGR0	Numeric	2	0	Age group of respondent's spouse/partner (groups of 10)	{1, 15 to 24 years}...	95 - 99	26	Right	Ordinal
7	AGEPRGRD	Numeric	2	0	Age difference between respondent and spouse/partner	{1, Respondent 11 + years older}...	95 - 99	36	Right	Ordinal
8	AGECRYGC	Numeric	2	0	Age of respondent's youngest single child in household	{1, 0 to 4}...	96 - 99	12	Right	Ordinal

Figure B.4 The Variable View

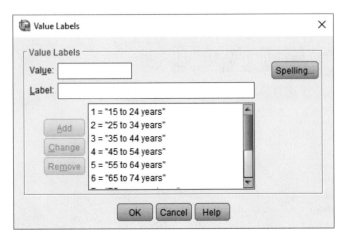
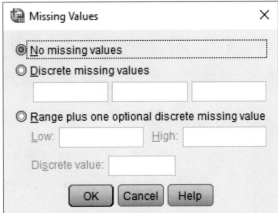

Figure B.5 The Value Labels and Missing Values Windows

"97" through "99"), with or without one additional non-contiguous missing value. To remove the missing values associated with a variable, select "No missing values."

The "Columns" and "Align" columns in the *Variable View* control how the information is displayed in the *Data View*. The former controls the width of each column, and the latter controls whether the cell is left-aligned, centred, or right-aligned. These controls are rarely used.

The "Measure" column in the *Variable View* is used to specify whether a variable is a nominal-level, ordinal-level, or ratio-level (scale) variable. Each of these levels of measurement is denoted using a different icon: a ruler for ratio-level (scale) variables, a bar graph for ordinal-level variables, and three circles for nominal-level variables. (See Figure B.6.) A variable's level of measurement affects which descriptive statistics are produced when they are requested through the pop-up menu that is accessed by right-clicking on a variable name. A variable's level of measurement also affects which types of graphs it can be used in and how it is treated in graphs. Because of this, be sure to assign each variable the correct level of measurement before graphing it.

Crucially, many datasets list an incorrect level of measurement for most variables; typically, all variables are listed as nominal-level (or are all listed as scale) by default. You must look at the values and value labels of each variable in order to determine whether it is actually nominal-level, ordinal-level, or ratio-level before using it in any statistical analyses or graphs.

Double-clicking on a variable name (in the *Data View*) or on the row number of a variable (in the *Variable View*) moves back and forth between the variable in the *Data View* and the *Variable View*.

Figure B.6 The Levels of Measurement Drop-Down Menu

The Output Viewer

Whenever a statistic is requested in SPSS, the results appear in a separate Output Viewer window. For each procedure, the Output Viewer shows the SPSS command syntax used to produce the statistic, and the resulting output. The tree diagram on the left side of the Output Viewer makes it easy to hide/show and navigate between groups of output.

The Output Viewer provides a running record of every statistical procedure that is produced in a session. It does not update retroactively; once a procedure is run, the information in the Output Viewer is fixed. So, if a variable is changed, the procedure needs to be re-run to see the effects of the change.

The appearance of tables and graphs in the Output Viewer can, however, be edited. To change the orientation of a table, double-click on the table to select it. Then, go to the **Pivot** menu, which only appears in the menu bar of the Output Viewer window when a table is selected. In the **Pivot** menu, selecting Transpose Rows and Columns will reverse the orientation of the table. Alternatively, choosing Pivoting Trays will launch a window like the one in Figure B.7. The Pivoting Trays window is used to arrange where variables and statistics appear in a table. Information can be nested in rows, in columns, or in layers. Arrange the elements by dragging-and-dropping them in the Pivoting Trays window, and then close the window to save the changes.

Tables can also be edited by double-clicking to select them and then right-clicking to access a pop-up editing menu. The pop-up editing menu provides access to items used to adjust many aspects of a table's appearance, including cell properties, table properties, and the table look (borders and shading). These items, as well as those used to add titles, captions, and footnotes, are also available in the **Insert** and **Format** menus that appear in the menu bar of the Output Viewer window when a table is selected.

SPSS output is easiest to work with when it shows both variable names and labels and the associated value numbers and labels. To display this information, select Options in the **Edit** menu, and navigate to the *Output* tab. (See Figure B.8.) Use the drop-down menus to select "Names and Labels" or "Values and Labels". for all four boxes (in the red squares), and then click the "OK" button to save the changes. I recommend that you develop the habit of doing this each time that you open SPSS.

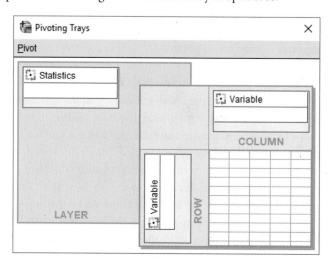

Figure B.7   **The Pivoting Trays Window**

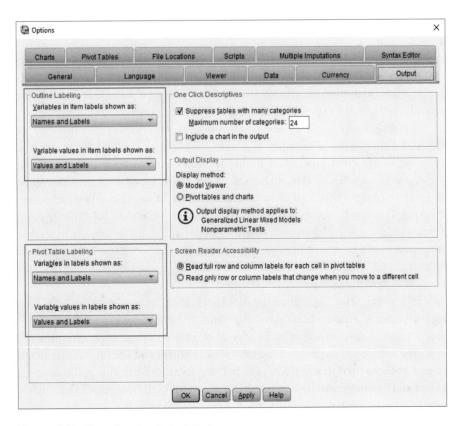

Figure B.8 **Changing the Output Options**

Many other default settings can be changed in the Options window, including where files are saved (using the *File Locations* tab), the appearance of tables (using the *Pivot Tables* tab), the appearance of charts (using the *Charts* tab), the appearance of output in the viewer (using the *Viewer* tab), and the language of the program menus (using the *Language* tab).

Copying and Pasting SPSS Output Researchers often copy output from SPSS and paste it into a word processor or spreadsheet. Tables, graphs, and any other output in the viewer can be copied by right-clicking the item and selecting Copy in the pop-up menu, or by selecting Copy in the **Edit** menu of the Output Viewer. It can then be pasted into another program.

By default, graphs are copied as images/pictures, and tables are copied as formatted text. It's sometimes useful, however, to copy tables as as images/pictures or into spreadsheets. This is done by selecting Copy As in the pop-up menu (see Figure B.9) which is accessed by right-clicking on any table in the viewer or by selecting Copy As in the **Edit** menu of the Output Viewer. When Excel Worksheet (BIFF) is selected the table structure is retained when it is pasted into a spreadsheet. When Image is selected the table is copied as a picture; you may need to use the

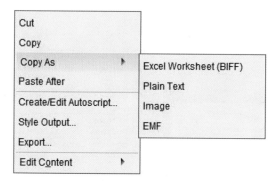

Figure B.9   The Copy As Pop-Up Menu

Paste Special option in your word processor for the table to be pasted as a picture. It can be particularly useful to copy wide tables as pictures, since they can become distorted when they are pasted in text form. When wide tables are pasted into documents as images, they can be resized without affecting the appearance of the table.

Exporting Output from SPSS

Instead of copying and pasting, output can also be exported from SPSS. This is particularly useful for transferring a large quantity of output to a spreadsheet, a word-processing document, or a PDF file. This is done using the Export Output window, which is accessed by right-clicking on any output in the viewer and selecting Export in the pop-up menu or by selecting Export in the **File** menu of the Output Viewer. In the Export Output window (in Figure B.10), you can select whether all of the output in the viewer, all of the visible (non-hidden) output in the viewer, or only the selected output will be exported. Use the drop-down menu to select the type of file that output will be saved as, specify the location where the file will be saved, and then click on the "OK" button.

Part II: Producing Statistics

One feature of SPSS is that there are often several ways to produce the same statistic. When this occurs, it usually does not matter which procedure is used since the results are identical. Most of the time, however, SPSS will produce any statistic for any variable, regardless of whether it is nominal-level, ordinal-level, or ratio-level. Thus, it often doesn't make sense to interpret all of the statistics that SPSS produces for a particular variable. As a data analyst, you are responsible for knowing which statistics are used with each type of variable and only reporting the results that are meaningful.

The windows used to access each statistical procedure in SPSS typically have five buttons across the bottom: "OK," "Paste," "Reset," "Cancel," and "Help."

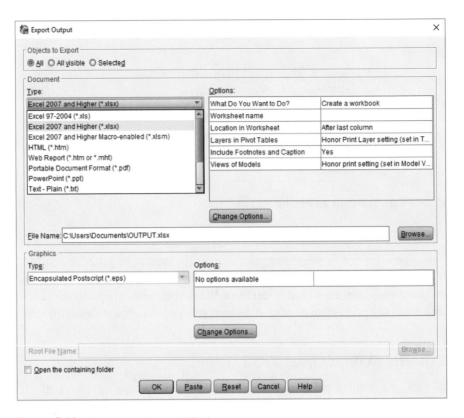

Figure B.10   **The Export Output Window**

- The "OK" button is used to run the procedure. For all of the procedures described in this section, click the "OK" button at the end of the instructions in order to produce the output.
- The "Paste" button is used to transfer the command syntax for each procedure into a syntax window so that it can be edited before the procedure is run.
- The "Reset" button resets the procedure window to its default state.
- The "Cancel" button closes the procedure window without producing any output.
- The "Help" button activates the default web browser in order to show context-specific help for the specific procedure or window.

Most SPSS procedure windows include additional buttons, typically located on the right, that are used to access options that are specific to each procedure. Clicking a button to access these options typically opens an additional window. Once you have made selections in the additional option window, click the "Continue" button to save the selections, close the window, and return to the main procedure window. (Clicking the "Cancel" button will close the additional window without saving the selections.) For all of the additional options that I describe in this section, click the "Continue" button to close the window after making selections.

For each statistical procedure, the variables used are specified by selecting them from a list of variables and putting them into a variable box in the procedure window by clicking on the arrow button, like the one in the red box in Figure B.11. Many procedure windows have more than one variable box and selection arrows that correspond to each one. As well, many procedures allow more than one variable to be put in the variable box. Variables can be removed from a variable box by selecting them and clicking on the arrow button (which will have reversed direction) to return them to the list of variables. Variables can also be dragged-and-dropped into or out of variable boxes.

There are several strategies that can make it easier to navigate the variables that are listed in each procedure window. First, most windows can be made wider by dragging the edge so that more of the variable label/name is visible. As well, right-clicking anywhere in the variable list provides access to the pop-up menu shown on the right of Figure B.11, which allows you to choose whether variable names (from the first column of the *Variable View*) or variable labels (from the fifth column of the *Variable View*) are displayed in the list. In addition, you can sort the variable names or labels alphabetically. If you click anywhere in the variable list, and press a letter on the keyboard, the highlighted variable will jump to the next variable name or label beginning with that letter. (Press the letter another time to jump to the next variable beginning with that letter.)

Descriptive Statistics

The **Analyze** menu provides access to Descriptive Statistics procedures. (See Figure B.12.) The first four procedures are commonly used: Frequencies, Descriptives, Explore, and Crosstabs.

The Frequencies Procedure The Frequencies procedure window has a single variable box. (See Figure B.13.) To produce a frequency distribution of one or more variables, put them in the variable box and click the "OK" button.

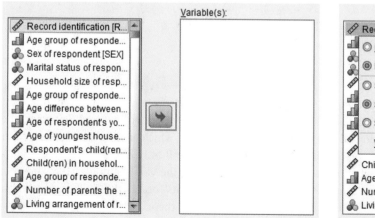
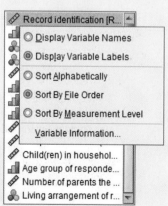

Figure B.11 Selecting Variables and Navigating the Variable List

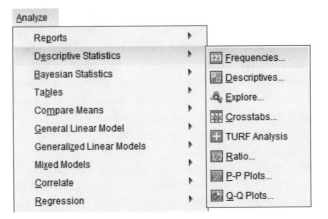

Figure B.12　**The Descriptive Statistics Procedures Menu**

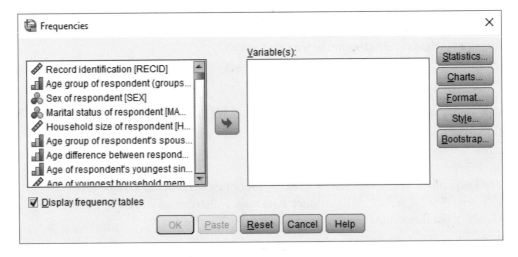

Figure B.13　**The Frequencies Window**

The Frequencies procedure options are listed in a series of buttons on the right side of the window. Two options are commonly used: Format and Statistics. The Format option allows you to specify the order in which attributes are displayed in a frequency distribution. (See the left of Figure B.14.) Selecting "Ascending/descending values" orders the attributes based on the value (number) that they are assigned. Selecting "Ascending/descending counts" orders the attributes based on their popularity or on how many people selected them.

The Statistics option allows you to select which statistics are produced for the variable(s). (See the right of Figure B.14.) Select the corresponding checkboxes

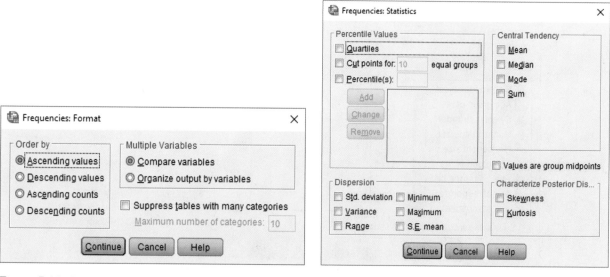

Figure B.14   The Frequencies Procedure Format and Statistics Windows

to produce statistics that show the central tendency ("Mean," "Median," and "Mode"), the dispersion ("Std. deviation," "Variance," "Range," "Minimum," and "Maximum"), and distribution ("Skew," "Kurtosis") of a variable, as well as percentile values (including "Quartiles"). To produce only the statistics but no frequency distribution, de-select the "Display frequency tables" checkbox in the main Frequencies window.

The Descriptives Procedure The Descriptives procedure window also has a single variable box. (See the left of Figure B.15.) To produce descriptive statistics for one or more variables, put them in the variable box and click the "OK" button. The "Options" button allows you to select which statistics are produced for the variable(s). (See the right of Figure B.15.) By default, the "Mean," "Std. deviation," "Minimum," and "Maximum" are shown. Select the corresponding checkboxes to also find the "Variance," "Range," "Kurtosis," and "Skewness."

The Descriptives procedure can also be used to create a new variable that captures the z-score of each case, based on the distribution of an existing ratio-level variable. In other words, the values on the original variable are mapped onto a standard normal distribution, and each case is assigned a value on the new variable that corresponds to its location in the standard normal distribution. The value "0" is assigned to cases with a value at the mean of the original variable, the value "1" is assigned to cases with a value that is one standard deviation above the mean of the original variable, the value "–1" is assigned to cases with a value that is one standard deviation below the mean of the original variable, and so on, for all of the cases. To create a variable that captures z-scores, select the "Save standardized values as variables" checkbox in the main Descriptives window. When the procedure is run, a new variable capturing the z-scores is added in the final row of the *Variable View*;

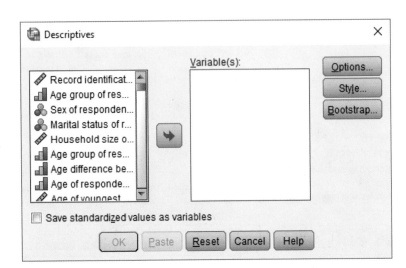

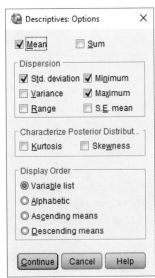

Figure B.15   The Descriptives Window and Its Options Window

it will have the same name as the original variable, prefaced with a *z*. Remember that it only makes sense to calculate z-scores for ratio-level variables.

The Explore Procedure   The Explore procedure produces summary statistics for a variable, including the mean, 95 per cent confidence interval for the mean, variance, standard deviation, minimum, maximum, range, interquartile range, skew, and kurtosis. To produce these statistics for one or more variable(s), put them in the "Dependent List" box. (See Figure B.16.) You can select whether to display "Statistics" only, "Plots" only, or "Both." If "Plots" or "Both" is selected, a stem-and-leaf plot and a box plot of each variable is displayed. For most analyses, I recommend selecting "Statistics" only to make the output more readable.

The Explore procedure can produce statistics and plots for all the cases together or divided by group. To divide the cases by group, put a categorical variable that defines the groups into the "Factor List" box.

The Explore procedure also provides access to Kolmogorov-Smirnov test, which shows whether the distribution of a variable is significantly different than normal. This test can be used to assess whether regression residuals are normally distributed. To request a Kolmogorov-Smirnov test, select the "Normality plots with tests" checkbox in the Plots window. (See the right of Figure B.16.) This produces both a Kolmogorov-Smirnov test and a quantile–quantile (Q-Q) plot for the variable(s) in the "Dependent List" box.

The Crosstabs Procedure   The Crosstabs procedure produces a cross-tabulation between two categorical variables: put the dependent variable into the "Row(s)" box and the independent variable into the "Column(s)" box. (See Figure B.17.)

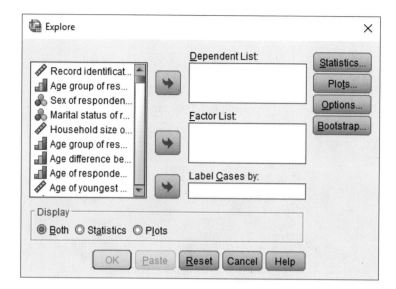
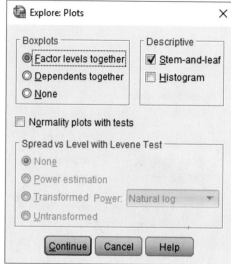

Figure B.16   **The Explore Window and Its Plots Window**

The Cells option allows you to select the information that is displayed in each cell of the cross-tabulation. By default, only the observed counts (or observed frequencies) are shown. To show the expected counts, select the "Expected" checkbox

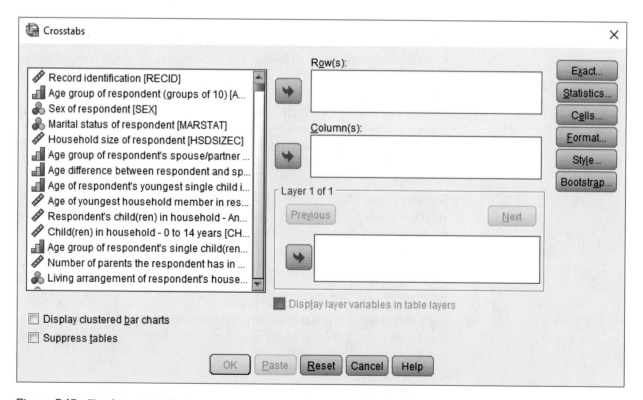

Figure B.17   **The Crosstabs Window**

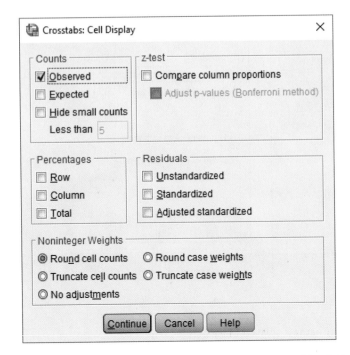
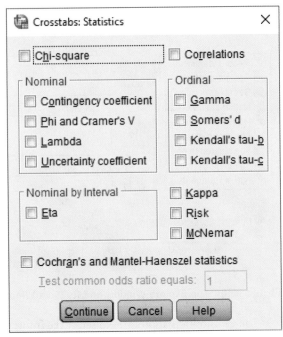

Figure B.18   The Crosstabs Procedure Cell Display and Statistics Windows

in the Cell Display window. (See the left of Figure B.18.) In addition, you can select whether "Row" percentages, "Column" percentages, and/or "Total" percentages are displayed. Row percentages show the percentage in each cell out of all of the cases in the same row that the cell is located in. Column percentages show the percentage in each cell out of all of the cases in the same column that the cell is located in. Total percentages show the percentage in each cell out of the total number of cases overall. If the independent variable is located in the columns of a cross-tabulation, it is most useful to produce and compare column percentages.

The Statistics option of the Crosstabs procedure allows you to request "Gamma," "Lambda," "Chi-square," and/or "Phi and Cramer's V" by selecting the appropriate checkboxes. Be sure to only request and interpret statistics that match the level of measurement (and the number of attributes) of the variables included in the cross-tabulation.

Comparing Means

The **Analyze** menu provides access to procedures that are used to Compare Means. (See Figure B.19.) Three procedures are commonly used: Means, Independent Samples T-Test, and One-Way ANOVA.

The Means Procedure The Means procedure produces statistics that show the centre, dispersion, and shape of a variable, either for all the cases together or

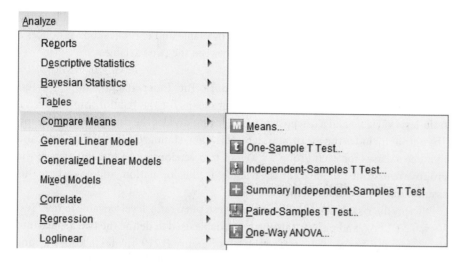

Figure B.19   **The Compare Means Procedures Menu**

divided by group. Begin by putting one or more ratio-level variable(s) in the "Dependent List" box. (See the left of Figure B.20.) To divide the cases by group, put a categorical variable that defines the groups into the "Independent List" box.

Use the "Options" button to launch a window where you can choose which statistics are produced for the variable(s) by selecting them from the list on the left and putting them in the "Cell Statistics" box. (See the right of Figure B.20.) By default, the "Mean," "Number of Cases," and "Standard Deviation" are displayed. Most other statistics used to describe the centre, dispersion, and shape of a variable are available in the Means procedure, with the exception of the mode and the

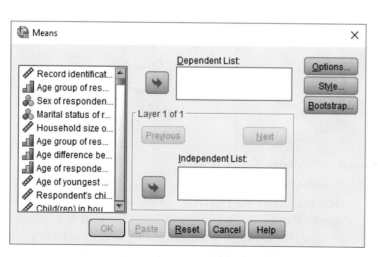

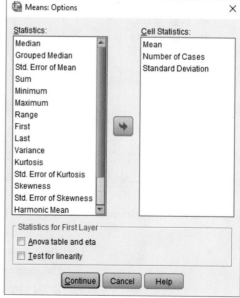

Figure B.20   **The Means Window and its Options Window**

inter-quartile range. If the cases are divided by group (by putting a variable in the "Independent List" box), selecting the "Anova table and eta" checkbox produces a one-way ANOVA test for the relationship between the two variables.

The Independent Samples T-Test Procedure The Independent Samples T-Test procedure produces a t-test of independent means for the relationship between a ratio-level variable and a categorical variable that divides the cases into two groups. The output includes the mean, standard deviation, standard error of the mean, and number of cases for each group, as well as two versions of a t-test (one where the groups are assumed to have equal variances in the population, and one where this assumption is not made).

To produce an independent samples t-test, put a ratio-level variable in the "Test Variable(s)" box, and put the categorical variable used to define the two groups into the "Grouping Variable" box. (See the left of Figure B.21.) Then, click the "Define Groups" button to provide SPSS with instructions on which two groups to compare. There are two ways to define the groups used in the t-test: either specify a single value that identifies Group 1 and a single value that identifies Group 2, or specify a value to use as a cut point. (See the right of Figure B.21.) When a "Cut point" value is specified, cases with a value greater than or equal to the cut-point value are treated as the first group, and cases with a value less than the cut-point value are treated as the second group.

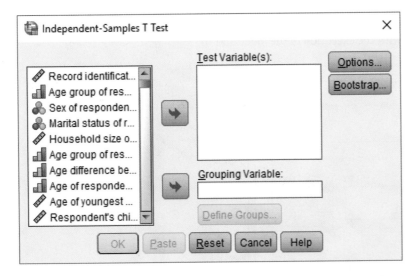

Figure B.21   **The Independent Samples T-Test Window and Its Define Groups Window**

The One-Way ANOVA Procedure The One-Way ANOVA procedure tests whether the mean of a ratio-level variable is likely to be different between two or more groups in a population. ANOVA tests are typically used with large samples or when a researcher is comparing three or more groups. To produce a one-way ANOVA test, put a ratio-level variable in the "Dependent List" box, and put the categorical variable used to define the groups into the "Factor" box. (See the left of Figure B.22.)

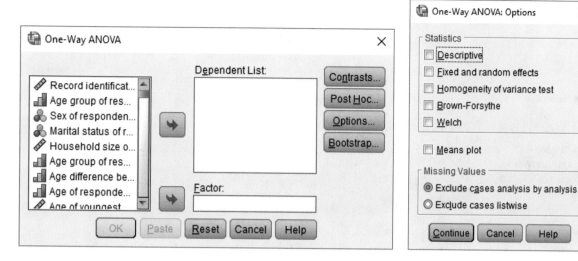

Figure B.22   The One-Way ANOVA Window and Its Options Window

By default, the procedure shows only the ANOVA test results. Use the "Options" button and select the "Descriptive" checkbox to also show the mean, 95 per cent confidence interval for the mean, standard deviation, standard error of the mean, minimum, maximum, and number of cases for each group and for all the cases together. (See the right of Figure B.22.)

The Post Hoc option is used to request post-hoc tests for each pair of groups included in a one-way ANOVA test. (See Figure B.23.) There are many different types

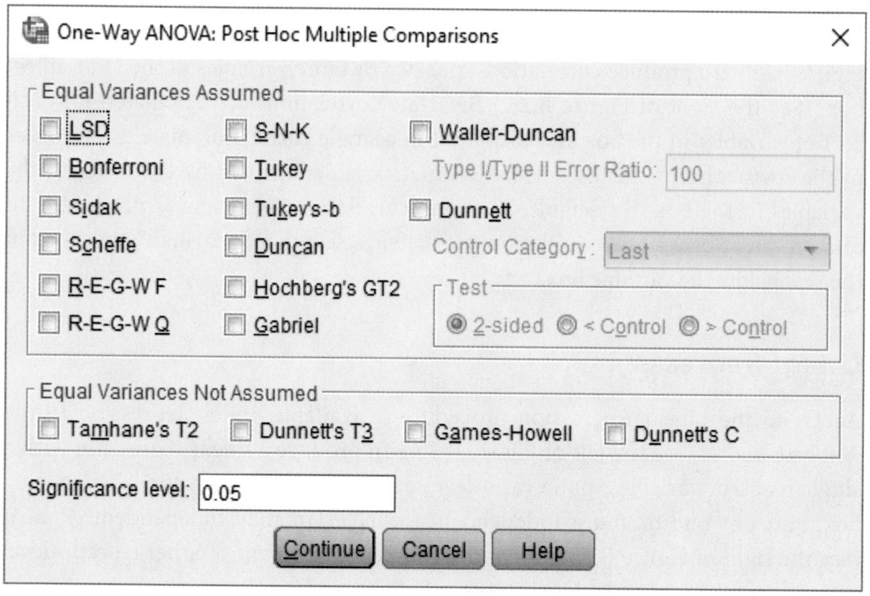

Figure B.23   The One-Way ANOVA Post-Hoc Multiple Comparisons Window

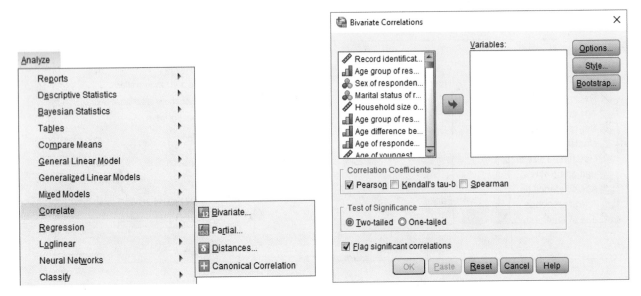

Figure B.24   The Correlate Procedures Menu and the Bivariate Correlations Window

of post-hoc tests; LSD and Bonferroni tests are commonly used. The "Help" button in the Post-Hoc window opens an SPSS help page that describes the different types of post-hoc tests and what makes each unique. Although you can change the alpha value used to determine statistical significance in the post-hoc tests, social science researchers typically leave this at 0.05.

Bivariate Correlations

The Bivariate Correlations procedure, available in the **Analyze** menu, is used to produce both Pearson's and Spearman's correlation coefficients. (See the left of Figure B.24.) To produce correlations, put two or more variables in the "Variables" box. (See the right of Figure B.24.) Bivariate correlations are calculated for each pair of variables in the box and displayed in a single correlation matrix. The order of the rows/columns in the correlation matrix is determined by the order of the variables in the box. By default, Pearson's correlation coefficient is produced. To also produce Spearman's correlation coefficient, select the "Spearman" checkbox in the row below the variable box.

Linear Regression

Access to the Linear Regression procedure is available under Regression in the **Analyze** menu. (See the left of Figure B.25.) To produce a linear regression with a single block of variables, put a ratio-level dependent variable in the "Dependent" box, and put one or more independent variables in the "Independent(s)" box. (See the right of Figure B.25.) The independent variables must either be ratio-level variables or dummy variables with the values "0" and "1".

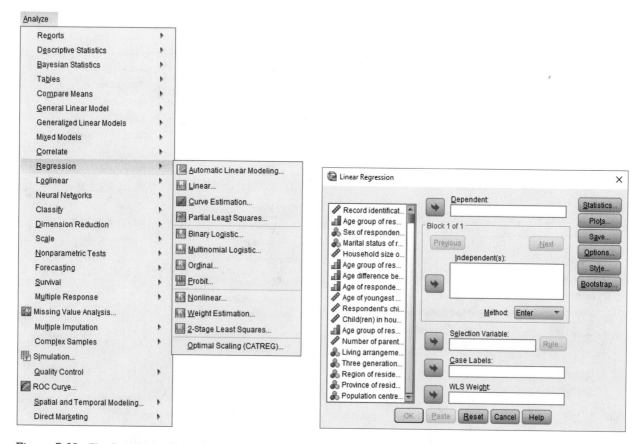

Figure B.25   **The Regression Procedure Menu and the Linear Regression Window**

Notice that the "Independent(s)" variable box is located inside a square labelled "Block 1 of 1" by default. To produce a linear regression with more than one block of variables, put the independent variables in the first block in the "Independent(s)" box, and then click the "Next" button above the box. The square's label will change to "Block 2 of 2" and the "Independent(s)" variable box will be empty again. Put the independent variables in the second block into the newly empty "Independent(s)" box. Repeat this process for each additional block of independent variables in the regression. To view or change the independent variables in previous blocks, click the "Previous" button located to the left of the "Next" button. An R^2 and an adjusted R^2 will be produced for each block in the regression.

The Statistics option provides access to additional regression statistics. (See the left of Figure B.26.) To show the confidence interval for each regression coefficient, select the "Confidence intervals" checkbox. By default, 95 per cent confidence intervals are produced. To display the tolerance and variance inflation factor of each independent variable used in the regression, select the "Collinearity diagnostics" checkbox.

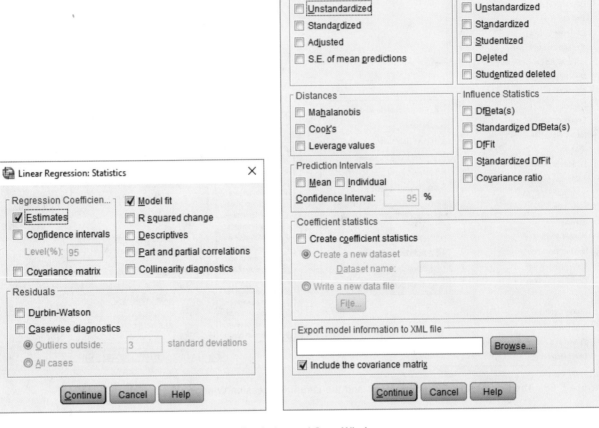

Figure B.26   The Linear Regression Procedure Statistics and Save Windows

The Save option allows you to save regression outcomes as new variables in the dataset. Notice that the Save window is divided into several sections. (See the right of Figure B.26.) In the "Predicted Values" section, select the "Unstandardized" checkbox to create a new variable that contains the value on the dependent variable that the regression predicts for each case. Similarly, in the "Residuals" section, select the "Unstandardized" checkbox to create a new variable that contains the difference between the value predicted by the regression and the actual value on the dependent variable for each case. New variables always appear in the final row of the *Variable View*. By default, for the first regression that is produced, the new variable capturing the unstandardized predicted values is assigned the name "PRE_1," and the variable capturing the unstandardized residuals is assigned the name "RES_1." For subsequent regressions, the number appended to the end of the variable name changes. (For instance, the saved variables for a second regression are named PRE_2 and RES_2.) The new, saved variables can be used like any other variable in SPSS procedures in order to assess how well a regression model fits the data and whether or not it is biased.

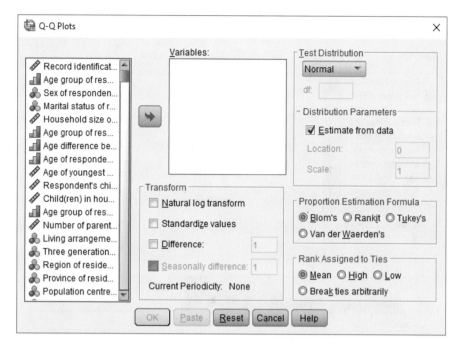

Figure B.27   **The Q-Q Plots Window**

Q-Q Plots The Q-Q Plot procedure is available under Descriptive Statistics in the **Analyze** menu. (Look back at Figure B.12.) To produce a Q-Q plot of regression residuals to test for normality, put the unstandardized residual variable (RES_1), saved from the Linear Regression procedure, into the "Variables" box. (See Figure B.27.) By default, variables are tested against a normal distribution. Once the Normal Q-Q plot has been produced, its axes and appearance can be edited in the same way as any other graph. (You'll learn how to do this in Part III of this appendix.)

Part III: Graphing

Using the Chart Builder Tool

The Chart Builder is a flexible tool that makes it easy to create graphs by dragging-and-dropping variables, which can be accessed in the **Graphs** menu. The Chart Builder is organized around a central canvas (the large white space) that you can put graphs and graph elements into. (See Figure B.28.) The easiest way to create a graph is by using the gallery. In the *Gallery* tab (located mid-way down the window), graphs are divided into categories in the "Choose from" list on the left. Each category in the list includes several pre-formatted graph templates, which are represented by a series of icons. Hover over each graph icon to see the name of the graph. To begin building a graph, either drag a graph icon into the canvas area or double-click the graph icon to make the template appear on the canvas.

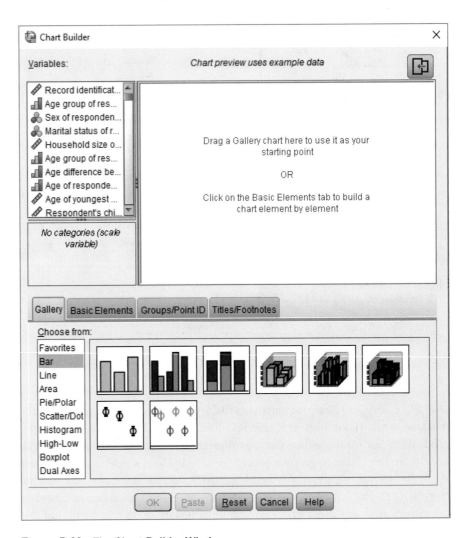

Figure B.28 **The Chart Builder Window**

For each graph template, the canvas shows one or more dashed boxes, called drop zones, for various elements. (See Figure B.29.) Drag variables from the list into each drop zone to include them in the graph. For instance, to select the variable that will be displayed on the x-axis, select a variable from the list and drag it into the "X-Axis" drop zone. Although using the chart builder is generally intuitive, remember that it relies on the level of measurement assigned to each variable to determine how the variable will be treated in graphs. If the level of measurement assigned to a variable in SPSS does not match the variable's actual level measurement (which often happens), you may get unexpected results. The drop zones for some graphs will only accept variables with a specific level of measurement. If you try to drag a variable into a graph and are not able to put it in a drop zone, it's likely because the variable has a level of measurement that does not match the requirements of the graph.

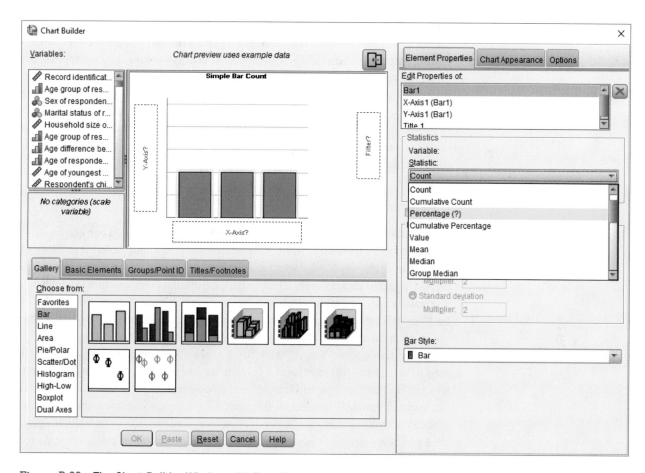

Figure B.29 The Chart Builder Window with Drop Zones and the Side Panel

To change a variable's level of measurement without exiting the Chart Builder tool, right-click on the variable name/label in the "Variables" list, and select the new level of measurement from the pop-up menu. Note, however, that when a variable's level of measurement is changed this way, any variables that have already been dropped into the canvas are not updated. The variable must be re-dragged into the canvas to use the updated level of measurement. (You can see which level of measurement SPSS is using for each variable by looking at the icon beside the variable name/label.) When a variable's level of measurement is changed inside the Chart Builder tool, the change is only temporary; it is *not* updated in the *Variable View*. For variables that are regularly used in graphs, permanently changing their level of measurement in the *Variable View* screen is more efficient.

When a graph template is put in the canvas, a side panel appears in the Chart Builder window (see Figure B.29). The side panel includes three tabs that let you alter various graph elements before a graph is produced; the elements that can be changed depend on the type of graph that is in the canvas. The *Element Properties* tab is used to change which statistics are displayed; for instance, to show percentages instead of counts on a graph, use the drop-down menu to change the "Statistic" to "Percentage." The *Element Properties* tab can also be used to assign a custom graph title and axis labels, and to make changes to how the axes appear. The *Chart Appearance* tab is used to alter the colours, borders, and grid lines used in the graph. Select the "Use custom color, border and grid line settings" checkbox to override the defaults, and then select (or de-select) the checkboxes corresponding to various graph elements (or use the colour picker) to make changes. The *Options* tab is used to alter how SPSS handles missing vales; do not change the default settings in this tab.

Clustered, Stacked, and Panelled Graphs

Some clustered and stacked graphs—such as clustered and stacked bar graphs—can be selected from the gallery in the Chart Builder window. But panelled graphs and other types of clustered and stacked graphs that are not available in the gallery can be created using the *Groups/Point ID* tab in the Chart Builder window. (See Figure B.30.) To create a clustered graph, select the "Clustering variable on X" checkbox. A "Cluster on X" drop zone will appear on the canvas, which is used to specify the variable that the graph is clustered on. Similarly, to create a stacked graph, select the "Grouping/ stacking variable" checkbox. A "Stack" drop zone will appear on the canvas, which is used to specify the variable that the graph is stacked on. To create a panelled graph, select either the "Rows panel variable" or the "Columns panel variable" checkbox. If "Rows panel variable" is selected, the panelled graphs are displayed above one another, as the rows in a single column. If "Column panel variable" is selected, the panelled graphs are displayed beside one another, as the columns in a single row. As for clustering and stacking, selecting either panelling checkbox adds a "Panel" drop zone to the canvas, which is used to specify the variable that defines the panels.

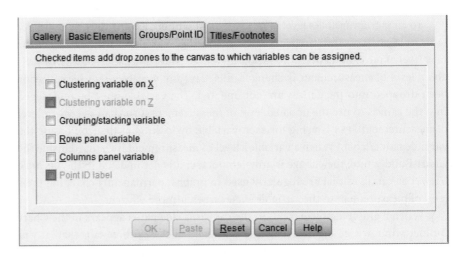

Figure B.30 The Groups/Point ID Tab in the Chart Builder Window

Building Common Graphs

Many different types of graphs can be constructed using the Chart Builder tool. In this section, I describe how to build the most commonly used graphs. Be prepared to experiment in order to develop graphs that effectively display the data, and explore the *Element Properties* and *Chart Appearance* tabs in the side panel to see what can be altered for each graph.

Bar graph Select the "Simple Bar" icon. (Look back at the left of Figure B.29.) Put a nominal- or ordinal-level variable in the "X-Axis" drop zone on the canvas. In the *Element Properties* tab in the side panel, change the bar statistic to "Percentage." (Look back at the right of Figure B.29.)

Histogram Select either the "Simple Bar" or the "Simple Histogram" icon. Put a scale variable in the "X-Axis" drop zone on the canvas. In the *Element Properties* tab in the side panel, change the bar statistic to "Histogram Percent."

   The same procedure is used to create bar graphs and histograms in SPSS: if a scale variable is used in a bar graph, SPSS will automatically produce a histogram; if a categorical variable is used in a histogram, SPSS will automatically produce a bar graph.

Pie graph Select the "Pie Chart" icon. Put a nominal- or ordinal-level variable in the "Slice by" drop zone on the canvas. In the *Element Properties* tab in the side panel, change the polar-interval statistic to "Percentage."

Line graph Select the "Simple Line" icon. Put a scale variable in the "X-Axis" drop zone on the canvas. In the *Element Properties* tab in the side panel, change the line statistic to "Histogram Percent."

Box plot Select the "Simple Boxplot" icon. Put an ordinal-level or scale variable into the "Y-Axis" drop zone on the canvas. To create box plots for different groups, put a nominal- or ordinal-level variable that defines the groups into the "X-Axis" drop zone. Note that in SPSS version 24 and on, only scale variables can be put into the "Y-Axis" drop zone for box plots.

Panelled pie graph or panelled bar graph Begin by building a pie graph or a bar graph of the dependent variable. Then, select either the "Rows panel variable" or the "Column panel variable" checkbox in the *Groups/Point ID* tab (look back at Figure B.30) to make a "Panel" drop zone appear on the canvas. Then, put a nominal- or ordinal-level independent variable into the "Panel" drop zone. The variable in the panel drop zone is used to define the groups that appear in each panel. In the *Element Properties* tab in the side panel, change the bar/polar interval statistic to "Percentage." In order to ensure that the percentages shown on the graph correspond with the column percentages in a cross-tabulation, click on the "Set

Parameters" button (located below the "Percentage" statistic in the *Element Properties* tab) to specify the denominator that will be used to calculate the percentages, and choose "Total for Panel".

Clustered bar graph Select the "Clustered Bar" icon. Put the nominal- or ordinal-level dependent variable in the "X-Axis" drop zone and put the nominal- or ordinal-level independent variable in the "Cluster on X: set color" drop zone that appears on the canvas. In the *Element Properties* tab in the side panel, change the bar statistic to "Percentage." In order to ensure that the percentages shown on the graph correspond with the column percentages in a cross-tabulation, click on the "Set Parameters" button (located below the "Percentage" statistic in the *Element Properties* tab) to specify the denominator that will be used to calculate the percentages, and choose "Total for Each Legend Variable Category (same fill color)."

Stacked bar graph Select the "Stacked Bar" icon. Put the nominal- or ordinal-level dependent variable in the "Stack: set color" drop zone that appears on the canvas and put the nominal- or ordinal-level independent variable in the "X-Axis" drop zone. In the *Element Properties* tab in the side panel, change the bar statistic to "Percentage." In order to ensure that the percentages shown in the graph correspond with the column percentages in a cross-tabulation, click on the "Set Parameters" button (located below the "Percentage" statistic in the *Element Properties* tab) to specify the denominator that will be used in the calculation of percentages, and choose "Total for each X-Axis category."

Error-bar graph Select the "Simple Error Bar" icon (from the "Bar" category in the gallery). Put a scale variable in the "Y-Axis" drop zone on the canvas. By default, the mean and the 95 per cent confidence interval are displayed in the error bars. To compare means and 95 per cent confidence intervals for different groups, put a nominal- or ordinal-level variable that defines the groups into the "X-Axis" drop zone.

Scatterplot Select the "Simple Scatter" icon. Put the independent scale variable in the "X-Axis" drop zone, and put the dependent scale variable in the "Y-Axis" drop zone on the canvas.

Editing Graphs

Once SPSS has produced a graph and it appears in the Output Viewer, you can edit it by double-clicking on it. This will launch the Chart Editor window, which is used to alter the appearance of graphs after they have been produced.

The Chart Editor window has several toolbars (shown in Figure B.31); you can select which ones are displayed in the **View** menu. Hovering over each toolbar icon provides information about what it does. The three icons in the red box on the left of the first toolbar row add reference lines to the graph: a straight line that serves

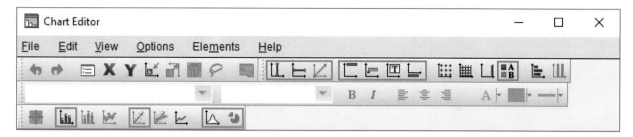

Figure B.31   **The Chart Editor Menus and Toolbars**

as a point of comparison for the viewer. The reference line can be a vertical line or a horizontal line, or it can be defined by a linear equation. The four icons in the red box in the middle of the first toolbar row add titles, annotations, textboxes, and footnotes to the graph. By default, SPSS shows variable labels on graphs, but if a variable has no label, the variable name is used instead; you can edit axis and legend labels, as well as graph titles, by double-clicking on them in the graph. The icon in the red box on the right of the first toolbar row shows/hides the graph legend. All of these items can also be accessed in the **Options** menu.

The bottom toolbar shows items from the **Elements** menu. The icon in the red box on the left shows data labels (i.e., percentages) on the graph. The icon in the red box in the middle adds a regression line to the graph. The first icon in the red box on the right adds a normal curve to a histogram to serve as a reference, and the second icon "explodes" a slice of a pie graph. If an item is not applicable to a specific type of graph, it will be greyed out on the toolbar.

Double-clicking anywhere on a graph in the Chart Editor window will launch a context-specific Properties window. The Properties window has tabs at the top, which you can navigate between in order to alter different graph elements. If there is something you want to change on a graph, double-click on it to see what is editable. For instance, double-clicking on various text elements will launch a Properties window like the one on the left of Figure B.32. The *Text Style* tab is used to change the font.

Double-click on a graph axis to edit the length and units (for scale variables) or the categories that are displayed (for categorical variables). Double-clicking on the axis of a scale variable will launch a Properties window like the one shown in the centre of Figure B.32. The *Scale* tab is used to designate the minimum and maximum values that appear on the graph, as well as the major increments (i.e., the distance between the numeric labels). Researchers often adjust the length of an axis so that outliers don't appear and so that the graph takes up more of the chart space. The *Number Format* tab is used to specify how many decimals are displayed on the graph axes, and the *Grid Lines* tab is used to control which grid lines appear. For axes that show categorical variables, a *Categories* tab is used to specify which groups appear (and which do not appear) and their order in the graph (See the right of Figure B.32).

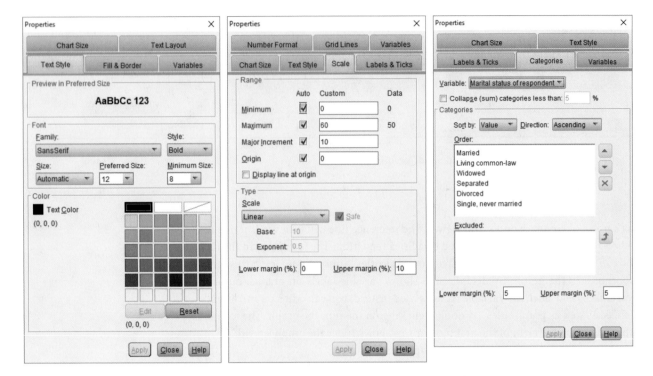

Figure B.32   Context-Specific Properties Windows in the Chart Editor

In addition to the text and the axes, almost every other aspect of a graph can be altered. Clicking on a dot or a marker (as in a scatterplot) launches a Properties window like the one on the left of Figure B.33. The *Marker* tab is used to change the shape, colour, and size of the marker. Similarly, clicking on a coloured graph element (such as a bar, or a pie slice, or the chart background) launches a Properties window like the one in the centre of Figure B.33. The *Fill & Border* tab is used to change colours. For histograms, clicking on the bars launches a Properties window like the one on the right of Figure B.33, with a *Binning* tab that is used to change the size of the bins, by specifying different interval widths (bin sizes) in the box. Always click on the "Apply" button after making any changes in the Properties window in order to ensure that the graph is updated.

Closing the Chart Editor window updates the appearance of the graph in the Output Viewer. Typically, researchers begin by producing basic graphs to explore patterns and relationships in the data. Then, they select key graphs to include in publications or reports and spend more time editing those graphs so that they are fully labelled and display all the relevant information.

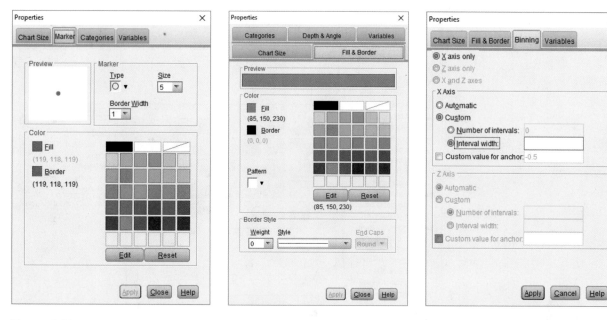

Figure B.33   **Additional Context-Specific Properties Windows in the Chart Editor**

Part IV: Variable and Data Management Tools

Computing a New Variable

The Compute Variable tool is available in the **Transform** menu. It can be used to create new variables or to change the values on existing variables. To create a new variable, type a new (unused) variable name in the "Target Variable" box. (See Figure B.34.) Optionally, the variable can be assigned a label using the "Type & Label" button below the "Target Variable" box. To change an existing variable, type the name of the existing variable in the "Target Variable" box. (Be sure to use the variable name, from the first column of the *Variable View*, and not the label.)

The "Numeric Expression" box is used to provide instructions about how to calculate the values on the target variable. For example, to make the values on the target variable equal to "1" for every case, put "1" in the "Numeric Expression" box. To make the values on the target variable equal to the values on an existing variable, put the existing variable (selected from the list) into the "Numeric Expression" box. To make the values on the target variable exactly double the values on an existing variable, put the existing variable (selected from the list) into the "Numeric Expression" box, followed by "*2" in order to multiply all of the values by 2. The calculator functions can be used to build more complex formulas that incorporate existing variables.

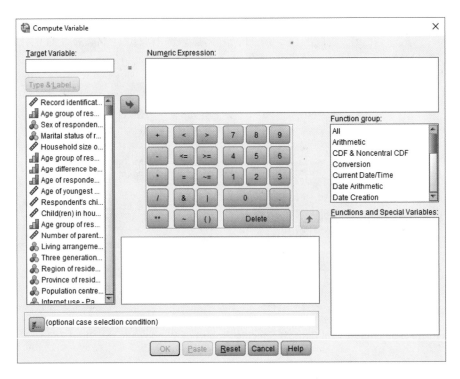

Figure B.34 **The Compute Variable Window**

The Compute Variable tool is commonly used to centre variables or to re-scale variables. For instance, to centre a variable called VARIABLE_A on the value 40:

- Open the Compute Variable tool and type "VARIABLE_A_CENTRED" into the "Target Variable" box.
- Put "VARIABLE_A – 40" in the "Numeric Expression" box.
- Click the "OK" button to compute the new variable; VARIABLE_A_CENTRED will appear in the final row of the *Variable View*.

Similarly, to rescale a variable that is measured in exact dollars (VARIABLE_A) so that it is measured in thousands of dollars:

- Open the Compute Variable tool and type "VARIABLE_A_RESCALED" into the "Target Variable" box.
- Put "VARIABLE_A / 1000" in the "Numeric Expression" box.
- Click the "OK" button to compute the new variable; VARIABLE_A_RESCALED will appear in the final row of the *Variable View*.

Sometimes, you only want to make changes to some cases in a dataset. To limit the cases affected by the Compute Variable tool, click the "If" button in the bottom left of the Compute Variable window to launch the If Cases window. The If Cases window is used to specify which cases a transformation applies to. To choose the cases that are affected, select "Include if case satisfies condition," and put one or more variables and their associated conditions in the box. (See Figure B.35.) If a case meets that condition, then it will be affected by the Compute Variable tool (and if it does not meet that condition, it will not be affected). This technique is often used to create a new variable that assigns different values to cases depending on their values on one or more existing variables.

For example, to create a new variable called MARRIED that equals "1" for all married people (who have the value "1" on an existing marital status variable, called MARSTAT) and that equals "0" for all non-married people (who have a value higher than "1" on the existing marital status variable):

- Open the Compute Variable tool and type "MARRIED" into the "Target Variable" box.
- Type "1" in the "Numeric Expression" box.
- Click the "If" button to open the If Cases window.
- Put "MARSTAT = 1" into the "Include if case satisfies condition" box.
- Click the "Continue" button to close the If Cases window and click the "OK" button to compute the new variable.

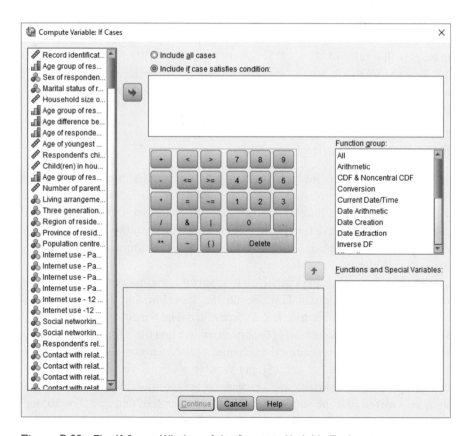

Figure B.35   **The If Cases Window of the Compute Variable Tool**

At this point, all married people are assigned the value "1" on the variable MARRIED. To assign the value "0" to the non-married people:

- Open the Compute Variable tool and type "MARRIED" into the "Target Variable" box.
- Type "0" in the "Numeric Expression" box.
- Click the "If" button to open the If Cases window.
- Put "MARSTAT > 1" into the "Include if case satisfies condition" box.
- Click the "Continue" button to close the If Cases window and click the "OK" button to compute the new variable.
- A warning message will pop up, asking for confirmation that an existing variable is being changed; click the "OK" button since the existing variable MARRIED is being changed.

The resulting variable, called MARRIED, appears in the final row of the *Variable View*. The MARRIED variable has the value "1" for all cases where MARSTAT is equal to "1", and has the value "0" for all cases where MARSTAT is greater than "1". You can produce a cross-tabulation between the variables MARRIED and MARSTAT to confirm that the new variable was created correctly.

It is also possible to combine two or more conditions, using the "&" symbol or the "|" symbol (which denotes "or"). The variable name must be repeated after each "and" or "or" symbol. In other words, to specify that a transformation should apply to cases with a value of "1" or "2" on a variable called VARIABLE_A, put "VARIABLE_A=1 | VARIABLE_A=2" (the variable equals "1" or the variable equals "2") in the "Include if case satisfies condition" box. The condition "VARIABLE_A=1 | 2" (the variable equals "1" or "2") will not be accepted.

Recoding an Existing Variable

SPSS includes two tools for recoding variables: the Recode into Same Variables tool and the Recode into Different Variables tool. Both tools are available in the **Transform** menu. The Recode into Same Variables tool overwrites the original variable, which can make it impossible to repair problems if the recoding process does not work as expected. Because of this, I strongly recommend using only the Recode into Different Variables tool.

The first step in recoding a variable is to specify the original, existing variable that will be recoded. Put the original variable into the "Input Variable -> Output Variable" box in the middle of the Recode into Different Variables window. (See Figure B.36.) Then, specify the name and label of the new variable, using the "Output Variable" box on the right of the window. Click the "Change" button to make the name of the new variable appear after the "->" in the "Input Variable -> Output Variable" box.

Once the original variable and new variable are specified, the recoding instructions are entered into the window that launches when the "Old and New Values" button is clicked. (See Figure B.37.) The Old and New Values window is used to provide instructions about what values on the new variable should be assigned to cases

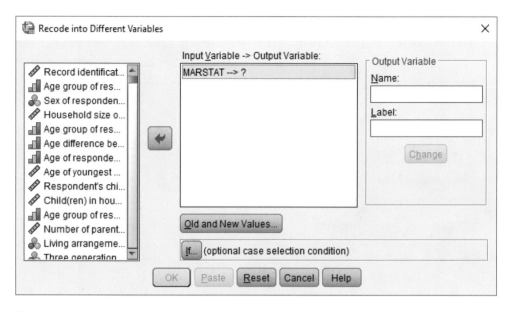

Figure B.36   **The Recode into Different Variables Window**

with each value on the original variable. The window is divided into two sections: the "Old Value" section is on the left, and the "New Value" section is on the right. List each value on the original variable in the "Old Value" section of the window, and assign it a new value in the "New Value" section of the window. After specifying each pairing, click the "Add" button, and the instruction will appear in the "Old ->New" box on the bottom right.

The values on the original variable can either be specified one at a time, or a range of values can be specified. You can also provide instructions about what to do with values/attributes that are designated as missing (user-missing) or blank cells (system-missing). The new values that are assigned can be a single numeric value that you specify, a "system-missing" value (i.e., make the cell blank), or a copy of the old value.

For example, to recode the variable that captures marital status (MARSTAT) into a new variable called MARRIED, that equals "1" for all married people (who have the value of "1" on the existing marital status variable) and that equals "0" for all non-married people (who have a value higher than "1" on the existing marital status variable):

- Open the Recode into Different Variables tool and put MARSTAT into the "Input Variable -> Output Variable" box.
- Type "MARRIED" into the Output Variable "Name" box, and assign a label to the new variable.
- Click the "Change" button, so that the "Input Variable -> Output Variable" box shows the expression "MARSTAT->MARRIED."
- Click the "Old and New Values" button to open the associated window.

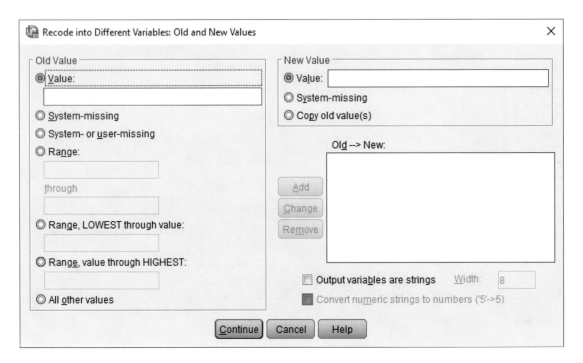

Figure B.37   The Old and New Values Window of the Recode into Different Variables Tool

- Put "1" into the "Value" box in the "Old Value" section of the window, and select "Copy old value(s)" in the "New Value" section of the window.
- Click the "Add" button so that the "Old ->New" box shows the expression "1-> Copy."
- Select "Range" in the "Old Value" section of the window, and put "2" through "6" in the boxes on the "Old Value" side of the window, and put "0" into the "Value" box on the "New Value" side of the window. This will assign all cases with the values "2" through "6" the value "0" on the new variable.
- Click the "Add" button so that the "Old ->New" box shows the expression "2 thru 6 ->0."
- Click the "Continue" button to close the Old and New Values window and click the "OK" button to create the new variable.

The resulting variable, called MARRIED, appears in the final row of the *Variable View*. The MARRIED variable has the value "1" for all cases where MAR-STAT is equal to "1", and has the value "0" for all cases where MARSTAT is greater than "1". You can produce a cross-tabulation between the variables MARRIED and MARSTAT to confirm that the new variable was created correctly.

As in the Compute Variable tool, the cases affected by the Recode into Different Variables tool can be limited by clicking the "If" button (located below the "Old and New Values" button) to launch the If Cases window. The If Cases window and the process for specifying which cases will be recoded are the same as in the Compute Variable tool.

Sorting Cases

The Sort Cases tool is used to sort the cases in the dataset based on their values on one or more variables. The Sort Cases tool is available in the **Data** menu. Put the variable(s) that you want to sort on in the "Sort by" box (See Figure B.38). You can specify whether cases should be sorted in ascending order (from the lowest to the highest values) or in descending order (from the highest to the lowest values).

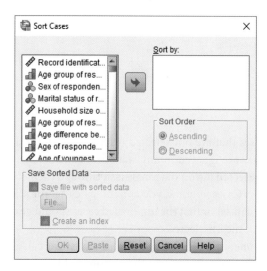

Figure B.38   **The Sort Cases Window**

Selecting or Filtering Cases

Sometimes researchers only want to use some cases in a statistical analysis. The Select Cases tool is used to choose which cases are used when SPSS calculates a statistic. The Select Cases tool is available in the **Data** menu.

There are several ways to select cases. It is most common to select cases using the "If condition is satisfied" method. (See the left of Figure B.39.) When this method is used, click the "If" button to open a window where you can specify which cases are used in the analyses, based on their values on an existing variable. (See the right of Figure B.39.) Specifying the conditions for selecting cases follows the same pattern as in the If Cases window of the Compute Variable tool.

Alternatively, you can use the Select Cases tool to select a random sample of cases. When the "Random sample of cases" method is selected, click on the "Sample" button to open a window where you can specify the percentage of cases that are randomly selected to use in the analysis. (Alternatively, you can randomly select "Exactly ___ cases from the first ___ cases.")

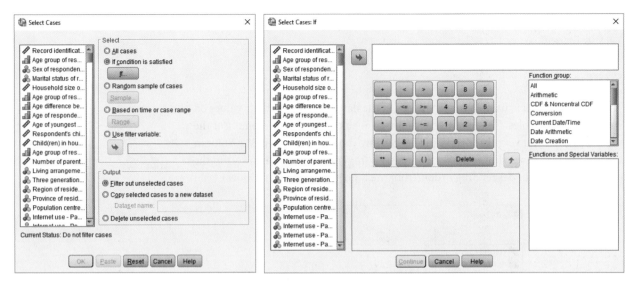

Figure B.39   **The Select Cases Window and Its If Cases Window**

By default, cases that are not selected (excluded from the analysis) remain in the dataset, but they are ignored in any statistical calculations. In the *Data View*, cases that are not selected have a diagonal slash through their row numbers. At the bottom of the Select Cases window, you can indicate that the selected cases should be copied to a new dataset or that the unselected cases should be deleted. It's usually best to just filter out the unselected cases.

When only some cases are being used in statistical calculations, the words "Filter on" appear in the status area in the bottom right corner of the Data Editor. Once the Select Cases tool has been used to filter the cases, all statistical calculations use only the selected cases until the filter is turned off. To turn off the filter, return to the Select Cases tool and choose "All cases."

Splitting a File

SPSS can also virtually divide a dataset into groups and produce statistical results for each group separately. This technique is particularly useful for exploring multivariate relationships. The Split File tool is available in the **Data** menu. When data are virtually divided into groups, there are two ways of displaying the results: "Compare groups" and "Organize output by groups." (See Figure B.40.) When "Compare groups" is selected, the results for each group appear sequentially within a single large table. (For example, three frequency distributions are printed in sequence within a single table.) When "Organize output by groups" is selected, the results for each group are printed in different tables. (For example, three separate frequency distributions are created.) Regardless of which way the results are displayed, the groups are defined by putting a nominal- or ordinal-level variable into the "Groups Based on" box. Before a variable can be used to split a file into groups, the data must be sorted on the values of that variable. (This is selected by default; do not change it.)

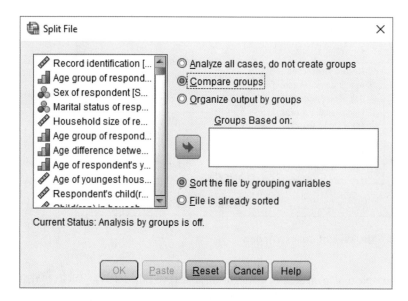

Figure B.40   **The Split File Window**

When a dataset is virtually divided into groups, the words "Split by" and the name of the grouping variable appear in the status area in the bottom right corner of the Data Editor. Once the Split File tool has been used to divide the dataset into groups, the results of all statistical calculations are divided into groups until the splitting is turned off. To turn the splitting off, return to the Split File tool and select "Analyze all cases, do not create groups."

Weighting Data

The Weight Cases tool, which is available in the **Data** menu, makes it easy to incorporate weights into statistical calculations. Simply select "Weight cases by" and put the variable that contains the weights into the "Frequency Variable" box. (See Figure B.41.) When weights are in effect, the words "Weight on" appear in the status area in the bottom right corner of the Data Editor. Once data are weighted, all statistical calculations incorporate the weighted contribution of each case, until the weights are turned off. To turn off the weights, return to the Weight Cases tool and select "Do not weight cases."

Creating a Standardized Weight Because SPSS does not effectively distinguish between the weighted number of cases and the actual number of cases used in statistical calculations, researchers often create and use standardized weights, instead of population weights, for statistical analyses using SPSS. The Compute Variable tool (in the **Transform** menu) is used to create a standardized weight variable. The first step in creating a standardized weight variable is to find the mean of the population weight variable (with the weights off). Copy or record the mean, including all of the decimals. Then, use the

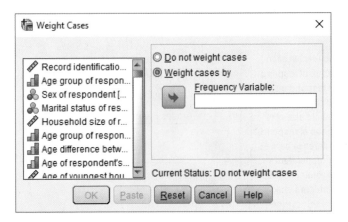

Figure B.41   **The Weight Cases Window**

Compute Variable tool to create a new weight variable (I've called it STD_WGHT) that is equal to the value on the population weight variable (called WGHT_PER in this dataset) divided by its mean (1058.034174 in this dataset). (See Figure B.42.)

To confirm that the standardized weight variable has been created correctly, find its mean (with the weights off), which should be equal to 1. Once you are confident that the standardized weight variable is correct, put it in the "Frequency Variable" box of the Weight Cases tool to use it to weight the data.

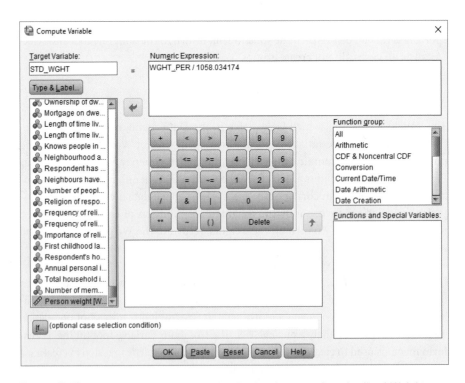

Figure B.42   **Using the Compute Variable Tool to Create a Standardized Weight**

Answers to Odd-Numbered "Practice What You Have Learned" Questions

Chapter 1

1. The two variables you will use are: (1) gender and (2) poverty status (which captures whether or not people's income is below a specific low-income threshold).

3. The group of people who are not in long-term relationships would include those with these attributes: "Widowed" (3), "Separated" (4), "Divorced" (5), and "Single, never married" (6). The group of people who are in long-term relationships would include those with these attributes: "Married" (1) and "Living common-law" (2).

5. Highest educational credential would be treated as the independent variable and annual personal income would be treated as the dependent variable.

7. It is a dichotomous variable.

9. It will be a dichotomous variable.

11. It will be a nominal-level variable.

13. It will be an ordinal-level variable.

15. It will be a ratio-level variable.

17. The value "97" indicates that people were not asked (skipped) the question because it doesn't apply to them (likely because they do not have a job). The value "98" indicates that people said they do not know how much they earn at their job each year. The value "99" indicates that people refused to report how much they earned at their job each year. All of these attributes should be designated as missing in statistical analyses and excluded from any calculations.

19. The unit of analysis is the country.

Chapter 2

1.

Does your family help pay your tuition costs?	Frequency	Percentage
Yes	38	76.00
No	12	24.00
Total	50	100.00

3. Overall, 760 out of every 1,000 students' families help pay their tuition costs.

5. The ratio of students whose families help pay their tuition costs to those whose families do not help pay their tuition costs is 38:12, or 3.17 (or 3.17:1). For every 3.17 students whose families help pay their tuition costs, one student's family does not help pay tuition costs.

7.

Does your family help pay your tuition costs?	Mature Student Status		
	Mature Student	Not a Mature Student	Total
Yes Count	6	32	38
Column %	37.50%	94.12%	76.00%
No Count	10	2	12
Column %	62.50%	5.88%	24.00%
Total Count	16	34	50
Column %	100.00%	100.00%	100.00%

9. a. It is a stacked bar graph.
   b. The two variables are (1) parents' highest level of schooling and (2) saving/not-saving for their children's post-secondary education.
   c. The independent variable is parents' highest level of schooling. The dependent variable is saving/not-saving for their children's post-secondary education.
   d. Parents' highest level of schooling is an ordinal-level variable. Saving/not-saving for their children's post-secondary education is a dichotomous variable.
   e. The general pattern shows that the higher the level of education that parents have, the more likely they are to be saving for their children's post-secondary education.

11. a.

Answer	Social Science Students Cumulative Percentage	Business Students Cumulative Percentage
$0 (no debt)	42.00%	56.67%
$1 to $4,999	48.00%	66.67%
$5,000 to $9,999	57.00%	76.67%
$10,000 to $24,999	81.00%	91.67%
$25,000 or more	100.00%	100.00%

b. Among social science students, 43 per cent graduate with $10,000 or more in student debt (100 − 57.00 = 43.00). Among business students, 23 per cent graduate with $10,000 or more in student debt (100 − 76.67 = 23.33). Thus, a higher percentage of social science students graduate with $10,000 or more in student debt.

13. a. Among social science students, 420 out of every 1,000 graduate without any student debt.

b. Among business students, 567 out of every 1,000 graduate without any student debt.

c. Comparing these two rates shows that business students are more likely to graduate without any student debt, and thus social science students are more likely to graduate with student debt.

15. a. The dependent variable is "Post-secondary education status:" attending college or university (or neither). The three independent variables are "Family income," "Immigrant status," and "Aboriginal status."

b. A third (33 per cent) of youth overall attended college and 42 per cent of youth overall attended university. Three-quarters (75 per cent) of youth overall attended either college or university.

c. Young people with family incomes below $50,000 are less likely to attend a post-secondary institution than young people with family incomes greater than $50,000. Young people with family incomes below $50,000 are more likely to attend college than university, whereas young people with family incomes greater than $50,000 are more likely to attend university than college.

d. First- and second-generation immigrants are more likely than non-immigrants to attend a post-secondary institution. Compared to non-immigrants, first- and second-generation immigrants are more likely attend university than college.

e. Aboriginal youth are less likely than non-Aboriginal youth to attend a post-secondary institution. Aboriginal youth are more likely to attend college than university, whereas non-Aboriginal youth are more likely to attend university than college.

Chapter 3

1. a. It will be a nominal-level variable.

b. The mode is used to describe the centre of a nominal-level variable.

c. The centre of this variable (the mode) is "Independent-living apartments."

3. a. It is an ordinal-level variable.

b. The mode and the median are used to describe the centre of an ordinal-level variable.

c. The centre of this variable is "$500 to $599." This is both the mode and the median of the variable.

5.

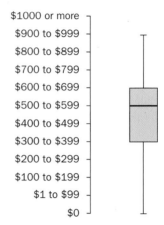

Students' Monthly Housing Costs

7. a. The twentieth percentile is "$200 to $299," the fortieth percentile is "$400 to $499," the sixtieth percentile is "$500 to $599," and the eightieth percentile is "$700 to $799."

b.

Quintile Group	Frequency	Percentage	Cumulative Percentage
Quintile Group 1 ($0 to $299)	269	21.3	21.3
Quintile Group 2 ($300 to $499)	312	24.7	46.0
Quintile Group 3 ($500 to $599)	256	20.3	66.3
Quintile Group 4 ($600 to $799)	286	22.7	89.0
Quintile Group 5 ($800 or more)	139	11.0	100.0
Total	1,262	100.0	

9. a. This range of the variable is from "Less than once a month" to "Every day."

b. The interquartile range of the variable is from "Less than once a month" to "A few times a week."

11. a. The fortieth percentile is "Once a month." This shows that 40 per cent of seniors worry about having enough money to pay for their home and utilities once a month or less often. The remaining 60 per cent of seniors worry about having enough money to pay for their home and utilities once a month or more often.

b. The eightieth percentile is "A few times a week." This shows that 80 per cent of seniors worry about having enough money to pay for their home and utilities a few times a week or less often. The remaining 20 per cent of seniors worry about having enough money to pay for their home and utilities a few times a week or more often.

13. a. The range of the variable is from 2 to 20 years (18 years).

b. The interquartile range of the variable is from 4 to 15 years (11 years).

15. a. It is a clustered bar graph.

b. The three variables are (1) "Housing tenure" (owner or renter), (2) "After-tax household income," and (3) "Year."

c. Some possible claims are:

- In every year from 2006 to 2014, renters had a substantially lower median after-tax household income than owners.

- From 2006 to 2014, there was a slight increase in the median after-tax household income of owners, but the same trend is not as evident among renters.

- Among households overall (regardless of housing tenure), there was only a slight increase in the median after-tax household income between 2006 and 2014.

Chapter 4

1. a. The median number of weeks it took these 10 clients to find a job is 5.25.

b. The average (mean) number of weeks it took these 10 clients to find a job is 5.30.

3. a. The median number of weeks it took these 11 clients to find a job is 5.50.

b. The average (mean) number of weeks it took these 11 clients to find a job is 8.00.

c. The median more accurately reflects how long it takes people to find a job using the agency's services. Only two clients took longer than the average number of weeks.

5. The standard deviation of clients' starting wage is $6.63.

7. a. It takes 68 per cent of the agency's clients between 7.50 and 12.50 weeks to find a job.

b. It takes 95 per cent of the agency's clients between 5.00 and 15.00 weeks to find a job.

9. It took people without a post-secondary educational credential 8.50 more weeks, on average, to find a job than people with a post-secondary educational credential—or about two months longer. There is also much more variation in the length of time it took people without a post-secondary credential to find a job, compared to people with a post-secondary credential. Since the number of weeks it takes to find a job is normally distributed, it took 95 per cent of people with a post-secondary credential between 1.50 and 5.50 weeks to find a job, whereas it took 95 per cent of people without a post-secondary credential between 4.00 and 20.00 weeks to find a job.

11. The histogram will have a right skew. Most people's starting wages will be clustered on the left/low-wage side of the graph and there will be a tail on the right-hand side, showing that a few people made high starting wages. The histogram will also be more peaked than the normal curve.

13. a. The mode shows that it is most common for post-secondary students aged 15 to 24 to earn $5,000 in wages each year. The median shows that half of these post-secondary students earn annual wages of $9,000 or less, and half earn annual wages of $9,000 or more. The average annual wage among these post-secondary students is $11,446. Overall, these wages seem low compared to students' costs; the Canadian Federation of Students found that in 2013, average undergraduate tuition fees for domestic students in Canada were $5,722. Students who pay their tuition fees themselves have relatively little left over to pay for housing, utilities, food, transportation, and other necessities; as a result, many students must rely on loans to cover their tuition and/or their living expenses.

b. Because the standard deviation is almost as large as the mean, it indicates that post-secondary students' wages are widely spread out, or widely dispersed.

c. Since the mode of this distribution is lower than the median, which is lower than the mean, the distribution of post-secondary students' annual wages will be right-skewed.

15. a. The group of young people with the highest unemployment rate is young people aged 15 to 19 who are visible minorities. The group of young people with the lowest unemployment rate is young people aged 20 to 24 who are not visible minorities.

b. In general, people aged 20 to 24 have lower unemployment rates than their counterparts aged 15 to 19. In both age groups, people who are Aboriginal have higher unemployment rates than people who are not Aboriginal, immigrants have higher unemployment rates than non-immigrants, people who are visible minorities have higher unemployment rates than people who are not visible minorities, and people with disabilities have higher unemployment rates than people who do not have disabilities. Overall, people who are perceived as being different or outside the social norm—because of their identity, appearance, place of birth, or ability—are more likely to be unemployed than those who are not.

Chapter 5

1. a. The probability of picking a red marble is 0.4.
    b. The probability of picking a red or green marble is 0.6.
3. a. The probability of drawing a seven is 0.08.
    b. The probability of drawing a face card is 0.23.
5. If you are not a mature student, then the probability that the classmate you are assigned to interview will not be a mature student is 0.82. If you are a mature student, then the probability that the classmate you are assigned to interview will not be a mature student is also 0.82.
7. The probability that a student in the course will be included in the sample is 0.17.
9. Statistics Canada uses a stratified sample for the National Graduates Survey. The population is divided into groups based on: the province/territory of the institution people graduated from (13 groups), the level of certification that people received (5 groups), and the field of study that the certification was in (12 groups). Overall, there were 636 possible combinations of groups, but only 434 of the groups had graduates in them. Once the groups were established, a type of random sampling called systematic sampling was used to randomly select people from each of the 434 groups to participate in the survey.
11. a. The probability that a randomly selected post-secondary student employee will work 7.5 hours a week or more is 0.84.
    b. The probability that a randomly selected post-secondary student employee will work 7.5 hours a week or less is 0.16.
13. a. In general, the household income of people living in First Nations communities is lower than the household income of the Canadian population. In 2006, 66 per cent of people in the Canadian population lived in households with an income of $50,000 or more, whereas in 2008/10, only 22 per cent of people in First Nations communities lived in households with an income of $50,000 or more. In contrast, in 2006, only 6 per cent of people in the Canadian

population lived in households with an income less than $20,000, whereas in 2008/10, 37 per cent of people in First Nations communities lived in households with an income less than $20,000.
    b. Surveys that exclude information from on-reserve and northern First Nations communities may be biased because they under-represent the number of people with low incomes. As a result, they may also under-represent the number of people with characteristics that are associated with being low income, such as unemployment, food insecurity, and poor physical or mental health.
15. a. The unweighted average number of hours worked per week is 21.67.
    b. The weighted average number of hours worked per week is 31.67.

Chapter 6

1. a. The standard error of the mean of time spent doing domestic labour is 3.32.
    b. The lower bound of the 95 per cent confidence interval for the mean of time spent doing domestic labour is 293.49, and the upper bound is 306.51.
    c. The 95 per cent confidence interval for the mean shows the range that the average is likely to be between in the population; there's a 95 per cent chance that the population parameter will be within this range. In the population of post-secondary students in Canada, the average amount of time spent doing domestic labour is likely between 293.49 and 306.51.
3. a. Among students who live with parents or relatives, the standard error of the mean of time spent doing domestic labour is 4.85. Among students who live independently, the standard error of the mean of time spent doing domestic labour is 4.23.
    b. Among students who live with parents or relatives, the 95 per cent confidence interval for the mean of time spent doing domestic labour is 252.49 to 271.51. Among students who live independently, the 95 per cent confidence interval for the mean of time spent doing domestic labour is 331.71 to 348.29.
    c. In the population of post-secondary students in Canada, there is likely a difference in the average amount of time spent doing domestic labour in each group. The two confidence intervals do not overlap one another; students who live independently are likely to spend more time doing domestic labour, on average, than students who live with parents or relatives.

5. a. The proportion of students in the sample who consider themselves feminists is 0.4535.

    b. The standard error of this proportion is 0.0173.

7. a. The proportion of social science students who consider themselves feminists is 0.3314. The proportion of natural science students who consider themselves feminists is 0.4800.

    b. The standard error of the proportion of social science students who consider themselves feminists is 0.0245. The standard error of the proportion of natural science students who consider themselves feminists is 0.0346.

9.

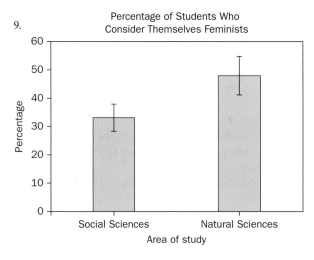

11. a. Caregivers are more likely to be women than men.

    b. Non-caregivers are just as likely to be women as they are men.

    c. In the population of people aged 45 and older, caregivers are more likely to be women than men, whereas this is not true of non-caregivers. In other words, women are over-represented among caregivers.

13. a. Compared to non-caregivers, a larger percentage of caregivers are in the two youngest age groups: 45 to 54 and 55 to 64. Compared to caregivers, a larger percentage of non-caregivers are in the two oldest age groups: 65 to 74 and 75 or older.

    b. Comparing these two distributions shows that in the population of people aged 45 and older, caregivers generally tend to be younger than non-caregivers. In other words, people aged 45 to 64 are over-represented among caregivers, whereas people aged 65 and older are over-represented among non-caregivers.

15. a. In Canada, among women who are employed full-time, the average daily leisure and personal care time is lower than in most other OECD countries, with the exception of Korea, Poland, and Slovenia. Researchers have attributed the reduced leisure time among Canadians—and among Canadian women in particular—to many factors, including the need to provide caregiving for children and seniors, employers' expectations for workers to remain "on call" or electronically available outside of scheduled working hours, non-standard work schedules, and long commute times.

    b. Although women in Canada who are employed full-time devote less time to leisure and personal care each day, on average, than their counterparts in most other OECD countries, the difference between men's and women's averages is relatively small compared to other countries, and is roughly on par with the OECD overall. All of the countries where women who are employed full-time devote less time, on average, to leisure and personal care than in Canada have a larger gender gap between men and women. In the United States, New Zealand, Sweden, and Denmark, men devote less time to leisure and personal care each day, on average, than women.

Chapter 7

1. a. "Employment status" (having paid employment or not) should be treated as the independent variable, and "Anxiety score" should be treated as the dependent variable.

    b. A non-directional hypothesis that the counsellor can test is this: "In the school population, there is a relationship between students' employment status and their anxiety scores." (Alternatively, the counsellor can test the non-directional hypothesis: "In the school population, the average anxiety score among students with paid employment is different than the average anxiety score among students who do not have paid employment.")

    c. A directional hypothesis that the counsellor can test is this: "In the school population, students with paid employment will have higher anxiety scores, on average, than students who do not have paid employment."

    d. The null hypothesis for the non-directional hypothesis in (b) is this: "In the school population, there is no relationship between students' employment status and their anxiety scores." (Alternatively, the null hypothesis is this: "In the school population, students with paid employment and students without paid employment have the same average anxiety score.") The null hypothesis for the directional hypothesis in (c) is this: "In the school population, students with paid employment will not have higher anxiety scores, on average, than students who do not have paid employment."

3. a. The t-statistic is 2.26 (or −2.26).

   b. The degrees of freedom of the t-statistic is 58.

   c. The t-statistic of 2.26 is beyond the critical value of +/−2.00 so you reject the null hypothesis. There is likely a relationship between students' employment status and their anxiety scores in the school population. In other words, in the school population, the mean anxiety score of students who have paid employment is likely to be different than the mean anxiety score of students who do not have paid employment.

5. a. Among all 18 first-year students, the mean anxiety score is 28.44 (s.d. = 9.76).

   b. Among the 10 first-year students who have paid employment, the mean anxiety score is 27.20 (s.d. = 8.55).

   c. Among the 8 first-year students who do not have paid employment, the mean anxiety score is 30.00 (s.d. = 11.51).

7. a. The t-statistic is −0.57 (or 0.57).

   b. The degrees of freedom of the t-statistic is 13.

   c. The t-statistic of −0.57 is not beyond the critical value of +/−2.16, so you fail to reject the null hypothesis. We are not confident that there is a relationship between employment status and anxiety scores in the population of first-year students at the school. In other words, in the population, the mean anxiety score of students who have paid employment might be exactly the same as the mean anxiety score of students who do not have paid employment.

9. a. Among people aged 20 to 24, there is not a big difference in the average mental-health scores of those who are currently attending school and those who are not. On average, people who are attending school have a positive mental-health score that is 1.6 points higher than people who are not currently attending school—a relatively small difference given that the scale of positive mental-health scores ranges from 0 to 70.

   b. Cohen's d is 0.14. This effect size is very small. The relationship between school attendance and positive mental-health scores is weak.

11. a. A non-directional hypothesis that the researchers can test is this: "Watching the DVD is related to people's scores on the OMS-HC scale in the population of health-care providers and students."

    b. The null hypothesis associated with the non-directional hypothesis in (a) is this: "Watching the DVD is not related to people's scores on the OMS-HC scale in the population of health-care providers and students."

13. a. Among the students in health care, the average post-test score was 0.3 points lower than the average pre-test score. This reduction indicates that, on average, people's stigma was reduced after watching the DVD. But the average follow-up score was 0.1 points higher than the average pre-test score and 0.4 points higher than the average post-test score. One month later, health-care students' stigma was higher than it was before participating in the Opening Minds initiative.

    b. The p-value of 0.943 is higher than the alpha value of 0.05. As a result, we are not confident that there is a relationship between watching the DVD and people's scores on the OMS-HC scale in the population of health-care students.

15. a. This infographic uses a (stacked) bar graph and a doughnut graph.

    b. In addition to the information in the graphs, the four other pieces of statistical information are as follows:
    - The median age of onset for mental illness
    - The median age of onset for anxiety disorders
    - The percentage of people with a moderate mental disorder who are in specialist treatment
    - How the mortality rate of people with bipolar disorder or schizophrenia compares to the general population

Chapter 8

1. a. The mean number of friends among the 40 students in the sample overall is 5.18.

   b. The mean number of friends among the 13 first-year students is 3. The mean number of friends among the 12 second-year students is 4. The mean number of friends among the 15 upper-year students is 8.

   c. On average, second-year students have one more friend at school than first-year students. But, on average, upper-year students have five more friends at school than first-year students and four more friends at school than second-year students.

3. a. The mean number of friends among the 25 first- and second-year students in the sample is 3.48.

   b. The total sum of squares is 78.23. The within group sum of squares is 72.00. The between group sum of squares is 6.23.

5. The total sum of squares is 415.68. The within group sum of squares is 218.00. The between group sum of squares is 197.68.

7. a. "Perceived job security" should be treated as the independent variable and "Positive mental-health score" should be treated as the dependent variable.

   b. A non-directional hypothesis that can be tested is this: "There is a relationship between perceived job security and people's positive mental-health scores in the population." (An alternative non-directional hypothesis is this: "In the population, people who agree that they have good job security, people who neither agree nor disagree they have good job security, and people who disagree that they have good job security have different positive mental-health scores, on average.")

   c. The null hypothesis associated with the hypothesis in (b) is this: "There is no relationship between perceived job security and people's positive mental health scores in the population." (An alternative null hypothesis is this: "In the population, people who agree that they have good job security, people who neither agree nor disagree they have good job security, and people who disagree that they have good job security have the same positive mental-health scores, on average.")

9. a. The within group sum of squares is 170,154.

   b. The between group degrees of freedom is 2. The within group degrees of freedom is 1,633.

   c. The F-statistic is 12.02.

   d. The F-statistic of 12.02 is higher than the critical value of 3.00, so you reject the null hypothesis. There is likely a relationship between perceptions of job security and positive mental-health scores among workers in the population. That is, people who agree that they have good job security, people who neither agree nor disagree they have good job security, and people who disagree that they have good job security are likely to have different mental-health scores, on average, in the population.

11. a. No, there is likely no difference between the percentage of men and the percentage of women with complete mental health in the Canadian population. The 95 per cent confidence intervals show that, in the population, the percentage of men with complete mental health is likely to be between 70.6 and 73.3, and the percentage of women with complete mental health is likely to be between 71.8 and 74.3. Thus, it's entirely possible that the same percentage of women and men in the Canadian population have complete mental health.

   b. Yes, there likely is a difference between the percentage of people with a post-secondary education and the percentage of people without a post-secondary education who have complete mental health in the Canadian population. This assessment can be made using the 95 per cent confidence intervals for each group or by using the results of the statistical significance test (designated by the **) in Table 8.8.

13. a. Yes, there likely is a difference between the percentage of males and females who have moderate-to-serious psychological distress in the population of Ontario grade 7 to 12 students. In this error-bar graph, the 95 per cent confidence interval for the proportion of males who have psychological distress does not overlap with the 95 per cent confidence interval for the proportion of females who have psychological distress. This indicates that each of the two groups are likely to have a different proportion of people who have moderate-to-serious psychological distress.

   b. Yes, there is a statistically significant difference between the percentage of males and females who have moderate-to-serious psychological distress. This is indicated by the footnote on the bottom of the graph, which states that there is a "significant difference by sex . . . ($p < 0.05$)."

15. a. No, in the population, there is likely no difference between the percentage of grade 7 to 12 students in each Ontario region who have moderate-to-serious psychological distress. In the sample, a higher proportion of students from Toronto and from northern Ontario have moderate-to-serious psychological distress, compared to other regions. However, as the 95 per cent confidence intervals for the percentages overlap for all of the regions, we are not confident that there is any difference between the regions in the population.

   b. No, there is not a statistically significant difference between the percentage of students in each region who have moderate to high psychological distress. This is indicated by the footnote on the bottom of the graph, which states that there is "no significant difference by region."

Chapter 9

1. a.   "Immigration status" (immigrant/born in Canada) should be treated as the independent variable, and "Trust in police" should be treated as the dependent variable.
    b.   A non-directional research hypothesis that can be tested is this: "There is a relationship between immigration status and trust in police in the community population."
    c.   The null hypothesis associated with the research hypothesis in (b) is this: "There is no relationship between immigration status and trust in police in the community population."

3. Lambda is equal to 0. The result shows that knowing whether or not people were born in Canada does not help you to reduce the error in predicting their trust in police.

5. Phi is 0.0173. It shows that the effect size is very small. The relationship between immigration status and trust in police is very weak.

7. a.

Overall, would you say that police in your community are doing:	Place of residence			
	Urban area	Suburban area	Rural area	Total
A poor job	11.90%	6.40%	1.49%	8.46%
A good job	50.40%	32.27%	8.96%	38.11%
An excellent job	37.70%	61.33%	89.55%	53.43%
Total	100.00%	100.00%	100.00%	100.00%

    b.   People who live in urban areas are more likely than people who live in other areas to say that police are doing a poor job: 11.90 per cent of people living in urban areas say police do a poor job, compared to only 1.49 per cent of people living in rural areas, a 10.41 percentage point difference. People who live in suburban and rural areas are more likely than people who live in urban areas to say that police are doing an excellent job. Whereas only 37.70 per cent of people living in urban areas say that police are doing an excellent job, 61.33 per cent of people living in suburban areas and 89.55 per cent of people living in rural areas say that police are doing an excellent job. In other words, the proportion of people who live in rural areas who say that police are doing an excellent job is more than double that of people who live in urban areas.

9. The total number of pairs of cases in Table 9.9 is 504,510. Only 34.37 per cent of pairs are used in the gamma calculation; the remaining 65.63 per cent are tied pairs that are excluded from the gamma calculation.

11. Cramér's V is 0.24. It shows that the effect size is small to medium. The relationship between people's place of residence and their overall perception of police is weak to moderate.

13. a.

Do you think your local police force does a good job, an average job, or a poor job of enforcing the laws?	Gender		
	Men	Women	Total
A poor or average job	41.81%	29.08%	34.57%
A good job	58.19%	70.92%	65.43%
Total	100.00%	100.00%	100.00%

    b.   This is an example of specification. In the sample overall, women are more likely than men to say police do a good job enforcing the laws: 70.92% of women say this, compared to 58.19% of men, for a difference of 12.73 percentage points. Among people who are visible minorities, the differences between women and men are larger: 79.62 per cent of women say police do a good job enforcing the laws, compared to 58.82 per cent of men, for a difference of 20.80 percentage points. Among people who are not visible minorities, 63.94 per cent of women say police do a good job enforcing the laws, compared to 57.86 per cent of men, for a difference of only 6.08 percentage points.

    c.   The relationship between gender and perceptions of how well police enforce laws is stronger among people who are visible minorities and weaker among people who are not visible minorities. A comparison of the results shows that this is primarily because of differences between women who are visible minorities and those who are not. Whereas roughly the same proportion of men who are visible minorities and men who are not visible minorities say that police do a good job enforcing the laws, this is not so for women. A much higher proportion of women who are visible minorities (79.62 per cent) say that police do a good job enforcing the laws than women who are not visible minorities (where only 63.94 per cent say the same).

15. Lambda is 0.03. This shows that you can reduce the error in predicting people's perceptions of police bias against Black people by 3 per cent if you know their racial identification.

17. a. The six different cross-tabulations displayed in the graph show the relationships between:
    - "Aboriginal status" and "Perception that police are approachable/easy to talk to"
    - "Aboriginal status" and "Perception that police ensure the safety of citizens"
    - "Aboriginal status" and "Perception that police treat people fairly"
    - "Aboriginal status" and "Perception that police enforce the laws"
    - "Aboriginal status" and "Perception that police promptly respond to calls"
    - "Aboriginal status" and "Perception that police provide information on ways to prevent crime"

    b. The * symbols indicate that there is a statistically significant relationship between the two variables in each cross-tabulation. In particular, they show that in the population of Canada's three territories, the percentage of non-Aboriginal people who say that police do a good job in each area is likely to be different than the percentage of Aboriginal people who say that police do a good job in each area.

    c. The general pattern that is evident in this chart is that non-Aboriginal people are more likely than Aboriginal people to say that police do a good job, in all six areas of police performance. Since all of these relationships are statistically significant, this pattern is likely to occur in the population of people living in Canada's three territories.

Chapter 10

1. Scatterplots (i) and (iv) show linear relationships. Scatterplots (ii) and (iii) show non-linear relationships.

3. Scatterplots (i), (ii), and (iv) show monotonic relationships. Scatterplot (iii) shows a non-monotonic relationship.

5. a. The sum of products is 143.
   b. The sum of squares of the "Satisfaction with life" variable is 146.
   c. The sum of squares of the "Number of close friends" variable is 285.

7. a. The t-statistic for Pearson's correlation coefficient is 3.10.

   b. The degrees of freedom of the t-statistic for Pearson's correlation coefficient is 10.
   c. The t-statistic of 3.10 is beyond the critical value of +/− 2.23, so you reject the null hypothesis. In the population, there is likely to be a relationship between recent immigrants' number of close friends and their satisfaction with life in the town.

9. a. A non-directional research hypothesis that can be tested using Spearman's correlation is this: "In the population of recent immigrants in the town, there is a rank-order relationship between people's number of close friends and their satisfaction with life in the town."
   b. The null hypothesis associated with this research hypothesis is this: "In the population of recent immigrants in the town, there is no rank-order relationship between people's number of close friends and their satisfaction with life in the town."

11. a. Spearman's correlation coefficient is 0.68.
    b. The Spearman's correlation coefficient shows that there is a moderate to strong rank-order relationship between these two variables.
    c. The rank-order relationship is positive. This indicates that people who are ranked as having more close friends are also ranked as being more satisfied with life in the town.

13. The Pearson's correlation coefficient for this relationship is 0.70, and the Spearman's correlation coefficient for this relationship is 0.68. This linear relationship between these two variables is about the same strength as the monotonic relationship. (Although technically, the linear relationship is slightly stronger.)

15. a. Six months after arrival, 91.2 per cent of immigrants would come to Canada if they had to make the decision again. Four years after arrival, only 86.5 per cent of immigrants would come to Canada if they had to make the decision again—a slightly smaller percentage.
    b. Six months after arrival, almost a quarter of immigrants (24.3 per cent) report that their life in Canada is somewhat or much worse than they expected. Four years after arrival, about the same proportion of immigrants (24.2 per cent) report that their life in Canada is somewhat or much worse than expected. The proportion of immigrants who report that their life is much worse than expected grew from 3.9 per cent six months after arrival to 5.0 per cent four years after arrival.

c. Overall, recent immigrants are more likely to be satisfied than dissatisfied with their life in Canada. Six months after arrival, 73.0 per cent of immigrants say that they are satisfied with life in Canada, and, similarly, four years after arrival 73.4 per cent of immigrants say the same. Six months after arrival, only 9.4 per cent of immigrants say that they are dissatisfied with life in Canada, and this drops to 7.1 per cent four years after arrival. Notably, though, the proportion of immigrants who are "Very/completely" dissatisfied four years after arrival (2.9 per cent) is higher than it is six months after arrival (1.4 per cent).

Chapter 11

1. a. "Months living in the town" should be treated as the independent variable, and "Hourly wage" should be treated as the dependent variable.
    b. A non-directional hypothesis that the agency worker can test is this: "There is a relationship between the number of months people have lived in the town and their hourly wages within the town's population of recent immigrants."
    c. The null hypothesis for the non-directional hypothesis in (b) is this: "There is no relationship between the number of months people have lived in the town and their hourly wages within the town's population of recent immigrants."
3. a. The constant coefficient is $15.56.
    b. The constant coefficient shows that recent immigrants who have just arrived in the town (i.e., they have lived there for 0 months) are predicted to earn $15.56 in hourly wages.
5. The residual of Alisha's case is –12.48.
7. a. The standard error of the estimate is 8.44.
    b. The standard error of the slope coefficient is 0.27.
9. a. "Years of work experience" should be treated as the independent variable, and "Hourly wage" should be treated as the dependent variable.
    b. A non-directional hypothesis that the agency worker can test is this: "There is a relationship between people's years of work experience and their hourly wages within the town's population of recent immigrants."
    c. The null hypothesis for the non-directional hypothesis in b) is this: "There is no relationship between

people's years of work experience and their hourly wages within the town's population of recent immigrants."
11. a. The constant coefficient is $28.30.
    b. The constant coefficient shows that recent immigrants who have 0 years of work experience are predicted to earn $28.30 in hourly wages.
13. a. The standard error of the estimate is 7.88.
    b. The standard error of the slope coefficient is 0.47.
15. The R^2 of this regression is 0.24. This result shows that 24 per cent of the variation in recent immigrants' hourly wages can be explained by their years of work experience.

Chapter 12

1. a. The partial slope coefficient is –0.9350.
    b. This partial slope coefficient shows that every one-year increase in age is associated with a decrease of $0.94 in the amount donated, controlling for weekly income.
3. a. The constant coefficient is 873.0750.
    b. The constant coefficient shows that people aged 0 who earn $0 per week are predicted to donate $873.08 to the agency.
5. a. The unstandardized slope coefficient of the "Age" variable shows that every one-year increase in age is associated with donating $5.04 more to charitable organizations in a 12-month period, controlling for annual personal income.
    b. The unstandardized slope coefficient of the "Annual personal income" variable shows that each additional $1,000 in annual personal income is associated with donating $3.33 more to charitable organizations in a 12-month period, controlling for age.
    c. The constant coefficient shows that people aged 45 who earn $0 in annual personal income are predicted to donate $187.07 to charitable organizations in a 12-month period.
7. a. $320.27
    b. $86.27
    c. $554.27
9. a. The unstandardized slope coefficient of the "Annual personal income" variable shows that each additional $1,000 in annual personal income is associated with donating $3.36 more to charitable organizations in a 12-month period, controlling for whether or not someone has a post-secondary education.

b. The unstandardized slope coefficient of the "Has a post-secondary education" dummy variable shows that people who have a post-secondary education are predicted to donate $105.36 more to charitable organizations in a 12-month period than people who do not have a post-secondary education, controlling for annual personal income.

c. The unstandardized slope coefficient of the "Has a post-secondary education" dummy variable in the regression in Table 12.16 is substantially smaller than in the regression in Table 12.15. This is because people who have a post-secondary education also typically have higher incomes than those who do not have a post-secondary education. In the regression in Table 12.15, the slope coefficient of the "Has a post-secondary education" variable was capturing both the influence of having a higher income *and* the influence of having a post-secondary education, whereas in the regression in Table 12.16, income is controlled for.

11. See graph below.

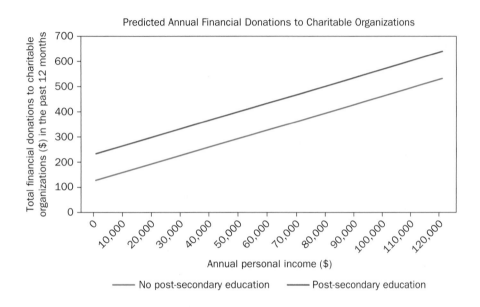

13.

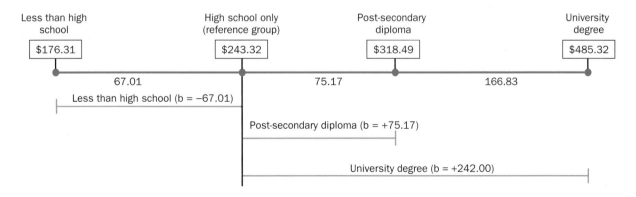

15. I would advise the agency that its marketing campaign should be designed to appeal to people who are older and who have higher personal incomes and a university education. The standardized slope coefficients show that age has the strongest association with the amount of money that people donate.

Chapter 13

1. a. The unstandardized slope coefficient of the "Age" variable in the first model shows that every one-year increase in age is associated with donating $5.87 more to charitable organizations in a 12-month period, controlling for sex/gender. The unstandardized slope coefficient of the "Women" dummy variable in the first model shows that women are predicted to donate $17.10 less than men to charitable organizations in a 12-month period, controlling for age. However, since this result is not statistically significant, we cannot be confident that in the population, women donate any more or less than men to charitable organizations.

    b. The constant coefficient in the first model shows that 45-year-old men are predicted to donate $327.08 to charitable organizations in a 12-month period.

    c. The R^2 and the adjusted R^2 of the first model show that age and sex/gender explain 2.7 per cent of the variation in the amount of financial donations to charitable organizations. Since the R^2 and the adjusted R^2 are the same, this suggests that both the independent variables in this model are good predictors of the dependent variable.

3. a. The unstandardized slope coefficient of the "Annual personal income" variable in the second model shows that each additional $1,000 in annual personal income is associated with donating $2.66 more to charitable organizations in a 12-month period, controlling for age, sex/gender, and highest educational credential.

    b. The unstandardized slope coefficient of the "Less than high school graduation" dummy variable shows that people who have less than a high school education are predicted to donate $58.89 less than people who have only a high school education to charitable organizations in a 12-month period, controlling for age, sex/gender, and annual personal income. People with a post-secondary diploma and people with a university degree are predicted to donate $38.47 and $183.27 more, respectively, than people who have

only a high school education, to charitable organizations in a 12-month period, controlling for age, sex/gender, and annual personal income. The general pattern of the relationship shows that the higher people's level of education, the more money they are predicted to donate to charitable organizations.

5. The constant coefficient in the first model is $327.08. It drops to $262.29 in the second model, and then to $106.40 in the third model. This change occurs because the constant coefficient shows the prediction for different types of people in each model. In the first model, the constant coefficient shows the prediction for 45-year-old men (regardless of their highest educational credential, annual personal income, volunteer status, or religious participation). In the second model, the constant coefficient shows the prediction for 45-year-old men who have only a high school education and who earn $40,000 per year (regardless of their volunteer status or religious participation). In the third model, the constant coefficient shows the prediction for 45-year-old men who have only a high school education, who earn $40,000 per year, who do not volunteer, and who participate in religious activities/services less than once a month.

7. a. The three variables that show the most signs of collinearity are the three education dummy variables ("Less than high school graduation" "Post-secondary diploma," and "University degree"). These three variables have the three lowest tolerances and the three highest variance inflation factors.

    b. No, none of the variables in this model show a level of collinearity that suggests a problem in this regression. All of the tolerances are above 0.1, and all of the variance inflation factors are below 10.

9. a. The Q-Q plot illustrates that the distribution of unstandardized residuals is right-skewed. This is evident because the tails of the residuals in the Q-Q plot curve downward, away from the diagonal line that represents a normal distribution.

    b. The Q-Q plot and the histogram in Figure 13.5 both show that the distribution of the unstandardized residuals is right-skewed. In both graphs, the residuals begin around –1,000. The trail of residuals moving away from the diagonal line, to the right of 2,000 on the Q-Q plot, corresponds to the tail of residuals to the right of 2,000 on the histogram.

11. a. The ideal relationship between regression residuals and an independent variable is the absence of a pattern. In other words, there should be no relationship

between regression residuals and an independent variable.

b. The box plot results do not show a strong relationship between the regression residuals and people's highest educational credential. The three box plots for people with less than a high school education, people with only a high school education, and people with a post-secondary diploma are all about the same height and all have a median near the ideal value of 0. This indicates that the regression model makes equally good predictions for people with these three levels of education. The box plot for people with a university degree has a median that is farther from 0 and is slightly taller than the box plots for the other levels of education. This indicates that the residuals for this group are more dispersed or spread out; in other words, the regression model makes slightly worse predictions for people with a university degree, compared to people with lower levels of education.

13. The Model 1 results show that women are predicted to have higher life satisfaction than men, older people are predicted to have lower life satisfaction than younger people, immigrants are predicted to have lower life satisfaction than non-immigrants, and Aboriginal people are predicted to have higher life satisfaction than non-Aboriginal people (after controlling for the other variables in the model).

15. a. People who have a feeling of belonging to their local community are predicted to have higher life satisfaction than people who do not have a feeling of belonging to their local community.

b. The slope coefficients associated with the variables "Women" and "Aboriginal persons" are larger in the second model than the first model. In other words, sex/gender and Aboriginal status are more strongly related to life satisfaction once people's feeling of belonging to their community is taken into account. The slope coefficients associated with the variables "Age" and "Immigrants," as well as the dummy variables capturing marital status and household income are smaller in the second model than the first model. In other words, age, immigrant status, marital status, and household income are more weakly related to life satisfaction once people's feeling of belonging to their community is taken into account. Some of these characteristics (age, immigrant status, marital status, and household income) are likely related to people's feeling of belonging to their community. There is not a consistent change in the other independent variables (capturing education, health status, and employment status) between the first model and the second model.

Glossary

95 per cent confidence interval (95% CI) Shows the range that a population parameter is estimated to be within, based on information from a *sample* selected from the *population*. There's a 95 per cent chance that the population parameter will actually be within this range.

adjusted R² A variation on *R²* that penalizes *linear regressions* that use independent variables that are poor predictors of the dependent variable. Although the R² will always increase when additional independent variables are added to a regression, this is not true of the adjusted R².

aggregation The process of collecting and summarizing many pieces of information (or observations) in order to develop conclusions. Quantitative researchers use aggregation when they summarize people's answers to survey questions in order to make a general claim.

alpha value (α) The threshold that researchers establish for their *p-value*. It shows the chance of making a *type I error* that a researcher is willing to accept.

antecedent variable A variable that occurs before both the independent variable and the dependent variable in time and, thus, may influence both of them. Antecedent control variables can be used in the *elaboration model*.

attributes Capture the potential range of variation within a single *variable*. Each variable must have two or more attributes. Attributes are divide cases into groups or categories, or represent a quantity or amount of something.

bar graph A graph that depicts the *proportion* or *percentage* of cases with each attribute using the relative height or length of a bar. Bar graphs are used to display the distribution of *ordinal-level variables* or *nominal-level variables* with many attributes.

beta coefficient (β) A common name for a *standardized slope coefficient*.

bias Systematic error, leading to the consistent overestimation or underestimation of a population parameter. Sampling bias occurs when a sample is systematically different from the population in some way.

box plot A graph that depicts the *median*, the *interquartile range*, and the *minimum* and *maximum* of a variable using a box with whiskers. Box plots are used to display the dispersion of *ordinal-* or *ratio-level variables*.

categorical variable A variable where the *attributes* divide cases into groups or categories. *Nominal-level* and *ordinal-level variables* are categorical variables.

causal relationship A statistical relationship where a causal variable occurs before an outcome variable in time, and where there are no other unaccounted for factors influencing both the causal and the outcome variables.

census A survey that collects information from every case in a *population*, rather than from a *sample* of cases. Census results are useful because they provide accurate information about population parameters instead of relying on statistical *estimation*. Census information is often used as the basis for *post-stratification* in sample surveys.

central limit theorem A theorem that states that, given a large enough number of random *samples* of the same size selected from a *population*, the distribution of a statistic calculated from each of those samples will be approximately normal, and centre on the population parameter. The central limit theorem is combined with the concept of a *sampling distribution* to form the basis for statistical *estimation*.

chi-square test of independence A *statistical significance test* used to assess the reliability of a relationship between two *categorical variables*. It is a *non-parametric test* that relies on comparing the observed frequencies in each cell of a cross-tabulation with the frequencies that are expected in each cell of a cross-tabulation if the null hypothesis is true.

class intervals Used to collapse all of the possible *attributes* of a variable into a smaller number of groups that each represent a range of attributes.

cluster sampling A sampling method that relies on dividing the *population* into groups, or clusters, and then randomly selecting among the clusters. This method is typically used when researchers are not able to generate or obtain a population list. Cluster sampling is often used in *multi-stage sampling* designs.

Cohen's d (d) A measure of effect size that standardizes the size (or magnitude) of an effect by dividing the difference between group *means* by the *standard deviation* of the variable in the sample overall.

collinearity Occurs when the *linear relationship* between one independent variable and the dependent variable in a regression is very similar to the linear relationship between another independent variable and the dependent variable. Collinearity is a concern because it becomes difficult to estimate how much of the variation in the dependent variable can be uniquely attributed to each independent variable.

concordant pair of cases Formed when one case ranks higher than another case on both the independent variable and the dependent variable (or a case ranks lower than another case on both the independent and dependent variables). Concordant and *discordant pairs of cases* are used to calculate *gamma*.

constant coefficient Provides information about where a *regression line* crosses the vertical axis. The constant coefficient is the predicted value on the dependent variable (y) when the value on the independent variable(s) (x) is/are 0. It is sometimes referred to as the intercept. The constant coefficient is reported in the same units as the dependent variable.

continuous variable A variable where the *attributes* are quantities or amounts; theoretically, continuous variables have an infinite number of attributes because any fraction of an amount is a legitimate attribute.

count variable A variable with *attributes* that are quantities or amounts that capture the number of times something occurs; the attributes usually include only whole numbers greater than or equal to 0.

covariance A measure of how two variables change (or vary) in relation to each other.

Cramér's V (V) A chi-square-based *measure of association* used to assess the magnitude of a relationship between two *categorical variables*, where one or both variables have more than two attributes. Cramér's V ranges from 0 to 1.

credibility interval The equivalent of a confidence interval or a *margin of error* for researchers using a Bayesian statistical approach. Credibility intervals are typically used to report the uncertainty associated with an estimate from a *non-probability sample*.

critical value For a test statistic, the cut-off point where the probability of making a *type I error* matches the *alpha value*. In many *statistical significance tests*, the critical value varies depending on the *degrees of freedom* of the probability distribution.

cross-tabulation Shows how the distribution of one variable is related to or is contingent on group membership (as defined by another variable). Cross-tabulations are used to show the relationship between two *nominal-level variables*, two *ordinal-level variables*, or one nominal-level and one ordinal-level variable. Cross-tabulations are sometimes referred to as crosstabs or contingency tables.

cumulative percentage Shows the *percentage* of cases with an attribute or one ranked below it (with a lower value). Cumulative percentages are used when reporting the distribution of *ordinal-* or *ratio-level variables*.

curvilinear relationship A *non-linear relationship* characterized by the shape of a curve or an arc.

data Information; often generated from empirical observations.

dataset A collection of *data* or information, usually focused on a main topic. Most statistical datasets come in electronic form and store information as a series of numbers.

degrees of freedom (df) Account for the number of free parameters in an equation. Mathematically, degrees of freedom account for the sample size and the number of groups in tests of *statistical significance*. The shape of some probability distributions vary depending on the degrees of freedom.

dependent variable (DV) A variable that captures the characteristic that is considered to be the "effect" or the result of whatever is captured in the *independent variable*. Many statistical procedures require researchers to conceptually designate one variable as a dependent variable.

descriptive statistics A series of techniques used to aggregate and summarize data; used to make claims about the people or cases that information was collected from.

dichotomous variable A special type of *nominal-level variable* that has only two attributes and values ("0" and "1") that indicate the presence or absence of something. A dichotomous variable can sometimes be used like a *ratio-level variable*.

directional hypothesis A statement of the expected relationship between variables that specifies the direction of the relationship.

discordant pair of cases Formed when one case ranks higher than another case on the independent variable and lower than it on the dependent variable, or vice versa. *Concordant* and discordant pairs of cases are used to calculate *gamma*.

distortion In the *elaboration model*, when the relationship between two variables in each subgroup of the sample is in the opposite direction of the relationship in the sample overall.

dummy variable A *dichotomous variable* used to incorporate a *categorical variable* as an independent variable in a regression. It has only two values, "0" and "1". Each attribute of the categorical variable is typically captured in a single dummy variable.

elaboration model An analytic approach based on assessing a relationship between two variables, and then investigating how the relationship changes after controlling for a third variable. There are six possible scenarios: *replication*, *specification*, *explanation*, *interpretation*, *suppression*, and *distortion*.

empirical research Research that relies on making direct observations of the world in order to generate knowledge; commonly used in the social sciences.

error-bar graph A graph that shows a sample statistic and the associated confidence interval for one or more groups (typically a *95 per cent confidence interval*). Error-bar graphs are particularly useful for comparing confidence intervals between groups.

estimation The process of making inferences about a *population* parameter using information collected from a random *sample* of that population. Researchers establish the amount of uncertainty associated with each estimate.

expected frequency (or expected count) The number of cases that researchers expect to be in a cell of a *cross-tabulation* if the *null hypothesis* is true; that is, if group membership (the independent variable) is not related to the dependent variable in the population. Expected frequencies are used to calculate the chi-square statistic.

explanation In the *elaboration model*, when the relationship between two variables in each subgroup of the sample is weaker than the relationship in the sample overall, as a result of controlling for an *antecedent variable*.

F-distribution A probability distribution used to determine the likelihood of randomly selecting a sample with the observed ratio of between group variation to within group variation (or a larger ratio), if the group means are equal in the population. The shape of the F-distribution varies depending on the sample size and on the number of groups being compared.

frequency distribution Shows how many cases in a dataset have each *attribute* of a variable, or how frequently each attribute occurs; a main tool of *descriptive statistics*. Frequency distributions are used to display the distribution of *nominal-* and *ordinal-level variables*.

gamma (γ) A *proportionate reduction in error measure* of the magnitude and direction of an association between two *ordinal-level variables*. It is a symmetric *measure of association*. Gamma ranges from −1 to +1.

heteroscedastic relationship A relationship in which the dispersion or spread of the dependent variable is different across different values on the independent variable; can sometimes be identified by a trumpet-shaped scatterplot.

histogram A graph that shows the number or percentage of cases that have values within equal-sized *class intervals* (called bins) using the relative height of a bar. Histograms are used to display the distribution of *ratio-level variables*.

homoscedastic relationship A relationship in which the dispersion or spread of the dependent variable is the same across all the values on the independent variable.

hypothesis A statement of the expected relationship between two or more *variables*. It is an educated guess that researchers make about what they expect to find, framed in the context of the specific variables available in the data.

independent variable (IV) A variable that captures the characteristic that is considered to be the "cause" of the outcome that is captured in the *dependent variable*. Many statistical procedures require researchers to conceptually designate one or more variables as independent variables.

inferential statistics A series of techniques used to make estimates or predictions about a population, using information collected from a *probability sample* of that population.

influential case A case that substantially affects the location or the direction of a *regression line*. Influential cases can be omitted so that the regression line better represents the general pattern in the majority of cases.

interpretation In the *elaboration model*, when the relationship between two variables in each subgroup of the sample is weaker than the relationship in the sample overall, as a result of controlling for an *intervening variable*.

interquartile range (IQR) The distance between the twenty-fifth *percentile* and the seventy-fifth percentile of a variable. The IQR shows the dispersion of the middle 50 per cent of cases for *ordinal-* and *ratio-level variables*. Although the IQR technically refers to the distance between the twenty-fifth and seventy-fifth percentiles, it is common to just report the upper and lower bounds of the IQR.

interval-level variable A variable that has *attributes* and *values* with an inherent order to them and an equal distance between them, but where the value "0" does not indicate the absence of something. A variable's level of measurement is used to determine which statistical techniques can be used with it. Most statistical techniques used with *ratio-level variables* can also be used with interval-level variables.

intervening variable A variable that occurs between the independent variable and dependent variable in time; it may be influenced by the independent variable and may also influence the dependent variable. Intervening control variables can be used in the *elaboration model*.

joint probability The chance that two or more outcomes will occur, either simultaneously or in sequence. The second (or subsequent) event can either be independent of or dependent on the result of the prior event(s).

kurtosis Indicates how peaked or flat the centre of a distribution is and how fat or skinny the tails of a distribution are, compared to a *normal distribution*. Distributions that are thinner and more peaked than a normal distribution are referred to as leptokurtic; distributions that are flatter and wider than a normal distribution are referred to as platykurtic.

lambda (λ) A *proportionate reduction in error measure* of the magnitude of an association between two *nominal-level variables* or between a nominal-level and an *ordinal-level variable*. It is an asymmetric *measure of association*. Lambda ranges from 0 to 1.

law of large numbers A law stating that the larger the number of trials used to establish the *observed probability* of an outcome, the closer the result will be to the *theoretical probability* of that outcome. The law of large numbers is useful for understanding how a *sampling distribution* relates to a population parameter.

linear regression A type of regression used to predict straight-line relationships between one or more independent variables and a ratio-level dependent variable. Simple linear regression uses only one independent variable whereas multiple linear regression uses more than one independent variable.

linear relationship A "straight-line" relationship between two ratio-level variables.

margin of error Specifies the distance above and below a statistic that a population parameter is likely to be within. The margin of error is typically reported in popular media and polling reports, and often corresponds to a *95 per cent confidence interval*.

mean ($\overline{X}$ or $\bar{x}$) The arithmetic average of a variable. The mean is used to describe the centre of *ratio-level variables*.

measures of association Measures that show the strength of a relationship between variables, usually summarized in a single number. Some measures of association also provide information about the direction of a relationship between variables. The choice of which measure of association to use depends on the level of measurement for each variable and, sometimes, on the number of groups or attributes in a *categorical variable*.

measures of effect size Measures that show how much of an effect an independent variable has on a dependent variable, often in a standardized way. They are used to indicate whether the independent variable has a small, medium, or large effect on the dependent variable, or to indicate whether there is a weak, moderate, or strong relationship between two variables.

median The attribute or value of the case located at the middle-most point of a variable when all of the cases are arranged in order from the lowest attribute or value to the highest attribute or value. Half of cases are at or above the median of a variable, and half of cases are at or below the median of a variable. The median is used to describe the centre of *ordinal-* and *ratio-level variables*.

minimum/maximum The lowest and highest attributes or values on a variable that occur in the data (and not just those that are theoretically possible). The minimum and maximum are used to calculate the *range* of *ordinal-* and *ratio-level variables*.

mode The attribute or value of a variable that occurs most frequently in the data. It is the most common answer. A variable can have more than one mode. The mode can be reported for variables with any level of measurement.

model specification The decisions that researchers make about which independent variables to use (or not use) in a regression model. Researchers strive to create regression models that accurately reflect social characteristics and processes and their relationship to a dependent variable.

monotonic relationship A relationship in which an increase in one variable is consistently associated with an increase in a second variable or consistently associated with a decrease in a second variable. All *linear relationships* are monotonic.

multi-stage sampling A process that involves sequentially sampling at different levels in order to select a final sample. For example, a random sample of clusters may be selected and then cases may be randomly selected within each cluster. Multi-stage sampling is typically used in research with large *populations* because it simplifies the sampling process.

negative relationship A relationship where higher values or attributes on one variable are associated with lower values or attributes on another variable (or vice versa). *Measures of association* sometimes show whether a relationship is positive or negative. Reporting the direction of a relationship makes sense only when both variables are measured at either the ordinal or ratio levels.

nested regressions Two or more regressions that use the same *dependent variable*, where each subsequent regression adds new *independent variables* without removing any of the previous independent variables.

nominal-level variable A variable with *attributes* that divide cases into groups or categories, but where the categories do not have an inherent order to them. A variable's level of measurement is used to determine which statistical techniques can be used with it.

non-directional hypothesis A statement of the expected relationship between variables that does not specify the direction of the relationship.

non-linear relationship A relationship between two ratio-level variables characterized by any other type of pattern besides a straight line.

non-monotonic relationship A relationship in which the direction of the relationship between two variables is not consistent across all of the values on a variable.

non-parametric test A test that does not rely on any assumptions about the underlying distribution of the variables being tested and that does not rely on estimating population parameters. The *chi-square test of independence* is a non-parametric test.

non-probability sample A sample where it is not possible to calculate the chance that a case from the *population* will be selected into the sample. Some cases in the population may have no chance of being selected into the sample. As a result, non-probability samples cannot be used to generate population estimates.

normal distribution A theoretical construct that is central to frequentist statistics. Along with the *central limit theorem*, it is central to the process of statistical *estimation*. The normal distribution is also used as a reference point for describing the shape of a distribution.

null hypothesis (H_0) A statement that there is no relationship between two or more variables in the *population* (or that the relationship is not in the expected direction, for *directional hypothesis*). In *inferential statistics*, it is the hypothesis that a researcher sets out to disprove with a *statistical significance test*.

null model A hypothetical regression model with only a dependent variable; that is, there are no independent variables. The mean of the dependent variable is predicted for every case. The null model is used in the calculation of R^2.

observed frequency (or observed count) The number of cases that are actually in a cell of a *cross-tabulation*. Observed frequencies are used to calculate the chi-square statistic.

observed probability A probability that is established by conducting multiple empirical trials to determine how often a specific outcome occurs. With a large enough number of trials, the observed probability is expected to converge with the *theoretical probability*. The concept of observed probability is useful for understanding how a *sampling distribution* relates to a population parameter.

omitted variable bias When an important independent variable is omitted from a regression and it affects the slope coefficients of one or more of the independent variables in the regression. Omitting an important independent variable can make the slope coefficients of other independent variables higher or lower than they would be if the variable were not left out.

one-tailed significance test Used to assess a *directional hypothesis*. For statistical significance tests that rely on symmetrical probability distributions, one-tailed tests take into account the proportion of samples that fall into only one tail of the distribution, resulting in a lower *critical value*. Researchers should use one-tailed significance tests only when there is a clear argument for why a relationship can go in only one direction.

one-way ANOVA test A *statistical significance test* used to assess whether the means of two or more groups are likely to be different from each other in the population, using sample data. It is used to assess the reliability of the relationship between a categorical independent variable and a ratio-level dependent variable. It is typically used with larger samples or when a researcher is comparing three or more groups.

ordinal-level variable A variable with *attributes* that divide cases into groups or categories, where the categories do have an inherent order to them. A variable's level of measurement is used to determine which statistical techniques can be used with it.

ordinary least squares (OLS) In *linear regression*, a method of determining the line of best fit, by finding the line that produces the smallest (least) number when the distance between every case and the line is squared and then summed.

outlier A case that does not fit into the general pattern of a distribution; often a case with an unusually high or unusually low value. Outliers affect the calculation of the *mean*, the *standard deviation*, and the *variance*.

parametric test A test that assumes that the variables being tested have an underlying *normal distribution* in the population. It relies on estimating population parameters. *T-tests* and *ANOVA tests* are parametric tests.

parsimonious The simplest and most efficient plausible explanation or model. Researchers strive to create parsimonious regression models that use just the right independent variables to make predictions about a dependent variable.

partial relationship In the *elaboration model*, the relationship between two variables in a subgroup of the sample, or after controlling for a third variable. Partial relationships are compared to the *zero-order relationship*.

Pearson's correlation coefficient (r) A *measure of association* that provides information about the strength and direction of the *linear relationship* between two *ratio-level variables*. Pearson's correlation coefficient ranges from −1 to +1.

percentage (%) Shows the *proportion* of people with an attribute, out of a base of 100. Percentages range from 0 to 100 and are used to make comparisons between groups of different sizes.

percentile A cut-off point used to divide cases based on the *percentage* of cases that are at or below some attribute or value on a variable. Percentiles are used to describe the dispersion of *ordinal-* and *ratio-level* variables.

phi (φ) A chi-square-based *measure of association* used to assess the magnitude of a relationship between two *categorical variables* that have only two attributes each. Phi ranges from 0 to 1.

pictograph A data visualization strategy that illustrates a *rate* using coloured/shaded icons.

pie graph A graph that depicts the *proportion* or *percentage* of cases with each attribute as a portion of the area in a circle. Pie graphs are used to display the distribution of *dichotomous variables* and *nominal-level variables* with relatively few attributes.

population The whole group that a researcher is interested in studying or making claims about. Researchers typically define a population at the outset of their study. The population is the group that the *sample* is selected from.

positive relationship A relationship where higher values or attributes on one variable are associated with higher values or attributes on another variable (or lower values or attributes on one variable are associated with lower values or attributes on another variable). *Measures of association* sometimes show whether a relationship is positive or negative. Reporting the direction of a relationship makes sense only when both variables are measured at either the ordinal or ratio levels.

post-hoc tests Tests conducted after a statistically significant relationship has been established. They typically provide more details about the relationship that was established.

post-stratification The process of assigning *weights* to a sample so that it matches the population on key characteristics, such as age and sex. The distribution of the key characteristics must be established by an external data source. Post-stratification is used to compensate for the fact that some people are less likely to participate in research.

primary data collection Collecting information for the explicit purpose of answering a specific research question. Primary data collection can be time consuming and expensive.

probability (p) The chance that an outcome or event will occur. Probabilities range from 0 to 1, where 0 indicates that there is no chance that the outcome will occur, and 1 indicates that the outcome will definitely occur. Researchers use probabilities to calculate the chance of a case being selected into a sample or to assess the chance of a sample being selected from a population.

probability sample A sample where each case in the *population* has a known, non-zero chance of being randomly selected. Researchers use data collected from probability samples to generate population estimates.

proportion (p) Shows the fraction of people with an attribute out of a whole (out of 1). Proportions range from 0 to 1 and are used to compare the relative frequency of different attributes.

proportionate reduction in error (PRE) measures Show how much the error in predicting the attributes of the dependent variable can be reduced if the attributes of the independent variable are known. *Lambda* and *gamma* are two common PRE measures.

p-value (p) Shows the *probability* of randomly selecting a sample with the observed relationship (or one of greater magnitude), if no relationship exists in the population the sample was selected from. Like all probabilities, the p-value theoretically ranges from 0 to 1 (although these extremes are never achieved in practice).

quantification The process of translating a concept, idea, behaviour, feeling, identity or something else into a number so that it can be used as *data* in statistical analyses. Quantification makes it possible for researchers to analyze societies using statistical techniques.

quantile A cut-off point used to divide cases into roughly equal-sized groups based on their attributes or values on a variable. Common quantiles are quartiles, which divide cases into four equal groups, and quintiles, which divide cases into five equal groups. Quantiles are used to describe the dispersion of *ordinal-* and *ratio-level variables*.

quantile–quantile plot (Q-Q plot) A special type of *scatterplot* that compares the value of each case on an observed variable to the value of a case located at the corresponding quantile in a perfect *normal distribution*. When an observed variable is normally distributed, all of the cases are located on the diagonal line in a Q-Q plot.

range The distance between the highest attribute or value and the lowest attribute or value that occur in the data. The range is used to describe the dispersion of ordinal- and ratio-level variables. For *ordinal-level variables*, this distance is usually reported in words, and for *ratio-level variables*, this distance is reported in the same unit as the variable is measured in.

rate Shows the number of times that something occurs, relative to the number of times that it could possibly occur. Rates are often used with standardized denominators that show the number of times that something occurs out of every 1,000 or every 100,000 instances.

ratio Shows how the frequencies of two attributes compare directly to each other. Ratios are used to illustrate how common or rare something is.

ratio-level variable A variable with *attributes* and *values* that are specific amounts or quantities of something, where the value "0" means having none of something. A variable's level of measurement is used to determine which statistical techniques can be used with it.

reference group The attribute or category corresponding to the *dummy variable* that must be omitted when a group of dummy variables (capturing the attributes of a *categorical variable*) are used as independent variables in a regression; the results associated with the remaining dummy variables are interpreted in relation to this omitted attribute or category.

regression line The line that best fits the pattern of an observed relationship between variables; it is the line that minimizes the sum of squared differences between the cases and the line.

replication In the *elaboration model*, when the relationship between two variables in each subgroup of the sample is roughly the same as the relationship between the two variables in the sample overall.

representative sample A sample that accurately represents the diversity of the *population* from which it was selected. Researchers strive to select a representative sample.

research hypothesis (H₁) A statement of an expected relationship between two or more variables in the *population*. In *inferential statistics*, it is the hypothesis that a researcher is ultimately trying to test or prove.

residual In regression, the difference between the predicted value on the dependent variable and the actual value, for each case. Residuals capture the unexplained variation in a regression model.

R-squared (R^2) Shows the proportion of the variation in the dependent variable that can be explained by the independent variable(s) in a *linear regression*. It ranges from 0 to 1 and is typically reported as a percentage. It is formally called the "coefficient of determination."

sample A group of people (or cases) who are selected from the larger *population* and about whom information is collected. The sample is a subset of the population. Researchers select a sample because it is usually too time consuming and expensive to collect information from the whole population.

sampling distribution The distribution of a statistic produced by every possible unique sample of the same size that can theoretically be selected from a population. The concept of a sampling distribution is combined with the *central limit theorem* to form the basis for statistical *estimation*.

sampling distribution of mean differences A theoretical distribution of the differences between the group means produced by every possible unique sample of the same size that can be selected from a population. Like all *sampling distributions*, it forms the basis for statistical *estimation*.

sampling error The error that inevitably results from collecting information from a *sample* of the *population* as opposed to the total population. Researchers strive to minimize sampling error by selecting large samples.

scatterplot A graph that shows the relationship between two variables by representing each case as a dot on the graph; typically, the independent variable is plotted on the x-axis and the dependent variable is plotted on the y-axis.

secondary data analysis The analysis of *data* that were not specifically collected to answer researchers' questions and were often not collected by the researchers themselves. The analysis of publicly available datasets, such as those produced by statistical agencies, is a form of secondary data analysis.

sigma (Σ) Statistical notation used to indicate that the calculation following the symbol should be completed for each case, and the results should be summed together. The sigma notation is used in many statistical formulas.

simple random sampling A sampling method that relies on listing each case in the *population*, assigning each case a number, and then randomly selecting numbers to select a sample. (This process is usually automated.) Simple random sampling is typically used in small-scale research studies or as the final stage in a *multi-stage sampling* design.

skew Indicates how asymmetrical a distribution is. The type of skew reflects the direction that the tail trails off in: a right-skewed distribution has a tail that trails off to the right, or in the positive direction on the horizontal axis; a left-skewed distribution has a tail that trails off to the left, or in the negative direction on the horizontal axis.

slope coefficient Provides information about the angle of a *regression line*. It indicates whether an increase in the independent variable is associated with an increase or a decrease in the dependent variable and what the size of that increase or decrease is. Unstandardized slope coefficients are reported in the same units as the dependent variable.

Spearman's rank-order correlation coefficient (ρ) A non-parametric *measure of association* that provides information about the strength and direction of the *monotonic relationship* between two variables. It is calculated using the rank of each case within a variable instead of the values on each variable. It can be used to assess the relationship between two *ordinal-level variables*, two *ratio-level variables*, or one ordinal-level and one ratio-level variable. Spearman's rank-order correlation coefficient ranges from −1 to +1.

specification In the *elaboration model*, when the relationship between two variables in one or more subgroups of the sample is stronger than the relationship in the sample overall and is weaker (or disappears entirely) in other subgroups.

spurious relationship A relationship between two variables that disappears once additional variables are controlled for. Researchers strive to avoid using independent variables that have a spurious relationship with the dependent variable in regressions.

standard deviation (S or s) The standardized, average deviation from the *mean*. It is the most common way of describing the dispersion of *ratio-level variables*.

standard error (se) The *standard deviation* of a *sampling distribution*. It is estimated using only information from a sample. The standard error is used to calculate confidence intervals.

standard error of the estimate A statistic that captures the overall accuracy of the predictions made by a *linear regression*. It is used primarily to calculate the *standard error* of regression coefficients.

standardized slope coefficient A regression *slope coefficient* that is reported in *standard deviations* instead of the original units of each variable; used to determine which independent variables have the strongest relationships with the dependent variable and which have the weakest relationships.

standardized weights Versions of *weights* that retain the relative contribution of each case but that have an average multiplier of 1 so that the weighted number of cases matches the unweighted sample size. Some statistical software requires standardized weights to accurately calculate statistical tests. They are also called relative weights, adjusted weights, or normalized weights.

statistical significance test Estimates the likelihood of randomly selecting a sample with the observed relationship (or one of greater magnitude), if no relationship exists in the population that the sample was selected from.

stratified sampling A sampling method that relies on dividing the *population* into analytically important strata or subgroups and then randomly

selecting a sample from each stratum. Stratified sampling is used to ensure that there are enough cases in the sample from each important subgroup.

suppression In the *elaboration model*, when the relationship between two variables in each subgroup of the sample is stronger than the relationship between the two variables in the sample overall.

t-distribution A probability distribution used to determine the likelihood of randomly selecting a sample with the observed difference between group means if the group means are equal in the population. The shape of the t-distribution varies depending on the sample size.

theoretical probability A probability that is established by dividing the number of outcomes of interest by the total number of possible outcomes. It is used to calculate the chance of each case being selected into a sample.

tolerance Shows how much of the variation in an independent variable cannot be explained by the other independent variables in a regression. The tolerance is the inverse of the *variance inflation factor (VIF)*. Tolerance ranges from 0 to 1; tolerances smaller than 0.1 indicate a potential *collinearity* problem.

t-test of independent means A *statistical significance test* used to determine whether the difference between two group means in a sample is large enough to assert that there is likely a difference between the group means in the population. It is used to assess the reliability of the relationship between an independent variable with two attributes and a ratio-level dependent variable.

two-tailed significance test Used to assess a *non-directional hypothesis*. For statistical significance tests that rely on symmetrical probability distributions, two-tailed tests take into account the proportion of samples that fall into both tails of the distribution. Researchers use two-tailed significance tests most of the time.

type I error When no relationship actually exists between two or more variables in the population, but a researcher claims that it likely does exist, based on an analysis of a random sample of cases from that population (a "false positive").

type II error When a relationship actually exists between two or more variables in the population, but a researcher claims it likely doesn't exist, based on an analysis of a random sample of cases from that population (a "false negative").

unit of analysis The basic "unit" that researchers treat as a single case in their analysis. Researchers use different units of analysis in order to make claims about different social configurations, such as individuals, households, families, cities, or countries.

value A number that is assigned to each *attribute* for the purpose of statistical manipulation. Sometimes values represent the group or category a case belongs to, and sometimes values represent a quantity or amount.

variable Captures some characteristic that varies across cases; such as across people, households, places, or time. Variables are often used to translate survey questions into data that can be analyzed using statistical techniques.

variance (S^2 or s^2) The square of the *standard deviation*. The variance is used in the calculation of many statistical tests, such as those in the analysis of variance (ANOVA) family.

variance inflation factor (VIF) Shows how much the variance of a slope coefficient is likely to be inflated as a result of an independent variable being correlated with the other independent variables in a regression. The VIF is the inverse of the *tolerance*. VIFs range from 1 to infinity; VIFs greater than 10 indicate a potential *collinearity* problem.

weights Multipliers assigned to each case that make the sample represent the larger population more accurately. Weight variables are typically included in large survey datasets, such as those produced by national statistical agencies.

zero-order relationship In the *elaboration model*, the original relationship between two variables in a sample overall. The zero-order relationship is compared to the *partial relationships*.

z-score A standardized score that shows how far an individual case is from the *mean* of a variable, using *standard deviations* as the unit of measurement. Researchers use z-scores to describe the position of a case within a single distribution and to compare the relative position of cases in different distributions.

Index